FLIM MICROSCOPY
IN
BIOLOGY AND MEDICINE

FLIM MICROSCOPY

IN

BIOLOGY AND MEDICINE

EDITED BY
AMMASI PERIASAMY
ROBERT M. CLEGG

CRC Press
Taylor & Francis Group
Boca Raton London New York

CRC Press is an imprint of the
Taylor & Francis Group an **informa** business

A CHAPMAN & HALL BOOK

The copyright to Chapter 10 is held by the National Institutes of Health.

Chapman & Hall/CRC
Taylor & Francis Group
6000 Broken Sound Parkway NW, Suite 300
Boca Raton, FL 33487-2742

First issued in paperback 2020

© 2010 by Taylor and Francis Group, LLC
Chapman & Hall/CRC is an imprint of Taylor & Francis Group, an Informa business

No claim to original U.S. Government works

ISBN-13: 978-0-367-57730-8 (pbk)
ISBN-13: 978-1-4200-7890-9 (hbk)

Library of Congress Cataloging-in-Publication Data

FLIM microscopy in biology and medicine / editors, Ammasi Periasamy, Robert M. Clegg.
 p. cm.
 Includes bibliographical references and index.
 ISBN 978-1-4200-7890-9 (hardcover : alk. paper)
 1. Fluorescence microscopy. I. Periasamy, Ammasi. II. Clegg, Robert M. III. Title.

QH212.F55F545 2009
570.28'2--dc22   2009015824

Visit the Taylor & Francis Web site at
http://www.taylorandfrancis.com

and the CRC Press Web site at
http://www.crcpress.com

We dedicate this book to past, present, and future FLIM enthusiasts, including those involved with instrumentation and software development as well as users mainly interested in FLIM applications.

Table of Contents

CHAPTER 2 ▪ Principles of Fluorescence for Quantitative
               Fluorescence Microscopy   35

NEIL ANTHONY, PENG GUO, AND KEITH BERLAND

CHAPTER 3 ▪ Visible Fluorescent Proteins for FRET-FLIM 65

RICHARD N. DAY

CHAPTER 7 ▪ Multiphoton Fluorescence Lifetime Imaging at the Dawn of Clinical Application

KARSTEN KÖNIG AND AISADA UCHUGONOVA

CHAPTER 8 ▪ FLIM Microscopy with a Streak Camera: Monitoring Metabolic Processes in Living Cells and Tissues

V. KRISHNAN RAMANUJAN, JAVIER A. JO, RAVI RANJAN, AND BRIAN A. HERMAN

Chapter 9 ■ Spectrally Resolved Fluorescence Lifetime Imaging Microscopy: SLIM/mwFLIM

Christoph Biskup, Birgit Hoffmann, Klaus Benndorf, and Angelika Rück

SECTION 3 **Data Analysis**

CHAPTER 12 ▪ Nonlinear Curve-Fitting Methods for Time-Resolved Data Analysis   341

IGNACY GRYCZYNSKI, RAFAL LUCHOWSKI, SHASHANK BHARILL, JULIAN BOREJDO, AND ZYGMUNT GRYCZYNSKI

SECTION 4 **Applications**

Preface

Fluorescence microscopy is an established tool for a variety of applications in biology and biomedical research. Recent advances leading to improved contrast and high sensitivity allow for the detection of signals at the single-molecule level. In conjunction with this platform, fluorescence lifetime imaging microscopy (FLIM) provides another dimension of contrast and sensitivity and also offers the additional benefit of independence from fluorophore concentration and excitation intensity. Moreover, the fluorescence lifetime is often sensitive to the physical and chemical environment of the fluorophore; as such, it is an excellent reporter of conformational changes and variations of the molecular surroundings of biological molecules.

These unique advantages of lifetime-resolved fluorescence measurements extend the information that is obtained from measuring only the intensity. The rationale for performing lifetime-resolved measurements in an imaging environment is to acquire, at every pixel of a fluorescence image, the critical information provided by dynamic fluorescence measurements, which has been available for decades from single-channel (cuvette) fluorescence dynamics measurements. The lifetime-resolved fluorescence parameters coupled with the spatial dimension provide valuable insight into the functioning of complex biological systems.

The primary objective of a FLIM investigation is usually quite different from that of single-channel measurements. One is still interested in determining the lifetime-resolved information as accurately, reproducibly, and robustly as possible; however, in FLIM, the structure/morphology of some object (e.g., a structure in a biological cell) under physiological conditions is often the investigation's target of major concern. Thus, the scientific questions asked are analogous to those for normal intensity fluorescence imaging; that is, one is interested in correlating the spectroscopic information with different locations in the imaged object. When lifetime-resolved fluorescence is acquired in addition to the intensity, the identification and quantitative differentiation of fluorophores are considerably improved.

For instance, in the case of FRET, if the lifetime of a donor fluorophore is known in the absence of an acceptor, it is relatively easy for FLIM to differentiate locations in an image with dissimilar lifetime decays; faster lifetimes indicate increased efficiency of energy transfer. Because lifetimes in FLIM are independent of the concentration, complicated control experiments and multiple wavelengths, which may be difficult to align in the image, are not required. FLIM is also an excellent way to discriminate objects that have similar emission wavelengths but different lifetimes; thus, the image contrast is improved. A common application of FLIM is the elimination of background fluorescence, such as intrinsic fluorescence, or unbound fluorophores, where the bound and unbound fluorophores exhibit different lifetimes. FLIM can be used for many quantitative determinations of ion concentrations, pH, oxygen content, protein–protein interactions, cell motility, and cancer diagnosis. All of these applications are discussed in the various chapters of this book.

It is difficult to say when and where the first FLIM images were observed. The "dawn of FLIM" took place in a few research laboratories with ready expertise in time-resolved fluorescence, fast electronics, and, usually, a strong interest in solving biological problems. The original developments made use of instrumentation already available in cuvette-based spectrofluorometers to acquire the fluorescent decay. The lifetime data were analyzed using on-hand fitting methods.

Once the power and broad applicability of FLIM became evident to the general scientific community, biological investigators' interest in exploring lifetime measurements developed quickly. Both time- and frequency-domain methodologies were rapidly improved and extended in many laboratories. Thanks to the development of various technologies, including optics, electronics, detectors, and the new discovery of visible fluorescent proteins, this development took place at a rapid pace. Not surprisingly, after the introduction of various commercial units, which simplified data acquisition and analysis and provided biological laboratories with ready-made instrumentation, publications covering FLIM microscopy have grown rapidly since 2000.

This book presents the fundamentals of FLIM so that a wider audience can appreciate the rapid advances and increasing applications reported in the literature. In this sense, the goal is pedagogical: In addition to reviewing the latest developments, applications, and approaches to data analysis, we want to convey the exciting future of FLIM and indicate the present state of the art in FLIM imaging as described in the instrumentation section. No measurement method is perfect; the authors have strived to present pros and cons of different methods and to give some indication of where improvements are necessary and desired. Each chapter critically compares FLIM measurements to other techniques.

The book also describes ancillary techniques related to the direct determination of lifetimes, including imaging fluorescence anisotropy for the study of molecular rotations. Moreover, in addition to discussions related directly to FLIM, we also address the fundamentals of dynamic fluorescence measurements and the basic pathways of de-excitation available to electronically excited molecules. An awareness of the diversity of pathways available to an excited fluorophore will assist potential users in recognizing the value of FLIM measurements, as well as inspire innovative experiments using lifetime-resolved imaging.

As time passes, more of the sophisticated methods used in photophysics and photochemistry, as well as new instrumentation, are being incorporated into FLIM. Novel features that apply exclusively to FLIM are being developed, including sophisticated image analysis. Our purpose has been to showcase the broad application of fluorescence lifetime-resolved imaging in biology. We include different aspects of FLIM data acquisition and applications, as well as discussions of FLIM data processing, in a separate section on data processing.

The discussion sections in all the chapters clearly show the challenges for implementing FLIM for various applications. Certain chapters discuss limits on the number of photons required for highly accurate lifetime determinations, as well as the accuracy with which multiple, closely associated lifetime components can reliably be determined. Highly accurate determinations of fluorescence lifetimes are sometimes necessary for answering certain specific, detailed questions concerning some molecular mechanisms. On the other hand, the change in lifetime-related parameters and their location in a cell are of primary concern for many investigations. Such considerations are important for the user when he or she is selecting the most advantageous method of FLIM to use for a particular application. These aspects are discussed in the various chapters. We hope that this book will be useful for experts in FLIM as well as for newcomers to this field.

We realize that the field of FLIM has grown rapidly in the recent past and that it is impossible to do justice to all those who have contributed unique FLIM applications, instrumentation, and data analysis. We acknowledge our indebtedness to all the FLIM enthusiasts who have made this such an exciting field. Most importantly, we wish to thank all the authors who have contributed chapters to this book and have strived to present the fundamentals so that novices can implement FLIM in their research and laboratories, as well as appreciate the uniqueness and usefulness of FLIM in their own research. It has been a great honor to work with them.

Ammasi Periasamy, PhD
Robert M. Clegg, PhD

Acknowledgments

We wish to acknowledge our respective universities' support in making this book possible. We also wish to acknowledge with gratitude the cover page illustrator, Mr. Hal Noakes of the University of Virginia. We would like to thank Luna Han, Judith M. Simon, and Amber Donley, Taylor and Francis–CRC Press, for all their help and support.

We also want to thank the following organizations for supporting this valuable book on FLIM:

Becker & Hickl GmbH

Intelligent Imaging Innovations

ISS

Lambert Instruments

PicoQuant GmbH

A. P.
R. M. C.

The Editors

Ammasi Periasamy is the director of the W.M. Keck Center for Cellular Imaging and a professor of biology and biomedical engineering at the University of Virginia. He received his doctorate in biomedical engineering from the Indian Institute of Technology, Madras, India, and performed postdoctoral research in biomedical imaging at the University of Washington, Seattle. Among his numerous research accomplishments has been the development of a steady-state, confocal, multiphoton, and FLIM-based Förster (fluorescence) resonance energy transfer (FRET) imaging system for protein localization. Dr. Periasamy's research focuses on advanced microscopy techniques, particularly molecular imaging in living cells and tissue. Dr. Periasamy serves on the editorial board of the *Journal of Biomedical Optics*.

Robert McDonald Clegg is a professor in the Departments of Physics and Bioengineering at the University of Illinois in Urbana (UIUC) and is presently the director of the Center for Biophysics and Computational Biology at UIUC. He is also an affiliate of the Biochemistry Department. Dr. Clegg received his doctorate in physical chemistry from Cornell University, followed by postdoctoral research at the Max Planck Institute for Biophysical Chemistry in Göttingen, Germany. His current research interests include the development and applications of fluorescence lifetime imaging microscopy (FLIM) apparatus and the development of unique dedicated software for analysis of FLIM data. He is well known for his research in fluorescence studies in a variety of complex biological systems, especially with FRET. His research involves rapid relaxation kinetics (T- and P-jump), the development of microsecond rapid mixing techniques, and high-pressure applications on biological systems, including nucleic acid conformational kinetics, multisubunit functional proteins, and photosynthetic systems.

Contributors

Sasha Agronskaia
Molecular Biophysics, Science Faculty
University of Utrecht
Utrecht, the Netherlands

Neil Anthony
Department of Physics
Emory University
Atlanta, Georgia

Arjen Bader
Molecular Biophysics, Science Faculty
University of Utrecht
Utrecht, the Netherlands

Klaus Benndorf
Universitätsklinikum Jena
Institut für Physiologie II
Jena, Germany

Keith Berland
Department of Physics
Emory University
Atlanta, Georgia

Shashank Bharill
Center for Commercialization of
    Fluorescence Technologies
Departments of Molecular Biology and
    Immunology and Cell Biology and
    Genetics
Health Science Center
University of North Texas
Fort Worth, Texas

Christoph Biskup
Universitätsklinikum Jena
Biomolecular Photonics Group
Jena, Germany

Paul S. Blank
National Institute of Child Health and
    Human Development
National Institutes of Health
Bethesda, Maryland

Julian Borejdo
Center for Commercialization of
    Fluorescence Technologies
Departments of Molecular Biology and
    Immunology and Cell Biology and
    Genetics
Health Science Center
University of North Texas
Fort Worth, Texas

Chittanon Buranachai
Department of Physics
Faculty of Science
Prince of Songkla University
Hatyai, Songkhla, Thailand

Yi-Chun Chen
Department of Bioengineering
University of Illinois at
    Urbana-Champaign
Urbana, Illinois

Robert M. Clegg
Department of Physics
Loomis Laboratory of Physics
University of Illinois at
Urbana-Champaign
Urbana, Illinois

Richard N. Day
Department of Cellular and
Integrative Physiology
Indiana University School of Medicine
Indianapolis, Indiana

James N. Demas
Department of Chemistry
University of Virginia
Charlottesville, Virginia

Hans C. Gerritsen
Molecular Biophysics, Science Faculty
University of Utrecht
Utrecht, the Netherlands

Hernan E. Grecco
Department of Systemic Cell Biology
Max Planck Institute of
Molecular Physiology
Dortmund, Germany

Ignacy Gryczynski
Center for Commercialization of
Fluorescence Technologies
Departments of Molecular Biology and
Immunology and Cell Biology and
Genetics
Health Science Center
University of North Texas
Fort Worth, Texas

Zygmunt Gryczynski
Center for Commercialization of
Fluorescence Technologies
Departments of Molecular Biology and
Immunology and Cell Biology and
Genetics
Health Science Center
University of North Texas
Fort Worth, Texas

Peng Guo
Department of Physics
Emory University
Atlanta, Georgia

Brian A. Herman
Department of Cellular and
Structural Biology
University of Texas Health
Science Center
San Antonio, Texas

Birgit Hoffmann
Universitätsklinikum Jena
Biomolecular Photonics Group
Jena, Germany

Javier A. Jo
Department of Biomedical Engineering
Texas A&M University
College Station, Texas

Karsten König
Faculty of Physics and Mechatronics
Saarland University
Saarbrücken, Germany

Srinagesh V. Koushik
National Institute on Alcohol
Abuse and Alcoholism
National Institutes of Health
Rockville, Maryland

Rafal Luchowski
Center for Commercialization of
    Fluorescence Technologies
Departments of Molecular Biology and
    Immunology and Cell Biology and
    Genetics
Health Science Center
University of North Texas
Fort Worth, Texas

George Malachowski
Bioscience Applications, Pty. Ltd.
Melbourne, Australia

Ammasi Periasamy
W. M. Keck Center for Cellular
    Imaging
University of Virginia
Charlottesville, Virginia

V. Krishnan Ramanujan
Minimally Invasive Surgical
    Technologies Institute
Department of Surgery
Cedars-Sinai Medical Center
Los Angeles, California

Ravi Ranjan
Department of Pharmacology
University of Texas Health
    Science Center
San Antonio, Texas

Angelika Rück
Institute for Laser Technologies in
    Medicine and Metrology
Ulm, Germany

Bryan Q. Spring
Center for Biophysics and
    Computational Biology
University of Illinois at
    Urbana-Champaign
Urbana, Illinois

Yuansheng Sun
W. M. Keck Center for Cellular Imaging
University of Virginia
Charlottesville, Virginia

Christopher Thaler
National Institute on Alcohol Abuse and
    Alcoholism
National Institutes of Health
Rockville, Maryland

Bianca Tong
Bioscience Applications, Pty. Ltd.
Melbourne, Australia

Aisada Uchugonova
Fraunhofer Institut of Biomedical
    Technology (IBMT)
Department of Biomedical Optics
St. Ingbert, Germany

Peter J. Verveer
Department of Systemic Cell Biology
Max Planck Institute of Molecular
    Physiology
Dortmund, Germany

Steven S. Vogel
National Institute on Alcohol Abuse and
    Alcoholism
National Institutes of Health
Rockville, Maryland

FLIM MICROSCOPY
IN
BIOLOGY AND MEDICINE

1

Introduction, Microscopy, Fluorophores

CHAPTER 1

Fluorescence Lifetime-Resolved Imaging

What, Why, How—A Prologue

Robert M. Clegg*

1.1 INTRODUCTION

Fluorescence lifetime-resolved imaging, FLI, acquires a fluorescence image whereby the dynamic response of the fluorescence decay is temporally resolved at every location (pixel) of the image. When specifically referring to measurements in a light microscope, the acronym is FLIM, where the "M" stands for microscopy. We will use the names interchangeably. FLI measurements are analogous to normal intensity fluorescence imaging measurements and are acquired on the same samples, except that information related to the fluorescence lifetime is recorded in addition to the normal measurement of the fluorescence intensity. One says that in FLI the fluorescence signal is "lifetime resolved" and "spatially resolved." The fluorescence lifetimes (or more often, the apparent fluorescence lifetime) can be determined with the temporal resolution of nanoseconds or less at every pixel of the recorded image. The spectroscopic lifetime-resolved information can be displayed at every pixel in image format. By considering the physical mechanisms that determine the life of an excited fluorophore, insight into the experimental possibilities afforded by FLIM can be better appreciated.

* The author wishes to express his gratitude and appreciation to the community of scientists who have been instrumental in the development of FLI. It is a definite pleasure to work within the "FLI community." Many aspects of FLI are not covered in this chapter, and no details are given of any particular study or instrument. This chapter is not a literature review. Therefore, justice is not afforded to the many innovative contributions of many research groups. The reader can find this information either in the original references that have been given or in the following chapters.

1.2 GOAL OF THIS CHAPTER

One aim of this chapter is to acquaint the reader with the information available by time-resolved fluorescence spectroscopy and to describe how the experimental measurements are related to the fundamental mechanism of fluorescence. Ultimately, an understanding of the different pathways of de-excitation available to a molecule in an excited state is necessary to interpret fluorescence data and leads naturally to an appreciation of the knowledge that can be gained by temporally resolving the emission of a fluorescence signal in FLIM. This is not a review of the many excellent publications and outstanding contributions that have been made in the last decade on different FLI instruments. We discuss the fundamental time-dependent mechanisms that play a role in fluorescence. Many different de-excitation pathways are available to an electronically excited fluorophore, and these independent pathways compete kinetically in parallel. The kinetic rate of each pathway of de-excitation is sensitive to the environment of the fluorophore, each through a different mechanism. The measured rate of fluorescence (which is the inverse of the measured lifetime) is a summation of the rates of all separate available pathways; therefore, the fluorescence lifetime bears witness to the rates of all contributing pathways of de-excitation.

A discussion of this inclusive property of the value of the fluorescence lifetime, which makes the fluorescence lifetime so valuable, will hopefully lead to new, innovative experiments on specific biological systems. This property of the lifetime is one of the major reasons for performing FLIM measurements. We also survey the present methods of FLIM instrumentation and discuss their comparative advantages. Detailed descriptions of FLIM are not the focus of this chapter. The reader should consult the other chapters for in-depth discussions of the different aspects and methods of FLIM.

1.3 WHY MEASURE FLUORESCENCE LIFETIMES?

Upon excitation of a molecule from the ground electronic state to a higher electronic state, a molecule will remain in its primarily excited electronic state only transiently. The residence time in the electronic excited state is usually in the range of picoseconds to tens of nanoseconds. The average time the molecule spends in the electronic excited state is referred to as the "fluorescence lifetime." The primary excitation event that boosts a molecule from its ground electronic state (S_0 state) to an excited state—a vibrationally excited S_1 electronic state— takes place in a femtosecond (fs; 1 fs = 10^{-15} s) or less. The initial vibrationally excited S_1 state rapidly loses its extra vibrational energy to the environment and decays in 10^{-14} to 10^{-12} s to a vibrationally relaxed excited state: the vibrationally relaxed S1 state. Also, through vibrational interactions, the surrounding solvent or other nearby molecules can interact through Coulomb or dipole interactions with the S_1 state, often in a time-dependent manner, and change the electronic energy level (and other properties) of the relaxed S_1 state.

The excited state is normally initiated by the absorption of a photon; but other means of excitation can produce the same relaxed excited state (such as energy transfer from another nearby excited molecule—Förster resonance energy transfer [FRET]—or through a chemical or biochemical reaction). Once in the S_1 vibrationally relaxed state, the molecule can undergo de-excitation through several different pathways (see the Perrin–Jablonski diagram in Box 1.2):

1. radiative emission of a photon (fluorescence), the *intrinsic* radiative rate, $k_{f,int}$ (this is not the measured rate of fluorescence and is considered to be an intrinsic property of the isolated molecule);

2. intersystem crossing to a triplet state, k_{isc};

3. nonradiative relaxation (usually losing the energy nonradiatively to the environment), which is termed internal conversion, k_{ic}, or nonradiative transitions, k_{nr};

4. dynamic quenching through collisions, k_q;

5. energy transfer to a nearby molecule in its ground state, k_{et};

6. *excited-state reactions* other than quenching, k_{er}, such as charge transfer, molecular isomerizations, and bimolecular reactions; and

7. photolysis (photodestruction of the excited molecule), usually by interaction of its triplet state with triplet oxygen, k_{ph}.

The notation, k_i, refers to the rate constants of the ith physical process. We also define k_{nf} as the sum of all the rate constants other than the rate constant for fluorescence, $k_{f,int}$. There are variations of these processes, but the preceding description is the usual case for most fluorophores used in FLIM. All these processes are dynamic, and there is a monomolecular rate constant for each pathway of de-excitation, depending on the conditions. Except for special situations, the "natural intrinsic rate constant," $k_{f,int}$, of radiative emission (fluorescence) is constant (e.g., not dependent on the temperature). $k_{f,int}$ can in principle be calculated from first principles; it is different for each molecule and depends on the details of the excited and ground-state electronic configurations. The rates of all other pathways are often sensitive to the molecular environment. All the separate pathways compete dynamically to first order.

The overall probability per unit time for the molecule to lose its excitation energy and pass to the ground state depends on the sum of the rate constants of all the different pathways (see Box 1.1). If we observe the fluorescence decay, the observed rate of the fluorescence relaxation will equal this sum of the rate constants of all the available pathways. The longest time the molecule can remain in the excited state is set by the rate constant of the fluorescence pathway (the intrinsic radiative lifetime, $\tau_{f,int} = 1/k_{f,int}$); however, as mentioned earlier, this is not the measured rate of fluorescence. The average time a molecule spends in the excited state will decrease as additional pathways become available for de-excitation. In this way, the excited fluorophore acts as a "spectroscopic spy," and the overall average lifetime of the fluorescence emission provides valuable information about the molecular environment of the excited fluorophore. A quantitative interpretation of the rate of emission is given in Box 1.1.

BOX 1.1: INTERDEPENDENCE OF THE PATHWAYS OF DE-EXCITATION AND THE MEASURED FLUORESCENCE LIFETIME OF A FLUOROPHORE, *D*

Discussing fluorescence from the natural point of view of competing kinetic rates emphasizes unequivocally that fluorescence is a convenient and very sensitive method for measuring kinetic mechanisms and molecular configurations on a molecular scale. Assume all the pathways of de-excitation are operative. The total rate of de-excitation from the excited-state D^* to the ground-state D can be depicted as a chemical reaction, $D^* \xrightarrow{\sum_i^{k_j}} D$. That is, the average time that the molecule stays in the excited state (the average lifetime of the excited-state τ_{D^*}) is inversely related to the sum of all the available different pathways of de-excitation, $1/\tau_{D^*} = \sum_i k_i$. The measured lifetime in the absence of pathway j is $(1/\tau_{D^*})_{i \neq j} = \sum_{i \neq j} k_i$. Thus, the rate of deactivation in the presence of pathway j is greater than in its absence; that is, $(1/\tau_{D^*})_{i \neq j} \leq 1/\tau_{D^*}$.

Whenever we allow an additional pathway of excitation, the rate of decay becomes faster. Usually, we choose to measure the rate of fluorescence decay, but the lifetime of D^* can be determined by measuring an experimental variable along *any* of the de-excitation pathways. Define the *measured* parameter to correspond to pathway f (where we have chosen the letter f because we usually measure fluorescence). The rate measured along pathway f in the presence and absence of pathway j will be $1/\tau_{f,meas} = \sum_i k_i$ and $(1/\tau_{f,meas})_{i \neq j} = \sum_{i \neq j} k_i$. Note that $k_{f,int}$ must be a member of both sums (because we are measuring fluorescence, this pathway cannot be the absent parameter). Also, note that $1/\tau_{f,meas}$ is the overall rate at which the excited state is depleted. $1/\tau_{f,meas}$ is *not* the intrinsic rate of de-excitation by way of the fluorescence pathway; that is, $1/\tau_{f,meas} \neq k_{f,int}$.

The intrinsic probability of fluorescence per unit time ($k_{f,int}$) is the same in both sums; however, the total pool of excited molecules becomes depleted faster than $k_{f,int}$ because other pathways for de-excitation are simultaneously actively available for depleting the excited state. The fluorescence decay signal mirrors the total decay of the excited-state population. For instance, in order to determine k_j, we simply subtract the two measured inverse decay times, $1/\tau_{f,meas} - (1/\tau_{f,meas})_{i \neq j} = \sum_i k_i - \sum_{i \neq j} k_i = k_j$.

Note that, in order to determine the rate of a pathway (j), we have measured fluorescence; however, pathway j has nothing to do with fluorescence. The obvious reason for choosing fluorescence to investigate all the other nonemissive pathways is because it is convenient and relatively easy to detect photons on the nanosecond time scale (which is the time window of the measurement). The lifetime of the measured fluorescence relaxation gives us direct insight into and quantitative estimates of the overall molecular dynamics of the nonfluorescence pathways. Through recent developments of FLI instrumentation, the considerable, powerful advantages of time-resolved fluorescence measurements are now available for imaging experiments.

The association between the measured fluorescence lifetime with the sum of all the rates of the competitive pathways of de-excitation is the primary motivation for carrying out fluorescence lifetime experiments. The average lifetime of the excited state, in the presence of all the available pathways of de-excitation, sets a limit on the temporal window of opportunity, during which the excited molecule can explore and report on its surroundings. As stated previously, the excited-state lifetime is usually in the nanosecond time region. Of course, by adding extra pathways for de-excitation from the excited state or by increasing the values of some of the rate constants, the probability per unit time that the molecule will emit a photon (fluorescence) will decrease; as a result, the intensity of fluorescence will decrease. Thus, the time-averaged fluorescence intensity also carries information on the rates of the different pathways taking place within this time window. However, as we will see, the lifetimes report directly on the molecular dynamics and the temporal information is much richer than simply the time-averaged intensity.

If the excited molecule has transferred from the primary excited singlet state to the triplet state by intersystem crossing (see later discussion), the emission from the triplet state is called phosphorescence. In the absence of triplet quenchers, such as triplet oxygen, the phosphorescence lifetime can be as long as many seconds because the transition from the triplet state to the ground singlet state is not allowed and requires a simultaneous spin flip. To first order, transitions between states with different spin values are forbidden.

There is an extensive history of fluorescence lifetime measurements in single-channel experiments (macroscopic samples in cuvettes) (Birks 1970; Birks and Dawson 1961; Cundall and Dale 1983; Gratton and Limkeman 1983; Grinvald and Steinberg 1974; Lakowicz 1999; Spencer and Weber 1969; Valeur 2002). Lifetime-resolved fluorescence measurements have provided a wealth of invaluable information about biological systems. Of course, directly measured spatial information (imaging) is not available in a cuvette-type spectroscopic measurement. On the other hand, routine fluorescence imaging, whereby one is measuring the time-averaged intensity of fluorescence, has become a familiar measurement in essentially every field of cellular biology and provides an enormous wealth of information in cellular biology. Until more recently, most fluorescence imaging measurements in an optical microscope were limited to spectrally resolved intensity measurements (where wavelengths are usually selected with a simple optical filter). Although normal fluorescence microscopes do not provide nanosecond temporal resolution of the fluorescence signal, sophisticated microscope instrumentation and powerful image analysis algorithms are available that reveal detailed morphological information with high spatial resolution. FLIM aspires to couple both these feature into a single measurement.

1.4 WHY MEASURE LIFETIME-RESOLVED IMAGES?

Due to advances in instrumentation in the last few decades, it has become possible to couple measurements of fluorescence lifetimes with the most common modes of fluorescence microscopy. As the FLI technique emerges and becomes available to more researchers, the enhanced and more refined information content of time-resolved fluorescence measurements will extend significantly the capability of the investigator to reveal physical details on the molecular scale in fluorescence images of biological samples. The dependence on the environment, the kinetic competition between different pathways of de-excitation, and the sensitivity of the measured fluorescence signal to physical events on the scale of microscopic dimensions make fluorescence imaging a valued and highly informative method of measurement. FLIM measures the kinetics of these dynamic processes directly, without "integrating over" the time-dependent information, as when the steady-state intensity is measured.

In addition to the mechanistic and molecular information available by fluorescence lifetime measurements, separating the fluorescence signal into its elementary lifetime components provides a practical way to increase image contrast and distinguish quantitatively the spatial distribution of multiple fluorophores with different lifetimes. FLIM can also help remove background fluorescence and discriminate intrinsic fluorescence in a biological cell. Reliable measurements can be made of the fraction of fluorophores in selected isomeric states, such as protonated and deprotenated forms in pH measurements (Carlsson et al. 2000; Rink, Tsien, and Pozzan 1982; Szmacinski and Lakowicz 1993). Measuring the fluorescence lifetime is also one of the best and most reliable ways to quantify FRET; this is the most common application of FLIM.

1.5 SPECIFIC FEATURES OF THE DIFFERENT PATHWAYS AND RATES OF DE-EXCITATION

In FLI we usually excite the molecules with light. As has been already mentioned, the total rate of leaving the excited state, which is the reciprocal of the measured excited-state lifetime, is the summation of the rates of all the possible pathways. This is usually depicted in the form of a Jablonski diagram (or perhaps this should also be termed a Perrin–Jablonski diagram; see Box 1.2; Birks 1970; Lakowicz 1999; Nickel 1996, 1997; Valeur 2002). We emphasize again that, when we measure fluorescence lifetimes, we are not measuring the intrinsic rate of emission proceeding only through the fluorescence pathway; on the contrary, we are measuring the total rate of leaving the excited state, which is the *sum* of all the rates for leaving the excited state.

During the time that a molecule is in an excited state, it interacts intimately and dynamically with its molecular environment. Spectroscopists have long taken advantage of the unique molecular information that is available from the emission from a molecule in an excited state when the dynamic decay is resolved by measuring it directly in the time domain (Birks 1970; Cundall and Dale 1983; Lakowicz 1999; Valeur 2002). Although a molecule spends only a very short time in the excited state—picoseconds to nanoseconds—the eventual emission of a photon bears the historical imprint of this sojourn in the excited state. Many events can happen during this short time. Understanding the dynamic characteristics of the major pathways of de-excitation is essential, and it is the starting point for interpreting lifetime measurements.

1.5.1 Intrinsic Rate of Emission (Fluorescence)

The *intrinsic spontaneous rate of fluorescence* defines the longest average time the molecule can stay in the S_1 excited state; it is defined wholly by the quantum mechanical nature of the excited and ground-state electric configurations of an isolated molecule. It can be thought of as the rate constant (that is, the probability per unit time) at which a molecule isolated from all other molecules and with no other pathways of de-excitation will leave the lowest (first) singlet excited state by emitting a photon and returning to its electric ground state. The intrinsic emission rate is the same as the Einstein spontaneous emission rate calculated in most spectroscopy and quantum mechanics textbooks (Atkins and Friedman 1997; Becker 1969; Chen and Kotlarchyk 1997; Craig and Thirunamachandran 1984; Förster 1951; Kauzmann 1957; Lakowicz 1999; Lippert and Macomber 1995; Parker 1968; Schiff 1968; Silfvast 1996). The intrinsic emission rate is also related to the uncertainty in the energy of a system that has a finite lifetime through the relation

BOX 1.2 PERRIN–JABLONSKI DIAGRAM

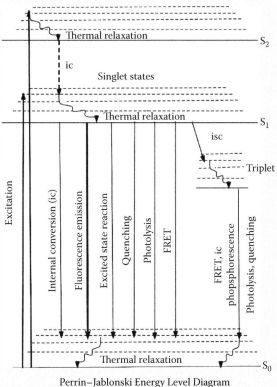

Perrin–Jablonski Energy Level Diagram
of a Fluorescent Molecule

Energy increases vertically up. The arrows represent the transition of a quantum change. The transitions begin at the lowest vibrational levels of each electronic state (S_0, S_1, and S_2) for both the absorption (upward arrows) and the de-excitation processes (downward arrows). The transition from the excited state to the ground state always happens from the lowest level of S_1 (first singlet excited state). The de-excitation transitions other than the emissive, internal conversion and intersystem crossing transitions are gathered together in a box at the right of the diagram. Via intersystem crossing (isc), the excited molecule passes from the singlet (S_1) to the triplet (T) state.

Emission from the triplet state is phosphorescence. The other de-excitation transitions from the triplet state are similar to those from the singlet state. The electronic transitions usually leave the molecule in an excited vibrational level of the end electronic state. This vibrational excitation energy relaxes thermally very rapidly, within picoseconds, to the lower vibrational levels of the corresponding electronic states. Two possible absorption (excitation) transitions to two different singlet excited states are shown (both contained within the dotted box). If the S_2 state becomes excited, the molecule immediately relaxes by internal conversion to the S_1 state.

$$\Delta E \Delta t = \Delta E \tau_f \geq \hbar$$

where $\hbar = h/2\pi$ and $h \equiv$ Planck's constant. This coupled uncertainty of the values of the energy of a system and the time that the system is in that state is not a true Heisenberg uncertainty relation (Atkins and Friedman 1997; Landau 1997; Schiff 1968).

Nevertheless, the lifetime of the intrinsic emission decay is related to the breadth of the energy of emission. This rate is never measured in FLIM (or any other spectroscopic experiment carried out in solution); neither is the extremely narrow spectral width of the intrinsic fluorescence rate. For this chapter, it is only important to recognize that the intrinsic rate of emission is a basic quantum mechanical property of an isolated molecule and can in principle be calculated from its energy levels and wave functions of its quantum states. The fundamental physical explanation for this intrinsic spontaneous emission is due to the coupling of the excited and ground states through the interaction with what is called fluctuations in the zero-point level of the photon modes of the vacuum radiation field.

Although the density of these photon modes is generally the same in most environments, it is possible that the density can change, and this will change the intrinsic rate of radiation emission. For instance, when a fluorophore is very close to a metal surface with a very sharp curvature, the intrinsic rate can be affected because the density of the photon modes can be increased. We do not discuss this interesting phenomenon, but refer the reader to recent literature where he or she can also find earlier references (Enderlein 2002; Fiuráek et al. 2001; Hamann et al. 2000, 2001; Sánchez, Novotny, and Xie 1999). This phenomenon has until now not been of much use for biological fluorescence measurements, but it is interesting for the future, especially from the point of view of lifetime measurements (it is a way to shorten the intrinsic lifetime). However, for our purposes, it is only necessary to know that this rate of spontaneous emission does not change unless the coupling between the ground and excited electronic structures of the molecule changes (e.g., a change in the positions of the nuclei, changing the molecular conformation).

1.5.2 Thermal Relaxation (Internal Conversion)

The process of internal conversion leads to a nonradiative transition from the excited to the ground state. Thermal interactions with the solvent surrounding the fluorophore or with the immediate surrounding molecular matrix reduce the time a molecule spends in the excited state by providing another pathway (other than the intrinsic photon emission) for leaving the excited state. The excited molecule is coupled to its environment through vibrations and collisions. These interactions are dependent on the temperature and the composition of the solvent. The vibronic coupling considerably broadens the spectrum in solution and reduces the fluorescence lifetime.

The intermolecular coupling of the thermal environment is not the only internal conversion pathway to the ground state. Intramolecular vibrational interactions also facilitate very rapid relaxations from higher vibrational levels of excited molecules to the lowest vibrational states of the first excited state (resulting in a Boltzmann distribution among the lowest vibrational levels of the first electronic excited state). At the normal temperatures of biology, the lowest energy vibrational state is by far the most populated vibrational state. Thermal coupling with the solvent molecules is the main reason why the energy of emission is always less than the energy of excitation (the Stokes shift: part of the excitation energy is lost to the environment).

This is clear by looking at the Perrin–Jablonski diagram in Box 1.1 and comparing the length of the excitation vector to that of the fluorescence emission vector. This vibrational coupling to the environment also broadens the fluorescence spectrum, so the individual vibrational bands can only rarely be observed. Biological samples are almost always in condensed media and usually measured at ambient temperatures or not far from them. Thus, the intrinsic rate of fluorescence plus the rate of thermal nonradiative deactivations will contribute to the apparent longest experimentally determined decay time observed for fluorophores in biophysical measurements.

1.5.3 Molecular Relaxation of the Solvent or Molecular Matrix Environment

In biophysical measurements, the fluorophores are almost always in a complex condensed matter environment, such as an aqueous environment or in an apolar surrounding such as in lipid membranes. Other biological components can also interact and couple strongly to the excited molecule. Coulomb and dipole interactions with the surrounding environment affect the position and the breadth of the emission. In a polar environment, the solvent molecules (or the neighboring molecular matrix) reorient around the excited molecule during the time window of the excited state (this happens especially in an aqueous solvent due to the very large dipole moment of water, 1.85 debye). This *solvent relaxation* will take place if the dipole moment of the excited molecule differs from the dipole moment of the ground state, changing the energy of the excited electronic state. This can often be observed in a polar aqueous environment and when internal charge transfer takes place after the molecule is excited into the excited state.

These local dipole relaxations, which take place subsequent to the much faster thermal relaxations to the lowest vibrational states, can lead to a further decrease in energy of the excited molecule before emission, shifting the emission toward lower energies—a red shift. The extent of this solvent relaxation is dependent on the temperature and the viscosity of the molecular environment. If the solvent relaxation occurs before or during the time when the molecule exits the excited state, it contributes significantly to the Stokes shift in a highly polar solvent environment; that is, the energy (wavelength) of the emission is less than (longer than) the energy (wavelength) of excitation.

Interestingly, if the "solvent" or "matrix" relaxation occurs on the same time scale of the fluorescence emission, the rate of the relaxation of the molecular matrix can be directly observed in the measured fluorescence lifetime. In this case, if the *measured* dynamic signal shifts out of (or into) the wavelength bandwidth of the emission optical filter, an extra component will enter into the measured signal. This would usually be classified as an artifact; however, it is clear that the time-dependent wavelength shift of the fluorescence emission contains valuable information on the molecular environment. Strong solvent interactions and charge transfer processes can also change the fluorescence lifetime by changing the overlap of the excited and ground-state wave functions.

The spectrum shifts can be observed in a steady-state fluorescence spectrum; spectral shifts of fluorophores have been used extensively in fluorescence imaging without temporal resolution. However, the emission spectrum can be recorded in a time-resolved mode, giving the shift of the spectrum as a function of the time. Such a time-resolved experiment would record directly the relaxation of the polar "solvent." Detailed measurements utilizing either of the solvent relaxation processes have not yet been extensively employed in FLI; however, considering the heterogeneous environment in a living cell, it is clear that this dynamic information is valuable. Detailed research of the polarity changes and micro-organization in the molecular environment of certain fluorophores—for example, laurydan (which has the very sensitive polarity-sensitive chromophore prodane)—has been carried out in biological membranes (Bagatolli and Gratton 1999; Dietrich et al. 2001).

1.5.4 Quenchers (Dynamic)

Dynamic quenchers collide with the excited molecule by random diffusion (Box 1.3). Because the excited-state lifetime is of the order of nanoseconds, only those quencher molecules very close to the excited molecule will be effective. Triplet oxygen is a major cause of dynamic quenching in normal biological milieu. However, other molecules are effective quenchers, such as Br^-, I^-, and acrylamide (nonpolymerized). Many effective "collisional quenchers" can increase the rate of transfer to the triplet state of the excited molecule (intersystem crossing; see later discussion), thus removing the singlet excited state and thereby decreasing the prompt fluorescence signal. This perturbation is usually accomplished through spin–orbit coupling (where the spins of electrons of the quencher perturb the orbital states of the fluorophore, increasing the rate of transfer from the singlet state to the triplet state of the fluorophore; see Section 1.5.7).

The spin–orbital coupling is especially effective if the collisional quencher has an unpaired, weakly held outer orbital electron (such as I^- and Br^-). Spin–orbit perturbation also takes place with the electrons of the excited molecule itself (see discussion of intersystem crossing in Section 1.5.7). In general, only smaller charged ions are effective quenchers due to their rapid diffusion; however, charged, or highly polarizable, groups on macromolecules can also effectively quench fluorophores that are attached, either covalently or simply bound, to the macromolecules.

For FLI, the principal effect of quenchers is that the fluorescence lifetime is shortened. Often in cellular imaging it is difficult to correlate the intensity of fluorescence with the concentration of the probe because a decrease in intensity could come from a smaller number of fluorescing molecules or from quenching. FLI can easily distinguish these two possibilities by determining the reduction in the quantum yield due to the dynamic quenching, thus allowing an accurate calculation of the concentrations. This is a very powerful application of FLI and is only possible if the fluorescence lifetime can be determined at every pixel of the image.

1.5.5 Excited-State Reactions

If the fluorophore in the excited state reacts with a reaction partner selectively, then lifetime-resolved imaging can be used to map the location of the reactive component. The analysis is the same as that discussed earlier for dynamic quenching (collisional quenching

BOX 1.3

Dynamic quenching involves diffusion of the quencher (and sometimes the diffusion of the fluorophore). We can measure the rate of dynamic quenching by simply measuring the rate of fluorescence decay in the presence and absence of quenching:

$$1/\tau_{f,meas}^{+quencher} - 1/\tau_{f,meas}^{-quencher} = \sum_i k_i - \sum_{i\neq quencher} k_i = k_{quencher} = k_q[Q]$$

If we assume that the encounter between the fluorophore and the quencher is diffusion controlled and neglect transient terms (Valeur 2002), then

> k_q is the rate constant of diffusional encounter quenching, $k_q = 4\pi N'D$, where the diffusion coefficient, D, is $D = kT/6\pi\eta a$ cm^2/s (usually about 10^{-5} cm^2/s);
> a is the radius of the quencher in centimeters (the distance of closest encounter);
> k is Boltzmann's constant;
> N' is Avagodro's constant divided by 1,000, and
> η is the viscosity.
> $[Q]$ is the concentration of the quencher in moles per liter.

This will give the rate constant k_q in liter mole^{-1} s^{-1}, and it usually has a value of about 10^9–10^{10} liter mole^{-1} s^{-1} for ion quenching.

If the quencher concentration is known, we can determine the effective viscosity of the environment. The rate of dynamic quenching is inversely dependent on the viscosity of the environment and representative dimensions of the quencher molecule and the fluorophore. If the quencher is spherical (with radius "a") and only the quencher diffuses, then $k_{quencher} \propto 1/6\pi\eta a$, where.

Thus, the lifetime not only can provide corrections to dynamic quenching, but also can furnish indications of the rigidity (effective viscosity) of the molecular environment of a fluorescence probe through the dynamics of quenching. (Actually, it provides information about the relative mobility of the fluorescence probe and the quencher molecule.) This could be deduced only with great difficulty from steady-state experiments in an image, but it is a simple experiment for FLI.

is essentially an excited-state reaction that does not destroy the fluorophore). For instance, pyrene forms eximers–dimers of pyrene, where one of the reaction partners is a molecule in the excited state and the other bimolecular partner is a pyrene in the ground state. The eximer emission is considerably red, shifted from the emission of the monomolecular species. However, the eximer is an independent chemical species and as such it is the product of an excited-state reaction and will shorten the emission lifetime (and lower the intensity) of the original independent monomolecular excited pyrene molecules. FRET can be considered to be an excited-state reaction.

1.5.6 Förster Resonance Energy Transfer (FRET)

FRET is one of the major applications of FLIM. In FRET the excitation energy of one molecule (called the donor, D) is transferred nonradiatively to a nearby molecular chromophore

BOX 1.4

The rate of energy transfer between single donor and acceptor molecules is proportional to $1/R^6$, where R is the distance between the centers of the two chromophores. Förster (1946, 1948, 1951) showed that the rate of energy transfer could be expressed as

$$k_{ET} = \frac{1}{\tau_{F_{D}^{-A}}}\left(\frac{R_0}{R}\right)^6 \tag{1.1}$$

R_0 is the value of R where the rate of energy transfer, k_{ET}, equals the rate of de-excitation from the excited state in the absence of the acceptor $(1/\tau_{F_{D}^{-A}})$. R_0 can be calculated from knowledge of the relative orientations between the transition dipoles and the spectral overlap of the emission spectrum of the donor and the absorption spectrum of the acceptor (Förster 1951).

in the ground state (called the acceptor, A). The probability of photon emission of D is thereby diminished, and D's lifetime in the excited state is shortened. FRET usually takes place over a D–A separation of 0.5–10 nm. A Coulomb charge–charge interaction (effectively, a dipole–dipole interaction) between the excited D molecule and the ground-state A molecule takes place through space; no photon is absorbed or emitted. Spectral overlap of the emission and absorption spectra of D and A is required. The quantum yield of D and the absorption coefficient of A must be great enough for a significant probability of transfer. FRET can be coupled to a wide variety of biological assays that yield specific information about the environments of the chromophores and distances between D and A. FRET is probably the major reason why many people want to make FLI measurements (see Chapters 2 and 9).

From the rate of energy transfer (see Box 1.4), we can gain *quantitative* information about the distance between D and A. We can sometimes learn about the relative orientation between D and A transition dipoles (the effectiveness of FRET depends on the orientation of the transition dipoles of D and A).

FLI overcomes difficulties in making reliable FRET measurements in an image using intensities. It is difficult to quantify FRET measurements in imaging experiments (fluorescence microscope) using steady-state fluorescence because standards must be used to calibrate the fluorescence signals; that is, we must compare the fluorescence intensity in the presence and absence of acceptor (Bright et al. 1989; Dunn and Maxfield 1998; Fan et al. 1999; Opitz 1998; Silver 1998). Usually, the variability of concentrations between different biological cells or the distribution within a biological cell is unknown. As can be seen from the equations in Box 1.5, there is no reason to calibrate intensities using FLIM; one only has to be able to measure the lifetimes accurately. This is the great advantage of lifetime measurements. Given that nowadays it does not take too much time to measure a lifetime-resolved image, FLI is the method of choice if one has the instrumentation. We defer further discussion of applications of FLI for FRET measurements until we have considered the methods (see also Chapter 2).

BOX 1.5

The efficiency of energy transfer can also be called the quantum yield of energy transfer because it is the fraction of times that excited molecules follow the ET pathway of de-excitation. Equivalently, it is the ratio of the rate of energy transfer (see Box 1.4) to the total rate (including energy transfer) of de-excitation from the excited state. The efficiency can be determined easily by measuring fluorescence lifetimes. One only has to measure (or know) the fluorescence lifetime of the donor in the absence of the acceptor, $1/\tau_{F_D^A}$, and measure the fluorescence lifetime in the presence of acceptor $1/\tau_{F_D^{+A}}$. The following ratio then determines the efficiency

$$E = \frac{\text{rate of energy transfer}}{\text{total de-excitation rate}} = \frac{\left(1/\tau_{F_D^{+A}}\right) - \left(1/\tau_{F_D^A}\right)}{\left(1/\tau_{F_D^{+A}}\right)}$$

$$= \frac{k_{ET}}{\tau_{F_D^{+A}}^{-1}} = \frac{k_{ET}}{\tau_{F_D^A}^{-1} + k_{ET}} = \frac{1}{\left(k_{ET}\tau_{F_D^A}\right)^{-1} + 1} = \frac{1}{1 + \left(\dfrac{R}{R_0}\right)^6} \qquad (1.2)$$

1.5.7 Intersystem Crossing and Delayed Emission

The first time-resolved imaging experiments were carried out on samples with delayed emission—phosphorescence and delayed fluorescence (Marriott et al. 1991). The time range of phosphorescence is microseconds to seconds. The ground states of most fluorophores are singlet states, and the first excited state is also a singlet state (the electrons in the highest occupied electronic levels, in the ground and the excited electron configurations, are paired with opposite spins). Thus, the transition from the excited state to the ground state (fluorescence) is "spin allowed" and therefore takes place in the nanosecond time scale. Due to spin–orbit coupling involving the electron that has been elevated to the excited state, there is a probability that the spin of the excited electron will flip, creating a triplet state (where two electrons have parallel spins) with lower energy (intersystem crossing, k_{isc}).

The transition between singlet and triplet states is not highly probable and is only partially allowed (not spin allowed). Therefore, the rate for the triplet molecule to deactivate (e.g., by emission of a photon) to the singlet ground state is much slower than the normal intrinsic rate of fluorescence or the deactivation by internal conversion. If the probability of a singlet–triplet transition (intersystem crossing) is high enough, then the formation of a triplet state will become a viable competitor with the other de-excitation processes of the original singlet state. Intersystem crossing then becomes a competing kinetic pathway, decreasing significantly the measured lifetime of the fluorescence emission from the singlet state. If the triplet can emit a photon, a long-lived emission decay is also observed. The triplet state is very short-lived unless oxygen is removed from solution; therefore, phosphorescence emission is usually not observed.

Triplet states have several important consequences, especially in the complex environment of a biological cell. Because of the unpaired available electron of the excited triplet molecule, the triplet is often highly reactive, and it can react with triplet scavengers such as triplet (ground state) oxygen. Oxygen is normally in solution at approximately 5 mM concentration. Upon reaction of triplet oxygen with the triplet excited state of the fluorophore, extremely reactive singlet oxygen is produced, which can then rapidly react with nearby molecules to produce a variety of highly destructive reactive oxygen species. These products (and the singlet oxygen) can lead to excited-state decomposition (photolysis) of the fluorophore itself.

If oxygen is rigorously removed, then some fluorophores, such as acridine orange and tryptophan, exhibit a high triplet emission quantum yield (phosphorescence) with very long lifetimes. *DLIM (delayed luminescence imaging microscopy)* makes use of this long delay to avoid completely the background fluorescence that is much faster than the phosphorescence. (Examples of lifetime-resolved delayed luminescence imaging can be found in Marriott et al. 1991 and Vereb et al. 1998.) A disadvantage of long-lived delayed luminescence is that the emission rate of photons is very slow, lowering the sensitivity (number of photons that can be counted per unit time).

1.5.8 Slow Luminescence without Intersystem Crossing

Lanthanides exhibit very long lifetime decays that are not due to triplet states (Chen and Selvin 1999; Xiao and Selvin 2001). The spectroscopic emission and absorption of lanthanides involve transitions between internal lower electronic orbitals that are well protected from the solvent environment; this leads to very long lifetimes of the excited states. The long lifetimes are accompanied by very narrow emission peaks (see earlier discussion concerning the energy–time uncertainty, $\Delta E \tau_f \sim \hbar$) that can very effectively be separated from other broad emissions with narrow band filters. The use of lanthanides has the great advantage that the delayed emission can be observed after the excitation has completely subsided, and this is easy to carry out (the lifetimes can be as long as a few milliseconds).

Time-resolved spectroscopy using lanthanides has found application in sensitive assay systems (Periasamy et al. 1995; Seveus et al. 1992, 1994) and has also been employed in imaging microscopy (Hennink et al. 1996; Marriott et al. 1991, 1994; Vereb et al. 1998). The absorption coefficients of lanthanides are extremely low by themselves (on the order of 1–10 $M^{-1}cm^{-1}$). However, it is possible to choose organic chelate structures with high absorption coefficients that very efficiently capture the ions (and hold them for very long times) and transfer essentially 100% of their excitation energy to the lanthanide ions (Xiao and Selvin 2001). This is a very effective way to excite the ions. The narrow emission energy bandwidth of the ions can be separated easily from other broad band fluorescence, and the long lifetime emission can be resolved temporally from prompt fluorescence. Selvin (1995) has conducted a review of measuring delayed emission from lanthanides. Imaging using the lanthanides is also an active development (De Haas et al. 1996; Vereb et al. 1998).

1.5.9 Photolysis (Process and Interpretation of Its Measurement)

The pathway of photolysis (irreversible destruction of the excited fluorophore) competes with the other pathways of de-excitation, usually via the triplet state. The result of photolysis is a product of an irreversible photochemical reaction, and the molecule is not returned to the pool of fluorescence-competent ground-state molecules. Every organic fluorophore will eventually undergo photolysis after an average number of excitations. This limits the number of photons that a single fluorescent molecule can emit; the number of photons is statistical distributed, is specific for each molecule (usually between 50,000 and 250,000), and depends on the environment (e.g., the oxygen concentration). Photolysis is a major inconvenience for all fluorescence microscopy experiments and is also a major problem for single-molecule experiments. In some cases, photolysis can be corrected for in FLIM (see Chapter 11).

Photolysis is not all bad. Photobleaching is a kinetic phenomenon in competition with the other kinetic deactivation processes in the excited state. Therefore, similar kinetic experiments such as that in Box 1.6 can be performed to determine other competitive rates (Clegg 1996) (see Box 1.5). The advantage is that the kinetics of photolysis is very slow (seconds to many minutes) compared to all the other rates (nanoseconds) and it is easy to carry out the kinetic experiments in this time range. This method, originally suggested and shown by Hirschfeld (1976), has been championed by Jovin and others (Jovin and Arndt-Jovin 1989a, 1989b; Kenworthy, Petranova, and Edidin 2000; Kubitscheck et al. 1993; Szabo et al. 1992; Young et al. 1994).

Box 1.6 shows how the kinetics of photolysis can be used to measure FRET. Again we see how simple it is to measure E_{ET} when measuring directly the kinetics of a dynamic process. This is also a time-resolved method of determining competitive rate constants directly, albeit very slowly, and the method eventually destroys the fluorescence of the sample and may be destructive for biological cells. Another simple way to measure FRET using photobleaching is to measure the fluorescence intensity of the donor in the presence of active

---

BOX 1.6

In the presence of other effective de-excitation pathways (such as FRET), the overall ensemble rate of photolysis will be diminished. Therefore, measuring the rate of photolysis of a donor in the presence and absence of an acceptor provides a method for determining the efficiency of energy transfer using equations similar to those describing the use of the fluorescence lifetime for determining the efficiency of FRET (Box 1.4). For instance,

$$E_{ET} = \frac{k_{PB}^{-A} - k_{PB}^{+A}}{k_{PB}^{-A}} = \frac{\left(\dfrac{1}{\tau_{PB-A}}\right) - \left(\dfrac{1}{\tau_{PB+A}}\right)}{\left(\dfrac{1}{\tau_{PB-A}}\right)} = 1 - \frac{\tau_{PB-A}}{\tau_{PB+A}} = \frac{1}{1 + \left(\dfrac{R}{R_0}\right)^6} \tag{1.3}$$

The subscript *PB* stands for "photobleaching" and, as before, the +A and −A refer to "in the presence and absence of the acceptor" (Clegg 1996).

acceptor and again after photobleaching the acceptor. After photobleaching of the acceptor, the donor fluorescence increases, and the fluorescence intensity of the donor is now that of the donor molecules in the absence of any acceptor (the acceptor is photolyzed and can no longer accept energy from the donor).

1.5.10 The Unifying Feature of Extracting Information from Excited-State Pathways

The unifying feature of the preceding mechanisms of de-excitation is that each pathway of de-excitation contributes individually and separately to the overall rate of excited state deactivation and every pathway of de-excitation competes with all the others. The separate rates of any individual pathway can be determined by measuring the relaxation rate of the fluorescence signal in the presence and absence of this pathway. The mechanism given in Box 1.5 is general. Any pathway could be used to make the measurement of any other pathway. Fluorescence is chosen in FLIM because it is relatively easy to measure. It is sensitive, versatile, and can be carried out at low concentrations with picosecond accuracy and high reproducibility.

1.6 OTHER PARAMETERS RELATED TO LIFETIME-RESOLVED FLUORESCENCE—DYNAMIC AND STEADY-STATE MEASUREMENTS

Other kinetic processes are united with the rate of deactivation of the excited state in addition to fluorescence decay. The time course of deactivation of the excited state is coupled to all other fluorescence measurements—steady state (integrated over time) or dynamic. We discuss two of these that are common. The experimenter should be aware of these effects because they can affect the FLIM measurements. They also have the possibility to provide independent estimates of the apparent fluorescence lifetimes, which can be compared to the direct FLIM measurements.

1.6.1 Anisotropy Decay

The decay of the extent of polarization is usually due to the physical molecular rotation of the excited molecule (Brownian rotation). Essentially, all fluorescence mechanisms operating in fluorescence and in FLIM involve dipole absorption and emission. That is, the transition from the ground to the excited state of a fluorophore is such that a well defined linear vector (the absorption dipole) within the molecular coordinate frame defines the direction in space that the oscillating electric field of the excitation light must have in order to display maximum absorption. Similarly, for emission, the electric vector of the emitted fluorescence (which is perpendicular to the direction of the outgoing light) is directed in space parallel to the emission dipole.

These dipole directions, which are often (but not necessarily) fairly parallel to each other in the molecular coordinate frame, are usually called the "absorption transition dipole" and the "emission transition dipole." They are not real dipoles, but rather represent the configurational transition of the electrons between states. One can picture these dipoles as the way the electric field vector of the excitation light can grab the electronic structural changes in the molecule to drive it into an excited state. The reverse can be pictured for the emission process. Of course, these transition dipoles are "dipole transitions" that are well defined by the dipole approximation of time-dependent quantum mechanical perturbation

BOX 1.7

The excited-state lifetime presents a time window to observe competing molecular relaxation processes. The extent of molecular rotation that takes place before fluorescence is emitted will depend on the length of this time window—that is, the statistical probability that the molecule remains in the excited state. The expression linking the fluorescence lifetime to the degree of polarization for the simple case of a freely rotating molecular sphere (where the fluorophore transition dipole is attached rigidly to the sphere) is

$$r_{\text{steady-state}} = \frac{I_{par} - I_{perp}}{I_{par} + 2I_{perp}} = r_0 \frac{1}{1 + 6D_{rot}\tau_F} \tag{1.4}$$

where

I_{par} and I_{perp} are the fluorescence intensity measured parallel and perpendicular to the excitation light polarization

r_0 is the maximum anisotropy (before rotation at zero time), which is 0.4 if the absorption and emission transition dipoles are exactly parallel

D_{rot} is the rotational diffusion constant of the molecular sphere

τ_F is the measured fluorescence lifetime of the chromophore

For a simple sphere (with diameter R_s), there is only one rotational diffusion constant,

$$D_{rot} = k_B T / 8\pi\eta R_s^3$$

where η is the viscosity and k_B is Boltzmann's constant. Such a simple relationship is only valid for a sphere, but this is often a good first approximation.

The rotational correlation time of a sphere is $\tau_{rot} = 1/6D_{rot}$, and the time dependent decay of anisotropy of a sphere is $r(t) = r_0 \exp(-t/\tau_{rot}) = r_0 \exp(-6t/D_{rot})$. More complex expressions correspond to nonspherical molecular structures and to cases of rotation of a covalently linked dye relative to the rotation of a macromolecule to which it is attached. The rotational correlation times become especially important for FRET (see Chapters 2, 10, and 12).

theory. For this discussion, we can think of them classically. As the molecule rotates, the statistical correlation of the emission molecular transition dipole with the original orientation of the transition dipole of the excited molecule decreases with time; that is, the ensemble of excited molecules loses "memory" of the original molecular orientations.

Because the electric field of the electrodynamic wave emitted by fluorophores is oriented along the molecular transition dipole axis, the measured anisotropy will tend to zero as the nonrandomly oriented photoselected ensemble of excited molecules undergoes free rotation. The time constant of anisotropy decay is related to the rotational diffusion coefficient of the molecule (see Box 1.7). Molecular rotations do not change the excited-state lifetime; only the extent of anisotropy is affected. If energy transfer takes place between identical molecules (homotransfer) the anisotropy will also change if the energy transfer is to a molecule with a different orientation of its transition dipole. However, the fluorescence lifetime is not changed. This is the only way to observe homotransfer.

If the excitation light is polarized, the measured rate of fluorescence decay will show signs of polarization *unless* the angle between the excitation and emission polarizers' axes is adjusted such that the *measured* polarization of any sample, no matter to what extent it is polarized, is zero (this angle is 54.72° and is called the "magic angle") (Lakowicz 1999; Spencer and Weber 1970). If the excitation and emission linear polarizers are adjusted to have their polarization directions at the magic angle to each other, the fluorescence decay will not be affected by changes in molecular orientations. Achieving magic angle conditions in a microscope is difficult and is usually not even attempted in FLIM experiments. But one should be aware of a possible artifact due to polarization, especially because samples in biology are often highly polarized. If the fluorophore randomizes its orientation much faster than the fluorescence emission, then the anisotropy is zero, and rotations also have no effect on the measured fluorescence decay.

One cannot avoid polarization by just using unpolarized excitation light. Incoming excitation light always polarizes the sample with cylindrical symmetry (along the axis of light propagation), even if the excitation light is unpolarized in the plane perpendicular to the direction of the light beam. Another important point to remember is that all optical components tend to partially polarize light propagating through. Of course, the usual sample in a fluorescence microscope is not randomly oriented because the dyes are attached to rigid particles. These problems have recently been analyzed and methods for correction given (Koshioka, Sasaki, and Masuhara 1995). General references to polarization can be found in any textbook on fluorescence (Lakowicz 1999).

Interestingly, if τ_F is known (which we can do with FLIM), then we can calculate D_{rot} by measuring $r_{\text{steady-state}}$ (see Box 1.7). On the other hand, as mentioned previously, if we know D_{rot}, we can calculate τ_F. Thus, if the rotational correlation times are known, we can even estimate fluorescence lifetimes from steady-state anisotropy measurements. These apparent lifetimes can be compared to the normal FLIM-measured lifetimes. Also, if we know τ_F and D_{rot}, then we can determine the effective viscosity of the solvent surrounding the rotating sphere. Anisotropy is a very powerful spectroscopic tool.

In general, following excitation, molecules will change their orientation by rotational diffusion (free rotation or hindered rotation). For small molecules and macromolecules similar in size to soluble proteins, the rotational correlation times are picoseconds to hundreds of nanoseconds. This is in the range that can be observed by fluorescence with nanosecond lifetimes. If we can measure the time dependence of the anisotropy decay (as with an FLI instrument with polarization optics), more accurate and reliable molecular information concerning molecular rotations can be determined.

Many studies and reviews of rotational diffusion have been conducted (Cherry 1979; Chuang and Eisenthal 1972; Ehrenberg and Rigler 1972; Lakowicz 1999; Lombardi and Dafforn 1966; Memming 1961; Perrin 1934; von Jena and Lessing 1979; Weber 1953; Williams 1978), but only recently have FLI enthusiasts turned their attention to this measurement (Clayton et al. 2002; Lidke et al. 2003). Molecular rotations' dynamic rotational movements tell us a great deal about the molecular environment of the macromolecules to which fluorescence probes are attached, and dynamic fluorescence measurements are necessary to extract this information.

1.6.2 Steady-State Quenching (Dynamic) Measurement

Another example of how steady-state measurements can be used to gather lifetime information is the measurement of the extent of dynamic quenching (Section 1.5.4). Dynamic quenching is due to molecular collision of the excited-state fluorophore with a diffusing quencher. The extent of quenching will be greater the longer fluorophore remains in the excited state. If a known concentration of quencher is calibrated for its quenching effectiveness with a certain fluorophore, then the intensity of fluorescence can be used as a measure of the fluorescence lifetime. This can be illustrated by inspecting a Stern–Volmer plot (Box 1.8).

BOX 1.8: STERN–VOLMER QUENCHING PLOT DESCRIPTION (VALEUR 2002)

For this calculation, we just calculate the quantum yield in the presence and absence of the quencher, using the equations in Box 1.3. Define τ_0 as the fluorescence lifetime in the absence of quencher. Then, using the methods outlined in Boxes 1.1 and 1.3, it is easy to show that $\Phi_0/\Phi = F_0/F = 1 + k_q\tau_0[Q]$, where Φ_0/Φ is the ratio of quantum yields in the absence, Φ_0, and presence, Φ, of the quencher, and F_0/F is the corresponding ratio of the measured steady-state fluorescence intensities. A plot of F_0/F versus $[Q]$ is a Stern–Volmer plot, and the slope is $k_q\tau_0$.

However, in contrast to the polarization measurements, where the number of excited species stays the same after transfer, dynamic quenching does compete directly with the basic deactivation pathways and reduces the number of excited species, changing the lifetime. Static quenching (forming a ground-state nonfluorescent species) does not affect the rate of deactivation of the activated state of those molecules not statically quenched. Therefore, it does not affect the fluorescence decay rate and cannot be used to determine the lifetime.

1.7 DATA ACQUISITION

In general, there are two optical configurations for gathering image data (scanning and full field) and two approaches to acquiring the dynamic information (time and frequency domains). As mentioned before, two other parameters mirror the fluorescence lifetimes. This section is only an overview of the following chapters, which explain the methods and analyses in detail.

1.7.1 Scanning and Full Field

Optical imaging measurements can be carried out in full-field (wide-field) or scanning modes. In *full-field mode,* all regions of an imaged object are illuminated simultaneously and all pixels are measured simultaneously. This parallel method of acquiring FLIM images captures the final images with a parallel array detector such as CCD (charge-coupled device) cameras, and the modulated excitation light illuminates the full field of the imaged sample. In *scanning mode,* a tightly focused beam of modulated excitation light scans the object (as in scanning confocal microscopy). The pixels of the recorded image are gathered serially, usually with a single-channel photodetector (photomultiplier or equivalent).

TABLE 1.1   Advantages of Scanning Two-Photon and Full-Field FLI

Scanning two-photon FLI	Full-field FLI
Spatial confinement of excitation-diffraction limited focusing	Simultaneous collection of all pixels
Confocal effect:	Even x–y illumination (low iris effect)
0.3 m × 1 m ($(c/h\nu)_{ex}$ = 700 nm$_{ex}$, NA = 1.3)	Simplicity of optical construction and operation; attach to any microscope
Little or no photo damage outside the 2-$h\nu$ region	FLIE (endoscopy)
Depth of penetration	Three-dimensional images possible with image deconvolution
Three-dimensional images possible without deconvolution	CCD data acquisition (long integration times possible without unreasonable total measurement time)
UV-excitation (localized) because of two-photon excitation	
PM detection—multifrequencies—Fourier spectrum	Phosphorescence (DLIM)
Detection straightforward—same as single-channel cuvette measurement	Real-time (video rate) lifetime-resolved image acquisition
Localized photo activation of caged compounds; localized rapid kinetic acquisition	Time resolution for image kinetics in millisecond range

Each of these methods has been reviewed (Bastiaens and Squire 1999; Clegg and Schneider 1996; Clegg, Schneider, and Jovin 1996; Dong et al. 1995; Dowling et al. 1998; French et al. 1998; Gadella 1999; Koenig and Schneckenburger 1994; Lakowicz and Szmacinski 1996; Phillips 1994; So et al. 1996; vande Ven and Gratton 1992; Wang et al. 1996) and the following chapters explain the fundamentals of the different methods. An overview of each method is given in Table 1.1. The list given in this table is to be considered only as a guideline. The following chapters in this book should be consulted for in-depth expositions; the chapters concerning the different methods are labeled in Table 1.1.

1.7.1.1 Scanning Modes

In essence, the scanning modes of data acquisition are similar to the classical one-channel, nonimaging techniques (Wilson and Sheppard 1984). The only difference is that the excitation light is scanned over the object and the lifetime is measured at different locations of the image (this is not meant to imply that the instrumentation is trivial). The scanning technique acquires data serially. Each point of the image is measured separately. For this reason, in order to acquire the images in reasonable times, the measurement at each pixel must be carried out rapidly. The critical limitation on the speed of point-by-point acquisition is the number of photons required to make a statistically sound lifetime measurement.

Scanning imaging methods can acquire well-resolved three-dimensional images by using confocal imaging, and the lifetime-resolved scanning imaging techniques have the same advantage. Another advantage of scanning techniques is that two-photon excitation can be employed. The excitation beam is highly focused, producing (in a little less than 1 μm^3 volume) a very high intensity locally, and the fluorophore can then be excited with two photons simultaneously. The modern pulsed Ti:sapphire lasers are ideal for the two-photon excitation, and the very short light pulses of this laser and the repetition frequency of the laser pulses are suitable for lifetime measurements (Gratton et al. 2001; So et al. 1998; vande Ven and Gratton 1992).

This laser is often used in both frequency- and time-domain measurements. The pulse is very short, which is needed for the time domain, and the repetitive frequency is in the right range for frequency-domain measurements. These scanning methods do not need to use spatial deconvolution techniques to remove out-of-focus fluorescence because the volume imaged through a pin hole in a confocal arrangement or the volume excited in a two-photon excitation arrangement is naturally limited in the z-direction. These confocal or two-photon scanning methods have also been extended to multifocus arrangements (using multilens objectives), thus allowing parallel lifetime-resolved data acquisition of many point-focused spots (Lakowicz et al. 1997; Schonle, Glatz, and Hell 2000).

1.7.1.2 Full-Field Modes

Full-field imaging has the great advantage that the measurement of the fluorescence response at every pixel in the image (and this can be up to a million pixels) is made simultaneously, saving considerable time (French et al. 1998; Gadella, Jovin, and Clegg 1993; Harpur, Wouters, and Bastiaens 2001; Squire, Verveer, and Bastiaens 2000). This makes full-field lifetime-resolved imaging the method of choice for real-time imaging applications. All full-field instruments require a high-frequency modulatable image detector, and this is usually accomplished by using a microchannel plate, high-frequency modulated intensifier (however, see Mitchell et al. 2002a, 2002b). The drawback of these devices is the cost and the noise of the intensifier (which has the same noise characteristics as a photomultiplier; see Chapters 5 and 12 for discussions of noise characteristics of the different detectors). But they are commercially available and can be operated at very high frequencies with almost no deleterious iris effects (for the frequency domain) or gating into the subnanosecond range (for the time domain).

The output of the intensifier (which is a phosphor screen onto which the accelerated electrons from the microchannel plate are focused, pixel by pixel) is focused onto a CCD camera; after averaging, the image is read into a computer for analysis. This whole process (data acquisition, image analysis, and image display) can now be made at video rates when the frequency domain is used (Clegg, Holub, and Gohlke 2003; Holub et al. 2000; Redford and Clegg 2005). Especially in recent years, a large selection of CCD cameras with excellent noise characteristics and very fast frame grabbers is available to transfer the data to the computer. PCs are now fast enough to control the experiment, gather the data, analyze the data to extract the lifetime information, and display the resultant images (Holub et al. 2000).

1.7.2 Time and Frequency Domains

In order to acquire nanosecond temporal resolution of the fluorescence decays, the excitation light must be modulated in a comparable time range, and the response of the fluorescence emission must be tracked in the nanosecond (and subnanosecond) time range. Other very rapid methods for exciting the fluorophore can also be employed, such as transferring energy from an excited donor molecule to an acceptor molecule (FRET) (see Chapter 11). The experimental techniques are usually subdivided into time- and frequency-domain methods and discussed separately; however, this is arbitrary. This artificial separation evolved historically by analogy to nomenclature in signal analysis. However, this notation is entrenched in the literature, so we will refer to this classification in this exposition.

All methods of lifetime determinations involve recording the time response of the fluorescence signal. The measured fluorescence response is a convolution of the "fundamental fluorescence response" with the time course of the excitation light. The fundamental fluorescence response is defined as the fluorescence signal that evolves following a "delta function excitation pulse" (we define the delta function excitation pulse as significantly shorter than all the fluorescence lifetimes). The fundamental fluorescence response is usually a weighted summation of exponential decays (see Chapter 11). The recorded time course of the fluorescence emission depends on the temporal form of the excitation event. Typically, the excitation consists of a train of identical repetitive pulses. The shapes of the individual repetitive pulses vary from very short repetitive pulses (pulse duration down to 50–100 fs) to continuous sinusoidal modulation. The pulses (modulation) are repeated as often as necessary and averaged to achieve a good signal-to-noise ratio.

In the time domain, the signal is recorded and analyzed directly as a function of time. If the excitation pulses are very short and provided that the repetition period is at least five times longer than the longest fluorescence lifetime, the fluorescence response can be analyzed directly as the fundamental fluorescence response (see Chapter 11). With the advent of pulse lasers with femtosecond pulse lengths, this is now routine. However, if the pulse is not sufficiently shorter than the lifetimes or the repetition frequency is too great, the fluorescence signal must be deconvolved from the temporal form of the excitation pulse before being analyzed as a weighted sum of exponentials corresponding to the fundamental fluorescence response of the system under study (see Box 1.1).

In the frequency domain, the time-dependent repetitive fluorescence response is expressed in terms of a phase lag and degree of demodulation relative to the phase and modulation depth of the excitation light. In essence, the excitation repetitive pulse signal is expressed as a Fourier series corresponding to the fundamental repetition frequency and all overtone frequencies of the excitation wave train. The fluorescence response is a repetitive signal with identical frequency components to those of the excitation light; however, the modulation depth and phase of each frequency component (relative to the corresponding component of the pulse train) are functions of the lifetime components of the fundamental fluorescence response of the system under study (see Chapter 11).

1.7.2.1 Time Domain

This method is easy to understand because the analysis takes place directly in the time domain. Time-domain measurements require that the data acquisition be fast enough to record the fluorescence intensity for several (sometimes hundreds of) consecutive time periods that are a fraction of the total decay time. The time progression of the fluorescence decay is recorded at specific times by delaying the period of observation following the start of each excitation pulse. The length of the delay before starting the measuring period is varied, and in this way the time course of the fluorescence decay is captured.

In a scanning mode, the time course of the fluorescence decay is recorded at every pixel separately. In the full-field mode, all pixels of the image are recorded simultaneously at every time period. This requires a means of time-gating the image; this is done using a fast gating image intensifier placed before the final image recorder, which is usually a CCD. The reader is referred to Chapters 6–9 and 12 and the literature for details of the instrumentation and analysis (Barzda et al. 2001; Carlsson et al. 2000; Cole et al. 2001; Cubeddu et al. 1999; Gerritsen, Vroom, and de Grauw 1999; Kohl et al. 1993; Minami and Hirayama 1990; Oida, Sako, and Kusumi 1993; Sanders et al. 1995; Smith et al. 1994; Van Der Oord et al. 1995; Wang et al. 1996, 1991). A technique for recording the lifetime information in the time domain that uses single-photon-counting techniques and a quadrant detection scheme to identify the location of fluorescence emission from an object is being developed based on new time- and space-correlated single-photon counting MCP-PMT instrumentation (Kemnitz 2001). This technique has a 10 ps resolution (see Chapter 14).

1.7.2.2 Frequency Domain

In the frequency domain, a continuous high-frequency (HF) repetitive train of pulses, which can be of any shape, excites the fluorophores (Birks 1970; Jameson, Gratton, and Hall 1984). Because this method is not as familiar as the direct time-domain measurements, we describe this in more detail. If the HF modulation is not purely sinusoidal (i.e., a sinusoid signal of one frequency), the repetitive pulse train will contain the frequency components required to define the repetitive pulse shape according to a Fourier analysis (see Chapters 5 and 11).

As the frequency is increased, the phase of the fluorescence is increasingly delayed relative to the phase of the excitation light modulation (0–90°) and the demodulation (the ratio of the AC to DC component of the fluorescence to the same AC/DC ratio of the excitation light) decreases from one to zero (Chapter 5). The range of high frequency over which the phase and demodulation changes are measured is determined by the inverse of the time decay of the fluorescence (usually between 10 and 200 MHz). The data are not recorded directly in the time frame of the 10–200 MHz high-frequency modulation. The detector amplification is modulated at a high frequency. The output of the mixing operation is a signal of the difference frequency between the frequencies of the excitation and fluorescence signals. The detection of the output signal involves either *heterodyne* techniques (where the two frequencies are different but close and the acquired signal is transferred by the mixing to the difference frequency) or *homodyne* techniques (where the frequencies are exactly the same and the acquired signal is transferred by the mixing to a DC signal).

The high-frequency phase and modulation information that contains the fluorescence lifetime parameters is preserved in both cases. In the homodyne measurement, the phase and demodulation of the fluorescence relative to the excitation light are determined by varying the phase between the high-frequency modulation of the excitation light and the detector, and recording the phase and modulation at each phase setting. These methods are described in Chapter 5 and have been described in detail in the literature (Birks 1970; Birks and Dawson 1961; Clegg and Schneider 1996; Clegg et al. 1996; Gadella et al. 1993; Gadella 1999; Holub et al. 2000; Lakowicz and Berndt 1991; Morgan, Mitchell, and Murray 1990, 1992; Squire et al. 2000). In the frequency domain, the fluorescence signal is described as a sum of sinusoids (which are the Fourier components of the excitation waveform) with characteristic phase and modulation amplitudes related to the fluorescence lifetimes. The fluorescence lifetimes are determined from the phase and demodulation values, which are functions of the modulation frequency (Chapters 5 and 11).

1.7.3 Equivalence of Time and Frequency Domains

The time and frequency domains are finite Fourier transforms of each other (Kraut 1967). The advantages and disadvantages of each technique are related more to the availability and cost of instrumentation, rather than to anything fundamental about the measurement process. As explained earlier, the basic physics of the time- and frequency-domain FLIM are essentially identical. In both cases (time and frequency domains), the time-dependent shape of the fluorescence signal arises from the convolution of the excitation waveform with the fundamental fluorescence response. The two methods of analysis portray the experiment and display the data in the two different Fourier spaces: time and frequency (Karut 1967).

1.7.4 Performance Goals and Comparisons

Several performance goals need to be considered in designing and constructing the lifetime-resolved imaging instruments. These are discussed in other chapters. Some of the major considerations are

1. accuracy of the lifetime determination;

2. sensitivity of the measurement;

3. time of data acquisition, data analysis, and display;

4. ease of making the measurement;

5. type and complexity of data analysis required; and

6. cost.

As usual, trade-offs must be made, depending on the most relevant requirements for a particular application. The basic components of the different measurement modes and the many variations are given in the corresponding chapters of this book, and the reader is also referred to the original literature. Specifically, Chapters 4, 5, 7, and 11–13 (especially Chapter 12) discuss the goals and performances of the different techniques. This is a rapidly changing field, and some of the later publications contain reference to earlier literature (Cubeddu et al. 1997; Gerritsen et al. 1999; Holub et al. 2000; Schneider and Clegg 1997; So et al. 1996; Squire et al. 2000).

1.8 DATA ANALYSIS

A variety of algorithms for analyzing lifetime-resolved data at every pixel of the recorded image have been proposed and implemented. Time-domain measurements are analyzed as exponentials and the frequency-domain measurements are analyzed by determining the phase and demodulation of the fluorescence relative to the excitation light modulation for all the frequency components (Clegg and Schneider 1996). Essentially, all the numerical methods developed for fitting single-channel fluorescence lifetime data (cuvette measurements) are also applicable; however, because the number of pixels varies from 10^3 to 10^6, the analysis of FLIM data can become very time consuming if many of these sophisticated methods are simply applied to FLIM data (see Chapters 11 and 12).

The history of data regression to determine the fluorescence lifetimes and their corresponding amplitudes is extensive. One major objective of these earlier programs was a detailed analysis of the required statistics to differentiate close lifetimes and to determine whether the fluorescence signal requires a discrete sum of exponentials or a continuous distribution of exponential time constants to interpret the fluorescence response (deconvoluted from the form of the excitation pulse). The sophistication and advanced technology of single-channel instrumentation generally delivers very high signal-to-noise data; consequently, fluorescence lifetime measurements have been very successful in aiding understanding of detailed mechanisms of macromolecular dynamics and biomechanisms (Cundall and Dale 1983; Istratova and Vyvenko 1999).

The key prerequisite for improving signal-to-noise ratio of any signal with random photon-counting noise is to increase the number of detected photons. This is an unavoidable constraint—whether the data are acquired in analogue or direct photon-counting fashion. The best one can do is to minimize as much as possible all other sources of noise arising from interference or directly from the equipment (dark noise, read-out noise, amplifier noise, etc.). If each pixel of a FLIM image is to be analyzed separately (not often the case), this poses a major problem if the goal is to acquire data that can be analyzed with the same statistical accuracy of the single-channel experiments. This problem is compounded because the number of fluorophores corresponding to a single pixel is usually many fewer than the number of molecules available to be excited in a cuvette lifetime experiment.

The chapters in this book discuss aspects of different approaches to analyze FLIM data. The methods chosen depend on the objective of each separate experiment, the experience of the research group, the quality of the data, and the availability of the requisite software—rather than on any theoretical motivation. Because of the large number of recorded picture elements, the challenge is to carry out the data analysis (and, of course, the data acquisition) on hundreds of thousands of pixels in a reasonably short time. Averaging (collecting the signal over a large number of repetitions) is always employed to reduce the statistical noise (no FLIM method uses only a single pulse of excitation). In FLIM, because the data are acquired in multiple pixels, the statistics can be improved by averaging temporal data over many pixels deemed to present identical lifetime data; for instance, global analysis can significantly improve the signal-to-noise ratio by assuming that identical fluorescence relaxation parameters pertain to many different pixels (see Chapter 13).

New opportunities for analysis (in addition to the obvious key advantage of having acquired the spatial distribution of fluorescence lifetimes in an extended sample) are presented by FLIM simply because they are collected in image format. This has been treated in detail in chapters in this book and elsewhere (Pepperkok et al. 1999; Squire and Bastiaens 1999; Verveer, Squire, and Bastiaens 2000, 2001). It is important to remember that the objectives of many FLIM applications are designed to detect differences of the overall fluorescence relaxation in various locations of the image, gather the image data rapidly enough to follow kinetic happenings in the biological sample, and track changes in time—rather than with ultra-accurate determinations of fluorescence lifetime parameters. The data analysis can be carried out in several different ways, depending on the objective and constraints of the experiment.

1.9 DISPLAY OF LIFETIME-RESOLVED IMAGES

FLI produces data sets that are more complex than the normal intensity images, especially as we add spectral and polarization capabilities to our repertoire for FLI imaging. Displaying the information for 10,000–100,000 pixels conveniently and informatively is demanding. This is especially challenging if the data are to be updated actively and displayed in real time. Commercial software is often not adequate for many needs of FLI because it is not fast enough or because the peculiar analysis employed for analyzing and displaying FLI data is not available.

However, FLI has matured significantly, and commercial instrumentation and analysis algorithms are rapidly becoming available. More sophisticated and informative routines for display will surely follow and are being developed in research laboratories. The judicious use of three-dimensional display, shading, lighting, color coding, and displaying contours can expediently and instructively accentuate critical elements of the data analysis to the user. In addition to showing where the lifetimes (or apparent lifetimes) are located, software for displaying FLIM data specifically tuned to disseminate the combination of all the spectroscopic information together with the information specific to FLIM easily, rapidly, and conveniently is still in its infancy compared to commercial programs available for image analysis and display of intensity images in cellular biology.

1.10 SUMMARY

The intention of this chapter is to present the fundamentals of fluorescence and to integrate the basics of fluorescence dynamics into a description of FLIM. The goal is pedagogical and is meant to orient readers less familiar with fluorescence mechanisms for the material presented in the following chapters. References to fluorescence can be found in popular textbooks (Becker 1969; Birks 1970; Lakowicz 1999; Valeur 2002); some of the general methods are also expanded and discussed in the following chapters, especially Chapters 2 and 3.

The probabilities of traversing different pathways of de-excitation from the excited state compete in a parallel kinetic fashion with the fluorescence pathway of deactivation; therefore, all the pathways other than the fluorescence pathway affect the measured fluorescence lifetime. It is this kinetic competition of de-excitation pathways through which we gain the information that lifetime-resolved fluorescence provides. The sensitivity of the dynamics of fluorescence emission to the surrounding environment of the fluorophore makes the direct knowledge of fluorescence lifetimes so valuable, especially for biological samples. The ability to acquire this dynamic information at every pixel of an image makes FLIM one of the most exciting emerging techniques in the field of optical microscopy.

REFERENCES

Atkins, P. W., and Friedman, R. S. 1997. *Molecular quantum mechanics*. Oxford: Oxford University Press.

Bagatolli, L. A., and Gratton, E. 1999. Two-photon fluorescence microscopy observation of shape changes at the phase transition in phospholipid's giant unilamellar vesicles. *Biophysical Journal* 77:2090–2101.

Barzda, V., de Grauw, C. J., Vroom, J. M., Kleima, F. J., van Grondelle, R., van Amerongen, H., et al. 2001. Fluorescence lifetime heterogeneity in aggregates of LHCII revealed by time-resolved microscopy. *Biophysical Journal* 81:538–546.

Bastiaens, P. I., and Squire, A. 1999. Fluorescence lifetime imaging microscopy: Spatial resolution of biochemical processes in the cell. *Trends in Cell Biology* 9:48–52.

Becker, R. S. 1969. *Theory and interpretation of fluorescence and phosphorescence*. New York: Wiley Interscience.

Birks, J. B. 1970. *Photophysics of aromatic molecules*. London: Wiley.

Birks, J. B., and Dawson, D. J. 1961. Phase and modulation fluorometer. *Journal of Scientific Instruments* 38:282–295.

Bright, G., Fisher, G., Rogowska, J., and Taylor, D. 1989. Fluorescence ratio imaging microscopy. *Methods in Cell Biology* 30:157–192.

Carlsson, K., Liljeborg, A., Andersson, R. M., and Brismar, H. 2000. Confocal pH imaging of microscopic specimens using fluorescence lifetimes and phase fluorometry: Influence of parameter choice on system performance. *Journal of Microscopy* 199:106–114.

Chen, J., and Selvin, P. R. 1999. Thiol-reactive luminescent chelates of terbium and europium. *Bioconjugate Chemistry* 10:311–315.

Chen, S.-H., and Kotlarchyk, M. 1997. *Interactions of photons and neutrons with matter—An introduction*. New Jersey: World Scientific.

Cherry, R. J. 1979. Rotational and lateral diffusion of membrane proteins. *Biochimica et Biophysica Acta* 559:289–327.

Chuang, T. J., and Eisenthal, K. B. 1972. Theory of fluorescence depolarization by anisotropic rotational diffusion. *Journal of Chemical Physics* 57:5094–5097.

Clayton, A. H. A., Hanley, Q. S., Arndt-Jovin, D. J., Subramaniam, V., and Jovin, T. M. 2002. Dynamic fluorescence anisotropy imaging microscopy in the frequency domain (rFLIM). *Biophysical Journal* 83:1631–1649.

Clegg, R. M. 1996. Fluorescence resonance energy transfer. In *Fluorescence imaging. Spectroscopy and microscopy*, ed. X. F. Wang and B. Herman, 137, 179–252. New York: John Wiley & Sons, Inc.

Clegg, R. M., Holub, O., and Gohlke, C. 2003. Fluorescence lifetime-resolved imaging: Measuring lifetimes in an image. *Methods in Enzymology* 360:509–542.

Clegg, R. M., and Schneider, P. C. 1996. Fluorescence lifetime-resolved imaging microscopy: A general description of the lifetime-resolved imaging measurements. In *Fluorescence microscopy and fluorescent probes,* ed. J. Slavik, 15–33. New York: Plenum Press.

Clegg, R. M., Schneider, P. C., and Jovin, T. M. 1996. Fluorescence lifetime-resolved imaging microscopy. In *Biomedical optical instrumentation and laser-assisted biotechnology,* ed. A. M. Verga Scheggi, S. Martellucci, A. N. Chester, and R. Pratesi, 325, 143–156. Dordrecht: Kluwer Academic Publishers.

Cole, M. J., Siegel, J., Webb, S. E. D., Jones, R., Dowling, K., Dayel, M. J., et al. 2001. Time-domain whole-field fluorescence lifetime imaging with optical sectioning. *Journal of Microscopy* 203:246–257.

Craig, D. P., and Thirunamachandran, T. 1984. *Molecular quantum electrodynamics. An introduction to radiation molecule interactions.* Mineola, MN: Dover Publications, Inc.

Cubeddu, R., Canti, G., Pifferi, A., Taroni, P., and Valentini, G. 1997. A real-time system for fluorescence lifetime imaging. *Proceedings of SPIE—the International Society for Optical Engineering, USA. Biomedical Sensing, Imaging, and Tracking Technologies II.* San Jose, CA. SPIE. IBOS. Feb. 11–13, 1997, 2976:98–104.

Cubeddu, R., Pifferi, A., Taroni, P., Torricelli, A., Valentini, G., Rinaldi, F., et al. 1999. Fluorescence lifetime imaging: An application to the detection of skin tumors. *IEEE Journal of Selected Topics in Quantum Electronics* 5:923–929.

Cundall, R. B., and Dale, R. E., eds. 1983. *Time-resolved fluorescence spectroscopy in biochemistry and biology.* NATO ASI series. Series A, life sciences. New York: Plenum.

De Haas, R. R., Verwoerd, N. P., Van der Corput, M. P., Van Gijlswijk, R. P., Sitari, H., and Yanke, H. J. 1996. The use of peroxidase-mediated deposition of biotin-tyramide in combination with time-resolved fluorescence imaging of europium chelate label in immunohistochemistry and in situ hybridization. *Journal of Histochemistry and Cytochemistry* 44:1091–1099.

Dietrich, C., Bagatolli, L., Volovyk, Z., Thompson, N., Levi, M., Jacobson, K., et al. 2001. Lipid rafts reconstituted in model membranes. *Biophysical Journal* 80:1417–1428.

Dong, C., So, P. T., French, T., and Gratton, E. 1995. Fluorescence lifetime imaging by asynchronous pump-probe microscopy. *Biophysical Journal* 69:2234–2242.

Dowling, K., Dayel, M. J., Hyde, S. C. W., Dainty, C., French, P. M. W., Vourdas, P., et al. 1998. Whole-field fluorescence lifetime imaging with picosecond resolution using ultrafast 10 kHz solid-state amplifier technology. *IEEE Journal of Selected Topics in Quantum Electronics* 4:370–375.

Dunn, K., and Maxfield, F. 1998. Ratio imaging instrumentation. *Methods in Cell Biology* 56:217–236.

Ehrenberg, M., and Rigler, R. 1972. Polarized fluorescence and rotational Brownian motion. *Chemistry and Physics Letters* 14:539–544.

Enderlein, J. 2002. Theoretical study of single molecule fluorescence in a metallic nanocavity. *Applied Physics Letters* 80:315.

Fan, G. Y., Fujisaki, H., Miyawaki, A., Tsay, R.-K., Tsien, R. Y., and Ellisman, M. H. 1999. Video-rate scanning two-photon excitation fluorescence microscopy and ratio imaging with cameleons. *Biophysical Journal* 76:2412–2420.

Fiuráek, J., Chernobrod, B., Prior, Y., and Sh., A. I. 2001. Coherent light scattering and resonant energy transfer in an apertureless scanning near-field optical microscope. *Physics Review B* 63:45420.

Förster, T. 1946. Energiewanderung und Fluoreszenz. *Naturwissenschaften* 6:166–175.

Förster, T. 1948. Zwischenmolekulare Energiewanderung und Fluoreszenz. *Annals of Physics* 2:55–75.

Förster, T. 1951. *Fluoreszenz organischer Verbindungen.* Göttingen: Vandenhoeck & Ruprecht.

French, T., So, P. T., Dong, C. Y., Berland, K. M., and Gratton, E. 1998. Fluorescence lifetime imaging techniques for microscopy. *Methods in Cell Biology* 56:277–304.

Gadella, T. W. J., Jr. 1999. Fluorescence lifetime imaging microscopy (FLIM): Instrumentation and applications. In *Fluorescent and luminescent probes,* 2nd ed., ed. W. T. Mason, 467–479, ISBN 0-12-447836-0. San Diego, CA: Academic Press.

Gadella, T. W. J., Jr., Jovin, T. M., and Clegg, R. M. 1993. Fluorescence lifetime imaging microscopy (FLIM): Spatial resolution of microstructures on the nanosecond time scale. *Biophysical Chemistry* 48:221–239.

Gerritsen, H. C., Vroom, J. M., and de Grauw, C. J. 1999. Combining two-photon excitation with fluorescence lifetime imaging. *IEEE Engineering in Medicine and Biology Magazine* 18:31–36.

Gratton, E., Barry, N. P., Beretta, S., and Celli, A. 2001. Multiphoton fluorescence microscopy. *Methods in Cell Biology* 25:103–110.

Gratton, E., and Limkeman, M. 1983. A continuously variable frequency cross-correlation phase fluorometer with picosecond resolution. *Biophysical Journal* 44:315–324.

Grinvald, A., and Steinberg, I. 1974. On the analysis of fluorescence decay kinetics by the method of least-squares. *Analytical Biochemistry* 59:583–598.

Hamann, H. F., Gallagher, A., and Nesbitt, J. D. 2000. Near-field fluorescence imaging by localized field enhancement near a sharp probe tip. *Applied Physics Letters* 76:1953–1955.

Hamann, H. F., Kuno, M., Gallagher, A., and Nesbitt, J. D. 2001. Molecular fluorescence in the vicinity of a nanoscopic probe. *Journal of Chemical Physics* 114:8596–8609.

Harpur, A. G., Wouters, F. S., and Bastiaens, P. I. 2001. Imaging FRET between spectrally similar GFP molecules in single cells. *Nature Biotechnology* 19:167–169.

Hennink, E. J., de Haas, R., Verwoerd, N. P., and Tanke, H. J. 1996. Evaluation of a time-resolved fluorescence microscope using a phosphorescent Pt-porphine model system. *Cytometry* 24:312–320.

Hirschfeld, T. 1976. Quantum efficiency independence of the time-integrated emission from a fluorescent molecule. *Applied Optics* 15:3135–3139.

Holub, O., Seufferheld, M., Gohlke, C., Govindjee, and Clegg, R. M. 2000. Fluorescence lifetime-resolved imaging (FLI) in real time—A new technique in photosynthetic research. *Photosynthetica* 38:581–599.

Istratova, A. A., and Vyvenko, O. F. 1999. Exponential analysis in physical phenomena. *Review of Scientific Instruments* 70:1233–1257.

Jameson, D. M., Gratton, E., and Hall, R. D. 1984. The measurement and analysis of heterogeneous emissions by multifrequency phase and modulation fluorometry. *Applied Spectroscopy Reviews* 20:55–106.

Jovin, T., and Arndt-Jovin, D. 1989a. FRET microscopy: Digital imaging of fluorescence resonance energy transfer. Application in cell biology. In *Cell Structure and Function by Microspectrofluorometry*, 99–117. Berlin: Springer.

Jovin, T., and Arndt-Jovin, D. 1989b. Luminescence digital imaging microscopy. *Annual Review of Biophysics and Biophysical Chemistry* 18:271–308.

Kauzmann, W. 1957. *Quantum chemistry. An introduction.* New York: Academic Press, Inc.

Kemnitz, K. 2001. Picosecond fluorescence lifetime imaging spectroscopy as a new tool for 3D structure determination of macromolecules in living systems. In *New trends in fluorescence spectroscopy. Applications to chemical and life sciences*, ed. B. Valeur and J.-C. Brochon, 381–410. Berlin: Springer.

Kenworthy, A. K., Petranova, N., and Edidin, M. 2000. High-resolution FRET microscopy of cholera toxin B-subunit and GPI-anchored proteins in cell plasma membranes. *Molecular Biology of the Cell* 11:1645–1655.

Koenig, K., and Schneckenburger, H. 1994. Laser-induced autofluorescence for medical diagnostics. *Journal of Fluorescence* 4:17–40.

Kohl, M., Neukammer, J., Sukowski, U., Rinneberg, H., Wohrle, D., Sinn, H. J., et al. 1993. Delayed observation of laser-induced fluorescence for imaging of tumors. *Applied Physics B—Photophysics & Laser Chemistry* B56:131–138.

Koshioka, M., Sasaki, K., and Masuhara, H. 1995. Time-dependent fluorescence depolarization analysis in three-dimensional microspectroscopy. *Applied Spectroscopy* 49:224–228.

Kraut, E. A. 1967. *Fundamentals of mathematical physics.* New York: McGraw–Hill.

Kubitscheck, U., Schweitzer-Stenner, R., Arndt-Jovin, D., Jovin, T., and Pecht, I. 1993. Distribution of type I Fc2-receptors on the surface of mast cells probed by fluorescence resonance energy transfer. *Biophysical Journal* 64:110–120.

Lakowicz, J. R. 1999. *Principles of fluorescence spectroscopy.* New York: Kluwer Academic/Plenum Publishers.

Lakowicz, J. R., and Berndt, K. W. 1991. Lifetime-selective fluorescence imaging using an RF phase-sensitive camera. *Review of Scientific Instruments* 62:1727–1734.

Lakowicz, J. R., Gryczynski, I., Malak, H., Schrader, M., Engelhardt, P., Kano, H., et al. 1997. Time-resolved fluorescence spectroscopy and imaging of DNA labeled with DAPI and Hoechst 33342 using three-photon excitation. *Biophysical Journal* 72:567–578.

Lakowicz, J. R., and Szmacinski, H. 1996. Imaging applications of time-resolved fluorescence spectroscopy. In *Fluorescence imaging spectroscopy and microscopy*, ed. X. F. Wang and B. Herman, 137, 273–311. New York: John Wiley & Sons, Inc.

Landau, L. D. 1997. *Quantum mechanics.* Oxford, England: Butterworth–Heinemann.

Lidke, D. S., Nagy, P., Barisas, B. G., Heintzmann, R., Post, J. N., Lidke, K. A., et al. 2003. Imaging molecular interactions in cells by dynamic and static fluorescence anisotropy (rFLIM and emFRET). *Biochemical Society Transactions* 31:1020–1027.

Lippert, E., and Macomber, J. D. 1995. *Dynamics of spectroscopic transitions. Basic concepts.* Berlin: Springer–Verlag.

Lombardi, J., and Dafforn, G. 1966. Anisotropic rotational relaxation in rigid media by polarized photoselection. *Journal of Chemical Physics* 44:3882–3887.

Marriott, G., Clegg, R. M., Arndt-Jovin, D. J., and Jovin, T. M. 1991. Time resolved imaging microscopy. Phosphorescence and delayed fluorescence imaging. *Biophysical Journal* 60:1374–1387.

Marriott, G., Heidecker, M., Diamandis, E. P., and Yan-Marriott, Y. 1994. Time-resolved delayed luminescence image microscopy using a europium ion chelate complex. *Biophysical Journal* 67:957–965.

Memming, R. 1961. Theorie der Fluoreszenzpolarisation für nicht kugelsymmetrische Moleküle. *Zeitschrift für Physikalische Chemie NF* 28:168–189.

Minami, T., and Hirayama, S. 1990. High quality fluorescence decay curves and lifetime imaging using an elliptical scan streak camera. *Journal of Photochemistry and Photobiology A* 53:11–21.

Mitchell, A. C., Wall, J. E., Murray, J. G., and Morgan, C. G. 2002a. Direct modulation of the effective sensitivity of a CCD detector: A new approach to time-resolved fluorescence imaging. *Journal of Microscopy* 206:225–232.

Mitchell, A. C., Wall, J. E., Murray, J. G., and Morgan, C. G. 2002b. Measurement of nanosecond time-resolved fluorescence with a directly gated interline CCD camera. *Journal of Microscopy* 206:233–238.

Morgan, C. G., Mitchell, A. C., and Murray, J. G. 1990. Nanosecond time-resolved fluorescence microscopy: Principles and practice. *Transactions of the Royal Microscopy Society* 1:463–466.

Morgan, C. G., Mitchell, A. C., and Murray, J. G. 1992. Prospects for confocal imaging on nanosecond fluorescence decay time. *Journal of Microscopy* 165:49–60.

Nickel, B. 1996. From the Perrin diagram to the Jablonski diagram. *EPA Newsletter* 58:9–38.

Nickel, B. 1997. From the Perrin diagram to the Jablonski diagram. Part 2. *EPA Newsletter* 61:27–60.

Oida, T., Sako, Y., and Kusumi, A. 1993. Fluorescence lifetime imaging microscopy (flimscopy). Methodology development and application to studies of endosome fusion in single cells. *Biophysical Journal* 64:676–685.

Opitz, N. 1998. Quantitative laser microscopy in single cells based on ratio-imaging of fluorescent indicators: A critical assessment. *Biomedizinische Technik (Berl)* 43 (suppl): 452–453.

Parker, C. A. 1968. *Photoluminescence of solutions.* Amsterdam: Elsevier.

Pepperkok, R., Squire, A., Geley, S., and Bastiaens, P. I. 1999. Simultaneous detection of multiple green fluorescent proteins in live cells by fluorescence lifetime imaging microscopy. *Current Biology* 9:269–272.

Periasamy, A., Siadat-Pajouh, M., Wodnicki, P., Wang, X. F., and Herman, B. 1995. Time-gated fluorescence microscopy in clinical imaging. Microscopy and analysis. *Microscopy Analysis* 11:33–35.

Perrin, F. 1934. Mouvement Brownien d'un ellipsoide (I). Dispersion dièlectrique pour des molécules ellipsoidales. *Journal de Physique et le Radium Ser VII* 5:497–511.

Phillips, D. 1994. Luminescence lifetimes in biological systems. *Analyst* 119:543–550.

Redford, G. I., and Clegg, R. M. 2005. Real-time fluorescence lifetime imaging and FRET using fast-gated image intensifiers. In *Molecular imaging: FRET microscopy and spectroscopy,* ed. A. Periasamy and R. N. Day, 193–226. New York: Oxford University Press.

Rink, T. J., Tsien, R. Y., and Pozzan, T. 1982. Cytoplasmic pH and free Mg2+ in lymphocytes. *Journal of Cell Biology* 95:189–196.

Sánchez, E. J., Novotny, L., and Xie, X. S. 1999. Near-field fluorescence microscopy based on two-photon excitation with metal tips. *Physical Review Letters* 82:4014–4017.

Sanders, R., Draaijer, A., Gerritsen, H. C., Houpt, P. M., and Levine, Y. K. 1995. Quantitative pH imaging in cells using confocal fluorescence lifetime imaging microscopy. *Analytical Biochemistry* 227:302–308.

Schiff, L. I. 1968. *Quantum mechanics.* New York: McGraw–Hill Book Co.

Schneider, P. C., and Clegg, R. M. 1997. Rapid acquisition, analysis, and display of fluorescence lifetime-resolved images for real-time applications. *Review of Scientific Instruments* 68:4107–4119.

Schonle, A., Glatz, M., and Hell, S. W. 2000. Four-dimensional multiphoton microscopy with time-correlated single-photon counting. *Applied Optics* 39:6306–6311.

Selvin, P. R. 1995. Fluorescence resonance energy transfer. *Methods in Enzymology* 246:300–334.

Seveus, L., Vaisala, M., Hemmila, I., Kojola, H., Roomans, G. M., and Soini, E. 1994. Use of fluorescent europium chelates as labels in microscopy allows glutaraldehyde fixation and permanent mounting and leads to reduced autofluorescence and good long-term stability. *Microscopy Research and Technique* 28:149–154.

Seveus, L., Vaisala, M., Syrjanen, S., Sandberg, M., Kuusisto, A., Harju, R., et al. 1992. Time-resolved fluorescence imaging of europium chelate label in immunohistochemistry and in situ hybridization. *Cytometry* 13:329–338.

Silfvast, W. T. 1996. *Laser fundamentals.* Cambridge: Cambridge University Press.

Silver, R. 1998. Ratio imaging: Practical considerations for measuring intracellular calcium and pH in living tissue. *Methods in Cell Biology* 56:237–251.

Smith, D. A. M., Williams, S. A., Miller, R. D., and Hochstrasser, R. M. 1994. Near-field time-resolved fluorescence studies of poly(phenylmethyl silane) with subwavelength resolution. *Journal of Fluorescence* 4:137–140.

So, P. T. C., French, T., Yu, W. M., Berland, K. M., Dong, C. Y., and Gratton, E. 1996. Two-photon fluorescence microscopy: Time-resolved and intensity imaging. In *Fluorescence imaging spectroscopy and microscopy*, ed. X. F. Wang, and B. Herman, 137, 351–373. New York: John Wiley & Sons, Inc.

So, P., Konig, K., Berland, K., Dong, C. Y., French, T., Buhler, C., et al. 1998. New time-resolved techniques in two-photon microscopy. *Cellular and Molecular Biology (Noisy-le-grand)* 44:771–793.

Spencer, R. D., and Weber, G. 1969. Measurement of subnanosecond fluorescence lifetimes with a cross-correlation phase fluorometer. *Annals of the New York Academy of Sciences* 158:361–376.

Spencer, R. D., and Weber, G. 1970. Influence of Brownian rotations and energy transfer upon measurements of fluorescence lifetimes. *Journal of Chemical Physics* 52:1654–1663.

Squire, A., and Bastiaens, P. I. 1999. Three-dimensional image restoration in fluorescence lifetime imaging microscopy. *Journal of Microscopy* 193:36–49.

Squire, A., Verveer, P. J., and Bastiaens, P. I. 2000. Multiple frequency fluorescence lifetime imaging microscopy. *Journal of Microscopy* 197:136–149.

Szabo, G., Pine, P., Weaver, J., Kasari, M., and Aszalos, A. 1992. Epitope mapping by photobleaching fluorescence resonance energy transfer measurements using a laser scanning microscope system. *Biophysical Journal* 61:661–670.

Szmacinski, H., and Lakowicz, J. R. 1993. Optical measurements of pH using fluorescence lifetimes and phase-modulation fluorometry. *Analytical Chemistry* 65:1668–1674.

Valeur, B. 2002. *Molecular fluorescence: Principles and applications.* Weinheim: Wiley–VCH.

Van Der Oord, C. J. R., Gerritsen, H. C., Rommerts, F. F. G., Shaw, D. A., Munro, I. H., and Levine, Y. K. 1995. Microvolume time-resolved fluorescence spectroscopy using confocal synchrotron radiation microscope. *Applied Spectroscopy* 49:1469–1473.

vande Ven, M., and Gratton, E. 1992. Time-resolved fluorescence lifetime imaging. In *Optical micros-copy: Emerging methods and applications*, ed. B. Herman, and J. J. Lemasters, 373–402. New York: Academic Press.

Vereb, G., Jares-Erijman, E., Selvin, P. R., and Jovin, T. M. 1998. Temporally and spectrally resolved imaging microscopy of lanthanide chelates. *Biophysical Journal* 74:2210–2222.

Verveer, P. J., Squire, A., and Bastiaens, P. I. 2000. Global analysis of fluorescence lifetime imaging microscopy data. *Biophysical Journal* 78:2127–2137.

Verveer, P. J., Squire, A., and Bastiaens, P. I. 2001. Improved spatial discrimination of protein reaction states in cells by global analysis and deconvolution of fluorescence lifetime imaging microscopy data. *Journal of Microscopy* 202:451–456.

von Jena, A., and Lessing, H. 1979. Rotational diffusion of prolate and oblate molecules from absorp-tion relaxation. *Berliner Bunsenges Physik Chemie* 83:181–191.

Wang, X. F., Periasamy, A., Wodnicki, P., Gordon, G. W., and Herman, B. 1996. Time-resolved fluorescence lifetime imaging microscopy: Instrumentation and biomedical applications. In *Fluorescence imaging spectroscopy and microscopy*, ed. X. F. Wang and B. Herman, 137, 313–350. New York: John Wiley & Sons, Inc.

Wang, X. F., Uchida, T., Maeshima, M., and Minami, S. 1991. Fluorescence pattern analysis based on the time-resolved ratio method. *Applied Spectroscopy* 45:560–565.

Weber, G. 1953. Rotational Brownian motion and polarization of the fluorescence of solutions. *Advances in Protein Chemistry* 8:415–459.

Williams, G. 1978. Time-correlation functions and molecular motions. *Chemistry Society Review* 7:89–131.

Wilson, T., and Sheppard, C. 1984. *Theory and practice of scanning optical microscopy*. London: Academic Press.

Xiao, M., and Selvin, P. R. 2001. Quantum yields of luminescent lanthanide chelates and far-red dyes measured by resonance energy transfer. *Journal of the American Chemical Society* 123:7067–7073.

Young, R., Arnette, J., Roess, D., and Barisas, B. 1994. Quantitation of fluorescence energy trans-fer between cell surface proteins is fluorescence donor photobleaching kinetics. *Biophysical Journal* 67:881–888.

CHAPTER 2

Principles of Fluorescence for Quantitative Fluorescence Microscopy

Neil Anthony, Peng Guo, and Keith Berland

2.1 INTRODUCTION

Continuing developments in fluorescence microscopy are providing increasingly sophisticated research tools that support detailed noninvasive investigations of molecular level functions in biological systems, including the capability to investigate molecular dynamics, interactions, and structure with exquisite sensitivity and specificity. Many of the important fluorescence microscopy-based research methods employ sophisticated data collection and analysis procedures, and a clear understanding of the fundamental principles of molecular fluorescence is a prerequisite for their successful implementation. This chapter therefore aims to introduce the basic photophysical properties of fluorescent molecules and to discuss some of the ways in which these properties influence experimental design and analysis in quantitative fluorescence microscopy. The emphasis is on fundamental principles rather than specific methods or applications.

2.2 WHAT IS FLUORESCENCE?

Fluorescence is a molecular luminescence process in which molecules spontaneously emit a photon as they relax from an excited electronic state to their ground state following absorption of energy—typically within a few nanoseconds or less following excitation. The characteristic electronic states and relaxation processes involved in fluorescence emission are illustrated in Figure 2.1, commonly referred to as a Jablonski diagram (Kasha 1987; Frackowiak 1988; Lakowicz 2007; see Chapter 1).

The nature and significance of the molecular transitions, shown as arrows in the diagram, are discussed in detail throughout this chapter. The states S_0 and S_1 are the ground state and lowest energy excited singlet electronic states, respectively, and T_1 is the lowest energy triplet state for the molecule. Singlet and triplet states are distinguished by the

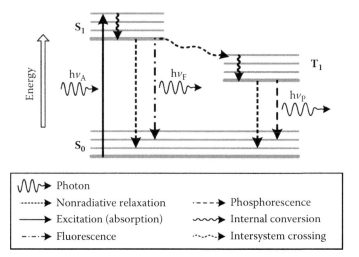

FIGURE 2.1 A Jablonski diagram representing the energy levels of a fluorescent molecule and several important transitions. The states S_0 and S_1 are the ground state and lowest energy excited singlet electronic states, and T_1 is the lowest energy triplet state for the molecule. Singlet and triplet states are distinguished by the orientation of their electron spins. The closely spaced levels within each electronic state (fine lines) represent the vibrational energy levels of the molecule, which are typically similar for the different electronic states. Each arrow represents a transition from one state to another, including absorption of light, internal conversion, nonradiative relaxation, fluorescence emission, intersystem crossing to the triplet state, and phosphorescence. Planck's constant ($h = 6.63 \times 10^{-34}$ J-s) relates a photon's energy, E, to its frequency, ν, as $E = h\nu$.

orientation of their electron spins.* The closely spaced levels within each electronic state represent the vibrational energy levels of the molecule, which are typically similar for the different electronic states. The different vibrational states are responsible for the characteristic shape of the absorption and emission spectral properties of a fluorescent molecule.

The absorbance and emission of light is a quantum process, and the energy is absorbed and emitted in discrete units called photons. A single photon has energy $h\nu$, where h is Planck's constant (6.63×10^{-34} J-s) and ν refers to the frequency of the radiation. The frequency and the wavelength, λ, of light are related parameters because their product is always the speed of light (i.e., $\lambda\nu = c$, where $c = 3 \times 10^8$ m s^{-1}). Thus, the photon energy can also be calculated as $E = hc/\lambda$. A convenient mnemonic for calculating the photon energy (in electron volts, 1 eV = 1.6×10^{-19} J) is $E = 1243/\lambda$, where λ is the wavelength in nanometers. The wavelengths of the visible spectrum run from approximately 400 (blue) to 700 (red) nm, corresponding to frequencies of 7.5×10^{14} and 4.3×10^{14} hertz (Hz), respectively.

2.3 ABSORPTION

Fluorescence is produced when molecules relax from an excited electronic state to the ground state; yet, at room temperature, virtually the entire population of any fluorescent molecules used in microscopy will occupy the ground-state S_0. Thus, fluorescence emission

---

* Singlet states have *antiparallel* electron spins, while triplet states have *parallel* electron spins.

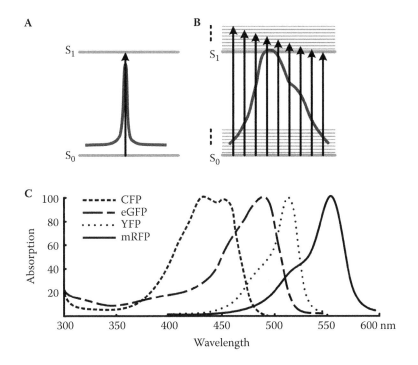

FIGURE 2.2   Absorption of light involves pumping the molecule from the ground electronic and vibrational states to an excited state. A single transition from one energy state to another would produce a sharp absorption peak as shown in (A). Molecules in solution have numerous closely spaced vibrational energy levels, and fluorescent molecules can be excited to many of these vibronic states, resulting in broad absorption spectra (B). The relative strength of each allowed energy level transition determines the shape of the absorption spectrum, with the most favorable transitions corresponding to the peak of the absorption spectrum. Typical absorption spectra for several fluorescent proteins commonly used in microscopy are shown in (C).

requires an energy source to pump the molecules into an excited state from which they can then emit fluorescence (Lakowicz 2007; Valeur 2002; Davidson and Abramowitz 2005). The energy source is typically an external light source such as a laser, lamp, or diode, which provides the excitation energy in the form of photons (Pawley 1995). The energy of the absorbed photons must correspond to the energy difference between the ground state and one of the excited states.

The Jablonski diagram in Figure 2.1 shows only a single absorption transition, which would produce a narrow spectrum (Figure 2.2A). For molecules in solution, there will generally be a wide range of allowed transitions of differing energy/frequency (Berlman 1971; Lakowicz 2007; Valeur 2002), corresponding to the different vibrational energy levels of the molecule, as shown in Figure 2.2(B). Some of the transitions are more favorable than others, and the relative strength of each transition can be found by measuring the absorption spectrum of a particular molecule. Photons with energies corresponding to the peaks of the absorption spectrum are more likely to interact with a fluorescent molecule than photons whose energies are not at the peak, although photons of many different wavelengths are capable of exciting the molecule. The absorption spectra of some common fluorescent proteins are shown in Figure 2.2(C). The spectra are normalized, so the figure only provides information about the spectral shape rather than the relative strength of the different absorbers.

Such normalized spectra can be misleading: One is naturally tempted to assume that the curve with a higher value at a particular wavelength is the stronger absorber at that wavelength, although this may not be accurate. For example, the extinction coefficient (Section 2.3.1.1) for the yellow fluorescent protein (YFP) is approximately two to three times larger than that for the cyan fluorescent protein (CFP) (Rizzo et al. 2004). There are thus wavelengths in the ~460–475 nm range for which YFP is the stronger absorber, even though the CFP spectrum appears to have the larger value in the figure.

In microscopy, it is very common for samples to contain multiple fluorescent species such as in experiments designed using multiple fluorescent probes to measure fluorescent signals from several different molecules of interest simultaneously. Even experiments designed with only a single fluorescent probe may still contain multiple emitting species because living systems can produce significant quantities of autofluorescing molecules (Aubin 1979; Andersson 1998; Benson et al. 1979). Because each type of fluorescent molecule has a unique absorption spectrum, varying the excitation wavelengths offers an opportunity to enhance or diminish the excitation of a particular fluorophore relative to the other fluorescent molecules present in the sample. One can therefore use excitation wavelength as a tool to optimize imaging conditions. Often the optimal excitation corresponds to the absorption peak, but it is not uncommon that one can optimize the signal to background and/or signal-to-noise ratios in a particular sample using off-peak excitation wavelengths.

The most commonly used excitation conditions for microscopy are designed to maximize excitation of a particular probe and minimize the excitation of other fluorophores present in the sample (Davidson and Abramowitz 2005; Reichman 2007; Haugland 2007). In modern automated digital imaging systems, an alternate approach is to image samples using multiple excitation wavelengths. When coupled with image processing algorithms, this approach can be used to separate the signals from multiple fluorescent emitters, provided the fluorescent probes have different excitation spectra, as they generally will. This approach is sometimes referred to as excitation fingerprinting (Dickinson et al. 2003; Lakowicz 2007) and can allow separation of fluorescence from multiple probes with similar or even unresolvable emission spectra. Selecting excitation wavelengths that efficiently excite multiple species and separating their signals with image processing can also improve the overall signal-to-noise ratios in imaging data (Thaler et al. 2005; Thaler and Vogel 2006).

2.3.1 Molecular Excitation Rates

A detailed description of the interaction between light and matter is beyond the scope of this chapter, although a basic physical description of molecular absorption is helpful for selecting appropriate excitation wavelengths and powers, as well as for understanding signal levels and artifacts in fluorescence microscopy. The total fluorescence signal generated from a given region of a microscopy sample will depend on how many fluorescent molecules are located at that position, their photophysical properties, and the rate at which those molecules are excited over time by the light source. The latter quantity is the focus of this section, in which we discuss absorption for both one-photon and two-photon excitation.

For each case, the rate at which individual fluorescent molecules in the sample are pumped from the ground state into an excited state will be specified by the quantity $w_i(\mathbf{r},t)$, which has units of inverse seconds and may be a function of time, t, and position, $\mathbf{r}$, within the sample. The index i specifies one- ($i = 1$) or two- ($i = 2$) photon excitation. The molecular excitation rate depends on both the absorption properties of the fluorescent molecule and the flux of exciting photons seen by each molecule. In most microscopy applications, the excitation source is not uniform throughout the sample, and therefore the fluorescence signals generated at any given time are also not spatially uniform. For example, confocal and two-photon laser scanning microscopy (Webb 1996; Conchello and Lichtman 2005; Wilson 1990; Pawley 1995; Masters 2006; Denk, Strickler, and Webb 1990; Williams, Piston, and Webb 1994; Masters and So 2008; Diaspro 2001; So et al. 2000) both employ focused laser excitation which excites molecules in the center of the focused beam more efficiently than those on the periphery of the focal spot.

Moreover, the focused laser only illuminates a small fraction of the sample at any given time. The focused beam is then raster scanned across the sample and the fluorescence signal from each successive position is measured sequentially and stored. The data are subsequently reconstructed into an image for visualization on a computer. The detailed functions describing the spatial profile of the focused laser beam for different microscope designs will not be discussed here (Born and Wolf 1997; Wilson 1990; Xu and Webb 1991). We will instead use generic notation with the dimensionless function $S(\mathbf{r})$ representing any normalized distribution function, with unit peak value, that models the three-dimensional spatial profile of the illumination source.

2.3.1.1 One-Photon Excitation

The most familiar parameter that specifies how strongly a molecule interacts with light is the molar extinction coefficient, $\varepsilon(\lambda)$. The extinction coefficient is often reported in terms of its peak value, although we here write it as a function of wavelength to emphasize that wavelength selection can be used advantageously in microscopy. The Beer–Lambert law states that for a beam of monochromatic light with initial intensity, I_0, traversing a path length l of an absorbing solution with molar concentration c, the extinction coefficient can be used to compute the transmitted intensity, I, using

$$I = I_0 10^{-\varepsilon(\lambda)cl} \tag{2.1}$$

A graphical representation of the Beer–Lambert law can be seen in Figure 2.3. A more fundamental measure of absorption on the molecular scale is a related parameter called the absorption cross section, $\sigma_1(\lambda)$. In the context of microscopy, $\sigma_1(\lambda)$ is the more natural quantity to work with because microscopy samples may not be spatially homogeneous and

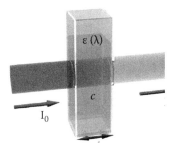

FIGURE 2.3   The Beer–Lambert law uses the extinction coefficient, $\varepsilon(\lambda)$, the molecular concentration, C, and the transmission path length, l, to compute the fraction of incident photons transmitted through a homogenous solution (Equation 2.1). In microscopy, the absorption cross section, $\sigma_1(\lambda)$, is a more natural measure of absorption, as described in the text.

transmission path lengths are in some cases not easily defined. Both $\sigma_1(\lambda)$ and $\varepsilon(\lambda)$ measure how strongly a particular molecule interacts with light of a given wavelength, and the two quantities are related by

$$\sigma_1\left(\lambda\right)=3.8\times10^{-21}\varepsilon\left(\lambda\right) \tag{2.2}$$

In Equation 2.2, the cross section $\sigma_1(\lambda)$ has units of square centimeters, and $\varepsilon(\lambda)$ has units of molar^{-1} centimeter^{-1} ($M^{-1}cm^{-1}$). Good absorbers, including many commonly used fluorescence probes, have cross sections on the order of 10^{-16} cm^2. For example, the widely used green fluorescent protein (eGFP) has a peak extinction coefficient of 57,000 M^{-1} cm^{-1}, which corresponds to a cross section of 2.2×10^{-16} cm^2 (Tsien 1998). Some other common fluorescent proteins have cross sections of similar magnitude, such as the cyan (CFP, 32,500 M^{-1} cm^{-1}, 1.2×10^{-16} cm^2) and yellow (YFP, 84,000 M^{-1} cm^{-1}, 3.2×10^{-16} cm^2) fluorescent proteins (Tsien 1998; Rizzo et al. 2004).

Computing the approximate number of molecular excitation events per unit time allows one to estimate how many fluorescence photons can be measured for a particular experimental system. For one-photon excitation, the rate that molecules are pumped from the ground to the excited state at position $\mathbf{r}$ is linearly proportional to the excitation flux and is computed as

$$w_1(\mathbf{r})=\sigma_1 I_0\left(t\right)S(\mathbf{r}) \tag{2.3}$$

The quantity $I_0(t)$ represents the peak photon flux of the laser excitation as a function of time with units of photons cm^{-2} s^{-1}.

Equation 2.3 has a useful geometric interpretation: The flux multiplied by the cross section yields a product of (1) the number of photons per second passing through an illuminated area A within the sample, and (2) the ratio of the absorption cross section to area A. In other words, the ratio of the cross section to the illuminated area specifies the fraction of the excitation photons passing through that area that will be absorbed (see Figure 2.4). Note that the only spatial dependence in Equation 2.3 comes from $S(\mathbf{r})$ itself, so in the

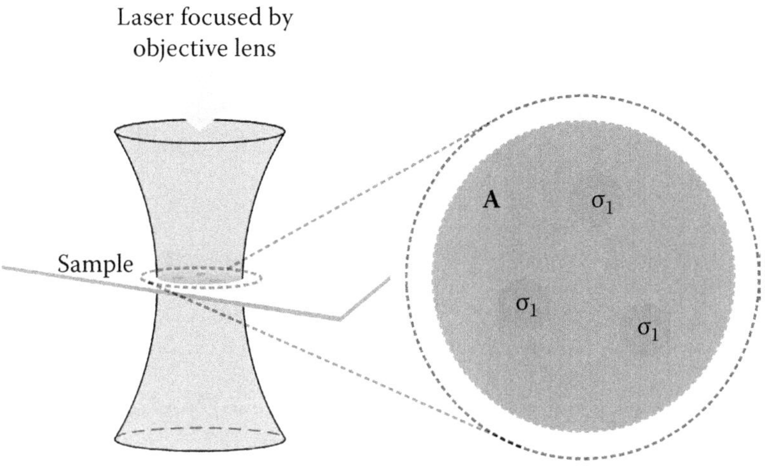

FIGURE 2.4 A geometrical interpretation of the absorption cross section. When a sample is illuminated with photon flux of I_0 photons cm^{-2} s^{-1}, this can be interpreted as a certain number of photons per second passing by area A defined by the physical size of the light source (dark blue shaded region). The cross section represents the effective area of the molecule (green circles) such that the quantity determines the fraction of photons passing through area A captured by each molecule. The cross section is an effective area because it depends on the electronic properties of the molecule and not directly on its physical size.

absence of saturation (Section 2.5.3), the spatial pattern of the excited molecules will match the profile of the excitation source.

Photon flux is difficult to measure directly, and in the laboratory one generally measures average power instead. The relationship between the photon flux at the sample and the average power of the source depends on the spatial pattern of the excitation source at the sample and how tightly it is focused in the sample. A precise description of the illumination pattern can be rather complex, although for reasonably accurate estimates of the molecular excitation rate it is often sufficient to assume the illumination profile is that of a focused Gaussian beam (Pawley 1995; Xu and Webb 1991; Nagy, Wu, and Berland 2005b).

In this case, assuming that constant power excitation is used (i.e., continuous wave [cw] nonpulsed sources), the peak intensity can be computed from the average power in watts using the expression $I_{0,cw} = 2\langle P\rangle/h\nu\,\pi\omega_0^2$, where ω_0 is the beam waist of the focused laser and $h\nu$ is the photon energy. The beam waist can be estimated by the expression $\omega_0 = \lambda/\pi NA$, where NA is the effective numerical aperture of the lens focusing the illumination source. Box 2.1 shows an example of how to compute the average molecular excitation rate for the green fluorescent protein for illumination with two different values of the average power and for objective lenses with different numerical apertures.

2.3.1.2 Two-Photon Excitation

Two-photon excitation has been demonstrated to offer several important advantages for many applications of fluorescence microscopy, such as enhanced depth penetration, reduced photobleaching, and reduced photodamage in living specimens (Denk et al. 1990; Zipfel, Williams, and Webb 2003; Squirrell et al. 1999; Masters and So 2008; So et al. 2000;

BOX 2.1

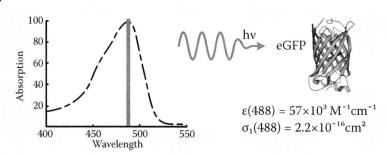

$\varepsilon(488) = 57 \times 10^3 \, M^{-1} cm^{-1}$
$\sigma_1(488) = 2.2 \times 10^{-16} cm^2$

The 488 nm excitation source is selected to match the peak of the eGFP absorption, with an absorption cross section of 2.2×10^{-16} cm² (extinction coefficient of 57,000 M⁻¹ cm⁻¹).

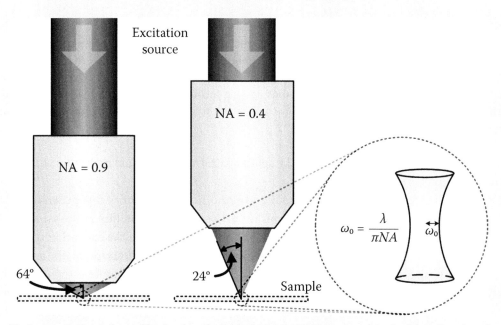

The laser excitation is focused by the objective lens, with higher NA lenses resulting in tighter focusing of the excitation source. The peak incidence angle is also a function of the numerical aperture, with $NA = n\sin\theta$ where θ is the half angle shown and n is the refractive index of the medium traversed by the focused light. The relationship between the incident power and intensity is defined in the text. Since the molecular excitation rate depends on intensity and not average power, the higher NA lens produces a higher molecular absorption rate, (w_1), for the same power, as shown in the table below.

	NA = 0.9		NA = 0.4	
$\langle P \rangle (\mu W)$	I_0(photons cm⁻² s⁻¹)	$\langle w_1 \rangle$(events s⁻¹)	I_0(photons cm⁻² s⁻¹)	$\langle w_1 \rangle$(events s⁻¹)
10	5.4×10^{22}	1.2×10^7	1.0×10^{22}	2.2×10^6
100	4.3×10^{23}	9.5×10^7	1.0×10^{23}	2.2×10^7

Computation of the average molecular excitation rate for the green fluorescent protein.

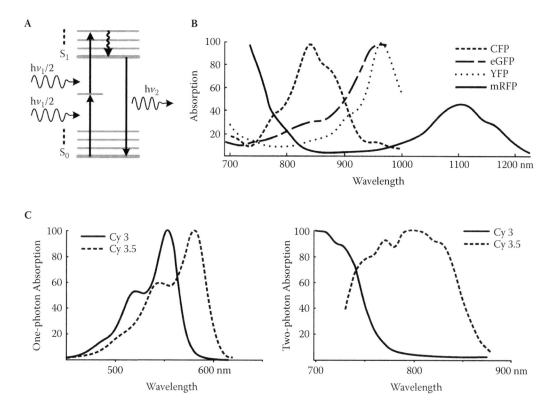

FIGURE 2.5 Two-photon excitation. (A) Jablonski diagram showing the molecular excitation transition in which two photons of equal energy provide the energy required to excite the molecule. (B) Two-photon excitation spectra of some commonly used fluorescent proteins, including the enhanced green fluorescent protein (eGFP), cyan fluorescent protein (CFP), yellow fluorescent protein (YFP), and monomeric red fluorescent protein (mRFP). (C) The one-photon (left) and two-photon (right) absorption spectra for the dyes Cy3 and Cy3.5. Two-photon absorption spectra sometimes have the same spectral shapes as their one-photon spectra; in other cases, they may be quite different, thus highlighting the importance of acquiring information about two-photon absorption spectra for probes used in two-photon microscopy.

Helmchen and Denk 2005). Two-photon excitation refers to the simultaneous absorption of two photons by a single fluorescent molecule, with the sum of the photon energies matching the molecular transition energy from ground to excited state, as shown in Figure 2.5(A).

Achieving significant fluorescence excitation by two-photon absorption requires high photon densities (Denk et al. 1990; So et al. 2000). Appreciable two-photon excitation in fluorescence microscopy thus generally requires ultrafast pulsed lasers focused using high numerical aperture lenses to concentrate photons in both space and time (Zipfel et al. 2003; Xu and Webb 1991). Unlike one-photon excitation, for which the molecular excitation rate is linearly proportional to the illumination intensity, two-photon absorptions rates are proportional to the square of the excitation intensity. This "power squared" dependence of two-photon absorption is what gives rise to the three-dimensional resolution of two-photon microscopes without the use of confocal pinholes (which are required for one-photon confocal microscopy to eliminate out-of-focus emission) (So et al. 2000; Denk et al. 1990). The rate that molecules are pumped from the ground to the excited state by two-photon absorption is given by

$$w_2(\mathbf{r}, t) = \frac{\sigma_2 I_0^2(t) S^2(\mathbf{r})}{2} \tag{2.4}$$

where σ_2 is the two-photon absorption cross section (Xu and Webb 1996; Xu et al. 1996). The preferred unit for two-photon cross sections is the Göppert-Mayer (GM), where 1 GM = 10^{-56} m^4 s. The unit is named after Nobel laureate Maria Goppert-Mayer, who first developed the theory of two-photon excitation (Göppert-Mayer 1931). Good two-photon absorbers typically have cross sections with orders of magnitude of ~100 GM (So et al. 2000; Xu and Webb 1991; Xu, Zipfel, Shear et al., 1996).

Typical laser systems for two-photon excitation have temporal pulse widths of ~100 fs and pulse repetition frequencies on the order of f_p = 80 MHz. The pulse width and repetition rate are generally significantly faster than the timescale of interest in a particular experiment; therefore, one is instead primarily interested in the average molecular excitation rate, which can be computed by integrating the total probability of molecular excitation per laser pulse and multiplying by the laser pulse repetition frequency. Thus, the average excitation rate is given by $\langle w_2(\mathbf{r}) \rangle = f_p \int w_2(\mathbf{r}, t) dt$, where the integral is computed over a single pulse and the angular brackets represent time averaging (Xu and Webb 1991; Nagy et al. 2005b).

As in the one-photon case, computing photon fluxes in terms of the easily measured average power requires knowledge about the temporal profile of the laser pulses as well as the spatial profile of the focused excitation source. If, for simplicity, one assumes a Gaussian profile in both space and time (Masters and So 2008; Xu and Webb 1991), then the photon flux can be computed in terms of the average power as

$$I_0(t) = \frac{\langle P \rangle}{h\nu} \frac{2}{\pi \omega_0^2} \frac{2\sqrt{\ln 2}}{\sqrt{\pi}} \frac{1}{f_p \alpha} e^{-\ln(2)\left(\frac{2(t-t_0)}{\alpha}\right)^2} = I_{0, peak} e^{-\ln(2)\left(\frac{2(t-t_0)}{\alpha}\right)^2} \tag{2.5}$$

where $\langle P \rangle$ is the time averaged laser power (in watts), ω_0 is the focused beam waist, and α is the temporal width (full width at half maximum) of the Gaussian pulse. One can then also compute the average molecular excitation rate as

$$\langle w_2(\mathbf{r}) \rangle = \sigma_2 I_{0, peak}^2 S^2(\mathbf{r}) \alpha f_p \frac{\sqrt{\pi}}{4\sqrt{\ln 4}} \tag{2.6}$$

Measurements of two-photon absorption spectra and cross sections are significantly more complicated than those of one-photon spectra and generally require specialized knowledge and equipment. Fortunately, two-photon absorption spectra of the most common fluorescent probes for microscopy are now available in the literature or from online sources (Najechalski et al. 2001; Yamaguchi and Tahara 2003; Heikal, Hess, and Webb 2001; Webb 2003). The two-photon excitation spectra for four commonly used fluorescent proteins are shown in Figure 2.5(B).

Many molecules have two-photon absorption spectra that essentially match their one-photon spectra in shape (with the wavelength axis of an absorption plot multiplied by two); other molecules have two-photon absorption spectra that are completely different from their one-photon spectra (the emission spectra are almost always the same for one- and two-photon excitation) (Lakowicz 2007; Spiess et al. 2005). An example of this is shown in Figure 2.5(C), which shows one- and two-photon absorption spectra for the dyes Cy3 and Cy3.5. Their one-photon absorption spectra are quite similar (left), yet the two-photon spectra are drastically different (right). Therefore, it is important to measure or look up the two-photon absorption spectrum whenever one is working with a new fluorophore.

2.4 FLUORESCENCE AND MOLECULAR RELAXATION PATHWAYS

2.4.1 Internal Conversion

Once a molecule is pumped into an excited state, it will relax to the ground vibrational level of electronic excited-state S_1 on a picoseconds (10^{-12} s) timescale by a process termed internal conversion (Lakowicz 2007). As a consequence of this rapid loss of vibrational energy, all subsequent molecular relaxation pathways proceed from the lowest vibrational level of the electronic excited state. Fluorescence emission spectra are thus typically independent of the excitation pathway (e.g., one- or two-photon excitation) and excitation wavelength.

The timescale of internal conversion is usually significantly faster than subsequent relaxation processes. Due to the separation of timescales between internal conversion and other pathways, it is reasonable to ignore the internal conversion process when interpreting fluorescence signals from the microscope. Following internal conversion, molecules will subsequently relax back to the ground state via fluorescence emission or by other competing relaxation pathways that are nonradiative.

These possibilities are summarized in Figure 2.6, where the constants k_1 and Γ represent the relaxation rates associated with internal conversion within the excited-state manifold and fluorescence emission, respectively. The rate constants k_{FRET} and k_{OTHER} represent the relaxation rates associated with Förster resonance energy transfer (FRET, Section 2.4.3.1) and the sum of all other nonradiative relaxation rates, respectively. From a physical perspective, k_{FRET} is simply one of the "other" nonradiative relaxation processes, but it has been noted separately here due to the high importance of FRET measurements for studying molecular interactions in microscopy (Jares-Erijman and Jovin 2003; Periasamy 2001; Clegg 1995; Periasamy and Day 2005).

2.4.2 Fluorescence Emission

For fluorescence microscopy, the fluorescence molecular relaxation process produces the photons that are detected; without fluorescence emission, there is no signal (excluding other types of nonfluorescent signals such as second harmonic generation) (Campagnola et

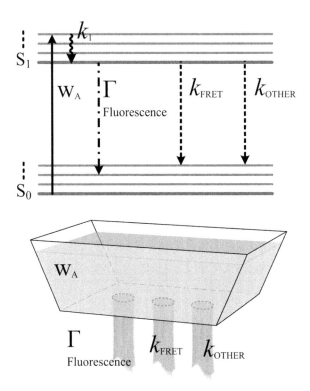

FIGURE 2.6 Molecular transitions are represented by arrows, and each arrow has an associated rate constant (units of s^{-1}). The rapid loss of vibrational energy (k_1) causes all further relaxation to begin in the ground vibrational state of S_1. The net rate at which the molecule relaxes back to the ground state is determined by the sum rates for all the available relaxation processes, which here include fluorescence, FRET, and other nonradiative energy loss mechanisms. A physical analogy that represents a similar process is to empty a trough of water through holes in the bottom. Molecular excitation (W_A) is like filling the trough. Then, each of the various relaxation pathways corresponds to a hole in the trough and the rate for each process is represented by the size of the hole. The time that the water remains in the trough is analogous to the lifetime of the excited state, and the fraction that drains through the fluorescence hole is analogous to the quantum yield.

al. 2002; Campagnola and Loew 2003). Due to competing mechanisms for an excited-state molecule to release its excess energy (see Figure 2.6), only a fraction of excited fluorescent molecules will relax via fluorescence emission.

The availability of different relaxation pathways depends not only on the specific fluorescent molecule, but also on its environment, which can reflect molecular interactions or solvent conditions (Lakowicz 2007). Therefore, imaging methods that quantitatively measure the dynamics of the molecular relaxation process have the potential to provide more information than is available from measurements of the fluorescence intensity alone (Lakowicz 2007; Becker 2007; Davidson and Abramowitz 2005; Periasamy and Day 2005; So et al. 1995; Clegg,

Holub, and Gohlke 2003). Two important parameters related to the prevalence or absence of competing relaxation mechanisms are the quantum yield and the excited-state lifetime.

2.4.2.1 Quantum Yield

The quantum yield of a fluorescent molecule is a measure of its fluorescence efficiency, defined as the probability that an excited molecule will emit a fluorescence photon. For populations of molecules, the quantum yield also specifies the fraction of excited molecules that relax by fluorescence emission. A water basin draining through multiple holes, as shown in Figure 2.6, provides a useful macroscopic analogy for the quantum yield. Each hole in the basin represents a specific relaxation mechanism and the size of each hole represents the relaxation rate for that pathway, with larger holes corresponding to faster relaxation rates. Although it is not possible to predict which hole any given water molecule will drain through, it is straightforward to determine the fraction of water molecules that will drain through each hole.

By analogy, the quantum yield represents the fraction of excited-state molecules that "drain" through the "fluorescence" hole in the basin. Quantitatively, the quantum yield is defined in terms of the rate constants as

$$Q = \frac{\Gamma}{\Gamma + k_{FRET} + k_{OTHER}} \tag{2.7}$$

It is apparent from this definition that the quantum yield has a maximum value of unity only when fluorescence is the single allowed relaxation process. Fluorescent molecules commonly used as probes in microscopy have quantum yields ranging from very low values (<0.05) to near unity.

Table 2.1 shows the quantum yield and spectral properties of some fluorescent proteins and other common fluorescent dyes. High quantum yield probes are generally desirable in most imaging applications because they are brighter than lower yield fluorescent molecules. However, as noted earlier, the availability of nonfluorescent relaxation pathways is sensitive to the environment of a fluorophore, and this sensitivity can be exploited to probe the nanoscale surroundings of a specific type of probe molecule. For example, FRET microscopy is widely used to investigate molecular interactions, and this capability arises from the different photophysical dynamics of the donor fluorophore in the presence or absence of a FRET acceptor (Periasamy and Day 2005; Clegg 1996). The quantum yield is difficult to measure directly in an experiment, but the quantum-yield-related information about the environment of a particular fluorophore is also available through measurements of the fluorescence lifetime.

2.4.2.2 Fluorescence Lifetimes

Fluorescence emission is a random process, and any particular fluorescent molecule will spend a random but finite amount of time in the excited state prior to releasing its energy to the environment as it relaxes to the ground state. However, each type of fluorescent molecule in a particular environment has a well defined *average* time that a molecule will spend in the excited state following excitation. This average time is referred to as the fluorescence

TABLE 2.1   Quantum Yield and Spectral Properties of Some Common Fluorescent Proteins and
Fluorescent Dyes

Probe	Quantum yield (QY)	Excitation wavelength	Emission wavelength
CFP	0.37	440	485
YFP	0.61	520	542
eGFP	0.6	488	535
Cerulean	0.62	440	485
Venus	0.6	520	542
DsRed	0.79	558	583
mRFP1	0.25	584	607
mHoneydew	0.12	487/504	537/562
mBanana	0.70	540	553
mOrange	0.69	548	562
tdTomato	0.69	554	581
mTangerine	0.30	568	585
mStrawberry	0.29	574	596
mCherry	0.22	587	610
mKO (Kusabira)	0.6	548	559
mKate (Katuska)	0.33	588	635
Alexa488	0.81	488	535
Alexa555 (Cy3)	0.13	555	575
Alexa647 (Cy5)	0.11	647	665

Note: Values separated by a forward slash denote the wavelengths of multiple peaks.

lifetime, τ. Using the previous water basin example, the lifetime is analogous to the average
time each water molecule poured into the basin (the pouring is analogous to molecular
excitation) will remain prior to draining out (molecules relaxing back to the ground state).
Clearly, the average time must be determined by the inverse of the sum of all the drainage
rates; in the case of fluorescence, the lifetime is given by

$$\tau = \frac{1}{\Gamma + k_{NF}} \tag{2.8}$$

where k_{NF} includes all nonfluorescent relaxation pathways. The sum of the relaxation rate
constants ($\Gamma + k_{NF} = 1/\tau$) specifies the rate at which excited-state molecules relax back to the
ground state. This rate is constant over time, and the temporal relaxation for a population
of excited molecules will therefore follow an exponential decay with time constant τ—that
is, $\exp(-t/\tau)$. The fluorescence signal is proportional to the excited-state population and
thus the fluorescence signal will decay with exponential kinetics (or possibly multiexponential kinetics if there are multiple lifetimes) following pulsed excitation.

The lifetimes of many common probes are on the order of a few nanoseconds but will, in
general, vary over a wide range depending on the specific fluorescent molecule, the solvent/
environment, and other factors (Lakowicz 2007; see Chapter 1). The lifetimes of several

TABLE 2.2   Lifetime of Fluorescent Proteins

Fluorescent protein	Lifetime (ns)	Source
GFP	2.45	Hillesheim et al., 2006
CFP	3.6 and 1.2[a]	Rizzo et al., 2004
YFP	2.9	Ganesan et al., 2006
Cerulean	3.0	Koushik et al., 2006

[a] Double component exponential decay.

common fluorescent proteins are listed in Table 2.2. Occasionally, it is useful to consider a quantity referred to as the natural lifetime, which refers to the lifetime a molecule would have in the absence of any nonfluorescent relaxation pathways (i.e., $k_{nf} = 0$). Thus, the natural lifetime is given by $\tau_n = 1/\Gamma$.

The versatile utility of fluorescence lifetime in microscopy arises because the lifetime is an accurate reporter of the relaxation pathways available to each molecule and thus of their local environment. The lifetime is an intrinsic property of a particular type of fluorescent molecule in a specific environmental condition. A very significant advantage of this fact for microscopy is that fluorescence lifetime measurements, unlike intensity measurements, are therefore not explicitly dependent upon molecular concentrations. This can be an important advantage in cells and tissue in which one is interested in probing molecular interactions or environmental surroundings but for which the concentrations of fluorescent reporter molecules may be far from uniform.

2.4.2.3 Fluorescence Emission Spectra

Due to internal conversion in the excited state following excitation, the energy of the excited state is lowered and the fluorescence emission is red shifted (i.e., lower energy and thus longer wavelength photons) relative to the excitation wavelength. In addition, the excited molecule usually relaxes from the lowest vibrational level of the excited state to a higher vibrational level of the ground electronic state, further lowering its energy compared with the excitation photon energy. This change in energy of the emission compared to the absorbed energy is called the Stokes shift. This shift in wavelength of the emission compared to the excitation is usually considerable (tens to hundreds of nanometers) and is very important for high-sensitivity fluorescence measurements because it allows for the use of optical filters that block the excitation source from reaching the detector. Fluorescence detection can therefore be measured against very low background signal, which allows even single fluorescent molecules to be detected under optimized conditions (Selvin and Ha 2008).

As for absorption, Figure 2.1 shows just one of the characteristic transitions involved in fluorescence emission. However, the complete emission spectrum is made up of multiple transitions from the lowest vibrational state of S_1 back to the various available vibrational levels of the ground state. Some transitions will be more favorable than others and together they determine the overall shape of the emission spectrum. In general, fluorescent probes in solution will have broad emission spectra. For example, Figure 2.7 shows the emission spectra of some common fluorescent protein variants. Broad emission spectra present a

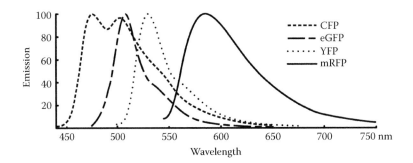

FIGURE 2.7 The emission spectra of some commonly used fluorescent proteins, including the cyan fluorescent protein (CFP), enhanced green fluorescent protein (eGFP), yellow fluorescent protein (YFP), and monomeric red fluorescent protein (mRFP).

challenge for fluorescence microscopy because they can make it difficult to separate the signals from multiple probes that are imaged simultaneously.

Both high-quality optical filters and image processing techniques are helpful for resolving multiple fluorophores with overlapping emission spectra. The recent introduction of spectrally resolved detection in microscopes has also greatly enhanced the capability to resolve fluorescence signals from multiple probes by using linear unmixing algorithms (Zimmermann, Rietdorf, and Pepperkok 2003; Thaler and Vogel 2006; Neher and Neher 2004; Kawata and Sasaki 1996). Fluorescence lifetime microscopy (FLIM) can offer another important tool to distinguish multiple probes through measurements of their different fluorescence lifetimes (Koushik et al. 2006; Wallrabe and Periasamy 2005; Kapusta et al. 2007; Kühnemuth and Seidel 2001; Becker 2007).

2.4.3 Nonradiative Relaxation Pathways

The availability of nonradiative relaxation pathways reduces the fluorescence signal compared to what it would be in the absence of nonfluorescent energy loss. For intensity-based fluorescence imaging, a reduction in quantum yield is undesirable; higher intensities allow high-quality images to be acquired in less time. The most commonly used fluorescent dyes and fluorescent proteins have therefore been selected for their bright fluorescence signals, a property that arises from both strong absorption coefficients and high quantum yields. However, as noted earlier, in some applications of fluorescence microscopy experimental conditions are intentionally designed to produce reductions in quantum yield for certain subpopulations of fluorescent molecules. The most widely used example of this is FRET microscopy (Wu and Brand 1994; Clegg 1996; Periasamy and Day 2005), which is widely used to measure molecular interactions by measuring the efficiency of nonradiative energy transfer between donor and acceptor fluorophores.

2.4.3.1 Basics of FRET

FRET is a process in which an excited-state molecule (the "donor") transfers its energy to a nearby molecule (the "acceptor") through near-field electromagnetic interactions. It falls in the category of nonradiative relaxation pathways because no fluorescence emission

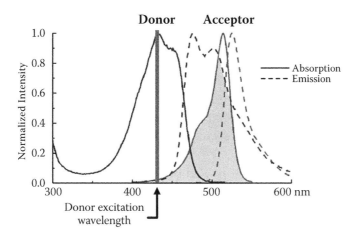

FIGURE 2.8 Efficient Förster resonance energy transfer (FRET) requires significant overlap between the donor emission and acceptor excitation spectra. This figure shows the absorption and emission spectra for both donor and acceptor molecules of an efficient FRET pair. The arrow indicates the excitation wavelength, and the shaded region exemplifies the spectral overlap required for FRET.

is associated with the energy loss for the donor molecule. The rate of FRET energy transfer is related to the overlap between the donor emission and acceptor absorption spectra (Figure 2.8), the distance between donor and acceptor molecules, and the relative positioning and orientation of the donor and acceptor transition dipoles (Förster 1965; Stryer 1978; Lakowicz 2007).

These factors go into computation of what is referred to as the Förster radius, R_0, which corresponds to the donor/acceptor separation at which FRET efficiency is 50% (Clegg 1996). The energy transfer efficiency has very strong dependence on the physical separation, r, between donor and acceptor molecules, with the transfer rate varying with the inverse sixth power of the separation:

$$k_{FRET} = \frac{1}{\tau_D^0} \left(\frac{R_0}{r} \right)^6 \tag{2.9}$$

Here, τ_D^0 is the lifetime of the donor in the absence of the acceptor molecule. The distances over which efficient FRET is achieved are typically below 100 Å. This short-length scale and the sixth-power dependence of k_{FRET} on separation distance are the factors that lead to the usefulness of FRET for measuring molecular interactions and distances between donor and acceptor molecules using FRET (Wallrabe and Periasamy 2005; Sekar and Periasamy 2003; Clegg 1996; Periasamy and Day 2005; see Chapters 3 and 9). The distance dependence and orientational sensitivity of FRET can also be exploited to probe conformational changes in biomolecules (Selvin and Ha 2008; Truong and Ikura 2001; Heyduk 2002; Rasnik, McKinney, and Ha 2005). Box 2.2 shows an example of how to use fluorescence lifetime measurements to determine energy transfer (FRET) efficiencies.

Spectral overlap is a requirement for FRET, and thus FRET measurements in the microscope must find ways to separate fluorescence signals quantitatively from donor and acceptor molecules, even though they have substantial spectral overlap. Moreover, accurate FRET measurements also require corrections for differences in the efficiency of the optics and detectors for the donor and acceptor fluorescence. A variety of data processing algorithms are used to correct for spectral cross-talk in FRET measurements (Tron et al. 1984; Gordon et al. 1998).

Alternatively, FRET efficiencies can also be accurately measured using FLIM because donor molecules that have FRET relaxation pathways available will have shorter lifetimes compared with donor molecules that do not relax *via* FRET (Koushik et al. 2006; Wallrabe and Periasamy 2005; Wang, Shyy, and Chien 2008; Chen, Mills, and Periasamy 2003; Periasamy and Day 2005; see Chapter 9). FRET standards based on cyan and yellow fluorescent proteins have recently been introduced, and they can be useful for calibrating FRET measurements in microscopy (Koushik et al. 2006).

2.4.3.2 Intersystem Crossing and Phosphorescence

Intersystem crossing occurs when an excited-state molecule relaxes to the triplet excited state, T_1. Molecules that reach the triplet state can relax *via* either photon emission or nonradiative relaxation. The radiative transition from the triplet state is called phosphorescence. The characteristic timescale for phosphorescence relaxation is much longer than that for fluorescence emission. For most applications of fluorescence microscopy, phosphorescence is not highly important. A possible exception is triplet saturation, in which a population of molecules builds up in the triplet state. This can reduce fluorescence signals because molecules in the excited triplet state may remain there for relatively long time periods (with typical average lifetimes of microseconds and longer) and are not able to produce fluorescence photons while they occupy the triplet state (Tsien and Waggoner 1995).

2.4.4 Photoselection and Anisotropy

The transition dipole moments for molecular absorption and fluorescence are vectors and have a spatial orientation that moves with the molecule. When a sample is illuminated with polarized light, molecules oriented such that their absorption dipole vectors are aligned with the electric field of the excitation light are preferentially excited. Polarized excitation thus photoselects a subpopulation from a randomly oriented sample, resulting in a partially oriented or anisotropic excited-state population. As a result, fluorescence emission is also partially polarized (i.e., anisotropic).

This effect is often ignored in microscopy without significant consequences. However, it can be important for several reasons. First, by measuring the extent of anisotropy, one can investigate molecular rotations (Steiner 1991; see Chapter 10). This is possible because molecular rotation during the time in which a molecule occupies the excited state causes a reorientation of the emission dipoles, thus reducing the anisotropy. Second, it is important

BOX 2.2

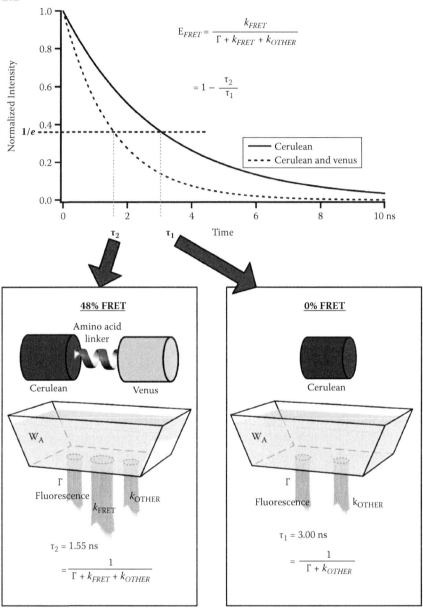

$$E_{FRET} = \frac{k_{FRET}}{\Gamma + k_{FRET} + k_{OTHER}}$$

$$= 1 - \frac{\tau_2}{\tau_1}$$

Fluorescence lifetimes can be used to measure quantum yields or FRET efficiencies (E_{FRET}). When a donor molecule (Cerulean) has no FRET acceptor available, it has a lifetime of $\tau_1 = 3.00$ ns, represented by the exponential decay (solid line, top). When a FRET acceptor (Venus) is held nearby—in this case, by a five amino acid linker—an additional relaxation pathway is opened up while the other relaxation rates remain unchanged. This new relaxation path has two effects: (1) a reduction in the fraction of excited state molecules that relax via fluorescence emission, and (2) a reduction in the average amount of time a molecule will spend in the excited state. This is seen in the reduced lifetime, with $\tau_2 = 1.55$ ns (dashed line, top). This corresponds to a 48% FRET efficiency, calculated by $E_{FRET} = 1 - \tau_2/\tau_1$.

to account for the effects of anisotropy changes (due to rotations) in order to measure fluorescence lifetimes accurately (Fixler et al. 2006; Lakowicz 2007; Becker 2007).

Two related parameters, anisotropy (r) and polarization (P), are commonly used to measure the degree of polarization. They are defined in terms of the fluorescence intensity polarized parallel (∥) and perpendicular (⊥) to the excitation polarization:

$$r = \frac{I_\parallel - I_\perp}{I_\parallel + 2I_\perp}, \qquad P = \frac{I_\parallel - I_\perp}{I_\parallel + I_\perp} \tag{2.10}$$

These two quantities are easily interconverted.*

2.5 MEASURING FLUORESCENCE IN THE MICROSCOPE

When measuring fluorescence in the microscope, one is often dealing with weak signals from relatively small molecular populations; this necessitates careful optimization of signal acquisition. In addition, fluorescent molecules may be spatially confined, immobile, or slowly moving, and thus photobleaching can become an important concern (Eggeling et al. 1998; Axelrod et al. 1976; Kenworthy 2005). Finally, because the spatial distribution of fluorescence signals is a primary concern in microscopy, it is useful to understand in detail how the fluorescence signal levels correspond to molecular populations and their spatial distribution. In this final section we consider each of these topics.

2.5.1 Sensitivity of Fluorescence Measurements

A variety of factors influence how effectively fluorescence signals are measured in the microscope. Most important among these are the collection efficiency of the objective lens, the throughput of the optics and filters, and the efficiency of the optical detectors. Because fluorescence emission for freely rotating molecules is isotropic (emitted in all directions), it is not possible for an objective lens to collect the entire emitted fluorescence signal. The collection efficiency (C.E.) for an objective lens is primarily a function of its numerical aperture (NA); higher numerical aperture lenses collect more light (Kazuko et al. 1999; Lindek, Swoger, and Stelzer 1999). If one assumes fluorescence emission is spatially isotropic, the collection efficiency of an objective lens (for an emitter at the focal point) is given by

$$C.E. = \frac{1 - \cos(\theta)}{2} \tag{2.11}$$

The angle θ is determined from the numerical aperture of the lens by $n \sin(\theta) = NA$, where n is the index of refraction ($n = 1$ for air lenses, $n = 1.3$ for water immersion lenses, and $n \sim 1.5$ for common immersion oil lenses). Thus, for example, an air lens with $NA = 0.9$ will collect about 28% of the fluorescence emission. This means that most of the fluorescence will never make it through the lens to the detector.

---

* $r = 2P/(3 - P)$. $P = 3r/(2 + r)$. These are easily derived from Equation 2.3.

On the other hand, high NA lenses collect substantially more light than low NA lenses and thus optimize signal levels. High NA lenses also maximize spatial resolution. One trade-off is that high NA lenses can cause problems in quantifying polarization as the steep focusing angles distort the polarization of the laser excitation (Axelrod 1979; Piston, Rizzo, and Kevin 2008). Another drawback of higher NA lenses is that they tend to have smaller working distances—defined as the space between the physical lens and the focal plane.

A second factor influencing the overall efficiency of fluorescence measurements in the microscope is the throughput (transmission) of the optics, including objective and other lenses, filters, windows, and mirrors. Throughput can vary substantially for different designs and materials, and careful selection of optics can dramatically enhance measurement sensitivity. For example, although many band-pass filters commonly used to measure fluorescence have transmission efficiencies in the range of 60–80%, specialized filters with efficiencies exceeding 90% are readily available from commercial sources. Similarly, objective lenses of different design, materials, and lens coatings can have widely different throughputs, ranging from very poor to over 90% transmission. Objective transmission can also vary substantially for different wavelengths. Given each of these factors, it is clear that measurements will benefit from selecting optimized optics and lenses with high efficiencies.

Finally, optical detectors only record a fraction of the photons incident upon them. This fraction is referred to as the quantum efficiency of the detector—not to be confused with the quantum yield of the fluorescent molecules relaxing from the excited state. Surprisingly, detector efficiencies can be as small as a few percent or less for many commonly used detectors. On the other hand, highly sensitive photon-counting detectors and cameras with much higher quantum efficiencies (up to ~90%) are now commercially available and provide an excellent resource to optimize fluorescence signal detection efficiency. Detector quantum efficiencies typically vary substantially for different wavelengths of light; this must be considered in selecting appropriate detectors (Hamamatsu Company 2008).

Taking each of these factors together, one can estimate the total efficiency with which emitted photons can be detected by the microscope. If one assumes 30% collection by the objective and 90% transmission efficiency for each of the other elements—including the dichroic mirror, emission filter, objective lens, and other microscope optics (e.g., mirrors and windows), then the overall efficiency of directing fluorescence emission to the detector would come out near 20%, as shown in Table 2.3. With a 50% efficient detector, this would

TABLE 2.3   Collection Efficiency of Lenses

Microscope elements	Optimized collection efficiency
Objective lens: numerical aperture	0.3
Objective lens: transmission	0.9
Dichroic mirror	0.9
Microscope throughput[a]	0.9
Filters	0.9
Total collection efficiency	~20%
Total detection efficiency (detector with 50% efficiency)	10%

[a] The light transfer efficiency from the objective to the detector, excluding other listed elements. This accounts for losses due to windows, mirrors and prisms, and apertures.

correspond to detecting ~10% of the fluorescence signal. This may seem small, but, in fact, it represents a significant improvement over what was achievable not long ago.

2.5.1.1 Fluorescence Signals

The fluorescence signal detected from any position within the sample will be proportional to the concentration of fluorescent molecules at that location, the rate at which the molecules are pumped into the excited state, and the detection efficiency. The fluorescence signal measured from a particular sample region can thus be written as $\kappa C(\mathbf{r},t)w(\mathbf{r})\Omega(\mathbf{r})$, where κ is a constant that accounts for both the quantum yield of the fluorophore and the fluorescence collection and detection efficiency of the microscope. The distribution function $\Omega(\mathbf{r})$ is included to describe any spatial filtering in the optical system, such as that in confocal microscopy, where pinholes are employed so that only fluorescence from certain sample regions reaches the detector. The function $\Omega(\mathbf{r})$ has unit value if there are no important apertures or pinholes, as is often but not always the case in two-photon microscopes. The total measured fluorescence signal can then be computed by integrating the contribution from different sample regions, with

$$\langle F \rangle = \kappa \int C(\mathbf{r}) \langle w(\mathbf{r}) \rangle \Omega(\mathbf{r}) d\mathbf{r} \tag{2.12}$$

For two-photon excitation, this integral has a finite value even without confocal pinholes, but for one-photon excitation, a confocal pinhole is required for the integral to have a finite value.

Equation 2.12 computes the average fluorescence signal from a particular spatial location within the sample (i.e., from the focused laser spot in a confocal or two-photon microscope). The laser is then scanned over various positions within the sample; other factors are otherwise held constant so that the fluorescence intensity from any given location is primarily determined by the local fluorophore concentration.

2.5.2 Observation Volumes and Molecular Brightness

In recent years, a variety of fluorescence fluctuation spectroscopy techniques have emerged that provide powerful capabilities for measuring molecular dynamics and interactions on the microscope and thus complement other fluorescence imaging methods (Rigler and Elson 2000; Schwille 2001; Hess et al. 2002; Webb 2001; Berland 2005; Hillesheim, Chen, and Muller 2006; Wu and Muller 1995; Muller 2004; Muller, Chen, and Gratton 2003; Chen et al. 1999, 2001). Detailed descriptions of fluctuation measurements are beyond the scope of this chapter; however, two important concepts have emerged from the development of fluctuation methods that are of significant value for understanding fluorescence signals in microscopes.

First, the concept of an observation volume has been introduced and provides a method to estimate the size and shape of the spatial region from which fluorescence signals are measured in two-photon and confocal microscopes. No rigorous definition for a physical volume (i.e., a container of some exact size) exists; however, one can compute the approximate measurement observation volume by dividing the total measured fluorescence signal (from a single position of a focused laser beam) by the amount of fluorescence per unit volume emitted when molecules are located in the center of the focal plane (Nagy et al. 2005b; Hess and Webb 2002). Formally, the volume is thus defined:

$$V = \int \frac{w(\mathbf{r})\Omega(\mathbf{r})}{w(0)\Omega(0)}d\mathbf{r} \tag{2.13}$$

The volume, V, estimates the approximate size of the physical region, which contributes substantially to the measured fluorescence signal when a focused laser beam illuminates a single position within the sample for confocal or two-photon microscopy. Images are created by moving the volume (i.e., the laser spot) across the sample in a raster pattern. The spatial dimensions of the observation volume can be useful to consider when selecting image pixel sizes. Pixels much larger than the volume will not optimize the available spatial resolution of the microscope optics.

Once the observation volume has been defined and assuming molecules are dispersed uniformly throughout the volume, Equation 2.12 can be rewritten as $\langle F \rangle = \varepsilon \langle C \rangle V$, where the parameter ε is referred to as the molecular brightness and is defined by

$$\varepsilon = \kappa \langle w(0) \rangle \Omega(0) = \kappa \langle w(0) \rangle \tag{2.14}$$

The molecular brightness specifies the average number of photons measured per unit time from a single molecule located at the center of the focused laser excitation. The molecular brightness is dependent upon the photophysical properties of the fluorescent molecule and the excitation conditions, as well as the microscope filters, optics, and detectors. It is therefore not a fundamental quantity, but it is measurable for any given experimental apparatus.

Both the volume and molecular brightness are valuable quantities in that they allow a fluorescence signal to be calibrated (by determining the volume and molecular brightness of any given fluorescent molecule on a particular instrument) so that the number of molecules responsible for a measured fluorescence signal can be determined. This has proved useful not only for measuring molecular concentrations, but also for investigating molecular interactions in the microscope (e.g., dimers, trimers, etc., as well as binding between two different molecular species such as between a labeled protein and a labeled ligand) (Hillesheim et al. 2006; Wu and Muller 1995; Muller 2004; Berland 2005; Schwille, Meyer-Almes, and Rigler 1997; Eigen and Rigler 1994).

2.5.3 Saturation

On average, a fluorescent molecule can emit no more than $1/\tau$ photons per second. With the tightly focused laser excitation sources used in microscopy, it is not unusual to illuminate fluorescent molecules with photon fluxes for which the nominal excitation rates for one- and two-photon excitation (defined previously) approach or exceed this maximum average fluorescence emission rate. When this occurs, the excitation is saturated and the fluorescence excitation will deviate from the linear (one-photon) or quadratic (two-photon) dependence on input power (Cianci, Wu, and Berland 2004; Wu and Berland 2007; Nagy et al. 2005b). When the excitation is saturated, further increases in laser power do not proportionally increase the signal from the molecules within the observation volume. Instead, saturation tends to increase the size of the effective observation volume (Cianci et al. 2004).

In the microscope, increasing illumination levels to increase signal levels can thus have the unintended effect of decreasing image resolution. Saturation can also introduce artifacts into quantitative analysis of fluorescence signal because the signals are not proportional to illumination levels and because one may be comparing signals measured from differently sized sample regions. For strongly absorbing molecules, saturation can be an important effect even at the average laser power levels that many researchers may consider "low power" excitation. Details of these effects are described in the literature (Cianci et al. 2004; Wu and Berland 2007; Nagy et al. 2005a, 2005b).

2.5.4 Photobleaching

Unfortunately, fluorescent molecules are often not terribly robust, and bleaching is frequently an important concern in fluorescence imaging. Upon repeated excitation, fluorescent molecules will undergo irreversible chemical reactions after which they no longer fluoresce. This is a statistical process and each fluorophore in a certain environment will photobleach, on average, after a certain number of excitations. (That is, there is a certain rate of photolysis once the molecule is in the excited state, and the photolysis pathway competes with the other pathways of de-excitation.) Photobleaching rates are highly dependent upon the illumination level, although the chemical mechanisms responsible for the bleaching process are often not fully understood.

One of the known bleaching mechanisms is due to interaction with molecular oxygen (Corbett, Cho, and Golan 1994; Rasnik, McKinney, and Ha 2006). The ground state of oxygen is a triplet, and after intersystem crossing of the excited fluorophore to the triplet state, the triplet fluorophore can interact with the triplet oxygen, thus creating the singlet ground state of the fluorophore and the singlet state of oxygen. Singlet oxygen is highly reactive and can chemically destroy the fluorophore as well as react with other surrounding molecules, creating reactive oxygen species that can also react with the fluorophore. Multiphoton absorption and excited-state absorption can also play a role in the bleaching of some molecular systems (Patterson and Piston 2000; Rubart 2004; Svoboda and Yasuda 2006).

The average number of emission cycles that occur before photobleaching depends on both the molecule in question and the environment (Patterson and Piston 2000; Hirschfeld 1976). Some molecules may bleach after emitting only a few photons, while other more robust probes can emit millions of fluorescence photons (on average) before bleaching. Photobleaching can cause artifacts in fluorescence microscopy; thus, for quantitative fluorescence measurements, it is always important to determine whether or not one observes significant photobleaching under illumination conditions. This is also true for FLIM microscopy because photobleaching can distort lifetime measurements when multiple lifetime components are involved.

It is sometimes possible to reduce the photobleaching rate chemically. For example, because bleaching is often highly dependent on oxygen concentration, various methods to remove oxygen from the sample can be very effective in stabilizing fluorescence emission (Rasnik et al. 2006). This, of course, assumes that the reduced oxygen concentration and the method used to remove the oxygen are not otherwise detrimental to the system under investigation.

It is worth noting that although bleaching can impede imaging applications, it can also be used to great advantage. For example, acceptor photobleaching is a useful tool for confirming molecular interactions as measured by FRET (Kenworthy 2001; Zal and Gascoigne 2004). Fluorescence photobleaching methods are also widely used to measure molecular dynamics (Axelrod et al. 1976).

2.6 SUMMARY

The aim of this chapter has been to introduce the fundamental concepts and parameters that characterize molecular fluorescence and some concepts relevant to measurement of fluorescence signals in microscopy. It is hoped that an understanding of this information will aid researchers in correctly applying fluorescence imaging technology and in interpreting experimental results. The concepts introduced can also serve as the foundation for understanding the sophisticated applications of fluorescence in microscopy.

REFERENCES

Andersson, H., Baechi, T., Hoechl, M., and Richter, C. 1998. Autofluorescence of living cells. *Journal of Microscopy* 191:1–7.

Aubin, J. E. 1979. Autofluorescence of viable cultured mammalian cells. *Journal of Histochemistry and Cytochemistry* 27:36–43.

Axelrod, D. 1979. Carbocyanine dye orientation in red cell membrane studied by microscopic fluorescence polarization. *Biophysical Journal* 26:557–573.

Axelrod, D., Koppel, D. E., Schlessinger, J., Elson, E., and Webb, W. W. 1976. Mobility measurement by analysis of fluorescence photobleaching recovery kinetics. *Biophysical Journal* 16:1055–1069.

Becker, W. 2007. *The Becker–Hickl TCSPC handbook.* Berlin: Becker & Hickl Gmbh.

Benson, R., Meyer, R., Zaruba, M., and McKhann, G. 1979. Cellular autofluorescence—Is it due to flavins? *Journal of Histochemistry and Cytochemistry* 27:44–48.

Berland, K. M. 2005. Quantifying molecular interactions with fluorescence correlation spectroscopy. In *Molecular imaging: FRET microscopy and spectroscopy,* ed. A. Periasamy and R. N. Day, 272–283. New York: Oxford University Press.

Berlman, I. B. 1971. *Handbook of fluorescence spectra of aromatic molecules.* New York: Academic Press.

Born, M., and Wolf, E. 1997. *Principles of optics.* New York: Cambridge University Press.

Campagnola, P. J., and Loew, L. M. 2003. Second-harmonic imaging microscopy for visualizing biomolecular arrays in cells, tissues and organisms. *Nature Biotechnology* 21:1356–1360.

Campagnola, P. J., Millard, A. C., Terasaki, M., et al. 2002. Three-dimensional high-resolution second-harmonic generation imaging of endogenous structural proteins in biological tissues. *Biophysical Journal* 82:493–508.

Chen, Y., Mills, J. D., and Periasamy, A. 2003. Protein localization in living cells and tissues using FRET and FLIM. *Differentiation* 71:528–541.

Chen, Y., Muller, J. D., Berland, K. M., and Gratton, E. 1999. Fluorescence fluctuation spectroscopy. *Methods* 19:234–252.

Chen, Y., Muller, J. D., Eid, J. S., and Gratton, E. 2001. Two-photon fluorescence fluctuation spectroscopy. In *New trends in fluorescence spectroscopy: Applications to chemical and life sciences,* ed. B. Valeur and J.-C. Brochon, 277–296. Berlin: Springer–Verlag.

Cianci, C. G., Wu, J. and Berland, K. M. 2004. Saturation modified point spread functions in two-photon microscopy. *Microscopy Research and Technique* 64:135–141.

Clegg, R. M. 1995. Fluorescence resonance energy transfer. *Current Opinion in Biotechnology* 6:103–110.

Clegg, R. M. 1996. Fluorescence resonance energy transfer. In *Fluorescence imaging spectroscopy and microscopy,* ed. X. F. Wang and B. Herman, 179–252. New York: John Wiley & Sons.

Clegg, R. M., Holub, O., and Gohlke, C. 2003. Fluorescence lifetime-resolved imaging: Measuring lifetimes in an image. *Methods in Enzymology* 360:509–542.

Conchello, J.-A., and Lichtman, J. W. 2005. Optical sectioning microscopy. *Nature Methods* 2:920–931.

Corbett, J. D., Cho, M. R., and Golan, D. E. 1994. Deoxygenation affects fluorescence photobleaching recovery measurements of red cell membrane protein lateral mobility. *Biophysical Journal* 66:25–30.

Davidson, M. W., and Abramowitz, M. 2008. *Molecular expressions optical microscopy primer,* Olympus America Inc. http://micro.magnet.fsu.edu/primer/index.html

Denk, W., Strickler, J. H., and Webb, W. W. 1990. Two-photon laser scanning fluorescence microscopy. *Science* 248:73–76.

Diaspro, A. 2001. *Confocal and two-photon microscopy: Foundations, applications and advances.* Wilmington, DE: Wiley–Liss.

Dickinson, M. E., Simbuerger, E., Zimmermann, B., Waters, C. W., and Fraser, S. E. 2003. Multiphoton excitation spectra in biological samples. *Journal of Biomedical Optics* 8:329–338.

Eggeling, C., Widengren, J., Rigler, R., and Seidel, C. A. M. 1998. Photobleaching of fluorescent dyes under conditions used for single-molecule detection: evidence of two-step photolysis. *Analytical Chemistry* 70:2651–2659.

Eigen, M., and Rigler, R. 1994. Sorting single molecules—Application to diagnostics and evolutionary biotechnology. *Proceedings of the National Academy of Sciences USA* 91:5740–5747.

Fixler, D., Namer, Y., Yishay, Y., and Deutsch, M. 2006. Influence of fluorescence anisotropy on fluorescence intensity and lifetime measurement: theory. Simulations and experiments. *Biomedical Engineering IEEE Transactions* 53:1141–1152.

Förster, T. 1965. Delocalized excitation and excitation transfer. In *Modern Quantum Chemistry,* ed. O. Sinanoglu, 93–137. New York: Academic Press.

Frackowiak, D. 1988. The Jablonski diagram. *Journal of Photochemistry and Photobiology B: Biology* 2:399–399.

Ganesan, S., Ameer-Beg, S. M., Ng, T. T. C., Vojnovic, B., and Wouters, F. S. 2006. A dark yellow fluorescent protein (YFP)-based resonance energy-accepting chromoprotein (REACh) for Forster resonance energy transfer with GFP. *Proceedings of the National Academy of Sciences of the United States of America* 103:4089–4094.

Göppert-Mayer, M. 1931. Über Elementarakte mit zwei Quantensprüngen. *Annalen der Physik* 401:273–294.

Gordon, G. W., Berry, G., Liang, X., B. L., and Herman, B. 1998. Quantitative fluorescence resonance energy transfer measurements using fluorescence microscopy. *Biophysical Journal* 74:2702–2713.

Hamamatsu Company. 2008. Quantum efficiency. Concepts in digital imaging technology. Hamamatsu Company.

Haugland, R. P. 2007. *Handbook of fluorescent probes and research chemicals. Molecular probes.* http://www.probes.com/handbook

Heikal, A. A., Hess, S. T., and Webb, W. W. 2001. Multiphoton molecular spectroscopy and excited-state dynamics of enhanced green fluorescent protein (EGFP): Acid-base specificity. *Chemical Physics* 274:37–55.

Helmchen, F., and Denk, W. 2005. Deep tissue two-photon microscopy. *Nature Methods* 2:932–940.

Hess, S. T., Huang, S. H., Heikal, A. A., and Webb, W. W. 2002. Biological and chemical applications of fluorescence correlation spectroscopy: A review. *Biochemistry* 41:697–705.

Hess, S. T., and Webb, W. W. 2002. Focal volume optics and experimental artifacts in confocal fluorescence correlation spectroscopy. *Biophysical Journal* 83:2300–2317.

Heyduk, T. 2002. Measuring protein conformational changes by FRET/LRET. *Current Opinion in Biotechnology* 13:292–296.

Hillesheim, L. N., Chen, Y., and Muller, J. D. 2006. Dual-color photon counting histogram analysis of mRFP1 and EGFP in living cells. *Biophysical Journal* 91:4273–4284.

Hirschfeld, T. 1976. Quantum efficiency independence of the time integrated emission from a fluorescent molecule. *Applied Optics* 15:3135–3139.

Jares-Erijman, E. A., and Jovin, T. M. 2003. FRET imaging. *Nature Biotechnology* 21:1387–1395.

Kapusta, P., Wahl, M., Benda, A., Hof, M., and Enderlein, J. 2007. Fluorescence lifetime correlation spectroscopy. *Journal of Fluorescence* 17:43–48.

Kasha, M. 1987. 50 Years of the Jablonski diagram. *Acta Physica Polonica A* 71:661–670.

Kawata, S., and Sasaki, K. 1996. Multispectral image processing for component analysis. In *Fluorescence imaging spectroscopy and microscopy,* ed. X. F. Wang and B. Herman, 55–86. New York: John Wiley & Sons.

Kazuko, K., Masahiro, Y., Motoyoshi, B., Tohru, S., and Hidefumi, A. 1999. High collection efficiency in fluorescence microscopy with a solid immersion lens. *Applied Physics Letters* 75:1667–1669.

Kenworthy, A. K. 2001. Imaging protein–protein interactions using fluorescence resonance energy transfer microscopy. *Methods* 24:289–296.

Kenworthy, A. K. 2005. Photobleaching FRET microscopy. In *Molecular imaging: FRET microscopy and spectroscopy,* ed. A. Periasamy and R. N. Day, 146–164. New York: Oxford University Press.

Koushik, S. V., Chen, H., Thaler, C., Puhl, H. L. I. and Vogel, S. S. 2006. Cerulean, Venus, and VenusY67C FRET reference standards. *Biophysical Journal* 91:L99–101.

Kühnemuth, R., and Seidel, A. M. C. 2001. Principles of single molecule multiparameter fluorescence spectroscopy. *Single Molecules* 2:251–254.

Lakowicz, J. R. 2007. *Principles of fluorescence spectroscopy,* 3rd ed. New York: Plenum Press.

Lindek, S., Swoger, J., and Stelzer, E. H. K. 1999. Single-lens theta microscopy: Resolution, efficiency and working distance. *Journal of Modern Optics* 46:843–858.

Masters, B. R. 2006. *Confocal microscopy and multiphoton excitation microscopy: The genesis of live cell imaging.* Bellingham, WA: SPIE Publications.

Masters, B. R., and So, P. T. C. 2008. *Handbook of biomedical nonlinear optical microscopy.* New York: Oxford University Press.

Muller, J. D. 2004. Cumulant analysis in fluorescence fluctuation spectroscopy. *Biophysical Journal* 86:3981–3992.

Muller, J. D., Chen, Y., and Gratton, E. 2003. Fluorescence correlation spectroscopy. *Methods in Enzymology* 361:69–92.

Nagy, A., Wu, J., and Berland, K. M. 2005a. Characterizing observation volumes and the role of photophysical dynamics in one-photon fluorescence fluctuation spectroscopy. *Journal of Biomedical Optics* 10:1–9.

Nagy, A., Wu, J., and Berland, K. M. 2005b. Observation volumes and {gamma}-factors in two-photon fluorescence fluctuation spectroscopy. *Biophysical Journal* 89:2077–2090.

Najechalski, P., Morel, Y., Stephan, O., and Baldeck, P. L. 2001. Two-photon absorption spectrum of poly(fluorene). *Chemical Physics Letters* 343:44–48.

Neher, R. A., and Neher, E. 2004. Applying spectral fingerprinting to the analysis of FRET images. *Microscopy Research and Technique* 64:185–195.

Patterson, G. H., and Piston, D. W. 2000. Photobleaching in two-photon excitation microscopy. *Biophysical Journal* 78:2159–2162.

Pawley, J. B. 1995. *Handbook of biological confocal microscopy.* New York: Plenum.

Periasamy, A. 2001. *Methods in cellular imaging.* New York: Oxford University Press.

Periasamy, A., and Day, R. N. 2005. *Molecular imaging: FRET microscopy and spectroscopy.* New York: Oxford University Press.

Piston, D. W., Rizzo, M. A., and Kevin, F. S. 2008. FRET by fluorescence polarization microscopy. *Methods in Cell Biology* 85:415–430.

Rasnik, I., McKinney, S. A., and Ha, T. 2005. Surfaces and orientations: Much to FRET about? *Accounts of Chemical Research* 38:542–548.

Rasnik, I., McKinney, S. A., and Ha, T. 2006. Nonblinking and long-lasting single-molecule fluorescence imaging. *Nature Methods* 3:891–893.

Reichman, J. 2007. Handbook of optical filters for fluorescence microscopy. Chroma Technology. http://www.chroma.com/resources/filter-handbook

Rigler, R., and Elson, E. 2000. *Fluorescence correlation spectroscopy theory and applications.* New York: Springer.

Rizzo, M. A., Springer, G. H., Granada, B. and Piston, D. W. 2004. An improved cyan fluorescent protein variant useful for FRET. *Nature Biotechnology* 22:445–449.

Rubart, M. 2004. Two-photon microscopy of cells and tissue. *Circulation Research* 95:1154–1166.

Schwille, P. 2001. Fluorescence correlation spectroscopy and its potential for intracellular applications. *Cell Biochemistry and Biophysics* 34:383–408.

Schwille, P., Meyer-Almes, F. J., and Rigler, R. 1997. Dual-color fluorescence cross-correlation spectroscopy for multicomponent diffusional analysis in solution. *Biophysical Journal* 72:1878–1886.

Sekar, R. B., and Periasamy, A. 2003. Fluorescence resonance energy transfer (FRET) microscopy imaging of live cell protein localizations. *Journal of Cell Biology* 160:629–633.

Selvin, P. R., and Ha, T. 2008. *Single molecule techniques: A laboratory manual.* Cold Spring Harbor, NY: Cold Spring Harbor Laboratory Press.

So, P. T. C., Dong, C. Y., Masters, B. R., and Berland, K. M. 2000. Two-photon excitation fluorescence microscopy. *Annual Review of Biomedical Engineering* 2:399–429.

So, P. T. C., French, T., Yu, W. M., et al. 1995. Time-resolved fluorescence microscopy using two-photon excitation. *Bioimaging* 3:49–63.

Spiess, E., Bestvater, F. A., Toth, K., et al. 2005. Two-photon excitation and emission spectra of the green fluorescent protein variants ECFP, EGFP and EYFP. *Journal of Microscopy* 217:200–204.

Squirrell, J. M., Wokosin, D. L., White, J. G., and Bavister, B. D. 1999. Long-term two-photon fluorescence imaging of mammalian embryos without compromising viability. *Nature Biotechnology* 17:763–767.

Steiner, R. F. 1991. Fluorescence anisotropy: Theory and applications. In *Topics in fluorescence spectroscopy: Techniques,* vol. 2, ed. J. R. Lakowicz, 1–52. New York: Springer.

Stryer, L. 1978. Fluorescence energy transfer as a spectroscopic ruler. *Annual Review of Biochemistry* 47:819–846.

Svoboda, K., and Yasuda, R. 2006. Principles of two-photon excitation microscopy and its applications to neuroscience. *Neuron* 50:823–839.

Thaler, C., Koushik, S., Blank, P. S., and Vogel, S. S. 2005. Quantitative multiphoton spectral imaging and its use for measuring resonance energy transfer. *Biophysical Journal* 89:2736–2749.

Thaler, C., and Vogel, S. S. 2006. Quantitative linear unmixing of CFP and YFP from spectral images acquired with two-photon excitation. *Cytometry Part A* 69A:904–911.

Tron, L., Szollosi, J., Damjanovich, S., et al. 1984. Flow cytometric measurement of fluorescence resonance energy transfer on cell surfaces. Quantitative evaluation of the transfer efficiency on a cell-by-cell basis. *Biophysical Journal* 45:939–946.

Truong, K., and Ikura, M. 2001. The use of FRET imaging microscopy to detect protein–protein interactions and protein conformational changes in vivo. *Current Opinion in Structural Biology* 11:573–578.

Tsien, R. Y. 1998. The green fluorescent protein. *Annual Review of Biochemistry* 67:509–544.

Tsien, R. Y., and Waggoner, A. 1995. Fluorophores for confocal microscopy. In *Handbook of biological confocal microscopy,* 2nd ed., ed. J. B. Pawley, 267–279. New York: Plenum.

Valeur, B. 2002. *Molecular fluorescence: Principles and applications.* New York: John Wiley & Sons.

Wallrabe, H., and Periasamy, A. 2005. Imaging protein molecules using FRET and FLIM microscopy. *Current Opinion in Biotechnology* 16:19–27.

Wang, Y., Shyy, J. Y. J. and Chien, S. 2008. Fluorescence proteins, live-cell imaging, and mechanobiology: Seeing is believing. *Annual Review of Biomedical Engineering* 10:1–38.

Webb, R. H. 1996. Confocal optical microscopy. *Reports on Progress in Physics* 3:427–471.

Webb, W. W. 2001. Fluorescence correlation spectroscopy: Inception, biophysical experimentations, and prospectus. *Applied Optics* 40:3969–3983.

Webb, W. W. 2003. Multiphoton excitation (MPE) index, DRBIO Research Webb Group. www.drbio. cornell.edu/MPE/mpe.html

Williams, R. M., Piston, D. W., and Webb, W. W. 1994. Two-photon molecular excitation provides intrinsic three-dimensional resolution for laser-based microscopy and microphotochemistry. *FASEB Journal* 8:804–813.

Wilson, T. 1990. *Confocal microscopy.* Oxford: Academic Press.

Wu, B., and Muller, J. D. 2005. Time-integrated fluorescence cumulant analysis in fluorescence fluctuation spectroscopy. *Biophysical Journal* 89:2721–2735.

Wu, J., and Berland, K. 2007. Fluorescence intensity is a poor predictor of saturation effects in two-photon microscopy: Artifacts in fluorescence correlation spectroscopy. *Microscopy Research and Technique* 70:682–686.

Wu, P. G., and Brand, L. 1994. Resonance energy transfer: Methods and applications. *Analytical Biochemistry* 218:1–13.

Xu, C. and Webb, W. W. 1991. Multiphoton excitation of molecular fluorophores and nonlinear laser microscopy. In *Topics in fluorescence spectroscopy: Nonlinear and two-photon induced fluorescence,* ed. J. R. Lakowicz and C. D. Geddes, 471–540. New York: Springer.

Xu, C. and Webb, W. W. 1996. Measurement of two-photon excitation cross sections of molecular fluorophores with data from 690 to 1050 nm. *Journal of the Optical Society of America B* 13:481–491.

Xu, C., Zipfel, W. R., Shear, J. B., Williams, R. M., and Webb, W. W. 1996. Multiphoton fluorescence excitation: New spectral windows for biological nonlinear microscopy. *Proceedings of the National Academy of Sciences USA* 93:10763–10768.

Yamaguchi, S., and Tahara, T. 2003. Two-photon absorption spectrum of all-trans retinal. *Chemical Physics Letters* 376:237–243.

Zal, T., and Gascoigne, N. R. J. 2004. Photobleaching-corrected FRET efficiency imaging of live cells. *Biophysical Journal* 86:3923–3939.

Zimmermann, T., Rietdorf, J., and Pepperkok, R. 2003. Spectral imaging and its applications in live cell microscopy. *FEBS Letters* 546:87–92.

Zipfel, W. R., Williams, R. M., and Webb, W. W. 2003. Nonlinear magic: Multiphoton microscopy in the biosciences. *Nature Biotechnology* 21:1369–1377.

CHAPTER 3

Visible Fluorescent Proteins for FRET-FLIM

Richard N. Day

3.1 INTRODUCTION

The use of noninvasive approaches such as fluorescence microscopy to detect events as they occur inside living cells is providing remarkable insight into molecular processes. Just in the last decade, the development of new genetically encoded probes, coupled with advances in digital image acquisition and analysis, has dramatically improved our ability to obtain quantitative measurements from living cells. Specifically, the cloning of the jellyfish *Aequorea victoria* green fluorescent protein (GFP; Prasher et al. 1992) sparked a revolution in studies of cell biology and physiology. For the first time, it became possible to produce genetically encoded fluorescent markers inside living cells and organisms (Chalfie et al. 1994; Inouye and Tsuji 1994). The utility of the fluorescent proteins (FPs) as noninvasive probes has been repeatedly proven by their integration into a variety of different living systems (reviewed by Hadjantonakis et al. 2003; Stewart 2006).

In the years since its cloning, the sequence encoding the *Aequorea* GFP has been engineered to yield new FPs emitting light from the blue to yellowish green range of the visible spectrum (Tsien 1998; Cubitt, Woollenweber, and Heim 1999; Nagai et al. 2002; Rizzo et al. 2004; Ai et al. 2007). Furthermore, it is now well appreciated that many marine organisms produce FPs that are homologous to the *Aequorea* GFP (Matz, Lukyanov, and Lukyanov 2002; Labas et al. 2002; Shagin et al. 2004). Recently, some of the GFP-like proteins that are responsible for the bright colors we see in reef corals have become available for live-cell imaging applications (Matz et al. 1999; Karasawa et al. 2004; Shcherbo et al. 2007).

The FP palette now spans the visible spectrum from deep blue to deep red, giving investigators a wide choice of genetically encoded markers for studies in cell biology (Shaner, Patterson, and Davidson 2007; Day and Schaufele 2008). The new colors expanded the repertoire of uses of the FPs to include multicolor imaging of protein co-localization and behavior inside living cells or to detect changes in intracellular activities, such as pH or ion concentration. However, their use for Förster resonance energy transfer (FRET) microscopy in living cells has generated the most interest in these probes (Tsien 1998; Lippincott-Schwartz, Snapp, and Kenworthy 2001; Zhang et al. 2002; Giepmans et al. 2006; Shaner et al. 2007).

Chapters 1 and 2 introduced FRET and microscopy and how to detect the transfer of excited-state energy nonradiatively from a donor fluorophore and nearby acceptor probes. This chapter will discuss the spectral properties of the different FPs that allow them to be used as donor and acceptor probes in FRET microscopy. Because energy transfer is limited to the scale of less than 100 Å, FRET provides unique information about the spatial relationships of proteins inside the living cell that is beyond the optical resolution limit of the conventional light microscope.

Although there are the many different methods for detecting FRET (reviewed by Jares-Erijman and Jovin 2003; Periasamy and Day 2005), the measurement of the donor fluorescence lifetime is considered to be the most rigorous method (see Chapters 1 and 9; Day and Piston 1999; Yasuda 2006; Piston and Kremers 2007). Intensity-based imaging measures a time-averaged fluorescent signal, and it is sometimes difficult to distinguish spectrally overlapping fluorophores and determine concentrations. Fluorescence lifetime imaging microscopy (FLIM) can map the spatial distribution of lifetimes in a sample to reveal heterogeneity in the probe environments. FLIM is particularly useful for biological applications because measurements made in the time domain are independent of variations in the probe concentration, excitation intensity, and other factors that can limit steady-state intensity-based measurements (see Chapters 1 and 2). Importantly, FRET is a dynamic process that nonradiatively depopulates the excited state of the donor fluorophore. This lowers the fluorescence lifetime of the donor, which is detected by FLIM, and provides a direct measurement of FRET.

This chapter presents selected practical guidelines for using the FPs as labels for FRET-FLIM studies. The objective is to relate the important features and photophysical properties of different FPs to their use as probes for FRET-FLIM measurements in biological systems. The intent is not to provide a comprehensive listing of all the FPs currently available, but rather to highlight a few exceptional probes and discuss their application for FRET-FLIM studies. The goal is to introduce the properties of the newer FPs in sufficient depth to be useful to those interested in pursuing these techniques in more detail.

The limitations, pitfalls, and critical considerations for FRET studies using these probes will be demonstrated using simplified theoretical frameworks, as well as experimental results. These considerations are especially relevant when making FRET-FLIM measurements in living cells because awareness of the potential pitfalls will help avoid complications and increase the reliability of these fluorescence measurements. To gain an appreciation of the applications of FRET and FLIM, the reader can then explore these topics in more depth in the references.

3.2 BACKGROUND

3.2.1 Overview of the Fluorescent Proteins

In 1962, Shimomura, Johnson, and Saiga purified the blue-light-emitting photoprotein, aequorin, from the jellyfish *Aequorea victoria*. They also reported the presence of an autofluorescent protein in extracts from the jellyfish. The protein was GFP, and it was later isolated and shown to be a companion protein for aequorin, absorbing its blue light emission and then reemitting it as green light (Morise et al. 1974). Prasher et al. (1992) cloned the gene encoding the *Aequorea* GFP, and the utility of this new probe for in vivo fluorescence labeling was proven by its expression in bacteria, mammalian and plant cells, and transgenic organisms (Chalfie et al. 1994; Inouye and Tsuji 1994; Plautz et al. 1996; van Roessel and Brand 2001; reviewed by Hadjantonakis et al. 2003; Stewart 2006).

To be fluorescent, GFP must fold into a tightly woven 11-strand beta-barrel structure (see Figure 3.1). Nearly the entire 238 amino acid sequence is required for its proper folding. The tripeptide sequence serine[65]–tyrosine[66]–glycine[67] is positioned at the core of the

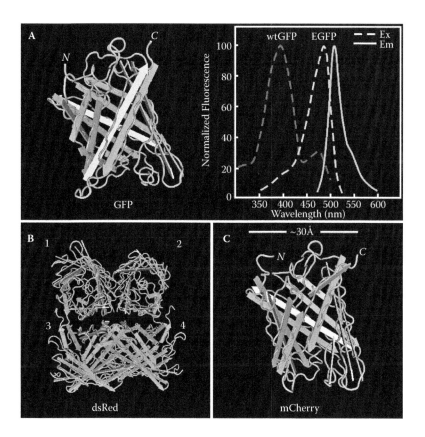

FIGURE 3.1 **(See color insert following page 288.)** The β-barrel structure of the FPs. (A) The structure of GFP is shown illustrating the 11-strand β-barrel that surrounds the chromophore. The excitation (Ex) and emission (Em) spectra for wtGFP and EGFP are plotted, showing how the S65T mutation in EGFP shifts the spectrum to a single peak excitation at 489 nm. (B) The tetrameric structure of the DsRed FP. (C) The structure of its monomeric derivative, mCherry. The illustration of DsRed looks down the barrel of two subunits (1,2) and views the other two subunits from the side. The approximate dimension of the β-barrel for the monomer mCherry is shown. (Data from the National Center for Biotechnology Information Molecular Modeling DataBase. The structures were rendered using the Cn3D software.)

beta-barrel as the protein matures, and this drives the cyclization and dehydration reaction that forms the chromophore (reviewed by Tsien 1998; Ward 1998). Once formed, the wild type (wt) *Aequorea* GFP displays a complex absorption spectrum, with maximal excitation occurring at 397 nm and a minor secondary peak of excitation at 476 nm (Figure 3.1A). Chattoraj et al. (1996) examined the excited-state dynamics of the wtGFP protein and showed that green fluorescence resulted from deprotonation within the chromophore, populating the state favoring excitation at 397 nm. In addition, a charged intermediate state is thought to relax slowly to a second stable state that is only rarely formed, accounting for the minor secondary excitation peak at 476 nm (Chattoraj et al. 1996).

The measurement of the fluorescence lifetime of the wtGFP revealed a multiexponential decay with components that corresponded to the different excited-state species (Striker et al. 1999). A mutated wtGFP, called Sapphire, was generated in which a chromophore contacting amino acid in the beta-barrel, threonine[203], was changed to isoleucine (Tsien 1998). As we will see in Section 3.2.2, this amino acid in the beta-barrel is a critical determinant of the photophysical properties of the chromophore. Changing the threonine[203] to isoleucine abolished the secondary excitation peak at 475 nm, generating a GFP with an exceptionally large Stokes shift. The Sapphire FP was further optimized for expression in mammalian cells and was shown to be a useful FRET donor for the red FPs described later (Zapata-Hommer and Griesbeck 2003).

3.2.2 Spectral Variants from the *Aequorea* GFP

Over the last decade, both targeted and random mutagenesis strategies have been used to modify the spectral and physical characteristics of the wtGFP, yielding new, enhanced (E) FPs ranging in color from blue to yellow-green (reviewed in Patterson, Day, and Piston 2001; Zhang et al. 2002; see Table 3.1). The sequences encoding the E FPs incorporate preferred human codon usage and silent mutations that improved the efficiency of production and maturation of the proteins in mammalian cells. In addition, many laboratories have contributed mutant variants that have altered spectral properties, providing different color fluorescent probes for live-cell imaging.

For example, one of the earliest color variants of the wtGFP is a blue FP (BFP) that results from substitution of tyrosine[66] with histidine (Heim and Tsien 1996; Cubitt et al. 1999). The original BFP, however, has a low quantum yield and is very susceptible to photobleaching. Recently, several groups used mutagenesis strategies to develop new blue FP variants with much higher quantum yields and photostabilities, greatly improving the utility of these deep blue probes (Ai et al. 2007; Kremers et al. 2007; Mena et al. 2006). The EBFP2, generated by Ai et al (2007), is the brightest and most stable of the blue FPs available and has been shown to be an excellent donor for FRET studies (see Table 3.1; Shaner et al. 2007). This new EBFP should be useful for long-term imaging in living cells where a blue probe is required, especially when two-photon excitation is used, which avoids cellular damage by near UV excitation (Wallrabe et al. 2003).

The development of cyan color variants (CFP) provided an early alternative to the blue FP. ECFP resulted from substitution of tyrosine[66] with tryptophan in combination with mutations in several other residues within the surrounding beta-barrel structure (Heim,

TABLE 3.1   Selected FP Color Palette Including Proteins Based on the *Aequorea* GFP, DsRed, and FPs Cloned from Other Marine Organisms

Fluorescent protein	Color	Peak Ex (nm)	Peak Em (nm)	Brightness[a]	Photostability[b]	Ref.
EBFP2	Blue	383	448	18	++	Ai et al. 2007
Cerulean	Cyan	433–445	475–503	27	++	Rizzo et al. 2004
mTFP	Teal	462	492	54	+++	Ai et al. 2006
EmGFP	Green	487	509	39	++++	Cubitt et al. 1995
Venus	Yellow-green	515	528	53	+[c]	Nagai et al. 2002
mKO (Kusabira)	Orange	548	559	31	+++	Karasawa et al. 2004
mOrange2	Orange	549	565	35	++++	Shaner et al. 2008
TagRFP-T	Orange	555	584	33	++++	Shaner et al. 2008
tdTomato	Orange	554	581	95	+++	Shaner et al. 2004
mCherry	Red	587	610	17	++	Shaner et al. 2004
mKate (Katushka)	Deep red	588	635	15[d]	++++	Shcherbo et al. 2007
REACh	Weak yellow	514	527	1[e]	nd	Ganesan et al. 2006; Murakoshi et al. 2008

[a] Intrinsic brightness is the product of quantum yield and extinction coefficient.

[b] Adapted from Shaner, N. C. et al. 2005. *Nature Methods* 2:905–909; Shaner, N. C. et al. 2007. *Journal of Cell Science* 120:4247–4260; Shaner, N. C. et al. 2008. *Nature Methods* 5:545–551.

[c] Proven useful for applications that involve photobleaching.

[d] Cellular autofluorescence is low at longer wavelengths, improving the signal-to-noise ratio.

[e] Dark probe useful for FRET-FLIM; see Section 3.3.6.

Prasher, and Tsien 1994; Cubitt et al. 1999). Although ECFP is optimally excited at 433 nm, its absorption spectrum is broad, with a second excitation peak near 445 nm. As had been observed for the wtGFP, the complex excitation spectrum of ECFP indicates more than one excited-state species. This was confirmed by fluorescence lifetime measurements (Tramier et al. 2002).

Efforts to improve the characteristics of ECFP yielded a mutant variant, called Cerulean, which results from substitutions on the solvent-exposed surface of ECFP (tyrosine[145] and histidine[148]). Cerulean has an increased quantum yield compared to ECFP, and Cerulean has been reported to have a single excited state (Table 3.1; Rizzo et al. 2004). However, when expressed in living cells, the fluorescence decay for Cerulean, similarly to ECFP, still indicates the presence of more than one excited-state species (Yasuda et al. 2006; Millington et al. 2007).

Improved green color variants result from changing the central chromophore, serine[65], to threonine (S65T). This stabilizes the chromophore in a permanently ionized form with a single peak absorbance at 489 nm and a peak emission at 507 nm (see Figure 3.1A; Heim et al. 1994; Brejc et al. 1997; Cubitt et al. 1999). Importantly, this simplifies the lifetime decay kinetics to a monoexponential (Traimer et al. 2006; Yasuda et al. 2006). Currently, the EGFP with the best imaging characteristics is called Emerald, which incorporates several additional point mutations that improve folding and increase its brightness (Table 3.1; Cubitt et al. 1999). The longest wavelength emission variants of *Aequorea* GFP result from again targeting the chromophore contacting amino acid in the beta-barrel, threonine[203]—in this case, changing it to tyrosine (Ormö et al. 1996). This generates a bright yellow green FP (YFP) that is optimally excited at 514 nm and has a peak emission at 527 nm.

The EYFP variant, however, is sensitive to both pH and halides, limiting its usefulness for studies in living cells. Starting with EYFP, Nagai et al. (2002) discovered that the substitution of phenylalanine[46] with leucine improves the maturation efficiency and reduces the halide sensitivity of YFP. Combined with several additional mutations, this results in a variant called Venus, which is among the brightest and most red shifted of the mutant variants based on the *Aequorea* GFP currently available (Table 3.1).

3.2.3 *Aequorea* Fluorescent Proteins and Dimer Formation

Most of the FPs from marine organisms that have been characterized were isolated as dimers, tetramers, or part of higher order complexes (see Figure 3.1B, for example; Shagin et al. 2004). Although the *Aequorea* GFP was isolated as a monomer, it can form dimers when the protein is highly concentrated (Ormö et al. 1996; Brejc et al. 1997). This self-association is not typically observed when the FPs diffuses freely within the cell, but dimers have a tendency to form when diffusion is restricted, such as in the two-dimensional space of biological membranes. Here, the formation of *Aequorea* FP dimers can cause the proteins that they label to form atypical complexes (Kenworthy 2002). To overcome this problem, Zacharias et al. (2002) developed monomeric forms of the *Aequorea*-based FPs by substitution of the alanine[206] with lysine, which blocked the dimer formation without altering the fluorescence characteristics of the *Aequorea*-based FPs.

3.2.4 New Fluorescent Proteins from Corals

Much of the color diversity in reef corals results from GFP-like proteins (Matz et al. 2002; Labas et al. 2002; Shagin et al. 2004). It is thought that these proteins evolved to fulfill a photoprotective function (Leutenegger et al. 2007) or, alternatively, to support symbiotic relationships between the corals and algae (Field et al. 2006). Over the past decade, some of these GFP-like proteins have been characterized, cloned, and optimized for imaging applications (Matz et al. 1999; Karasawa et al. 2004; Shcherbo et al. 2007). Among the first of the new FPs from coral is a protein called DsRed that was isolated from the mushroom anemone *Discosoma striata* (Matz et al. 1999). DsRed is the first red FP (RFP) to become available, with a peak absorbance at 558 nm and a maximum emission at 583 nm.

Unfortunately, the characteristics of DsRed do not particularly lend themselves to live cell imaging. For instance, DsRed is a very slowly maturing protein that generates a green intermediate as it matures. Even more problematic is that it is an obligate tetramer (Figure 3.1B) with a strong tendency to form oligomers when produced inside cells (Baird, Zacharias, and Tsien 2000). The fluorescence lifetime of the native complex has been determined to be about 3.6 ns (Heikal et al. 2000), but continued illumination causes photoconversion events, leading to more heterogeneous lifetimes ranging from 1.5 to 3.6 ns (Cotlet et al. 2001).

To overcome these problems, both random and directed mutagenesis strategies have been used to improve this novel red FP. Bevis and Glick (2002) used this strategy to address the problem of slow maturation, and they generated a rapid maturing variant called DsRedT.1. Starting with the DsRedT.1 variant, Campbell et al. (2002) applied directed mutagenesis to break the tetramer, but this generated a nonfluorescent species. They then used many rounds of random mutagenesis and selected for proteins with improved red fluorescence (an approach called directed evolution). This yielded a rapid maturing monomeric RFP (mRFP) that overcame the critical problems associated with DsRed, and also shifted the fluorescence emission deeper into the red spectrum (Campbell et al. 2002).

However, this new mRFP still has problems that limit its use as a probe for quantitative imaging studies. As expected, the fluorescence quantum yield is only about 25% of that of the DsRed tetramer. Further, there is an absorbance peak at 503 nm; however, this species is nonfluorescent, which might indicate a fraction of the protein that never fully matures. To generate additional FPs with improved characteristics, mRFP was subjected to many rounds of directed evolution using both the error-prone polymerase chain reaction and somatic hypermutation in B-lymphocytes (Wang et al. 2004; Shaner et al. 2004). When combined with cell-based screening methods, these approaches yielded a variety of new FPs. This new crop of FPs included mCherry (Figure 3.1C), which is a rapidly maturing, brighter, more stable version of mRFP (Table 3.1; Shaner, Steinbach, and Tsien 2005). In addition, a dimeric FP called tdTomato has been generated that is currently the brightest of the available FPs (Table 3.1). This probe is useful for applications that require minimal exposure to excitation illumination to maintain cell viability, but it is limited by its larger size (54 kDa).

Still other novel FPs have recently been cloned from corals and engineered to improve their utility for live-cell imaging. For example, the sequence encoding a cyan-colored protein from the coral *Clavularia* was modified by directed mutagenesis to generate a monomeric teal FP (mTFP) with remarkable brightness and photostability (Ai et al. 2006; see Table 3.1). Unlike CFP, which has a tryptophan residue at the central chromophore position, mTFP has a tyrosine residue in this position, which is typical of the GFPs. Indeed, both the excitation and emission spectra of the mTFP are shifted to the green wavelengths when compared to CFP. The mTFP protein has a high quantum yield and displays a relatively narrow emission spectrum that strongly overlaps the excitation spectrum of the yellow and orange FPs. The mTFP is also noteworthy because it fills the spectral gap between the cyan and green FPs and is optimally excited by the 457 nm laser line that is available on most confocal microscopes. This new blue green protein is also an excellent donor fluorophore for FRET studies using the Venus FP (Day, Booker, and Periasamy 2008; discussed later).

Several orange and red FPs are currently available that share significant spectral overlap with other commonly used FPs, providing alternative acceptor fluorophores. For example, a protein called Kusabira orange (KO) was isolated from the mushroom coral *Fungia concinna*. This FP was cloned and engineered to a bright, photostable monomeric protein called mKO (Karasawa et al. 2004; see Table 3.1). Another monomeric orange (mOrange) protein was generated during the directed evolution of mRFP (Shaner et al. 2004, 2005), but its use was limited by problems with photostability. Recently, an improved variant, mOrange2, was evolved and selected for increased photostability (Shaner et al. 2008; Table 3.1).

This same approach was applied to another orange FP called TagRFP, which was originally cloned from the sea anemone *Entacmaea quadricolor* (Merzlyak et al. 2007). The directed evolution yielded the TagRFP-T protein, with much improved photostability (Shaner et al. 2008; Table 3.1). Shcherbo et al. (2007) also applied this approach to the *Entacmaea* FP, but instead selected for deep red FPs. This yielded a dimeric RFP called Katushka, which was then engineered to the monomeric protein called mKate. The mKate protein is currently the brightest and most photostable of the deep RFPs (Shcherbo et al. 2007; Table 3.1). This makes mKate potentially useful for studies that combine FRET imaging in the deep red spectral window with blue or cyan probes for other protein activities.

3.3 METHODS

3.3.1 Visible Fluorescent Proteins for FRET Measurements

As we have seen in the earlier chapters in this book, the Förster distance (R_0) for a fluorophore pair depends on the quantum yield of the donor and the spectral overlap integral ($J\lambda$) of the donor emission with the absorption of the acceptor. The importance of the spectral overlap is illustrated by comparing the absorption and emission spectra for Cerulean CFP and Venus YFP (Figure 3.2), which share significant spectral overlap (shaded area), making them efficient FRET partners. When energy is transferred from Cerulean to Venus, the Cerulean emission detected in the donor channel is quenched and emission from the acceptor is increased (sensitized); this can be detected in the FRET channel (Figure 3.2). However, the strong spectral overlap also produces spectral

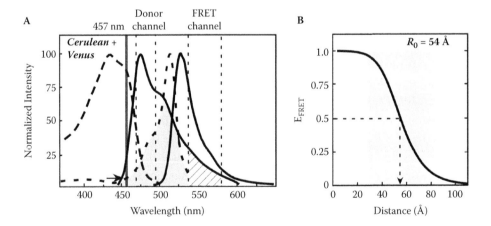

FIGURE 3.2   FRET efficiency as a function of distance separating Cerulean and Venus. (A) Cerulean CFP and Venus YFP are commonly used in FRET studies, and the excitation (- - -) and emission (——) spectra are illustrated here. The spectral overlap is illustrated by shading. Excitation with the 457 nm laser line is shown, and typical band-pass filters for detection of the donor and the acceptor are illustrated (vertical dotted lines). The spectral bleed-through caused by the direct excitation of the acceptor (arrow) and donor bleed-through fluorescence into the FRET channel (hatching) are also illustrated. (B) The Förster distance, R_0, for this fluorophore pair is 54 Å, and the efficiency of energy transfer, E_{FRET}, is plotted as a function of the separation distance (r). The distance spanning the range of 0.5–1.5 R_0 is shaded.

bleed-through signals that hinder the accurate measurement of FRET using filter-based methods.

For intensity-based measurements, the spectral bleed-through signals result from the direct excitation of the acceptor (arrow, Figure 3.2) and the donor emission that bleeds into the FRET channel (hatching, Figure 3.2). Therefore, the accurate measurement of FRET signals requires correction methods that detect and remove the different spectral bleed-through components (see Periasamy and Day 2005). Significantly, the accuracy of these spectral bleed-through correction methods is degraded as the spectral overlap between the FPs is increased to the point where spectral bleed-through components overwhelm the FRET signal (discussed later; see Berney and Danuser 2003).

The alternative is to measure the effect of FRET on the donor fluorophore, which requires only measurements in donor channel and therefore is less prone to bleed-through artifacts (see Figure 3.2). The FRET-FLIM approach uses optical filtering to isolate the donor fluorescence emission signal and then measure the fluorescence lifetime (for example, see Figure 3.2; this is discussed further later in the chapter). When FRET occurs, FLIM will detect at least two donor populations: the unquenched donors (free donor) and the donors that are quenched by the acceptors (bound donor). These different populations will be reflected in the donor fluorescence decay kinetics, which will be described by at least two exponential components.

Here, the accurate assignment of the donor populations will be improved if the donor fluorophore has simple decay kinetics. In addition, it is important to choose an optical filter that efficiently collects the donor emission signal while eliminating the acceptor emission bleed-through (discussed in Section 3.3.5); however, this may be at the expense of photon counts (Bastiaens and Squire 1999; Peter et al. 2005). If these requirements are met, the measurements of the donor lifetime provide a robust method to quantify FRET (Yasuda 2006; Piston and Kremers 2007).

3.3.2 Standards for Live-Cell FRET Imaging

As mentioned in Chapter 1 and other chapters in this book, numerous methods can be used to measure FRET (for examples, see Jares-Erijman and Jovin 2003 and Periasamy and Day 2005). However, it has been difficult to compare the accuracies of the different methods. An elegant solution to this problem is the development of genetically encoded FRET "standard" proteins. The direct coupling of donor and acceptor FPs to one another by a protein linker has been used to develop a variety of different FRET-based biosensor probes (reviewed by Zhang et al. 2002; Giepmans et al. 2006).

The Vogel laboratory adopted this strategy and developed a set of genetic constructs encoding fusion proteins containing donor and acceptor FPs separated by protein linkers of defined length (Thaler et al. 2005; Koushik et al. 2006). These fusion proteins were then produced in living cells, and two different intensity-based methods (sensitized acceptor emission and emission spectra measurements) and FRET-FLIM were used to measure the FRET efficiency. For each of the fusion proteins tested there was consensus in the results obtained by the different FRET methods, demonstrating that the genetic constructs could serve as FRET standards.

More importantly, other laboratories can verify and calibrate their FRET measurements using these same genetic constructs. For example, Thaler et al. (2005) showed that the Cerulean-5aa-Venus fusion protein yielded an average FRET efficiency of approximately 45%, and we confirmed these results in our laboratory using spectral FRET measurements (Chen et al. 2007). Furthermore, we used the FRET standard approach to compare another FP—the new Teal colored variant mTFP (see Section 3.2.4)—directly to Cerulean to determine its utility as a FRET donor for Venus (Day et al. 2008).

A genetic construct substituting the sequence coding for Cerulean with the cDNA for mTFP was generated to produce the mTFP-5aa-Venus fusion protein. The FRET-FLIM method was then used to acquire fluorescence lifetime measurements from cells that expressed Cerulean, mTFP, or the two different FRET standard constructs. The fluorescence lifetime measurements for Cerulean and mTFP expressed living cells shows that they have similar mean lifetimes (Figure 3.3 and Table 3.2). The FRET-FLIM measurements

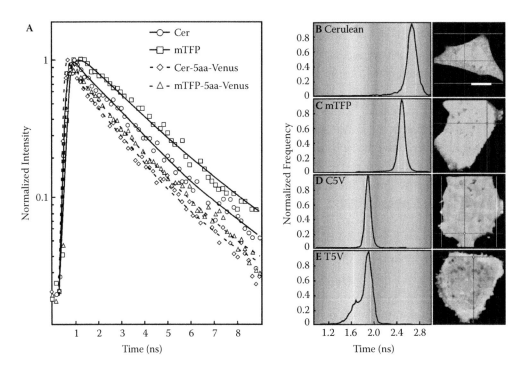

FIGURE 3.3 **(See color insert following page 288.)** Donor lifetime measurements for the fusion proteins consisting of Cerulean or mTFP linked to Venus. (A) Cells expressing Cerulean or mTFP alone, or the FRET standards Cer-5aa-Venus or mTFP-5aa-Venus were used to acquire fluorescence lifetime measurements as described in the text. The fluorescence lifetime decay kinetics for donor fluorophores alone or in the presence of Venus were determined by fitting the data to a double exponential decay. The lifetime distributions are shown for representative cells expressing (B) Cerulean (the calibration bar indicates 10 μm), (C) mTFP, (D) Cer-5aa-Venus, or (E) mTFP-5aa-Venus. The results of the fluorescence lifetime analysis are summarized in Table 3.2. (Adapted from Day, R. N. et al. 2008. *Journal of Biomedical Optics* 13:031203.)

TABLE 3.2   FRET Measurements of Cerulean
or mTFP Fusions to Venus

	Fluorescence lifetime	
Fusion protein	τ_{DA} (ns)	E_{FRET}[b]
Cerulean	2.7 ± 0.08[a]	
Cerulean-5aa-Venus	1.21 ± 0.12	51
mTFP	2.65 ± 0.12	
mTFP-5aa-Venus	1.11 ± 0.06	55

[a] ±SD; n = five to six cells.
[b] Determined by $E_{FRET} = 1 - \tau_{DA}/\tau_D$.

from cells expressing the FRET standard proteins consisting of either Cerulean or mTFP tethered to Venus are shown in Figure 3.3(B–E).

For both Cer-5aa-Venus and mTFP-5aa-Venus, strong quenching of the donor results in a shortening of the mean donor lifetime (Table 3.2 and Figure 3.3A). These measurements can be used to determine the FRET efficiency (see Table 3.2), and the results are in good agreement with measurements obtained using other methods (Day et al. 2008). Thus, mTFP, with its increased brightness and photostability and optimal excitation using the standard laser line, is an excellent donor fluorophore for FRET studies.

3.3.3 Using FRET-FLIM to Detect Protein Interactions in Living Cells

Once the instrumentation and FRET measurement methods have been verified using the standard protein, the goal is then to use the system to address biological questions. For example, we have used FRET-FLIM to characterize the protein interactions involving the transcription factor CCAAT/enhancer binding protein alpha (C/EBPα). The C/EBP family proteins bind to specific DNA elements as obligate dimers and function to regulate genes involved in energy metabolism and programs of cell differentiation (Wedel and Ziegler-Heitbrock 1995). Immunocytochemical staining in differentiated mouse adipocyte cells shows that the endogenous C/EBPα protein preferentially bound to repeated DNA sequences located in regions of centromeric heterochromatin (Tang and Lane 1999).

Similarly, when the FP-labeled C/EBPα is produced in cells originating from the mouse, it is also preferentially localized to the regions of centromeric heterochromatin (Schaufele et al. 2001; Enwright et al. 2003). In earlier studies, we showed that only the basic-region leucine zipper (BZIP) domain, which is sufficient for dimerization and specific DNA-binding, is necessary for this subnuclear positioning (Day et al. 2003). This positioning of C/EBPα BZIP domain dimers in discrete heterochromatin islands within the cell nucleus is clearly seen in the images shown in Figure 3.4.

The fluorescence lifetime distribution for the CFP-labeled BZIP proteins has been measured by the time-correlated single-photon counting (TCSPC) method (see Chapter 7 for method details), and the fluorescence decay histograms of photon emission times relative to the laser excitation pulse were generated from the distribution of interpulse intervals at each pixel in the image. This method was used to determine a mean lifetime (τ_m) for CFP-BZIP of 2.57 ns (Figure 3.4A). Time-resolved images were then acquired from cells that

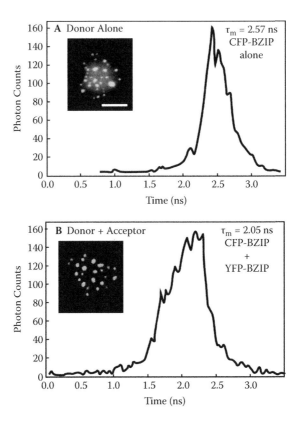

FIGURE 3.4 **(See color insert following page 288.)** FRET-FLIM of cells coexpressing donor and acceptor. (A) FLIM measurements from a mouse pituitary GHFT1-5 cell that expressed the CFP-BZIP protein. The images in (A) and (B) show the cell nuclei, where the BZIP proteins accumulate in regions of heterochromatin; the calibration bar in (A) is 10 μm. The lifetime distribution was determined for the entire nucleus, and the mean lifetime (τ_m) was 2.57 ns. (B) The FLIM measurements were repeated on a cell that coexpressed CFP-BZIP and YFP-BZIP proteins, which form dimers when bound to DNA. The donor (CFP) lifetime distribution was determined for the entire nucleus, and the mean donor lifetime (τ_m) was 2.05 ns, with slow (τ_{DA2}) and fast (τ_{DA1}) lifetime components of 2.65 and 1.05 ns, respectively. (Adapted from Day, R. N. et al. 2004. *Proceedings of SPIE* 5323:36–43.)

coexpressed the CFP- and YFP-labeled BZIP proteins. The BZIP proteins must dimerize to bind to the DNA elements in regions of centromeric heterochromatin. Therefore, if the donor- and acceptor-labeled proteins are produced at similar levels (discussed later) and the fluorophores are favorably positioned, it should be possible to detect dimerization events as a decrease in the mean donor lifetime (τ_m).

For the cells that coexpress the CFP- and YFP-labeled BZIP, the FLIM results show that lifetime decay kinetics for the CFP-labeled proteins are best fitted by a double-exponential decay, and the τ_m is 2.05 ns in the presence of the acceptor (Figure 3.4B). A slow lifetime component (τ_{DA2}) was measured at 2.65 ns, which is similar to that measured for the donor alone (Figure 3.4A) and represents the population of donor labeled proteins not quenched by the acceptor. A fast lifetime component (τ_{DA1}), representing the donor population quenched by FRET, is 1.05 ns (Figure 3.4B). The relative amplitudes for the τ_{DA2} and τ_{DA1} decay components are 71 and 29%, respectively, yielding a distribution ratio of quenched to unquenched donor molecules of 0.4. These results illustrate how FRET-FLIM can be used to detect the association of donor- and acceptor-labeled proteins and importantly, provide an estimate of the molar ratio of quenched donor to donor not associated with the acceptor.

3.3.4 Verifying Protein Interactions Using Acceptor Photobleaching FRET

If a shortening of the donor lifetime results from energy transfer, then destroying the acceptor fluorophore will eliminate FRET, and the donor signal should increase (dequench). The technique of acceptor photobleaching FRET exploits this effect and measures the dequenching of the donor signal in the regions of the cell where FRET has occurred (Bastiaens and Jovin 1996; Kenworthy and Edidin et al. 1998). The photobleaching approach requires that bleaching of the acceptor is selective because any bleaching of the donor fluorophore will lead to an underestimation of the dequenching. Further, the bleaching of the acceptor must be nearly complete because any remaining acceptor will still be available for FRET—again resulting in an underestimation of the donor dequenching.

The acceptor photobleaching method is used here to demonstrate that the fast donor lifetime component observed by FLIM was the result of FRET. The 514 nm laser line was used selectively to photobleach the YFP labeling the BZIP proteins. The donor lifetime was then reacquired, and the measurements revealed that the fast decay component disappeared after acceptor photobleaching, leaving only the unquenched donor population (Figure 3.5). The donor τ_m was found to be 2.49 ns, with over 90% of the population falling into the slow lifetime distribution.

These results provide independent verification of the FLIM measurements and clearly demonstrate that the average distance separating the fluorophores attached to the BZIP proteins bound to the heterochromatin is less than 80 Å. Although acceptor photobleaching is an end-point assay that cannot be repeated on the same cells, it does provide a straightforward method to verify FRET measurements made using other techniques, including FLIM.

3.3.5 Alternative Fluorophore Pairs for FRET-FLIM

Earlier chapters in this book describe how the spectral overlap, Jλ, between the donor and acceptor fluorophores is a critical determinant of FRET efficiency. By using fluorophores that share more spectral overlap, it is possible to increase the distance over which FRET can be detected (Patterson, Piston, and Barisas 2000). For example, Emerald is a bright, stable GFP that shares substantial spectral overlap with Venus (Table 3.1), making it an effective donor (see Figure 3.6). However, the strong spectral overlap that improves the FRET efficiency will also cause a profound increase in the background spectral bleed-through signals (Section 3.3.1). As the spectral bleed-through components become larger, they overwhelm the FRET signal, severely limiting the accuracy of sensitized acceptor emission measurements (compare Figures 3.2 and 3.6). This problem is not limited to

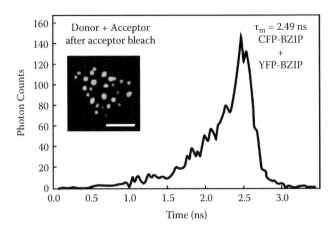

FIGURE 3.5 **(See color insert following page 288.)** FRET-FLIM of cells coexpressing donor and acceptor after acceptor photobleaching. The same cell shown in Figure 3.4B was used for acceptor photobleaching; the calibration bar is 10 μm. The YFP labeling the BZIP protein was bleached using the 514 nm laser line. The donor (CFP) lifetime distribution for the entire nucleus was reacquired after the acceptor photobleaching. The donor τ_m was shifted to 2.49 ns, which is similar to the lifetime of the donor alone (Table 3.2). (Adapted from Day, R. N. et al. 2004. *Proceedings of SPIE* 5323:36–43.)

intensity-based bleed-through correction methods, however, because the increased overlap also leads to acceptor signal bleed-through into the donor channel (back-bleed-through; see Figure 3.6).

This would appear to eliminate the GFP/YFP pair for FRET-FLIM measurements as well, but a strategy was developed that exploits the overlapping signals from the donor and acceptor fluorophores. Because energy transfer causes an increase in the acceptor lifetime (see Chapters 1 and 9), the lifetime of the combined donor and acceptor will also be increased, and this can be measured using FLIM (Harpur, Wouters, and Bastiaens 2001; Calleja et al. 2003). A broad band-pass filter is used to collect the emissions from both GFP and YFP simultaneously, allowing the combined lifetimes of the donor and acceptor to be determined. This method obviates the need for exclusive filtering to isolate the donor signal. However, the measurements of FRET using the combined spectral regions results in an increased number of lifetimes and will be less accurate than methods that are able to isolate the donor lifetime.

An alternative approach is to use acceptor probes with improved photophysical characteristics that do not fluoresce at donor emission wavelengths. For example, the red fluorescent cyanine dye Cy3 and the similar Alexa Fluor 555 are excellent acceptors for GFP (Bastiaens and Squire 1999; Ng et al. 1999). The Cy3 probe attached to antibodies or Fab fragments can be used in combination with expressed GFP-labeled proteins for FRET-FLIM experiments. Fixed cells expressing the GFP-labeled donor protein can be labeled with antibodies raised against a specific epitope or the entire interacting partner protein, thus increasing the chance of a favorable orientation of the tagged antibody to the donor fluorophore.

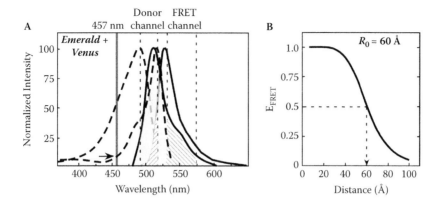

FIGURE 3.6 Spectral overlap and FRET efficiency for Emerald and Venus. (A) Emerald and Venus FPs share more spectral overlap (shaded area) than the commonly used CFP and YFP variants (see Figure 3.2 and legend for details). This, however, dramatically increases direct acceptor excitation (arrow) and donor bleed-through (right hatching) and leads to acceptor back-bleed-through (left hatching) detected in the donor channel. (B) The Förster distance, R_0, for this fluorophore pair is 60 Å, and the efficiency of energy transfer, E_{FRET}, is plotted as described for Figure 3.2.

This approach was used to study the epidermal growth factor receptor (EGFR), which becomes autophosphorylated when the ligand binds. Wouters and Bastiaens (1999) used FRET-FLIM to map the fluorescent lifetimes of GFP-tagged EGFR that was bound by Cy3-labeled antiphosphotyrosine antibodies in cells fixed after EGF stimulation. Their FRET results, acquired at different time points after EGF addition, provided evidence for the recruitment of signaling proteins to phosphorylated EGFRs that were internalized in endosomes. Their lifetime maps also showed differences between the antibody-labeled phosphotyrosine and the FRET distributions. It is important to point out that antibody staining used by these investigators does not provide a quantitative measure of the extent of phosphorylation. However, because FRET-FLIM only detected the donor fluorophore (here, GFP), the antibody labeled with acceptor could be used in excess because the non-specific interactions of the antibody would not be detected.

3.3.6 Fluorescent Proteins Designed Specifically for FLIM Applications

For live-cell imaging, using probes that are excited at longer wavelengths offers important advantages. The longer wavelength spectral windows decrease phototoxicity in the living specimens and also reduce the sample autofluorescence background. Here, we would expect the new orange and red FPs (Table 3.1) to have advantages for FRET studies; however, few published studies have used these probes for intensity-based FRET measurements. The reason for this is likely that sensitized acceptor emission measurements favor fluorophores with a high quantum yield, and most of the red FPs have low intrinsic brightness (Table 3.1). In stark contrast, the acceptor quantum yield is irrelevant if FLIM is used to detect the lifetime of the donor. Therefore, probes that are optimal for FLIM measurements have requirements different from those of probes that are best for intensity-based methods.

Ideally, the donor fluorophore used for FRET-FLIM should have simple lifetime decay kinetics to allow unambiguous assignment of quenched and unquenched fractions. In this regard, the complex decay kinetics of ECFP and Cerulean can be problematic for lifetime analysis (Traimer et al. 2004; Yasuda et al. 2006; Millington et al. 2007). Here, because of variations in the protonation of the chromophore (see Section 3.2.1, for example), multiple decay pathways are available to the donor fluorophore in the excited state. Because the fluorescence decay of the donor alone is already complex, this must be accounted for when extracting quantitative information from FRET-FLIM measurements.

Further, acceptor probes with high absorption coefficients are efficient in quenching the donor fluorophore, but the acceptor does not have to be fluorescent. An acceptor probe with optimal spectral overlap, but low quantum yield, will have decreased acceptor back-bleed-through detected in the donor channel (see Figure 3.6). Here, some FPs that have not been particularly useful for intensity-based FRET measurements have turned out to be most useful for FRET-FLIM studies.

For FRET-FLIM, there are several advantages to using GFP as a donor fluorophore. First, the donor quantum yield determines the R_0 for the FRET pair, and GFPs have a higher intrinsic brightness than the Cerulean CFP (Table 3.1). Second, GFP is excited in a spectral window that generates less autofluorescence than that for CFP. Third, as mentioned in Section 3.3.1, simple decay kinetics are an important characteristic of donor fluorophores for FLIM studies. The emission decay of GFP is monoexponential (Traimer et al. 2006; Yasuda et al. 2006), which allows the unambiguous assignment of quenched and unquenched fractions in FRET studies. These characteristics have prompted the search for optimal FRET acceptors for GFP.

Because of their spectral overlap with GFP, both mRFP and its variant, mCherry (Table 3.1), have been used as acceptors for EGFP in FLIM studies (Peter et al. 2005; Tramier et al. 2006; Yasuda et al. 2006). The low quantum yield of mRFP improves the signal-to-noise ratio because there is less bleed-through from the acceptor detected in the donor fluorescence channel (Yasuda et al. 2006). Still, acceptors that have increased spectral overlap with EGFP emission would improve the FRET efficiency (see Section 3.3.4). As we have seen, a substantial spectral overlap exists between GFP and YFP (Figure 3.6), but the strong back-bleed-through from the bright YFP variants prevented the use of this pair for donor-based measurements.

Recently, novel YFPs have been developed that have a high absorbance coefficient, but extremely low quantum efficiency. This class of chromophore, called resonance energy-accepting chromoproteins (REACh; see Table 3.1), permits the optimal use of GFP as a donor for FRET-FLIM (Ganesan et al. 2006; Murakoshi, Lee, and Yasuda 2008). Their very low quantum yield overcomes the problem of acceptor back-bleed-through emission into the donor channel. This allows the use of filters with a wider donor spectral window to collect optimally the donor signal. The measurement of a double-exponential fluorescence lifetime decay curve for EGFP in the presence of the dark chromoproteins will now accurately reflect the populations of free donor and donor quenched by the REACh probe (Ganesan et al. 2006).

What is more, the absence of fluorescence from REACh probes means that the spectral window normally occupied by the acceptor is now available for the detection of another probe. This opens the possibility of correlating the protein–protein interactions detected by FRET with the behavior of another labeled protein expressed inside the same living cells—the cellular biochemical network (Ganesan et al. 2006; Murakoshi et al. 2008).

3.4 CRITICAL DISCUSSION

3.4.1 General Considerations and Limitations

As we have seen in Section 3.3.2, the genetically encoded FRET standard proteins are useful tools for checking cell culture conditions for protein expression and for calibrating imaging systems for the detection of FRET. The standards developed by Thaler et al. (2005) also included FRET standards with low FRET efficiency. These low FRET efficiency standards are especially useful for assessing the background noise in the system. In addition, FRET standards were developed that included a mutant form of Venus, called Amber, in which the chromophore tyrosine[67] was changed to cysteine. The Amber protein folds correctly, but does not act as a FRET acceptor (Koushik et al. 2006). For some methods, such as FLIM and anisotropy approaches (see Chapter 10), the Amber-mutated FRET standard constructs may be more suitable for obtaining the donor-alone measurements because the donor-Amber fusion protein will have the same size, geometry, and mobility as the intact donor–acceptor fusion protein.

The linked Cerulean–Venus or mTFP-Venus constructs described here (see Section 3.3.2) also serve as a starting point for the design of biosensor probes. The biosensor proteins use a bioactive linker peptide to separate the donor and acceptor fluorophores. They use FRET to report changes in the conformation of the linker resulting from its modification or the binding of a substrate (reviewed by Zhang et al. 2002). These intramolecular FRET-based indicator proteins have been used to measure diverse intracellular events, including changes in intracellular calcium or protein kinase activity (Nagai et al. 2000; Miyawaki and Tsien 2000; Ting et al. 2001; Zhang et al. 2001). The direct tethering of the donor to the acceptor fixes the ratio of the expressed fluorophore pair at 1:1 and allows simple ratio imaging of acceptor-to-donor fluorescence.

The major limitation to the intramolecular FRET approach has been the poor dynamic range of the FRET sensor probes. To improve the range of response, a directed mutagenesis strategy was applied to a linked CFP-YFP construct, and mutant sensor proteins with increased FRET efficiency were selected (Nguyen and Daugherty 2002). This approach yielded sensor probes with substantially enhanced FRET signals, even though the coevolved FPs had similar spectral characteristics and slightly decreased quantum yields when compared to the original FPs. This paradox was recently resolved when it was shown that the enhanced FRET signals resulted from mutations that promoted the stabilization of an intramolecular complex formed between the linked FPs (Ohashi et al. 2007; Vinkenborg et al. 2007).

Although the self-association of the FPs is typically an unwanted characteristic for most experiments, the increased dynamic range of these probes can be an advantage for high-throughput screening methods. For example, the FRET sensor probes with markedly improved dynamic range were recently used in large-scale screening approaches (You et al. 2006). There are concerns, however, that the stabilized intramolecular association of the FPs will also contribute to increased false positives (Section 3.4.4). Therefore, the biosensor probes based on Cerulean or Teal coupled to Venus might still be preferable, and FRET-FLIM can be particularly useful for the biosensor measurements. Here, variations in the biosensor probe lifetimes can be accurately mapped throughout the cell to reveal localized changes in probe activity. Further, the dark chromoproteins such as REACh have the added advantage of allowing the biosensor probes to be used in combination with a separate probe for protein activity.

In contrast to the intramolecular FRET measurements, intermolecular FRET experiments are designed to detect the association of independently produced proteins that are labeled with either the donor or acceptor FPs. In this case, the ratio of donor to acceptor is not fixed, and the ratio can be highly variable between individual cells within the transfected population. Adjustments in the amount of the input plasmid DNAs in the cell transfections will only influence the average relative expression levels of donor- and acceptor-labeled proteins within the population. Furthermore, because the donor–acceptor ratio varies from cell to cell, the spectral bleed-through background signal will also be different for each transfected cell. Therefore, intensity-based measurements of FRET are highly dependent upon the donor-to-acceptor ratio and will work best over a limited range of ratios (Berney and Danuser 2003). FRET-FLIM is less restricted in this respect because the determination of donor lifetimes will also provide an estimate of the molar ratio of quenched donor to donor not involved in FRET (Section 3.3.3).

3.4.2 Overexpression Artifacts

The intermolecular FRET experiments in living cells typically involve transfection methods to introduce DNA constructs encoding the proteins labeled with the donor or acceptor FPs. This type of approach offers great flexibility for the analysis of protein interactions using FRET. However, it is important to recognize that any amount of exogenous protein that is produced in a cell is, by definition, overexpressed relative to its endogenous counterpart. The transfection approach can yield very high levels of the fusion proteins in the target cells, especially when strong promoters are used. This can result in improper protein distribution and protein dysfunction that could lead to erroneous interpretations of protein activities.

It is critical that immunostaining approaches be used to demonstrate that the subcellular distribution of the expressed proteins is the same as that for the endogenous protein. Moreover, several different approaches should be used to verify that FP-fusion proteins produced in living cells retain proper functions. Only a small percentage of the expressed protein detected by fluorescence microscopy might actually be involved in the cellular function that is being monitored. Therefore, it is important to work with cells expressing near-physiological ranges of the fusion proteins and to quantify the effect of protein concentration on the FRET results carefully.

3.4.3 Factors Limiting FRET-FLIM

The major limitation to FRET-FLIM studies has been the complexity of the imaging systems. Recently, however, commercial user-friendly systems have become available, making the FLIM approach much more accessible. Another limitation of FLIM is that the acquisition of the data is typically slow. For example, acquiring sufficient photon counts to assign lifetimes using the TCSPC method described earlier in Section 3.3.3 required 2 minutes for each image, which limits its utility for monitoring dynamic events. Some commercial systems use the frequency domain method, which can be faster and more photon efficient (see Chapter 5). Finally, fluorescence lifetime provides detailed information about any local environmental event that influences the excited state. Environmental factors, such as changing pH or collisional quenching, will also shorten the measured fluorescence lifetime. Therefore, care must be taken in interpreting FRET-FLIM data from living cells.

3.4.4 False Positives and False Negatives

Just as with any other imaging method, it is critical to identify the sources of noise in FRET-FLIM measurements in order to determine the reliability of the data analysis and to avoid overinterpretation of data. Instrumental errors are always a possible source of erroneous results. Here, the FRET standard proteins (Section 3.3.2) are valuable tools because they should report the same range of FRET signals each time they are used, and they will effectively reveal problems in the imaging system. A potential source of false-positive results, mentioned in Sections 3.2.3 and 3.4.1, is interactions driven by the association of the FPs themselves. This can be avoided by using the monomeric versions of the FPs (see Section 3.2.3). Further, false-positive signals can arise because of the overexpression artifacts mentioned previously. This can be avoided by choosing cells during imaging that express the labeled proteins at low levels.

In contrast, false-negative results are common because it is often difficult to achieve the required spatial relationships for FRET between the FPs that tag proteins—even when the proteins are interacting. For this reason, a negative FRET result provides information only in cases where the donor and acceptor are physically linked, such as the biosensor probes. Here, it is also important to recognize that the detection of FRET provides information about the spatial relationship of the fluorophores themselves and does not necessarily indicate the direct interactions between the proteins that they label. Rather, the FPs serve as surrogates for the relative spatial relationships between the specific protein domains to which they are attached.

Thus, FRET measurements alone are not sufficient to prove direct protein–protein interactions. Additional biochemical approaches are required to demonstrate that the protein partners that are being studied are actually in physical contact with each other. However, intermolecular FRET measurements with the FPs do provide direct evidence of protein associations within the natural environment of the living cell at less than 100 Å, revealing information about the structure of protein complexes that form at specific subcellular sites.

3.4.5 Analysis in the Cell Population

FRET results from single cells are, by themselves, not sufficient to characterize the associations between proteins in living cells. Although the FRET measurements, when collected and quantified properly, are remarkably robust, there is still heterogeneity in the measurements. Furthermore, substantial cell-to-cell heterogeneity may also be present for some types of interactions, and it is possible that only a subpopulation of cells responds to a particular stimulus. Therefore, data must be collected and statistically analyzed from multiple cells to prevent the user from reaching false conclusions from a nonrepresentative measurement.

3.5 SUMMARY

This chapter has presented some of the characteristics of the FPs that make them useful labels for FRET-FLIM studies in biological systems. The results presented here illustrated how the FRET standard proteins can be used to calibrate measurements and characterize new fluorophore pairs. The results also showed how FRET-FLIM can be used to detect the intermolecular interactions between proteins produced in living cells and how these measurements can be verified by acceptor photobleaching. In addition, the chapter outlined how some of the FPs that have not been useful for intensity-based FRET measurements are finding great utility in FRET-FLIM studies. Furthermore, FPs designed specifically for FLIM applications, such as the REACh probes, are overcoming some of the limitations to the FRET-FLIM approach in living cells. The reader is encouraged to explore these topics in greater depth using the references provided.

3.6 FUTURE PERSPECTIVE

The combination of the FPs and lifetime imaging provides a powerful tool for studies in cell biology. This approach will certainly come into the mainstream as user-friendly and less expensive commercial systems become available. The measurements of probe lifetimes can enhance the contrast in biological and medical imaging and have the potential to be used for large-scale screening applications in living cells. In addition, as more FPs with unique properties are discovered, the palette of useful probes for FLIM will expand. In this regard, new FPs have been identified that can be switched on by light or that change their emission characteristics when illuminated at specific wavelengths (reviewed by Day and Schaufele 2008). The photoswitching properties of these novel FPs are being used to measure the dynamic behaviors of proteins inside living cells (for example, see Demarco et al. 2006), and it may be possible to exploit these characteristics for FLIM studies as well.

REFERENCES

Ai, H. W., Hazelwood, K. L., Davidson, M. W., and Campbell, R. E. 2008. Fluorescent protein FRET pairs for ratiometric imaging of dual biosensors. *Nature Methods* 5:401–403.

Ai, H. W., Henderson, N. J., Remington S. J., and Campbell R. E. 2006. Directed evolution of a monomeric, bright and photostable version of Clavularia cyan fluorescent protein: Structural characterization and applications in fluorescence imaging. *Biochemical Journal* 400:531–540.

Ai, H. W., Shaner, N. C., Cheng, Z., Tsien, R. Y., and Campbell, R. E. 2007. Exploration of new chromophore structures leads to the identification of improved blue fluorescent proteins. *Biochemistry* 46:5904–5910.

Baird, G. S., Zacharias, D. A., and Tsien, R. Y. 2000. Biochemistry, mutagenesis, and oligomerization of DsRed, a red fluorescent protein from coral. *Proceedings of the National Academy of Sciences USA* 97:11984–11989.

Bastiaens, P. I., and Jovin, T. M. 1996. Microspectroscopic imaging tracks the intracellular processing of a signal transduction protein: Fluorescent-labeled protein kinase C beta I. *Proceedings of the National Academy of Sciences USA* 93:8407–8412.

Bastiaens, P. I., and Squire, A. 1999. Fluorescence lifetime imaging microscopy: Spatial resolution of biochemical processes in the cell. *Trends in Cell Biology* 9:48–52.

Berney, C., and Danuser, G. 2003. FRET or no FRET: A quantitative comparison. *Biophysical Journal* 84:3992–4010.

Bevis, B. J., and Glick, B. S. 2002. Rapidly maturing variants of the *Discosoma* red fluorescent protein (DsRed). *Nature Biotechnology* 20:83–87.

Brejc, K., Sixma, T. K., Kitts, P. A., et al. 1997. Structural basis for dual excitation and photoisomerization of the *Aequorea victoria* green fluorescent protein. *Proceedings of the National Academy of Sciences USA* 94:2306–2311.

Calleja, V., Ameer-Beg, S. M., Vojnovic, B., Woscholski, R., Downward, J., and Larijani, B. 2003. Monitoring conformational changes of proteins in cells by fluorescence lifetime imaging microscopy. *Biochemical Journal* 372:33–40.

Campbell, R. E., Tour, O., Palmer, A. E., et al. 2002. A monomeric red fluorescent protein. *Proceedings of the National Academy of Sciences USA* 99:7877–7882.

Chalfie, M., Tu, Y., Euskirchen, G., Ward, W. W., and Prasher, D. C. 1994. Green fluorescent protein as a marker for gene expression. *Science* 263:802–805.

Chattoraj, M., King, B. A., Bublitz, G. U., and Boxer, S. G. 1996. Ultra-fast excited state dynamics in green fluorescent protein: Multiple states and proton transfer. *Proceedings of the National Academy of Sciences USA* 93:8362–8367.

Chen, Y., Mauldin, J. P., Day, R. N., and Periasamy, A. 2007. Characterization of spectral FRET imaging microscopy for monitoring nuclear protein interactions. *Journal of Microscopy* 228:139–152.

Cotlet, M., Hofkens, J., Habuchi, S., et al. 2001. Identification of different emitting species in the red fluorescent protein DsRed by means of ensemble and single-molecule spectroscopy. *Proceedings of the National Academy of Sciences USA* 98:14398–14403.

Cubitt, A. B., Heim, R., Adams, S. R., Boyd, A. E., Gross, L. A., and Tsien, R. Y. 1995. Understanding, improving and using green fluorescent proteins. *Trends in Biochemical Science* 20:448–455.

Cubitt, A. B., Woollenweber, L. A., and Heim, R. 1999. Understanding structure–function relationships in the *Aequorea victoria* green fluorescent protein. *Methods in Cell Biology* 58:19–30.

Day, R. N., Booker, C. F., and Periasamy, A. 2008. The characterization of an improved donor fluorescent protein for Förster resonance energy transfer microscopy. *Journal of Biomedical Optics* 13:031203.

Day, R. N., Demarco, I. A., Voss, T. C., Chen Y., and Periasamy, A. 2004. FLIM-FRET microscopy to visualize transcription factor interactions in the nucleus of the living cell. *Proceedings of SPIE* 5323:36–43.

Day, R. N., and Piston, D. W. 1999. Spying on the hidden lives of proteins. *Nature Biotechnology* 17:425–426.

Day, R. N., and Schaufele, F. 2008. Fluorescent protein tools for studying protein dynamics in living cells: A review. *Journal of Biomedical Optics* 13:031202.

Day, R. N., Voss, T. C., Enwright, J. F., III, Booker, C. F., Periasamy, A., and Schaufele F. 2003. Imaging the localized protein interactions between Pit-1 and the CCAAT/enhancer binding protein alpha in the living pituitary cell nucleus. *Molecular Endocrinology* 17:333–345.

Demarco, I. A., Periasamy, A., Booker, C. F., and Day, R. N. 2006. Monitoring dynamic protein interactions with photoquenching FRET. *Nature Methods* 3:519–524.

Enwright, J. F., III, Kawecki-Crook, M. A., Voss, T. C., Schaufele, F., and Day, R. N. 2003. A PIT-1 homeodomain mutant blocks the intranuclear recruitment of the CCAAT/enhancer binding protein alpha required for prolactin gene transcription. *Molecular Endocrinology* 17:209–222.

Field, S. F., Bulina, M. Y., Kelmanson, I. V., Bielawski, J. P., and Matz, M. V. 2006. Adaptive evolution of multicolored fluorescent proteins in reef-building corals. *Journal of Molecular Evolution* 62:332–339.

Ganesan, S., Ameer-Beg, S. M., Ng, T. T., Vojnovic, B., and Wouters, F. S. 2006. A dark yellow fluorescent protein (YFP)-based resonance energy-accepting chromoprotein (REACh) for Förster resonance energy transfer with GFP. *Proceedings of the National Academy of Sciences USA* 103:4089–4094.

Giepmans, B. N., Adams, S. R., Ellisman, M.H., and Tsien, R. Y. 2006. The fluorescent toolbox for assessing protein location and function. *Science* 312:217–224.

Hadjantonakis, A. K., Dickinson, M. E., Fraser, S. E., and Papaioannou, V. E. 2003. Technicolor transgenics: Imaging tools for functional genomics in the mouse. *National Review of Genetics* 4:613–625.

Harpur, A. G., Wouters, F. S., and Bastiaens, P. I. 2001. Imaging FRET between spectrally similar GFP molecules in single cells. *Nature Biotechnology* 19:167–169.

Heikal, A. A., Hess, S. T., Baird, G. S., Tsien, R. Y., and Webb, W. W. 2000. Molecular spectroscopy and dynamics of intrinsically fluorescent proteins: Coral red (dsRed) and yellow (Citrine). *Proceedings of the National Academy of Sciences USA* 97:11996–12001.

Heim, R., Cubitt, A. B., and Tsien, R. Y. 1995. Improved green fluorescence. *Nature* 373:663–664.

Heim, R., Prasher, D. C., and Tsien, R. Y. 1994. Wavelength mutations and posttranslational autooxidation of green fluorescent protein. *Proceedings of the National Academy of Sciences USA* 91:12501–12504.

Heim, R., and Tsien, R. Y. 1996. Engineering green fluorescent protein for improved brightness, longer wavelengths and fluorescence resonance energy transfer. *Current Biology* 6:178–182.

Inouye, S., and Tsuji, F. I. 1994. Aequorea green fluorescent protein. Expression of the gene and fluorescence characteristics of the recombinant protein. *FEBS Letters* 341:277–280.

Jares-Erijman, E. A., and Jovin, T. M. 2003. FRET imaging. *Nature Biotechnology* 21:1387–1395.

Karasawa, S., Araki, T., Nagai, T., Mizuno, H., and Miyawaki, A. 2004. Cyan-emitting and orange-emitting fluorescent proteins as a donor/acceptor pair for fluorescence resonance energy transfer. *Biochemical Journal* 381:307–312.

Kenworthy, A. 2002. Peering inside lipid rafts and caveolae. *Trends in Biochemical Science* 27:435–437.

Kenworthy, A. K., and Edidin, M. 1999. Imaging fluorescence resonance energy transfer as probe of membrane organization and molecular associations of GPI-anchored proteins. *Methods in Molecular Biology* 116:37–49.

Koushik, S. V., Chen, H., Thaler, C., Puhl, H. L., III, and Vogel, S. S. 2006. Cerulean, Venus, and VenusY67C FRET reference standards. *Biophysical Journal* 91:L99–L101.

Kremers, G. J., Goedhart, J., van den Heuvel, D. J., Gerritsen, H. C., and Gadella, T. W., Jr. 2007. Improved green and blue fluorescent proteins for expression in bacteria and mammalian cells. *Biochemistry* 46:3775–3783.

Labas, Y. A., Gurskaya, N. G., Yanushevich, Y. G., et al. 2002. Diversity and evolution of the green fluorescent protein family. *Proceedings of the National Academy of Sciences USA* 99:4256–4261.

Leutenegger, A., D'Angelo, C., Matz, M. V., et al. 2007. It's cheap to be colorful. Anthozoans show a slow turnover of GFP-like proteins. *FEBS Journal* 274:2496–2505.

Lippincott-Schwartz, J. E., Snapp, E., and Kenworthy, A. 2001. Studying protein dynamics in living cells. *Nature Reviews. Molecular Cell Biology* 2:444–456.

Matz, M. V., Fradkov, A. F., Labas, Y. A., et al. 1999. Fluorescent proteins from nonbioluminescent *Anthozoa* species. *Nature Biotechnology* 17:969–973.

Matz, M. V., Lukyanov, K. A., and Lukyanov, S. A. 2002. Family of the green fluorescent protein: journey to the end of the rainbow. *Bioessays* 24:953–959.

Mena, M. A., Treynor, T. P., Mayo, S. L., and Daugherty, P. S. 2006. Blue fluorescent proteins with enhanced brightness and photostability from a structurally targeted library. *Nature Biotechnology* 24:1569–1571.

Merzlyak, E. M., Goedhart, J., Shcherbo, D., Bulina, M. E., Shcheglov, A. S., et al. 2007. Bright monomeric red fluorescent protein with an extended fluorescence lifetime. *Nature Methods* 4:555–557.

Millington, M., Grindlay, G. J., Altenbach, K., et al. 2007. High-precision FLIM-FRET in fixed and living cells reveals heterogeneity in a simple CFP-YFP fusion protein. *Biophysical Chemistry* 127:155–164.

Miyawaki, A., and Tsien, R. Y. 2000. Monitoring protein conformations and interactions by fluorescence resonance energy transfer between mutants of green fluorescent protein. *Methods in Enzymology* 327:472–500.

Morise, H., Shimomura, O., Johnson, F. H., and Winant, J. 1974. Intermolecular energy transfer in the bioluminescent system of Aequorea. *Biochemistry* 13:2656–2662.

Murakoshi, H., Lee, S. J., and Yasuda, R. 2008. Highly sensitive and quantitative FRET-FLIM imaging in single dendritic spines using improved nonradiative YFP. *Brain Cell Biology* May 30 [Epub ahead of print].

Nagai, T., Ibata, K., Park, E. S., Kubota, M., Mikoshiba, K., and Miyawaki, A. 2002. A variant of yellow fluorescent protein with fast and efficient maturation for cell-biological applications. *Nature Biotechnology* 20:87–90.

Nagai, Y., Miyazaki, M., Aoki, R., et al. 2000. A fluorescent indicator for visualizing cAMP-induced phosphorylation in vivo. *Nature Biotechnology* 18:313–316.

Ng, T., Squire, A., Hansra, G., et al. 1999. Imaging protein kinase Calpha activation in cells. *Science* 283: 2085–2089.

Nguyen, A. W., and Daugherty, P. S. 2002. Evolutionary optimization of fluorescent proteins for intracellular FRET. *Nature Biotechnology* 23:355–360.

Ohashi, T., Galiacy, S. D., Briscoe, G., and Erickson, H. P. 2007. An experimental study of GFP-based FRET, with application to intrinsically unstructured proteins. *Protein Science* 16:1429–1438.

Ormö, M., Cubitt, A. B., Kallio, K., Gross, L. A., Tsien, R. Y., and Remington, S. J. 1996. Crystal structure of the *Aequorea victoria* green fluorescent protein. *Science* 273:1392–1395.

Patterson, G., Day, R. N., and Piston, D. 2001. Fluorescent protein spectra. *Journal of Cell Science* 114:837–838.

Patterson, G. H., Piston, D. W., and Barisas, B. G. 2000. Forster distances between green fluorescent protein pairs. *Analytical Biochemistry* 284:438–440.

Periasamy, A., and Day, R. N. 2005. *Molecular imaging: FRET microscopy and spectroscopy.* New York: Oxford University Press.

Peter, M., Ameer-Beg, S. M., Hughes, M. K., et al. 2005. Multiphoton-FLIM quantification of the EGFP-mRFP1 FRET pair for localization of membrane receptor-kinase interactions. *Biophysical Journal* 88:1224–1237.

Piston, D. W., and Kremers, G. J. 2007. Fluorescent protein FRET: The good, the bad and the ugly. *Trends in Biochemical Science* 32:407–414.

Plautz, J. D., Day, R. N., Dailey, G. M., et al. 1996. Green fluorescent protein and its derivatives as versatile markers for gene expression in living *Drosophila melanogaster,* plant and mammalian cells. *Gene* 173:83–87.

Prasher, D. C., Eckenrode, V. K., Ward, W. W., Prendergast, F. G., and Cormier, M. J. 1992. Primary structure of the *Aequorea victoria* green-fluorescent protein. *Gene* 111:229–233.

Rizzo, M. A., Springer, G. H., Granada, B., and Piston, D. W. 2004. An improved cyan fluorescent protein variant useful for FRET. *Nature Biotechnology* 22:445–449.

Schaufele, F., Enwright, J. F., III, Wang, X., et al. 2001. CCAAT/enhancer binding protein alpha assembles essential cooperating factors in common subnuclear domains. *Molecular Endocrinology* 15:1665–1676.

Shagin, D. A., Barsova, E. V., Yanushevich, Y. G., et al. 2004. GFP-like proteins as ubiquitous meta-zoan superfamily: Evolution of functional features and structural complexity. *Molecular Biology and Evolution* 21:841–850.

Shaner, N. C., Campbell, R. E., Steinbach, P. A., Giepmans, B. N., Palmer, A. E., and Tsien, R. Y. 2004. Improved monomeric red, orange and yellow fluorescent proteins derived from *Discosoma* sp. red fluorescent protein. *Nature Biotechnology* 22:1567–1572.

Shaner, N. C., Lin, M. Z., McKeown, M. R., et al. 2008. Improving the photostability of bright mono-meric orange and red fluorescent proteins. *Nature Methods* 5:545–551.

Shaner, N. C., Patterson, G. H., and Davidson, M. W. 2007. Advances in fluorescent protein technol-ogy. *Journal Cell Science* 120:4247–4260.

Shaner, N. C., Steinbach, P. A., and Tsien, R. Y. 2005. A guide to choosing fluorescent proteins. *Nature Methods* 2:905–909.

Shcherbo, D., Merzlyak, E. M., Chepurnykh, T. V., et al. 2007. Bright far-red fluorescent protein for whole-body imaging. *Nature Methods* 4:741–746.

Shimomura, O., Johnson, F. H., and Saiga, Y. 1962. Extraction, purification and properties of aequorin, a bioluminescent protein from the luminous hydromedusan, *Aequorea. Journal of Cellular and Comparative Physiology* 59:223–239.

Stewart, C. N., Jr. 2006. Go with the glow: Fluorescent proteins to light transgenic organisms. *Trends in Biotechnology* 24:155–162.

Striker, G., Subramaniam, V., Seidel, C. A., and Volkmer, A. 1999. Photochromicity and fluorescence lifetimes of green fluorescent protein. *Journal of Physical Chemistry B* 103:8612–8617.

Tang, Q. Q., and Lane, M. D. 1999. Activation and centromeric localization of CCAAT/enhancer-binding proteins during the mitotic clonal expansion of adipocyte differentiation. *Genes and Development* 13:2231–2241.

Thaler, C., Koushik, S. V., Blank, P. S., and Vogel, S. S. 2005. Quantitative multiphoton spectral imag-ing and its use for measuring resonance energy transfer. *Biophysical Journal* 89:2736–2749.

Ting, A. Y., Kain, K. H., Klemke, R. L., and Tsien, R. Y. 2001. Genetically encoded fluorescent report-ers of protein tyrosine kinase activities in living cells. *Proceedings of the National Academy of Sciences USA* 98:15003–15008.

Tramier, M., Kemnitz, K., Durieux, C., and Coppey-Moisan, M. 2004. Picosecond time-resolved microspectrofluorometry in live cells exemplified by complex fluorescence dynamics of popu-lar probes ethidium and cyan fluorescent protein. *Journal of Microscopy* 213:110–118.

Tramier, M., Zahid, M., Mevel, J. C., Masse, M. J., and Coppey-Moisan, M. 2006. Sensitivity of CFP/YFP and GFP/mCherry pairs to donor photobleaching on FRET determination by fluorescence lifetime imaging microscopy in living cells. *Microscopy Research and Technique* 69:933–939.

Tsien, R. Y. 1998. The green fluorescent protein. *Annual Review of Biochemistry* 67:509–544.

van Roessel, P., and Brand, A. H. 2002. Imaging into the future: Visualizing gene expression and protein interactions with fluorescent proteins. *Nature Cell Biology* 4:E15–20.

Vinkenborg, J. L., Evers, T. H., Reulen, S. W., Meijer, E. W., and Merkx, M. 2007. Enhanced sensitivity of FRET-based protease sensors by redesign of the GFP dimerization interface. *ChemBioChem* 8:1119–1121.

Wallrabe, H., Stanley, M., Periasamy, A., and Barroso, M. 2003. One- and two-photon fluorescence resonance energy transfer microscopy to establish a clustered distribution of receptor-ligand complexes in endocytic membranes. *Journal of Biomedical Optics* 8:339–346.

Wang, L., Jackson, W. C., Steinbach, P. A., and Tsien, R. Y. 2004. Evolution of new nonantibody pro-teins via iterative somatic hypermutation. *Proceedings of the National Academy of Sciences USA* 101:16745–16749.

Ward, W. W. 1998. Biochemical and physical properties of green fluorescent protein. In *Green fluo-rescent protein: Properties, applications and protocols*, ed. M. Chalfie and S. Kain, 45–75. New York: Wiley-Liss.

Wedel, A., and Ziegler-Heitbrock, H. W. 1995. The C/EBP family of transcription factors. *Immunobiology* 193:171–185.

Wouters, F. S., and Bastiaens, P. I. 1999. Fluorescence lifetime imaging of receptor tyrosine kinase activity in cells. *Current Biology* 9:1127–1130.

Yasuda, R. 2006. Imaging spatiotemporal dynamics of neuronal signaling using fluorescence resonance energy transfer and fluorescence lifetime imaging microscopy. *Current Opinion in Neurobiology* 16:551–561.

Yasuda, R., Harvey, C. D., Zhong, H., Sobczyk, A., van Aelst, L., and Svoboda, K. 2006. Supersensitive Ras activation in dendrites and spines revealed by two-photon fluorescence lifetime imaging. *Nature Neuroscience* 9:283–291.

You, X., Nguyen, A. W., Jabaiah, A., Sheff, M. A., Thorn, K. S., and Daugherty, P. S. 2006. Intracellular protein interaction mapping with FRET hybrids. *Proceedings of the National Academy of Sciences USA* 103:18458–18463.

Zacharias, D. A., Violin, J. D., Newton, A. C., and Tsien, R. Y. 2002. Partitioning of lipid-modified monomeric GFPs into membrane microdomains of live cells. *Science* 296:913–916.

Zapata-Hommer, O., and Griesbeck, O. 2003. Efficiently folding and circularly permuted variants of the Sapphire mutant of GFP. *BMC Biotechnology* 3:5.

Zhang, J., Campbell, R. E., Ting, A. Y., and Tsien, R. Y. 2002. Creating new fluorescent probes for cell biology. *Nature Reviews. Molecular Cell Biology* 3:906–918.

Zhang, J., Ma, Y., Taylor, S. S., and Tsien, R. Y. 2001. Genetically encoded reporters of protein kinase A activity reveal impact of substrate tethering. *Proceedings of the National Academy of Sciences* 98:14997–15002.

2

Instrumentation

Wide-Field Fluorescence Lifetime Imaging Microscopy Using a Gated Image Intensifier Camera

Yuansheng Sun, James N. Demas, and Ammasi Periasamy

4.1 INTRODUCTION

Fluorescence lifetime imaging microscopy (FLIM) is an important tool in the investigation of biological events. Existing fluorescence microscopy techniques do not allow imaging of dynamic molecular interactions between cellular components on a precise spatial and temporal scale. Lifetime imaging modality visualizes or monitors spatial and temporal information of environmental changes in living specimens. Instrumental methods for measuring fluorescence lifetimes are divided into two major categories: frequency domain (Lakowicz 2007; Chapter 5) and time domain (O'Connor and Phillips 1984; Becker 2005; Lakowicz 2007; Chapter 7).

Frequency-domain fluorometers generate the fluorescence with light, which is sinusoidal and modulated at different frequencies depending on the lifetime values to be measured (for nanosecond decays, megahertz), and then they measure the phase shift and amplitude attenuation of the fluorescence emission relative to the phase and amplitude of the exciting light (see Figure 4.1). Thus, each lifetime value will cause a specific phase shift and attenuation at a given frequency. For single-lifetime samples, the lifetime may be calculated directly from either the phase shift or the magnitude of the attenuation (or both because both are available from a single measurement). For multiple lifetimes, many measurements are required over a range of excitation frequencies.

In time-domain methods, pulsed light is used as the excitation source, and fluorescence lifetimes are measured from the fluorescence signal directly or by using single-photon counting (Periasamy et al. 1996). It is possible to obtain fluorescence lifetime images by a combination of the lifetime determination techniques (time or frequency domains) with high-speed two-dimensional detectors and scanning techniques, such as laser beam

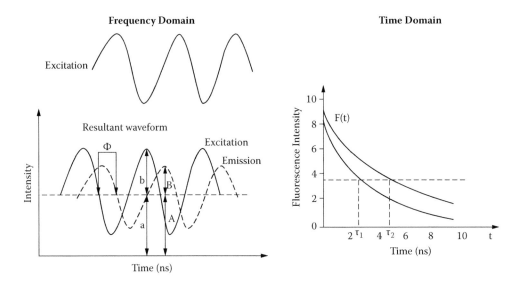

FIGURE 4.1 Illustration of frequency- and time-domain measurement. The dependence of phase angle Φ and modulation m $\{m = (B/A)/(b/a)\}$ on the light modulation frequency is used to recover the intensity decay of the sample.

scanning (point scanning or spinning disk), gated image intensifier, or electronic scanning methods (image dissector or streak camera).

The first paper published on lifetime imaging in 1989 was based on the use of a high-speed, sensitive image dissector tube (IDT) coupled with frequency-domain lifetime detection (Wang, Uchida, and Minami 1989). High-frequency response of up to 1 GHz and high resolution of images were obtained by heterodyning the modulated fluorescence from the specimen and reference signals at the blanking electrode of the IDT. Later, in 1990, a multi-channel photon-counting detector was coupled to a fluorescence microscope to obtain life-time images (Wang et al. 1990). The multichannel photon-detection methodology allowed enhancing the rate of data collection and improved the temporal resolution for two-dimensional lifetime images compared to IDT. The interested reader may review references that reported spatially mapping the excited-state lifetimes of dye molecules as well as various biological applications (Becker et al. 2004; Buurman et al. 1992; Chen and Periasamy 2004; Clegg et al. 1991; Gadella, van Hoek, and Visser 1997; Gerritsen et al. 2006; König et al. 1996; Krishnan et al. 2003; Lakowicz and Brendt 1991; Lakowicz et al. 1994; Marriot et al. 1991; Morgan, Mitchell, and Murray 1990; Oida, Sako, and Kusumi 1993; Periasamy et al. 1995, 1996; Straub and Hell 1998; Wang et al. 1992; Wouters and Bastiaens 1999).

In this chapter, we describe a wide-field lifetime imaging mode using a gated image intensifier camera. We also describe a methodology used for single- and double-exponential decay analysis. The time-domain-based gating camera method allows investigators to collect time-resolved images within a few seconds compared to time-correlated single-photon counting (TCSPC), which can take several minutes. The rapid lifetime determination (RLD) method, based on a wide-field instrument, is the focus of this chapter. Because many life-science departments have access to a wide-field system, retrofitting the system for RLD lifetime may be the least expensive approach to buying an alternative system. Moreover, the gating camera methodology can be used to measure the lifetime in the range of nanosecond to microseconds.

4.2 BACKGROUND

The fluorescence exponential decay $I(t) = k \exp(-t/\tau)$ describes a broad range of physical processes essential for understanding many areas of the physical sciences. In particular, luminescence decays are part of a number of analytical methods that contain molecular and environmental information (Lakowicz 2007). The pre-exponential factor (k) in the preceding equation is directly related to the concentration of the luminescent species and is independent of quenching errors (Demas, Jones, and Keller 1986). The method of evaluating k and τ for exponential decays is to fit ($I(t)$) versus t by linear least-squares or other methods described in other chapters in this book (see Chapter 12). These methods apply to decay with a zero base line or where the measured or known base line is subtracted.

On the other hand, the RLD method described in this chapter can be applied to single- or multiple-exponential decays with or without a known baseline (Woods et al. 1984). RLD can substantially reduce the computational time compared to the curve fitting or global analysis methods. An analog version of the RLD for single-exponential decay without a baseline was implemented in a luminescence decay time thermometer (Sholes and Small 1980).

The RLD method was theoretically evaluated by Ballew and Demas in 1989. In their paper, they compared the RLD method with the weighted linear least-squares (WLLS) method. At optimum conditions of total counts at 10^4, the standard deviations in the RLD method were found to be around 30% compared with WLLS; however, the calculations of lifetime and pre-exponential factors were hundreds of times faster (Ballew and Demas 1989). Later, this method was experimentally demonstrated using a gated image intensifier for biological applications (Wang et al. 1991, 1996; Periasamy et al. 1996). Moreover, the RLD method has been experimentally calibrated using fluorescence lifetime standards and the resultant lifetime values are within acceptable error ranges (<1%) of published values (Periasamy et al. 1996).

Sharman et al. (1999) evaluated theoretically the RLD method for double-exponential decay and applied different gating schemes. It appeared that the overlapped gating scheme provided more photons per pixel and reduced the error considerably in the calculations of pre-exponential factors and the lifetimes. Later, this gating scheme was demonstrated experimentally using a high-speed gating image intensifier (110 MHz repetition rate; 300 ps gating width) coupled to a CCD (charge-coupled device) camera to monitor protein–protein interactions in live specimens (Elangovan, Day, and Periasamy 2002; Periasamy et al. 2001, 2002). Here we provide the RLD theory, instrumentation, and step-by-step instructions for acquiring and processing images from biological samples.

4.3 METHODS

4.3.1 Theory behind the RLD Method

The RLD method allows us to calculate the decay parameters using the areas under different regions of the decay rather than recording a complete multipoint curve and analyzing the decay by the traditional least-squares methods (Demas 1983). For quantitative analysis, it is important to estimate measurement precision in the presence of noise. The RLD method has been evaluated over a wide range of experimental conditions to assess the optimum conditions and the theoretical limitations for contiguous and overlapped gating procedures for single- and double-exponential decays using Monte Carlo simulations (Sharman et al. 1999).

4.3.1.1 Single-Exponential Decay

For single-exponential decay, only two areas are calculated because there are only two unknowns: k and τ. The integrated areas D_0 and D_1 are shown in Box 4.1 (a) for equal gating and in Box 4.1 (b) for the 50% overlapped gating. The total counts would equal $k\tau$. The data analysis for the RLD method consists of substituting the integrals D_n in the following equations. The D_n are obtained by summing the data that have been acquired at equal time intervals δt. For example,

$$D_0 = \sum_{i=1}^{i=n/2} \left(I_i \delta t \right) \tag{4.1}$$

where I_i is the data point acquired at the ith time ($i\,\delta t$) and n is the total number of points acquired. In photon counting, the integrals are simply the total counts recorded in the interval Δt. The equations for τ and k are given in Box 4.1, with Δt the time interval of integration.

BOX 4.1 SINGLE-EXPONENTIAL DECAY.

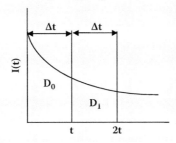

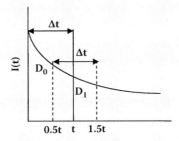

A Contiguous gating

Equations for calculating lifetime (τ) and pre-exponential factor (k) using two contiguous gates D_0 and D_1 with the gate width Δt - the time interval of integration, are given below:

$\tau = \Delta t / \ln(D_0/D_1)$ (4.2)

$k = D_0^2 \ln(D_0/D_1)/[(D_0-D_1)\,\Delta t]$ (4.3)

B Overlapped gating (50% overlap)

Equations for calculating lifetime (τ) and pre-exponential factor (k) using two overlapped gates D_0 and D_1 with the gate width Δt - the time interval of integration, are given below:

$\tau = -\Delta t / \ln(D_1^2/D_0^2)$ (4.4)

$k = 2D_0^3 \ln(D_1/D_0)/[(D_1^2-D_0^2)\,\Delta t]$ (4.5)

Shown are the graphs of integrated areas for the single-exponential decays with contiguous gates (A) and overlapped gates (B) and the corresponding equations to calculate lifetime (τ) and pre-exponential factor (k) (Sharman et al 1999).

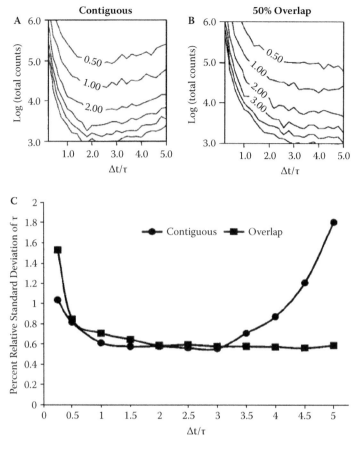

FIGURE 4.2 Percent relative standard deviation of τ for single exponential decays with contiguous gates (A) and overlapping gates (B) using Monte Carlo simulations. (C) $Ru(bpy)_3^{2+}$ decay comparing contiguous gates to overlapped gates using the RLD method. In (C), the standard deviation is almost stable for overlap gating starting at $\Delta t/\tau = 1$. It demonstrates that one can obtain reasonable results at lower lifetime values. (Adapted from Sharman, K. K. et al. 1999. *Analytical Chemistry* 71:947–952.)

As demonstrated in Figure 4.2, for very small values of dt/τ, the contiguous gates give a smaller relative standard deviation, whereas the overlapped gating case provides good precision over a much wider range of dt/τ. These results are consistent with Monte Carlo simulations (see Appendix 4.2) as shown in Figure 4.2. The overlapping case should therefore be used when little is known about the system. The RLD precision for τ (0.6–1.2%) is comparable to, or only slightly less than, that for nonlinear least-squares fitting (0.6%).

4.3.1.2 Double-Exponential Decay
Double-exponential decays were computed from (Sharman et al. 1999):

$$I(t) = k_1 \exp(-1/\tau_1) + k_2 \exp(-1/\tau_2) \tag{4.6}$$

τ_1 is the shorter lifetime.

BOX 4.2 DOUBLE-EXPONENTIAL DECAY.

A Contiguous gating

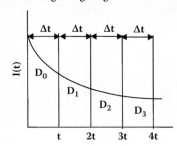

B Overlapped gating (50% overlap)

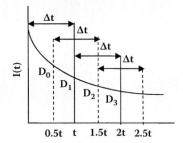

Equations for calculating lifetimes (τ_1, τ_2) and pre-exponential factors (k_1, k_2) using four contiguous gates D_0, D_1, D_2 and D_3 with the gate width Δt - the time interval of integration, are given below:

$$\tau_1 = -\Delta t/\ln(y) \quad (4.7)$$

$$\tau_2 = -\Delta t/\ln(x) \quad (4.8)$$

$$k_1 = (xD_0 - D_1)^3 \ln(y)/[(xD_0 - xD_1 - D_1 + D_2)(x^2D_0 - 2xD_1 + D_2)\Delta t] \quad (4.9)$$

$$k_2 = -R\,[\ln(yD_1 - D_2) - \ln(D_0y - D_1)]/[(x-1)(x^2D_0 - 2xD_1 + D_2)\Delta t] \quad (4.10)$$

Equations for calculating lifetimes (τ_1, τ_2) and pre-exponential factors (k_1, k_2) using four overlapped gates D_0, D_1, D_2 and D_3 with the gate width Δt - the time interval of integration, are given below:

$$\tau_1 = -\Delta t/\ln(y^2) \quad (4.11)$$

$$\tau_2 = -\Delta t/\ln(x^2) \quad (4.12)$$

$$k_1 = -2(xD_0 - D_1)^4 \ln(y)/[(xD_0 + xD_1 - D_1 - D_2)(x^2D_0 - 2xD_1 + D_2)\Delta t] \quad (4.13)$$

$$k_2 = -2R\,\ln(x)/[(x^2-1)(x^2D_0 - 2xD_1 + D_2)\,\Delta t] \quad (4.14)$$

Intermediate calculations for all equations above:

$$R = D_1D_1 - D_0D_2,\ P = D_0D_3 - D_1D_2,\ Q = D_2D_2 - D_1D_3$$

$$DISC = PP - 4RQ,\ x = (-P - \sqrt{DISC})/(2R),\ y = (-P + \sqrt{DISC})/(2R)$$

Shown are the graphs of integrated areas for the unconstrained double-exponential decays with contiguous gates (A) and overlapped gates (B) and the corresponding equations to calculate lifetimes (τ_1 and τ_2) and pre-exponential factors (k_1 and k_2) (Sharman et al 1999).

Four gated images are required for the unconstrained double-exponential decays for an unknown lifetime because there are four unknowns (k_1, k_2, τ_1, and τ_2), as shown in the equations in Box 4.2. It is advantageous to use the overlapping method so that the photons per pixel are increased to reduce the error in the lifetime calculations. The percent relative standard deviation from a representative set of Monte Carlo simulations can be seen in the literature (Sharman et al. 1999; Appendix 4.1). Plots of the best fits calculated by nonlinear least-squares and RLD methods are essentially indistinguishable except at the starting times. When computational times were compared, the RLD was in submilliseconds whereas the nonlinear least-squares method required several seconds (Sharman et al. 1999).

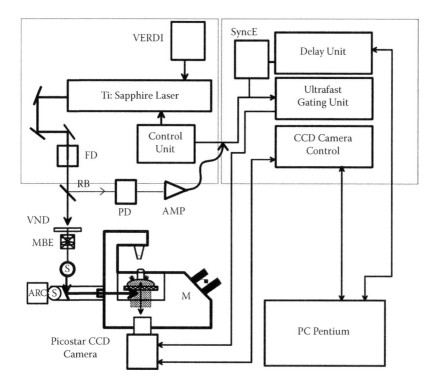

FIGURE 4.3 Schematic of the RLD-based wide-field fluorescence lifetime imaging microscopy. The high-speed gated camera system (PicoStar HR; LaVision, GMBH, Goettingen, Germany) consists of camera head (CCD and image intensifier) and its control units, TTL-I/O synchronization board, and A/D converter. This intensified gated/modulated camera gate pulse driver has a bandwidth of 1 GHz and it has internal pulse-forming circuitry to provide gate widths less than 300 ps at trigger rates from single shot to 110 MHz. The camera features an internal microcontroller with a front panel LCD display and keypad for all functions. FD: frequency doubler; VND: variable neutral density filter; MBE: motorized beam expander; M: microscope; ARC: arc lamp; S: shutter; SyncE: synchronization electronics; Rb: 4–5% of the excitation light as reference beam; PD: high-speed photodiode; AMP: amplifier. (Adapted from Elangovan, M. et al. 2002. *Journal of Microscopy* 205:3–14.)

4.3.2 Components Required for RLD-Based Lifetime Imaging

An RLD-based fluorescence lifetime imaging microscope system is shown in Figure 4.3. The following list discusses various components involved in setting up the lifetime imaging system.

1. In the time-domain method, a high-speed pulsed laser is required. In this case, a tunable Verdi (power: 10 W) pumped Ti:sapphire laser system providing 150 fs excitation pulses at repetition rate of up to 76 MHz (www.Coherent.com) was tuned to 880 nm.

2. This tuned frequency was doubled using a Coherent frequency doubler to 440 nm to excite the cyan fluorescent protein (CFP). Currently, a 440 nm pulsed diode laser is commercially available for RLD imaging.

3. This pulsed 440 nm blue laser line was coupled to a Nikon inverted epi-fluorescent microscope through beam steering optics.

4. A 60× water-immersion, 1.2 numerical aperture (NA) objective lens was used for the experiments. For laser coupling to a microscope, one should consider a nonphase objective lens to minimize any light scattering.

5. The output side port of the epi-fluorescent inverted microscope was coupled to a high-speed gated image intensifier to acquire the time-resolved fluorescent images.

6. This high-speed gated image intensifier camera is provided in a remote housing with a flexible connection to the respective control units. The gate width of the intensifier is variable from 300 ps to milliseconds. The quantum efficiency (QE) of the photocathode of this intensifier was about 20% at 540 nm. The output of this intensifier was coupled by fiber optics or lens to a 16 bit CCD chip at 12.5 MHz readout (640 × 480 pixels). After synchronizing the gated camera to the excitation laser pulses, time-resolved nanosecond images were acquired, as shown in Figure 4.5.

7. Four to five percent of the high-speed excitation laser light was allowed to enter on a high-speed photodiode (PD) before entering the microscope port. The signal from the photodiode was used to synchronize the gating of the high-speed camera with the fluorescent decay.

8. An appropriate filter cube was used to reflect the 440 nm laser beam to the specimen and transmits the 485/40 nm cyan emission signal to the camera coupled to the side port of the microscope.

9. The whole optical, camera, and other electronics components were assembled on a vibration-free table.

10. A four-channel oscilloscope was used to synchronize the excitation light source, gating pulse, and generator signal to trigger the gating camera and the camera gating pulse.

11. A temperature-controlled chamber was used to maintain the cell specimen under physiological conditions. The lifetime number will change depending on the environmental condition of the specimen or room temperature; thus, it is important to maintain the specimen temperature while collecting the lifetime images.

4.3.3 How Data Were Acquired Using the RLD Method

4.3.3.1 Calibration of the System with a Known Fluorophore (Single-Exponential Decay)
The RLD method uses two windows for single-exponential decay and four windows for double-exponential decays to calculate the lifetime as shown in Boxes 4.1 and 4.3. Here we explain step by step how to acquire the images and how to calibrate the system using the single-exponential decay approach. Table 4.1 shows the fluorophore standards that can be used for the lifetime system calibration.

BOX 4.3

The energy transfer efficiency (E), the rate of energy transfer (k_T), and the distance between donor and acceptor molecule (r) are calculated using the following equations (Lakowicz 1999):

$$E = R_0^6/(R_0^6 + r^6) \tag{4.15}$$

$$E = 1 - (\tau_{DA}/\tau_D) \tag{4.16}$$

$$k_T = (1/\tau_D)\,(R_0/r)^6 \tag{4.17}$$

$$r = R_0\{(1/E) - 1\}^{1/6} \tag{4.18}$$

$$R_0 = 0.211\{\kappa^2\,n^{-4}\,Q_D\,J(\lambda)\}^{1/6} \tag{4.19}$$

where τ_D and τ_{DA} are the donor excited-state lifetime in the absence and presence of the acceptor; R_0 is the Förster distance—that is, the distance between the donor and the acceptor at which half the excitation energy of the donor is transferred to the acceptor while the other half is dissipated by all other processes, including light emission; n is the refractive index; Q_D is the quantum yield of the donor; and κ^2 is a factor describing the relative dipole orientation.

The overlap integral $J(\lambda)$ (see Chapters 1 and 2) expresses the degree of spectral overlap between the donor emission and the acceptor absorption:

$$J(\lambda) = \int_0^\infty f_D(\lambda)\varepsilon_A(\lambda)\lambda^4\,d\lambda \Big/ \int_0^\infty f_D(\lambda)d\lambda \tag{4.20}$$

where $f_D(\lambda)$ is the corrected fluorescence intensity of the donor wavelength in the range λ and $\lambda + d\lambda$, with the total intensity normalized to unity; $\varepsilon_A(\lambda)$ is the extinction coefficient of the acceptor at λ, which is in units of M^{-1} cm^{-1}.

1. A 1 μM solution of rhodamineB in ethanol was prepared in a microscope chamber and placed on an inverted Nikon epi-fluorescence microscope stage.

2. The average power of the laser *at the specimen plane* needs to be measured carefully to reduce photobleaching of the specimen. The average power at the specimen plane for the described calibration in this chapter was about 1 mW at 76 MHz repetition rate.

3. Synchronize the camera gating to the laser pulse excitation of the specimen.

4. Following delivery of an excitation pulse, the camera is turned on for a very brief interval (Δt) at some delay time (t_1) after the pulse and the emitted fluorescence signal from the specimen is acquired on the photocathode of the gated image intensifier. This intensifier is coupled to a cooled CCD camera. The signal from the CCD camera is integrated and digitized as a gated or time-resolved image.

TABLE 4.1   Comparison of Literature and Measured Lifetimes Using RLD, TCSPC, and FM Techniques[a]

Fluorophores	RLD[b] (ns)	TCSPC[c] (ns)	FM[d] (ns)	Literature (ns)
RhodamineB in ethanol	2.9	3.0	2.9	2.88[e]
Rose Bengal (RsB) in acetone	2.5	2.4	2.6	2.57[f]
RsB in N,N,dimethyl formamide	2.2	2.2	2.3	2.19[f]
RsB in 2-propanol	1.0	1.1	1.1	0.975[f]
RsB in ethanol	0.8	0.8	0.8	0.739[f]

[a] Periasamy, A. et al. 1996. *Review of Scientific Instruments* 67:3722–3731.
[b] RLD: rapid lifetime determination
[c] TCSPC: time-correlated single-photon counting
[d] FM: frequency modulation
[e] Lakowicz, J. R. 2007. *Principles of Fluorescence Spectroscopy,* 3rd ed. New York: Plenum.
[f] Cramer, L. E., and Spears, K. G. 1978. *Journal of the American Chemical Society* 100:221–227.

5. The photocathode of the image intensifier is activated (opened or gated) for every excitation laser pulse. The specimen is excited at about 76×10^6 pulses per second. At each pulse excitation, there is fluorescent decay. The time between each laser pulse is about 12.5 ns. The images were accumulated at the same time frame and gate width, resulting in approximately 76×10^6 images per second.

6. The image is accumulated for a second (i.e., 76×10^6 images are integrated into a single image). The first gate image is D_0 (see Figures 4.4 and 4.5).

7. For the second gate image, the trigger pulse was shifted to t_2 $(t_2 > t_1)$. The camera is ready for photon collection after the laser pulse excites the specimen.

8. For example, if $t_1 = 1$ ns, after laser pulse excitation the second gate, t_2, was shifted to 3 ns (contiguous gating), assuming that the gate width Δt is 2 ns.

9. One cannot move the t_2 time beyond 12 ns because the time between each excitation laser pulse is around 13 ns. On the other hand, the camera gate is variable, depending on the excitation laser repetition rate. It is possible to collect a series of images and the image intensity will be decreasing as we move the gate time t_2 along the fluorescence decay, as shown in Figure 4.5

10. The exposure time or collection time of the images depends on the specimen's brightness or fluorophore labeling condition. Usually 1–3 s are required for collecting the images.

Thus, the fluorescence lifetimes and pre-exponential factors can be calculated directly from four parameters $(D_0, D_1, t_1, \text{ and } t_2)$ without fitting a large number of data points, as is required by the conventional least-squares method (see Figure 4.4).

4.3.3.2 Double-Exponential Decays: Biological Examples
Here we use a biological sample to demonstrate the double-exponential decay to localize protein–protein interactions in living cells. As explained in this book in many chapters, to

determine the energy transfer process from a donor to an acceptor molecule, one should follow the change in lifetime of the donor molecule in the presence of the acceptor and compare this with the lifetime of the donor in the absence of an acceptor. The protein inter-action occurs in the excited state at a nanosecond time scale. Lifetime imaging techniques provide the ability to record data at the time of interaction, at nanoseconds. In this RLD method, even though the image acquisition time is a few seconds, the photons are collected at the time of interactions. Here we offer a step-by-step explanation of how this method can be used for any biological or clinical samples.

With the RLD method, we acquired time-resolved images of CFP-C/EBPα (donor) in the absence and presence of the acceptor YFP-C/EBPα. Mouse pituitary GHFT1-5 cells were used to express with CFP-C/EBPα and YFP-C/EBPα. We need two cover slips to evaluate the protein–protein interactions: one CFP alone and another cover slip with cells expressing CFP and YFP. The expression and other details of this protein are described in Appendix 4.2.

4.3.3.2.1 Donor in the Absence of the Acceptor

1. CFP-C/EBPα was expressed alone on cover slips and placed on the microscope stage; cells to be imaged are selected through the ocular using an arc lamp light source.

2. As described in Section 4.3.2, the 440 nm laser line was used as an excitation wave-length to illuminate the specimen and a high-speed camera was used to acquire time-resolved images using 480/30 nm emission filters.

3. The camera gating is synchronized (see Figure 4.5) to the excitation laser pulse. Following the delivery of an excitation laser pulse, at some delay time $t_1 = 1$ ns, the photocathode of the image intensifier was turned on (gating, ΔT) for about 2 ns to accumulate the photons (see Figure 4.5). The amplified signal at the image intensifier was then accumulated on the coupled CCD chip.

4. The accumulation time varied from 500 ms to 1 s depending on the signal level from the labeled cells. This accumulation time is possible because many stable laser pulses (76×10^6 pulses/second) cause excitation during that time of data acquisition.

5. After the first gate image acquisition, the second gate image (D_1) is acquired by mov-ing the timing of turning the photocathode about one nanosecond ($t_2, t_2 > t_1$) from the previous gate window. This allows overlapping the gating time about 50% of the first gate, as shown in Figure 4.5. The overlapped gating scheme provides more photon counts than the contiguous gating method (Sharman et al. 1999).

6. Processing the images requires another 500 ms for 256×256 pixels with 12 bit resolu-tion. Thus, a maximum of 3 s is necessary to acquire, process, and display the images on a PC platform.

7. The decay equations (Equation 4.4) are used to process the acquired images to obtain a single-exponential decay value of the donor image (τ_D).

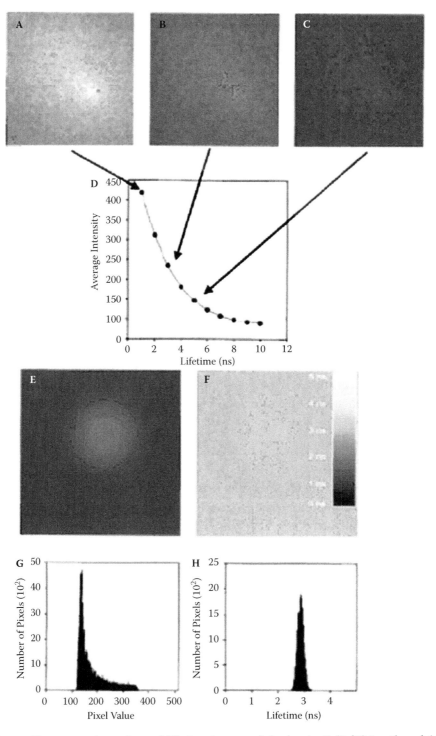

FIGURE 4.4 Fluorescent intensity and lifetime images of rhodamineB (RdB) in ethanol (simulation). A 1 μM solution of RdB was prepared in ethanol and imaged using RLD at room temperature. (A)–(C): Time-resolved fluorescence intensity images of RdB at various times (1, 3, and 6 ns) after the laser pulse excitation. The intensity of the images decreases as the time between the excitation pulse and turning on the gated image intensifier increases. (D): The decay of fluorescence intensity

with time following excitation ($t = 0$). Typical intensity (E) of lifetime (F) images and their respective histograms (G) and (H); lifetime images of RdB were derived from two time-resolved images using the contiguous gating equation as shown in Box 4.1(A). The RdB lifetime value of 2.88 ns was determined from the position of the peak of the histogram. Note the different shape and character of the histograms of the intensity image and the lifetime image. The intensity histogram has a peak that corresponds to a large number of background pixels; intensities of other pixels are quite uniformly distributed over a range of values. In contrast, the lifetime image histogram consists only of a well defined, symmetrical, and narrow peak with the shape of the Gaussian distribution. (Original data can be seen in Periasamy et al. 1996.)

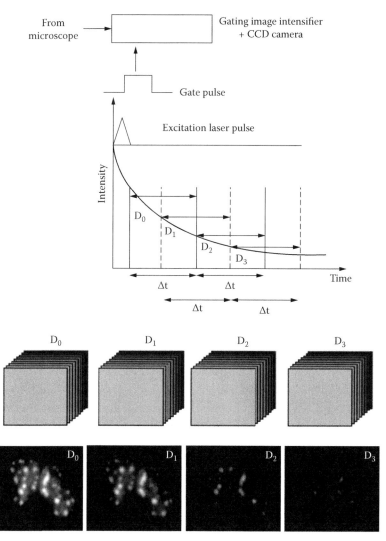

FIGURE 4.5   Schematic illustration for the acquisition of time-resolved FLIM images. After laser pulse excitation, the reference pulse from laser is used as a trigger pulse to gate the camera. The first gate image (D_0) is acquired with a gate window (or exposure time) of ΔT. This signal is accumulated for certain duration of time t. Likewise, other gated images (D_1, D_2, and D_3) are acquired. The respective cellular images (CFP-YFP-C/EBPα) are shown. (Adapted from Elangovan, M. et al. 2002. *Journal of Microscopy* 205:3–14.)

4.3.3.2.2 Donor in the Presence of the Acceptor

1. As the next step, we replaced the chamber with cells expressing the combination of CFP-C/EBPα and YFP-C/EBPα.

2. Time-resolved images are acquired using the same excitation (440 nm laser line) and emission filter (480/30 nm) used for donor-alone time-resolved image acquisition (www.chromatech.com).

3. The previously described procedure was repeated to collect four overlap-gated (50%) time-resolved images (D_0, D_1, D_2, and D_3; see Figure 4.5).

4. The double-exponential fluorescence lifetimes (τ_{DA1}, τ_{DA2}) were processed and calculated for overlapped gating by using Equations 4.11 and 4.12 without having to fit a large number of data points as required by conventional least-squares methods (O'Connor and Phillips 1984).

The lifetime values of donor images in the absence (τ_D) and in the presence of the acceptor (τ_{DA1}, τ_{DA2}) were calculated to determine the energy transfer efficiency and the distance between donor and acceptor molecules using Equations 4.16 and 4.18 (see Box 4.3). All images were acquired at room temperature (74°F). We did not observe any detectable autofluorescence signals imaging unlabeled cells in the same media. Moreover, the fluorescence signal was stronger than the autofluorescence signal and there was negligible involvement of autofluorescence in the processed lifetime images.

4.4 CRITICAL DISCUSSION

Lifetime imaging in general affords high temporal and spatial resolution; the double-exponential decay processing option adds another dimension by discriminating the binding and nonbinding conditions of the biological proteins, calcium, pH, etc. (Day et al. 2008; Lakowicz et al. 1994; Chen and Day 2005; Munro et al. 2005; Wallrabe and Periasamy 2005; Sanders et al. 1995). The RLD method is ideal to monitor protein–protein interactions in living specimens. This method also allows acquiring lifetime time-lapse images because it requires only a couple of seconds for data collection. This lifetime technique allows following the change in lifetime of the donor molecule in the absence and presence of the acceptor. Only two lifetime values are required to estimate the energy transfer efficiency between donor and acceptor molecules.

The minimum lifetime temporal resolution of this RLD imaging system is about 50–100 ps, depending upon the signal-to-noise ratio of the time-resolved images. The energy transfer efficiency (E) and distance (r) between the donor and acceptor are calculated using Equations 4.15–4.18 (see Box 4.3). The available literature value for the refractive index ($n = 1.4$), the assumed dipole orientation for the random movement ($\kappa^2 = 2/3$), and the quantum yield of the donor ($Q_D = 0.4$) have been used in the equations to calculate the R_0 value (Lakowicz 2007). Förster distance R_0 value was calculated ($R_0 = 52.77$) for the CFP-YFP fluorophore and Cerulean–Venus ($R_0 = 57$) (Elangovan et al. 2003; Day et al. 2008).

This detailed spatial and temporal distribution of protein–protein interaction cannot be ordinarily obtained using an intensity-based FRET (Förster resonance energy transfer) imaging system. Intensity-based imaging requires mathematical approaches to remove the contamination generated by spectral bleed-through and autofluorescence (Chen, Elangovan, and Periasamy 2005; Chen et al. 2007; Chen and Periasamy 2006). This clearly demonstrates that the RLD-based FRET-FLIM imaging system delineates the regional variations of protein–protein interactions at a nanometer scale. It is important to point out that less error is involved in distance calculation using $\kappa^2 = 2/3$ in the FRET-FLIM imaging mode compared to steady-state FRET imaging. For example, for the CFP/YFP pair, the steady-state (wide-field) FRET measurements provide efficiency $E = 55\%$ and distance $r = 5.1$ nm. The FRET-FLIM-based measurement values were $E = 26\%$ (± 4.1) and $r = 6.0$ nm (± 2.3) (Elangovan et al. 2002; see Figure 4.6). Reasonable agreement was found when we compared the FRET-FLIM-based distance measurement with the distance calculated using the molecular size and other parameters.

The number of gates that can be acquired is controlled by the strength of the signal levels from the biological specimen, the repetition rate of the laser pulse, and the sensitivity of the photocathode of the gated image intensifier and the CCD camera chip. So far, we have derived the decay equations for double-exponential decays, which required four gated images whether the specimens were single-, double-, or triple-labeled cells.

4.5 PITFALLS

Here we describe the issues involved in determination of the lifetime values using the RLD method.

1. In an exponential decay situation, the image intensity will decrease at different times of decay, as shown in Figure 4.4 (also see Figure 4.7). In double-exponential decay, it is important to determine whether the collected images (D_0, D_1, D_2, and D_3) are in an exponential pattern. If the collected images are in an exponential pattern, the ratio of the consecutive gate images should increase. For example, in panel B in Figure 4.7, we used C/EBPα dimerization proteins, which were expressed with CFP and YFP in the GHFT1 cell nucleus; four gate images were collected. As shown, the ratio values D_1/D_0, D_2/D_1, and D_3/D_2 are not increasing. In this kind of situation, the τ_{DA2} is less than τ_{DA1}. According to the RLD theory as mentioned in Equation 4.6, τ_1 should be less than τ_2. This demonstrates that the data were not acquired properly or photobleaching of the specimen might have occurred during the acquisition of these images.

   In the other set of data (panel A) expressed with FRET standards mTFP(Teal) linked with Venus by five amino acids (see the cell preparation in Appendix 4.2), the ratios of the gated image ratios are increasing. As expected by the theoretical simulation, the τ_{DA1} value is less than the τ_{DA2} value. It is important that one should verify the data after data acquisition. There could be a number of reasons for $\tau_{DA1} > \tau_{DA2}$. The main culprit could be a photobleaching problem of the specimen. If the specimens are in good condition, one should suspect the average power of the excitation laser light at the specimen plane to be too high. Too much excitation power could photobleach the fluorophore because each gate (image) takes 1 s to collect.

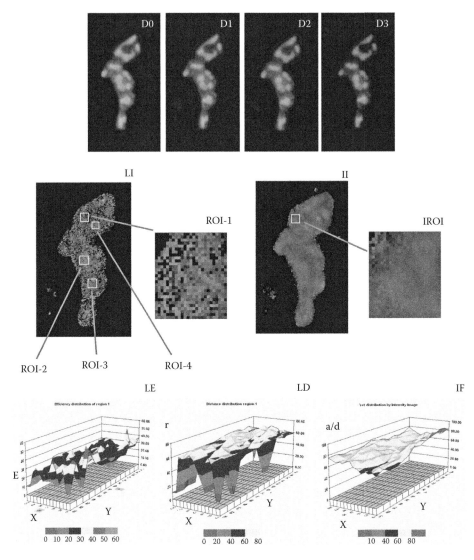

FIGURE 4.6 **(See color insert following page 288.)** Demonstration of CFP-YFP-C/EBPα protein behavior in the foci of a living cell nucleus during the protein interaction process and its comparison with the intensity-based FRET image. Time-resolved images (D_0, D_1, D_2, and D_3) of overlapped gating of donor in the presence of acceptor were acquired according to the scheme illustrated in Figure 4.4. Using the double exponential decay equations (Equations 4.11–4.14), we processed these images for one- (Figure 4.6LI) and two- (Figure 4.4E) component FRET-FLIM images. The spatially resolved protein image was obtained (LI) using the nanosecond FRET-FLIM microscopy compared to a patch of protein complex as shown in the ROI of an intensity-based FRET image (II or IROI). The three-dimensional plots of the efficiency (LE) and the distance (LD) distribution of the ROI-1 are shown. The various colors represent the distribution in efficiency and distance compared to the intensity-based ratioed (A/D) image (IF). Note that this high-speed imaging system demonstrated the protein interactions process with better details than the intensity-based imaging for the same cell as shown in Figure 4.6(IF). (Adapted from Elangovan, M. et al. 2002. *Journal of Microscopy* 205:3–14.)

Panel A

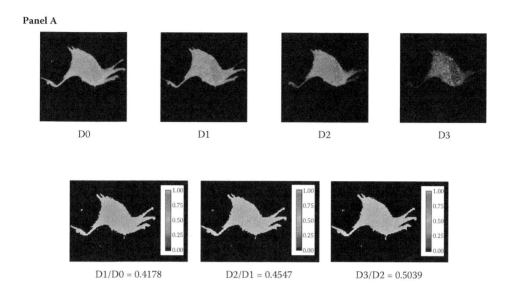

Panel B

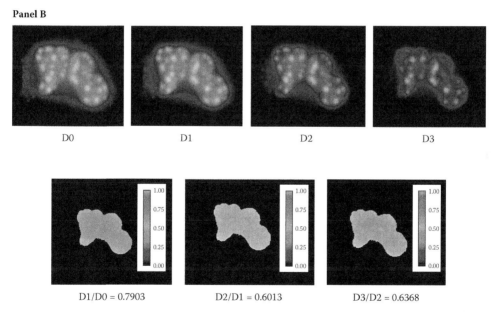

FIGURE 4.7   Issues in acquisition of images in RLD method. The ratios of collected images are in increasing order in panel A compared to in panel B. The images in panel A clearly demonstrate that the images collected are in exponential decay and the lifetime values are τ_{DA1} = 0.932 ns and τ_{DA2} = 2.526 ns (mTFP-5aa-Venus). In panel B, the specimen might have been photobleached during data collection and the lifetime values are not meaningful (CFP-YFP-C/EBPα).

2. Another issue is that of obtaining enough photons at short exposure time to avoid the issues described previously. More photons per pixel will provide lifetime values with good statistics. Achieving high numbers of photons per pixel depends on the amount of fluorophore expression in the specimen, the illumination of laser light to

the specimen (exposure time), and selection of appropriate gate width (Δt) to collect the images. Thus, the RLD method currently provides two component analyses.

3. The precision of RLD measurement is still very good (see Table 4.1) and not all instrumentation lends itself to recording the complete multipoint decay curves required of least-squares methods. On the other hand, the disadvantage of RLD is that it gives no warning of more complex decays, and the method should be applied to systems that have already been fully characterized as to their kinetics.

4.6 SUMMARY

In this chapter we explained step-by-step development of the gated image intensifier in a wide-field microscopy system to measure the lifetime distribution in a single living cell. Our results show that the new gating scheme for single-exponential and double-exponential decays provides a simple and extremely rapid method of lifetime evaluation.

It has been generally described in the literature that energy transfer is primarily considered a single donor–acceptor event or a situation in which there was a single acceptor near each donor. However, in many situations, multiple donor–acceptor interactions take place, such as in solutions, macromolecular protein assemblies, clustered receptor-based complexes, and focal adhesion proteins to allow the possibility of studying many proteins at one event. Steady-state FRET imaging is probably good for localizing single donor–acceptor interactions, but it is not possible to monitor the multiple pairs of interactions at one time.

Multiple pairs of interactions have been measured in solutions (Lakowicz 2007) but not in any biological specimen. RLD technology would help to provide more significant and useful information in complex situations involving protein–protein interactions in living cells. Moreover, the RLD method can very well be applied to time-lapse lifetime imaging because the data acquisition takes seconds compared to 2 min by the TCSPC method. The reader should also visit the other group Web sites (*French 2008; Gerritsen 2008*) to learn more about the usage of the gated image intensifier for lifetime imaging in the area of biological and clinical applications.

It is possible to obtain a three-dimensional (x, y, τ) distribution of the lifetimes in biological systems using the RLD method. It is also possible to collect different optical sections of the specimen by coupling this gated image intensifier to a spinning disk confocal and using a spot-scanning photomultiplier tube (PMT) as a detector in a confocal or multiphoton microscopy system; this may require different hardware and electronics configurations (Gerritsen et al. 2006). Currently used photocathode sensitivity is about 23%; increasing the photocathode sensitivity and using a fiber-coupled CCD camera would improve the number of photons per pixel. Moreover, the RLD software can also be used to process the data collected from the TCSPC method.

APPENDIX 4.1: MONTE CARLO SIMULATION

Monte Carlo simulations were used to judge the precision and accuracy of the data reduction methods. The parameter values selected to generate data corresponded to expected experimental values. For each set of selected decay parameters, a noise-free data set was generated from the model equation. Noise obeying Poisson statistics was added to the decays, which were then reduced by the appropriate algorithm. This process was repeated several times with different noise sets to determine the statistics for the fitting algorithm. The parameters from the reduction were compared with initial parameters used to generate the noise-free data. The calculations were then repeated with different sets of parameters for the modeling equation. Thus, the theoretical calculations allowed testing of the extremes over which RLD will provide reliable results and direct the design of experiments to achieve a desired accuracy.

Poisson noise was selected because it corresponds to noise that is frequently encountered experimentally (e.g., time-correlated single-photon-counting instrument). Even analog instruments based on photon detectors can have noise distributions that follow amplitude dependencies similar to Poisson statistics. In all cases, we base our calculation on the total number of photons that would be detected during the entire sample decay from zero time to infinity. We chose not to use error propagation for two reasons. First, it becomes exceptionally complex and cumbersome for the double-exponential case. Second, it is based on infinitesimal errors, which can distort results for the larger errors present in our simulations. Thus, while the plots are noisier, they are more reliable.

All Monte Carlo simulations were performed using Mathcad 7.0 (Waterloo Scientific, Toronto, Canada, 1997). Typically, 100 simulations were performed at each set of conditions to determine the uncertainty in the parameters. The built-in random number generator gives no repeats over >107 trials, which far exceeds the number involved in our calculations, and it gives a good Gaussian distribution for hundreds of trials. After each decay was scaled to give a total area equal to the specified number of photons, noise was added using a Poisson noise generator. Results are displayed as contour plots of fractional percent standard deviation $100\ \sigma_x/\bar{A}$, where σ_x is the standard deviation and $\bar{A}$ is the mean of the parameters determined from all simulations. Equations were derived by analytically evaluating the areas under the different portions of the decay curves. The decay parameters were determined by solving the system of equations (adapted from Sharman et al. 1999).

APPENDIX 4.2: PREPARATION OF CELLS

The recent resurgence of interest in FRET microscopy has been driven by the availability of the different color fluorescent proteins (Cubitt et al. 1999; Ellenberg et al. 1998; Heim and Tsien 1996; Tsien 1998). Mutagenesis of the *Aequoria victoria* green fluorescent protein (GFP) has yielded proteins that fluoresce from blue to yellowish green, and some of these expressed protein tags have proven to be suitable as donor and acceptor pairs for FRET microscopy (Periasamy and Day 2005). A cyan (blue green) color variant was generated that is resistant to photobleaching and shares an extensive spectral overlap with the yellowish fluorescent protein (YFP), thus allowing this combination to be used in FRET studies (Miyawaki 1999).

For the studies described here, the sequence encoding the DNA binding and dimerization domain of the transcription factor C/EBPα (Lincoln 1994) was fused in-frame to the commercially available CFP or YFP color variants (www.clontech.com) to generate CFP-C/EBPα and YFP-C/EBPα (Day et al. 2001). For transfection, mouse pituitary GHFT1-5 cells

(Lew 1992) were harvested and transfected with the indicated plasmid DNA(s) by electroporation (Day 1998; Schaufele et al. 2001). The total input DNA was kept constant using empty vector DNA. Cell extracts from transfected cells were analyzed by western blot to verify that the tagged proteins were of the appropriate size as described previously (Day 1998). For imaging, the cells were inoculated drop wise onto a sterile cover glass in 35 mm culture dishes, allowed to attach before gently flooding the culture dish with media, and maintained for 18–36 hours before imaging. The cover glass with cells attached was inserted into a chamber containing the appropriate medium and the chamber was then placed on the microscope stage.

mTFP-5aa-Venus: The plasmid vector encoding the mTFP (Gene Bank accession DQ676819) was obtained from Dr. Mike Davidson (Florida State University) and is commercially available from Allele Biotechnology & Pharmaceuticals (www.allelebiotech.com). The cDNA for Venus was obtained from Dr. Atsushi Miyawaki (RIKEN, Japan) (Nagai et al. 2002). The Cerulean FP was generated by mutagenesis of the sequence encoding the enhanced cyan FP (ECFP) to incorporate the S72A, Y145A, and H148D changes (Rizzo et al. 2004). The plasmid encoding the FRET standard fusion protein consisting of Cerulean tethered to Venus by a five-amino acid (aa) linker (Cer-5aa-Venus) was a gift from Dr. Steven Vogel (NIH) and was described earlier (Thaler et al. 2005). The plasmid encoding a similar mTFP FRET standard fusion protein (mTFP-5aa-Venus) was made by substitution of the coding sequence for Cerulean with the cDNA for mTFP, generated by PCR with primers incorporating suitable restriction enzyme sites.

REFERENCES

Ballew, R. M., and Demas, J. N. 1989. An error analysis of the rapid lifetime determination method for the evaluation of single exponential decays. *Analytical Chemistry* 81:30–33.

Becker, W. 2005. *Advanced time-correlated single photon counting techniques.* Berlin: Springer–Verlag.

Becker, W., Bergmann, A., Hink, M. A., König, K., Benndorf, K., and Biskup, C. 2004. Fluorescence lifetime imaging by time-correlated single-photon counting. *Microscopy Research and Technique* 63:58–66.

Buurman, E. P., Sanders, R., Draaijer, A., Van Veen, J. J. F., Houpt, P. M., and Levine, Y. K. 1992. Fluorescence lifetime imaging using a confocal laser scanning microscope. *Scanning* 14:155–159.

Chen, Y., Elangovan, M., and Periasamy, A. 2005. FRET data analysis—The algorithm. In *Molecular imaging: FRET microscopy and spectroscopy,* ed. A. Periasamy and R. N. Day, 126–145. New York: Oxford University Press.

Chen, Y., Mauldin, J. P., Day, R. N., and Periasamy, A. 2007. Characterization of spectral FRET imaging microscopy for monitoring the nuclear protein interactions. *Journal of Microscopy* 228:139–152.

Chen, Y., and Periasamy, A. 2004. Characterization of two-photon excitation fluorescence lifetime imaging microscopy for protein localization. *Microscopy Research and Technique* 63:72–80.

Chen, Y., and Periasamy, A. 2005. Time-correlated single photon counting (TCSPC) FLIM-FRET microscopy for protein localization. In *Molecular imaging: FRET microscopy and spectroscopy,* eds. A. Periasamy and R. N. Day, 239–259. New York: Oxford University Press.

Chen, Y., and Periasamy, A. 2006. Intensity range-based quantitative FRET data analysis to localize the protein molecules in living cell nucleus. *Journal of Fluorescence* 16:95–104.

Clegg, R. M., Feddersen, B., Gratton, E., and Jovin, T. M. 1991. Time-resolved imaging microscopy. *Proceedings of SPIE—International Society for Optical Engineering* 1640:448–460.

Cramer, L. E., and Spears, K. G. 1978. Hydrogen bond strengths from solvent-dependent lifetimes of rose Bengal dye. *Journal of the American Chemical Society* 100:221–227.

Day, R. N., Booker, C. F., and Periasamy, A. 2008. Characterization of an improved donor fluorescent protein for Förster resonance energy transfer microscopy. *Journal of Biomedical Optics* 13:031203.

Demas, J. N. 1983 *Excited-state lifetime* measurements. New York: Academic Press.

Demas, J. N., Jones, W. M., and Keller, R. A. 1986. Elimination of quenching effects in luminescence spectrometry by phase resolution. *Analytical Chemistry* 58:1717–1721.

Elangovan, M., Wallrabe, H., Chen, Y., Day, R. N., Barroso, M., Periasamy, A. 2003. Characterization of one- and two-photon excitation fluorescence resonance energy transfer microscopy. *Methods* 1:58–73.

Elangovan, M., Day, R. N., and Periasamy, A. 2002. A novel nanosecond FRET-FLIM microscopy to quantitate the protein interactions in a single living cell. *Journal of Microscopy* 205:3–14.

French, P. 2008. http://www3.imperial.ac.uk/people/paul.french

Gadella, T. W. J., Jr., van Hoek, A., and Visser, A. J. W. G. 1997. Construction and characterization of a frequency-domain fluorescence lifetime imaging microscopy system. *Journal of Fluorescence* 7:35–43.

Geriston, H. C. 2008. http://www/1.phys.uu.ni/wwwmbf

Gerritsen, H. C., Draaijer, A., van den Heuvel, D. J., and Agronskaia, A. V. 2006. Fluorescence lifetime imaging in scanning microscopy. In *Handbook of biological confocal microscopy,* ed. J. B. Pawley, 516–534. Berlin: Springer.

Konig, K., So, P. T., Mantulin, W. W., Tromberg, B. J., and Gratton, E. 1996. Two-photon excited lifetime imaging of autofluorescence in cells during UVA and NIR photostress. *Journal of Microscopy* 183:197–204.

Krishnan, R. V., Saitoh, H., Terada, H., Centonze, V. E., and Herman, B. 2003. Development of a multiphoton fluorescence lifetime imaging microscopy system using a streak camera. *Review of Scientific Instruments* 74:2714–2721.

Lakowicz, J. R. 2007. *Principles of fluorescence spectroscopy,* 3rd ed. New York: Plenum.

Lakowicz, J. R. 1999. *Principles of fluorescence spectroscopy,* 2nd ed. New York: Plenum.

Lakowicz, J. R., and Brendt, K. W. 1991. Lifetime-selective fluorescence imaging using an RF phase-sensitive camera. *Review of Scientific Instruments* 62:3653–3657.

Lakowicz, J. R., Szmacinski, H., Nowaczyk, K., Ledrer, W. J., Kirby, M. S., and Johnson, M. L. 1994. Fluorescence lifetime imaging of intracellular calcium in COS cells using Quin-2. *Cell Calcium* 15:7–27.

Marriott, G., Clegg, R. M., Arndt-Jovin, D. J., and Jovin, T. M. 1991. Time resolved imaging microscopy: Phosphorescence and delayed fluorescence imaging. *Biophysical Journal* 60:1374–1387.

Morgan, C. G., Mitchell, A. C., and Murray, J. G. 1990. Nanosecond time-resolved fluorescence microscopy: Principle and practice. *Transactions of the Royal Microscopical Society* (Micro '90):463–466.

Munro, I., McGinty, J., Galletly, N., et al. 2005. Towards the clinical application of time-domain fluorescence lifetime imaging. *Journal of Biomedical Optics* 10:051403.

Nagai, T., Ibata, K., Park, E. S., Kubota, M., Mikoshiba, K., and Miyawaki, A. 2002. A variant of yellow fluorescent protein with fast and efficient maturation for cell-biological applications. *Nature Biotechnology* 20:87–90.

O'Connor, D. V., and Phillips, D. 1984. *Time-correlated single photon counting.* London: Academic Press.

Oida, T., Sako, Y., and Kusumi, A. 1993. Fluorescence lifetime imaging microscopy (FLIMscopy): Methodology development and application to studies to endosome fusion in single cells. *Biophysical Journal* 64:676–685.

Periasamy, A., Elangovan, M., Elliott, E., and Brautigan, D. L. 2002. Fluorescence lifetime imaging of green fluorescent fusion proteins in living cell. In *Methods in molecular biology,* vol. 183, *Green fluorescent protein: Applications and protocols,* ed. B. Hicks, 89–100. Totowa, NJ: Humana Press.

Periasamy, A., Elangovan, M., Wallrabe, H., Demas, J. N., Barroso, M., Brautigan, D. L., and Day, R. N. .2001. Wide-field, confocal, two-photon and lifetime resonance energy transfer imaging microscopy. In *Methods in cellular imaging,* ed. A. Periasamy, 295–308. New York: Oxford University Press.

Periasamy, A., Siadat-Pajouh, M., Wodnicki, P., Wang, X. F., and Herman, B. 1995. Time-gated fluorescence microscopy for clinical imaging. *Microscopy and Microanalysis* March:19–21.

Periasamy, A., Wodnicki, P., Wang, X. F., Kwon, S., Gordon, G. W., and Herman, B. 1996. Time-resolved fluorescence lifetime imaging microscopy using picosecond pulsed tunable dye laser system. *Review of Scientific Instruments* 67:3722–3731.

Rizzo, M. A., Springer, G. H., Granada, B., and Piston, D. W. 2004. An improved cyan fluorescent protein variant useful for FRET. *Nature Biotechnology* 22:445–449.

Sanders, R., Draaijer, A., Gerritsen, H. C., Houpt, P. M., and Levine, Y. K. 1995. Quantitative pH imaging in cells using confocal fluorescence lifetime imaging microscopy. *Analytical Biochemistry* 227:302–308.

Sharman, K. K., Asworth, H., Snow, N. H., Demas, J. N., and Periasamy, A. 1999. Error analysis of the rapid lifetime determination method for double-exponential decays and new windowing schemes. *Analytical Chemistry* 71:947–952.

Sholes, R. R., and Small, J. G. 1980. Fluorescent decay thermometer with biological applications. *Review of Scientific Instruments* 51:882–884.

Straub, M., and Hell, S. W. 1998. Fluorescence lifetime three-dimensional microscopy with picoseconds precision using a multifocal multiphoton microscope. *Applied Physics Letters* 73:1769–1771.

Thaler, C., Koushik, S. V., Blank, P. S., and Vogel, S. S. 2005. Quantitative multiphoton spectral imaging and its use for measuring resonance energy transfer. *Biophysical Journal* 89:2736–2749.

Wallrabe, H., and Periasamy, A. 2005. Imaging protein molecules using FRET-FLIM microscopy. *Current Opinion in Biotechnology* 16:19–27.

Wang, X. F., Kitajima, S., Uchida, T., Coleman, D. M., and Minami, S. 1990. Time-resolved fluorescence microscopy using multichannel photon counting. *Applied Spectroscopy* 44:25–29.

Wang, X. F., Periasamy, A., Coleman, D. M., and Herman, B. 1992. Fluorescence lifetime imaging microscopy: Instrumentation and applications. *CRC Critical Reviews in Analytical Chemistry* 23:369–395.

Wang, X. F., Periasamy, A., Wodnicki, P., Gordon, G. W., and Herman, B. 1996. Time-resolved fluorescence lifetime imaging microscopy: Instrumentation and biomedical applications. In *Fluorescence imaging spectroscopy and microscopy,* ed. X. F. Wang and B. Herman, Chemical Analysis Series, vol. 137, 313–350. New York: John Wiley & Sons.

Wang, X. F., Uchida, T., Coleman, D. M., and Minami, S. 1991. A two-dimensional fluorescence lifetime imaging system using a gated image-intensifier. *Applied Spectroscopy* 45:360–367.

Wang, X. F., Uchida, T., and Minami, S. 1989. A fluorescent lifetime distribution measurement system based on phase-resolved detection using an image dissector tube. *Applied Spectroscopy* 43:840–845.

Woods, R. J., Scypinski, S., Cline Love, L. J., and Asworth, H. A. 1984. Transient digitizer for the determination of microsecond luminescence lifetimes. *Analytical Chemistry* 56:1395–1400.

Wouters, F. S., and Bastiaens P. I. H. 1999. Fluorescence lifetime imaging of receptor tyrosine kinase activity in cells. *Current Biology* 9:1127–1130.

CHAPTER 5

Frequency-Domain FLIM

Bryan Q. Spring and Robert M. Clegg

5.1 INTRODUCTION TO FREQUENCY-DOMAIN METHODS

5.1.1 Overview

For frequency-domain FLIM (fluorescence lifetime imaging microscopy), the amplitude of the excitation light is modulated repetitively at high frequency (HF). Radio frequencies between 1 and 200 MHz are nominal and chosen in order for the fluorescence response (nanoseconds) to be sensitive to the frequency of repetition/modulation. The waveform of the HF modulation is often sinusoidal, but it can be any repetitive shape; such a repetitive waveform can be decomposed into multiple harmonics in a Fourier series, and each sinusoidal harmonic component is treated separately. Only the fundamental frequency is present if the excitation is a pure sinusoidal modulation; other waveforms (e.g., square waves or pulse trains from mode-locked lasers) contain multiple harmonics of the repetition frequency.

The time-dependent fluorescence emission of a fluorophore is the convolution of the impulse response of the fluorophore (which is an exponential decay in time following excitation by a short excitation pulse) with the actual repetitive excitation light waveform (see Equation 5.6). As we show later, the fluorescence emission signal oscillates at the same high frequencies as the corresponding Fourier components of the excitation light, but exhibits a demodulation and a phase lag at each frequency. Demodulation is the reduction in the amplitude of the oscillation (the AC component) relative to the mean signal (the DC component). The phase lag means that the fluorescence waveform is delayed in time relative to the excitation waveform. The fluorescence lifetime acts as an effective low-pass filter and removes the higher frequency harmonics with which it cannot keep pace.

The many different ways that the lifetimes can be measured (time and frequency domain) are discussed in a pedagogical paper (Chandler et al. 2006), to which the reader is referred. In this reference, the effect of decaying luminescence is demonstrated with a sample with long fluorescence (e.g., ruby crystals with a millisecond decay time) or a phosphorescent sample (microsecond decay time). The damping of the higher frequency tones that are demodulated by the low-pass filtering by the decay time is easily discerned.

Nanosecond fluorescence decays are probed using megahertz frequencies. For convenience and simplicity of the instrumentation and to avoid high-frequency noise, the resulting HF fluorescence signal is not measured directly. Rather, gain-modulated detectors are used to convert the megahertz HF signal to a low-frequency (LF)—10–10000 Hz—or constant (DC) signal, which can be more readily measured. This is accomplished by "frequency mixing" between the detector gain (or, equivalently, the detector amplification) and the excitation light waveforms. This frequency mixing phenomenon—the conversion of HF signals to LF signals—is well known and the basis of radio technology (Goldman 1948). There are two main approaches of frequency mixing to produce the desired LF signal. The detector gain is modulated at a frequency either very close to the excitation modulation frequency (the heterodyne method) or at the precise frequency of the excitation modulation (the homodyne method).

5.1.2 Heterodyne and Homodyne Methods for Measuring Fluorescence Lifetimes

Frequency-domain techniques for measuring fluorescence lifetimes have been used for a long time, and the technique has become standard (Alcala, Gratton, and Prendergast 1987; Bailey and Rollefson 1953; Birks 1970; Birks and Little 1953; Birks and Dawson 1961; Gratton and Limkeman 1983; Maercks 1938; Spencer and Weber 1969; Tumerman 1941; Tumerman and Szymanowski 1937; Tumerman and Sorokin 1967; Weber 1981a).

For the heterodyne method, the measured signal is a difference frequency oscillating at the frequency difference between the light modulation and the detector gain modulation. This difference frequency, or cross-correlation frequency, is usually chosen to be in the range between a few hundred to a few thousand hertz. For cuvette measurements, the heterodyne method is the method of choice for precision estimates of the fluorescence lifetime components. The modulation frequency of the excitation light, and detector amplification is scanned through a range of frequencies, keeping the cross-correlation frequency the same. The resulting frequency response of the phase shift and demodulation values can be iteratively fit numerically to extract the fluorescence lifetime components (see Equations 5.11–5.14).

Addressing multiple frequencies in a sequential fashion is a normal way to proceed for cuvette (single-channel) experiments, but because this is time consuming for many pixels, it is not normally applied for imaging applications. For FLIM, the heterodyne method is usually carried out in scanning mode with a single-channel detector and a single frequency (however, see Coyler, Lee, and Gratton 2008), such as a photomultiplier tube (PMT) or an avalanche photodiode (APD) for maximal photon collection efficiency. Only a single frequency is used to modulate the excitation light. This means that a focused excitation beam is scanned across the sample to collect the image points sequentially. This approach is compatible with scanning confocal microscopy and multiphoton excitation, which is convenient for many imaging applications.

In the homodyne method, the excitation light and detector are modulated at the same frequency. The advantage of this approach is that a gain-modulated image intensifier and CCD (charge-coupled device) detector can be used for full-field FLIM imaging. Full-field mode (such as in a normal full-field fluorescence microscope) implies that all locations of the sample are imaged simultaneously. Thus, rapid FLIM acquisition by parallel detection is enabled. Detection of multiple lifetime components is still possible with this approach (Clegg and Schneider 1996; Schneider and Clegg 1997).

Thus, the defining distinction between heterodyne and homodyne FLIM is the type of detection system used. In both cases, the demodulation and phase lag of the fluorescence at the high frequency are preserved in the integrated homodyne or cross-correlated heterodyne signal. After straightforward corrections for instrumental offsets, the demodulation and phase lag can then be used to estimate the lifetimes (see Equations 5.8–5.14).

5.1.3 A Few Preliminary Comments

The mathematical expressions for the measured signals discussed later correspond to each individual pixel. It is assumed that each pixel of the detector captures only the fluorescence signal emanating from the corresponding location of the sample plane (for scanning, the fluorescence response is acquired for every addressed location of the excitation focused beam). For three-dimensional confocal imaging, the excitation volume is referred to as a "voxel."

Reflected and scattered light (as well as fluorescence from planes above and below the focus) can create artifacts and contribute additional fluorescence lifetime components from neighboring (or even relatively distal) regions of the sample. In addition, each pixel can contain multiple fluorescent lifetime species, each with different lifetimes. These fluorescent lifetime species often have different emission and excitation spectra. Such artifacts must always be kept in mind when analyzing a FLIM image.

The focus in this chapter is rapid full-field FLIM. We discuss the instrumentation necessary to achieve video-rate confocal FLIM imaging (to reduce the contribution of reflected and scattered light and obtain three-dimensional sections of the sample). We also discuss the implementation of spectral FLIM imaging (to resolve individual fluorescence lifetimes from a mixture of fluorophores). Both the confocal and spectral FLIM modes are carried out with modular components that can be easily inserted into a full-field, homodyne FLIM instrument.

5.2 RELATIONSHIP BETWEEN OBSERVABLES AND FLUORESCENCE LIFETIMES

5.2.1 A Primer in Complex Analysis

The calculations and Fourier series in this section are carried out using complex notation. The general expressions for multiple lifetime species and for repetitive modulation of the excitation light with multiple harmonics are simpler in complex notation. The relationships given in Box 5.1 are helpful for following the mathematics in the following derivations; there are many excellent references for reviewing complex notation (Bracewell 1978; Brigham 1974; Butz 2006; Byron and Fuller 1969; Dern and Walsh 1963; Goldman 1948; Tolstov 1962).

5.2.2 A General Expression for the Fluorescence Signal

The time-resolved *fundamental fluorescence response* is defined as the fluorescence response to a delta function excitation pulse (it is an exponential decay). The fluorescence response

BOX 5.1: SHORT REVIEW OF COMPLEX NOTATION

The real part of a complex number can be found by applying the following expression (where θ is a real number; in this chapter, we define $i = \sqrt{-1}$):

$$2 \cdot \cos(\theta) = \left[\exp\left(i\theta\right) + \exp\left(-i\theta\right) \right] \tag{5.1}$$

This is easy to derive from Euler's identities, $e^{\pm i\theta} = \cos(\theta) \pm i \sin(\theta)$.

Complex numbers can be expressed as vectors in polar coordinates. The conversion from the Cartesian representation for a complex number (where a is the abscissa and b is the ordinate) to the polar notation is

$$
\begin{aligned}
a + ib &= \sqrt{a^2 + b^2}\, \exp\left(i \tan^{-1}\left(b/a\right)\right) \\
&= \sqrt{a^2 + b^2}\, \exp\left(i\theta\right) = \sqrt{a^2 + b^2}\left(\cos\left(\theta\right) + i \sin\left(\theta\right)\right)
\end{aligned}
\tag{5.2}
$$

$\sqrt{a^2 + b^2}$ is the magnitude and $\theta = \tan^{-1}(b/a)$ is the phase in polar coordinates.

The following relationship is very useful for extracting the real part of a complex Fourier series representation of a general time-varying function—$A(t)$—throughout the following discussion:

$$
\begin{aligned}
\mathrm{Re}\left\{A(t)\right\} &= \mathrm{Re}\left\{ A_0 + \sum_{\substack{n=-\infty \\ n\neq0}}^{\infty} A_n \cdot e^{i(2\pi nt/T)} \right\} \\[2mm]
&= \mathrm{Re}\left\{ A_0 + \sum_{n=1}^{\infty}\left[A_n \cdot e^{i(2\pi nt/T)} + A_{-n} \cdot e^{-i(2\pi nt/T)} \right] \right\} \\[2mm]
&= \mathrm{Re}\left\{ A_0 + \sum_{n=1}^{\infty}\left[\left|A_n\right| \cdot e^{i\varphi_n} \cdot e^{i(2\pi nt/T)} + \left|A_n\right| e^{-i\varphi_n} \cdot e^{-i(2\pi nt/T)} \right] \right\} \\[2mm]
&= A_0 + 2 \cdot \sum_{n=1}^{\infty} \left|A_n\right| \cdot \cos\left(2\pi nt / T + \varphi_n\right)
\end{aligned}
\tag{5.3}
$$

where $|A_n| \times \exp(i\phi_n)$ has been substituted for the complex Fourier coefficients, A_n.

to an arbitrary waveform of the excitation is the convolution of the fundamental fluorescence response with the excitation waveform (Box 5.2).

For multiple fluorescence lifetime species, the fluorescence response $F_\delta(t)$ to a delta function excitation pulse, $F_\delta(t)$, is composed of several exponentially decaying components (see Chapter 11):

$$F_\delta(t) = \sum_{s=1}^{S} a_s \cdot \exp\left(-t/\tau_s\right) \tag{5.7}$$

BOX 5.2: THE FOURIER EXPANSION OF THE EXCITATION REPETITIVE PULSE AND THE FUNDAMENTAL FLUORESCENCE RESPONSE OF A SINGLE-FLUORESCENCE SPECIES

The modulated excitation light signal $E(t)$ can be a train of laser pulses or any repetitive wave with fundamental period, T. $E(t)$ can be represented as a Fourier series in complex notation:

$$E(t) = E_0 + \sum_{\substack{n=-\infty \\ n \neq 0}}^{\infty} \left| E_{\omega,n} \right| \cdot \exp\left(i\left(\omega_n^E t + \varphi_n^E \right) \right) \tag{5.4}$$

where

$$E_{\omega,n} = \frac{1}{T} \int_0^T E(t) \exp\left(-i\omega_n^E t \right) dt \tag{5.5}$$

$E_{\omega,n}$ is the amplitude of the nth frequency component, ω_n. The Fourier series includes the fundamental frequency ($\omega = 2\pi/T$ and overtones ($\omega_n = n\omega = 2\pi n/T$, $|n| > 1$), as well as the time independent long time average, E_0. The *fundamental fluorescence response* of a single-fluorescence component to a delta function excitation, $E_\delta(t)$, is $F_\delta(t) = a \times \exp(-t/\tau)$. The expression for the fluorescence response with repetitively modulated excitation light of any waveform, $E(t)$, is the convolution in the following equation:

$$F(t) = Q \int_0^t E(t') F_\delta(t - t') dt' \tag{5.6}$$

The coefficients of the Fourier series, $E_{\omega,n}$, are complex numbers, and in Equation 5.4 we have used the polar form of a complex number, $E_{\omega,n} = \left| E_{\omega,n} \right| \cdot \exp(i\varphi_n^E)$, where $\left| E_{\omega,n} \right|$ is the magnitude and φ_n^E is the phase of $E_{\omega,n}$. Q is a factor to take care of all the instrumentation aspects (see Box 5.3).

where a_s is the fractional amplitude of the sth fluorescence component with a corresponding fluorescence lifetime of τ_s. The fractional amplitude of the sth fluorescence component is proportional to its concentration (c_s), extinction coefficient (ε_s), and natural radiative lifetime, $1/\tau_s^\circ$; that is, $a_s = c_s \times \varepsilon_s \times 1/\tau_s^\circ$, relative to that of all other contributing fluorescence components.

Finally, we can compute the measured fluorescence signal, $F(t)$, including all fluorescence species and all frequency components of the Fourier expansion of $E(t)$, by substituting Equations 5.4 and 5.7 into Equation 5.6 (Clegg and Schneider 1996) (see Box 5.3).

Thus, the fluorescence signal, Equation 5.8, is modulated at the same frequencies ω_n^E as the excitation light (the excitation light drives the fluorescence emission at the frequencies of the Fourier expansion of $E(t)$). But $F(t)$ exhibits a phase lag, ϕ_n (see Equation 5.11), and a demodulation (attenuation in the modulation depth relative to the modulation depth of the excitation waveform), $M_{\omega,n}$, defined in Equation 5.12:

BOX 5.3: MEASURED FLUORESCENCE SIGNAL FOR MULTIPLE FLUORESCENT SPECIES (S) AND MULTIPLE FREQUENCY FOURIER COMPONENTS (N) (CLEGG AND SCHNEIDER 1996)

$$F(t) = Q \cdot \left\{ F_0 + \sum_{\substack{n=-\infty \\ n \neq 0}}^{\infty} |E_{\omega,n}| \cdot F_{\omega,n} \cdot \exp\left(i(\omega_n^E t + \varphi_n^E - \varphi_n)\right) \right\} \tag{5.8}$$

where

$$F_0 = E_0 \cdot \sum_{s=1}^{S} a_s \cdot \tau_s \tag{5.9}$$

$$F_{\omega,n} = \left[\left(\sum_{s=1}^{S} \frac{a_s \cdot \tau_s}{1 + \left(\omega_n^E \tau_s\right)^2} \right)^2 + \left(\sum_{s=1}^{S} \frac{a_s \cdot \omega_n^E \cdot \tau_s^2}{1 + \left(\omega_n^E \tau_s\right)^2} \right)^2 \right]^{1/2} \tag{5.10}$$

and

$$\varphi_n = \tan^{-1}\left(\sum_{s=1}^{S} \frac{a_s \cdot \omega_n^E \cdot \tau_s^2}{1 + \left(\omega_n^E \tau_s\right)^2} \middle/ \sum_{s=1}^{S} \frac{a_s \cdot \tau_s}{1 + \left(\omega_n^E \tau_s\right)^2} \right) \tag{5.11}$$

The factor Q takes into account the efficiency of the instrumentation and optics in collecting the fluorescence signal such that $F(t)$ represents the actual fluorescence signal. This efficiency can have a spectral dependence that will be taken into account for spectral FLIM. $F_{\omega,n}$ and $\varphi_n^F (= \varphi_n^E - \varphi_n)$ are the measured amplitude and the measured phase of the nth harmonic of the measured fluorescence emission, respectively. Note that the values of $F_{\omega,n}$ and φ_n^F are the overall amplitudes and phase shifts, which include the sum over all the fluorescence lifetime species (s).

$$M_{\omega,n} = \frac{M_{F,\omega,n}}{M_{E,\omega,n}} = \frac{Q \cdot E_{\omega,n} \cdot F_{\omega,n} / Q \cdot F_0}{E_{\omega,n} / E_0} = \frac{E_0 \cdot F_{\omega,n}}{F_0}$$

$$= \left[\left(\sum_{s=1}^{S} \frac{a_s \cdot \tau_s \middle/ \sum_{s=1}^{S} a_s \cdot \tau_s}{1 + \left(\omega_n^E \tau_s\right)^2} \right)^2 + \left(\sum_{s=1}^{S} \frac{a_s \cdot \omega_n^E \cdot \tau_s^2 \middle/ \sum_{s=1}^{S} a_s \cdot \tau_s}{1 + \left(\omega_n^E \tau_s\right)^2} \right)^2 \right]^{1/2} \tag{5.12}$$

The values of the phase lag ϕ_n and demodulation $M_{\omega,n}$ are dependent on the residence time of all the different fluorescent species in their excited electronic states (their fluorescence lifetimes) and the frequencies ω_n^E. ϕ_n and $M_{\omega,n}$ are the *experimental values* that we directly measure, and from such measurements we can determine the values of the fluorescence lifetime components.

The often quoted demodulation and phase lag for the case of a single fluorescence lifetime species are derived by considering just a single fluorescence species from Equations 5.11 and 5.12:

$$M_{\omega,n} = 1 \Big/ \sqrt{1 + \left(\omega_n^E \tau\right)^2} \tag{5.13}$$

and

$$\varphi_n = \tan^{-1}\left(\omega_n^E \tau\right) \tag{5.14}$$

Equation 5.8 refers to the fluorescence signal impinging on a single pixel and corresponds to the fluorescence lifetime components contained in the sample at that particular location. Every individual pixel of the detector collects the fluorescence signal from its corresponding region of the sample in the general form of Equation 5.8. For complex biological systems, the fluorescence lifetimes vary from pixel to pixel, corresponding to the local environment of the fluorescent molecules and the different fluorescent species.

The fundamental frequency of the excitation light is chosen to resolve particular fluorescence lifetime components best. For instance, the contributions of faster decaying components to the total fluorescence signal (e.g., short-lifetime components due to quenched donor molecules participating in Förster resonance energy transfer [FRET]) are de-emphasized in FLIM, according to Equation 5.10, because the coefficient of each term is proportional to its fluorescence lifetime—$a_s \cdot \tau_s / (1 + (\omega_n^E \tau_s)^2)$). Also, by tuning the fundamental frequency of the excitation light waveform to a sufficiently high frequency, the slow lifetime components can be suppressed in order to resolve the fast lifetime components better (assuming that a_s is comparable for both the fast and slow components, which is often the case).

5.2.3 A General Expression for the Measured Homo-/Heterodyne Signal

In this section, we present the conversion of the HF signal to a more convenient and more easily handled LF or DC signal. To achieve this, the gain or amplification of the detector is also modulated at radio frequencies in order to convert the fluorescence signal to an LF or a DC signal—for example, frequency mixing (Clegg and Schneider 1996; Goldman 1948; Jameson, Gratton, and Hall 1984; Spencer and Weber 1969). This is, for instance, what is done in a radio receiver to lower the frequency of the carrier wave (which is in the megahertz region) to audio frequencies (1–20 kHz). The general form of the gain modulation may be expressed as a Fourier series in complex notation, as was done for the excitation light in Equation 5.4:

$$G(t) = G_0 + \sum_{\substack{m=-\infty \\ m\neq 0}}^{\infty} \left|G_{\omega,m}\right| \cdot \exp\left(i\left(\omega_m^G t + \varphi_m^G\right)\right) \tag{5.15}$$

where $G_{\omega,m}$ is the amplitude of the mth frequency component, ω_m. Note that the *frequency components of the gain are not necessarily equal to those of the excitation modulation;* hence, for clarity we use the subscript m for the gain.

The measured signal, S, is the real-time product of the fluorescence signal and the detector gain—$F(t) \times G(t)$. Many of the terms resulting from this multiplication are modulated at very HF and average to zero during the actual measurement due to LF filtering. Therefore, we eliminate these terms and only consider the LF terms of the signal that will pass the LF filter, $\{\ldots\}_{LF}$:

$$S = \left\{ F(t) \cdot G(t) \right\}_{LF}$$

$$= Q \cdot \left\{ F_0 \cdot G_0 + 2 \cdot \sum_{n=1}^{\infty} \left| G_{\omega,n} \right| \cdot F_0 \cdot M_{E,\omega,n} \cdot M_{\omega,n} \cdot \cos\left(\Delta\omega_n t - \varphi_n^F + \Delta\varphi_n^{EG} \right) \right\} \quad (5.16)$$

where $\Delta\varphi_n^{EG} = \varphi_n^E - \varphi_m^G$ and $\Delta\omega_n = \omega_n^E - \omega_m^G$. The demodulation and phase shift are preserved in the signal and only the terms in the sums where $0 < \Delta\omega_n / \omega_n^E \ll 1$ (i.e., LF heterodyne signal) or $\Delta\omega_n = 0$ (i.e., DC homodyne signal) survive the LF filter.

Equation 5.16 is the measured frequency-domain FLIM signal in its most generalized form for a single fluorophore that is directly excited by the repetitive light pulse train. The corresponding expression that includes multiple fluorescence lifetime species is found by substituting Equations 5.9, 5.11, and 5.12 into Equation 5.16 for F_0, φ_n^F, and $M_{\omega,n}$. Theoretically, all frequency overtones $\Delta\omega_n$ are present in the fluorescence signal; however, in practice only a finite number of harmonics can be present due to the frequency response of the measuring apparatus. It is possible to measure multiple harmonics simultaneously; they are all simultaneously present in the measured signal, S.

Multifrequency FLIM instrumentation is designed to identify and separate the fluorescence lifetime components and their relative contributions. If the excitation has a waveform with many higher harmonics, such as a short rectangular pulse, the harmonic components can be measured simultaneously, using Equation 5.16. Conversely, one can vary the repetition frequency (usually applying a relatively pure sinusoidal excitation) and measure the signal of the fundamental repetition frequency component at every different repetition frequency separately. Both methods are used. Single-frequency FLIM often refers to experiments carried out at a single fundamental frequency only. In the next section, we discuss single- and multifrequency modes of FLIM in more detail.

5.2.4 Single- Versus Multifrequency FLIM

5.2.4.1 Single-Frequency FLIM

In many applications of FLIM, only the modulation and phase of the fundamental frequency components are considered. The modulation of the excitation light can be sinusoidal, but the presence of higher harmonics is in no way detrimental to the measurement. Any of the frequency components that happen to be present in the fluorescence signal can be determined independently of all others due to the orthogonality of the cosine and sine functions. Thus, the overtones are simply ignored during the determination of the modulation and phase of the fundamental frequency signal. For homodyne detection (where $\Delta\omega_n = 0$), considering only the fundamental frequency in Equation 5.16 results in a DC homodyne signal, $S_{\text{homodyne},k}$:

$$S_{\text{homodyne},k} = S_{\text{homodyne}}\left(\Delta\varphi_k^{EG}\right) = Q \cdot \left\{F_0 \cdot G_0 + 2 \cdot |G_\omega| \cdot F_0 \cdot M_E \cdot M \cdot \cos\left(\Delta\varphi_k^{EG} - \varphi\right)\right\} \quad (5.17)$$

where M and φ are the modulation factor and phase lag of the fluorescence signal for the fundamental frequency.

Note that the factor of 2 in Equation 5.17 results from the way we have defined the real amplitudes. That is, we defined the complex amplitudes as $E_{\omega,n}$ and $G_{\omega,n}$ to be the Fourier coefficients of the complex Fourier expansion of the repetitive waveforms. Then, the real amplitudes are $2 \times |E_{\omega,n}|$ and $2 \times |G_{\omega,n}|$. This definition is convenient for working with complex notation.

Other homodyne FLIM reports (e.g., see Buranachai et al. 2008), which ignore the overtones for simplicity of presentation, derive Equation 5.17 with a factor of 0.5 rather than 2. The two derivations are reconciled by the fact that the magnitudes of the amplitudes of the real portion of the Fourier series in complex notation are defined as $2 \times |A_{\omega,n}|$. In Figure 5.1, we denote the magnitudes of the amplitudes as $2 \times |A_{\omega,n}|$ to be consistent with our chosen notation.

The DC values of $S_{\text{homodyne},k}$ are recorded for each pixel for each of k phase offsets—$\Delta\varphi_k^{EG}$. That is, the signal is integrated at K equally spaced phase offsets, $\Delta\varphi_k^{EG} = 2\pi k/K$ (where $k = 0, 1, 2\ldots$). Variation of $\Delta\varphi_k^{EG}$ is done by varying the phase of either the excitation light or detector gain modulation. The signal is integrated at each phase setting for a time period much longer than the period of the HF modulation, averaging the signal. The time constant of the acquisition electronics can have a very low frequency response because the signal at each phase setting is DC. Thus, the final data set for each pixel is a series of k phase-selected data points. The phase-selected DC measurements are fit at each pixel to calculate the demodulation and phase shift images (as explained later). It is not necessary that the phase settings be equally spaced, but this speeds up the analysis considerably (see Section 5.3).

For heterodyne FLIM, the detector is modulated at the frequency $\omega + \Delta\omega$ when the excitation is modulated at a frequency ω, where $\Delta\omega$ is a low frequency in the kilohertz (or less) range. The resulting signal—$S_{\text{heterodyne}}$—is modulated in time at the low frequency $\Delta\omega$:

$$S_{\text{heterodyne}}(t) = Q \cdot \left\{F_0 \cdot G_0 + 2 \cdot G_\omega \cdot F_0 \cdot M_E \cdot M \cdot \cos\left(\Delta\omega t - \varphi + \Delta\varphi^{EG}\right)\right\} \quad (5.18)$$

Thus, the detector output can be sampled at the lower cross-correlation frequency, $\Delta\omega$, independently of the excitation frequency. The LF signal can be sampled by a data acquisition system with sliding time windows ($\Delta t \ll 2\pi/\Delta\omega_n$) at various time delays in a boxcar fashion.

5.2.4.2 Multifrequency FLIM

Multiple fluorescence lifetime values can be obtained by scanning the fundamental frequency; that is, the single-frequency FLIM measurement is repeated for a number of frequencies. Such sequential measurements are time consuming and a minimum of M

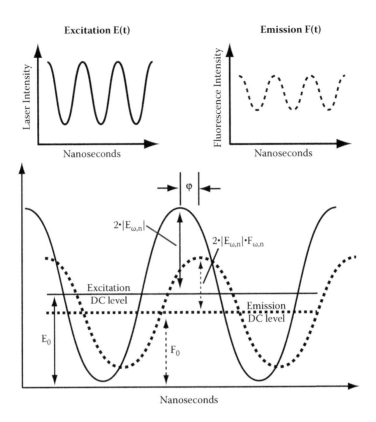

FIGURE 5.1 The frequency-domain FLIM signal. The fluorescence emission is demodulated and phase lagged relative to the phase and modulation depth of the excitation light waveform.

frequency measurements is required to discern *M* lifetime components (Jameson et al. 1984; Weber 1981b). Nevertheless, frequency scanning can be employed to generate a frequency dispersion curve (measured demodulation and phase values vs. frequency), which can be analyzed to extract the fluorescence lifetimes using multifrequency fitting techniques (Gratton et al. 1984; Jameson and Gratton 1983; Jameson et al. 1984; Lakowicz et al. 1984; Piston et al. 1989).

Acquiring multiple frequencies sequentially slows down the image acquisition, and the sample is exposed to longer periods of excitation light, which can lead to significant photobleaching of microscopic samples. Alternatively, multiple frequencies can be recorded simultaneously by taking advantage of multiple harmonics when the excitation light is not modulated as a single sinusoidal wave. Such "multifrequency FLIM" measurements circumvent the need for sequential scanning of the frequency. In the following, we briefly discuss homodyne and heterodyne FLIM measurements utilizing multiple harmonics.

5.2.4.3 Homodyne Multifrequency FLIM
In the *homodyne* implementation of multifrequency FLIM, the specific harmonic frequencies present in the image intensifier match those of the excitation light (Squire, Verveer, and Bastiaens 2000). Due to the nonlinear gain of the image intensifier, the light modulation

passing through it contains multiple higher harmonics even when the modulated photocathode voltage is sinusoidal (it is approximately a square wave; Squire et al. 2000). In the work by Squire and colleagues, the excitation laser light is passed through a pair of acousto-optic modulators in series, and harmonics are introduced into the laser intensity modulation to match that of the image intensifier gain. Just as for a single-frequency homodyne experiment, the DC fluorescence signal is recorded for k phase offsets, $\Delta\varphi_{n,k}^{EG}$, in order to extract the demodulation and phase shift. The multifrequency homodyne signal—$S_{\text{MF-homodyne},k}$—is then

$$S_{\text{MF-homodyne},k} = \left\{F(t) \cdot G(t)\right\}_{LF}$$

$$= Q \cdot \left\{F_0 \cdot G_0 + 2 \cdot \sum_{n=1}^{\infty} |G_{\omega,n}| \cdot F_0 \cdot M_{E,\omega,n} \cdot M_{\omega,n} \cdot \cos\left(\Delta\varphi_{n,k}^{EG} - \varphi_n\right)\right\} \tag{5.19}$$

The resulting signal includes multiple harmonics (at least four harmonics; Squire et al. 2000) and the k phase-selected measurements are used to form a frequency dispersion curve. Although the multifrequency, homodyne FLIM measurement is much faster than carrying out sequential frequency scans, it is obviously more time consuming than single-frequency FLIM. The resolution of the multiple harmonics requires that the density of phase points be sufficient to satisfy the Nyquist criterion for the highest harmonic, which means longer times to collect these additional phase offsets in comparison to single-frequency homodyne FLIM.

Thus, there is always a trade-off between data acquisition speed and resolution of the fluorescence lifetime components. Nevertheless, multifrequency FLIM produces additional information, and if one can afford the decrease in image acquisition speed, the acquisition of the extra frequencies can be a valuable asset for understanding the system under study.

5.2.4.4 Heterodyne Multifrequency FLIM

The *heterodyne FLIM* multifrequency signal, $S_{\text{MF-heterodyne}}$, contains n harmonic HF frequencies at the cross-correlation frequency $\Delta\omega_n$:

$$S_{\text{MF-heterodyne}}(t) = \left\{F(t) \cdot G(t)\right\}_{LF}$$

$$= Q \cdot \left\{F_0 \cdot G_0 + 2 \cdot \sum_{n=1}^{\infty} |G_{\omega,n}| \cdot F_0 \cdot M_{E,\omega,n} \cdot M_{\omega,n} \cdot \cos\left(\Delta\omega_n t - \varphi_n + \Delta\varphi_n^{EG}\right)\right\} \tag{5.20}$$

Multifrequency heterodyne fluorescence lifetime measurements can be carried out with a digital-acquisition system to digitize the voltage signal immediately following the PMT/APD output. The frequency filtering is then done using software rather than hardware. Analog heterodyne systems were used for the first time-resolved fluorometers and digital systems were introduced later as a convenient method for multifrequency fluorometry (Feddersen, Piston, and Gratton 1989). A digital, multifrequency heterodyne FLIM method has recently been developed by Coyler and colleagues (2008). The digital FLIM system acquires multiple harmonics in parallel from a single point of the sample using a PMT or APD detector, and it has a photon efficiency and accuracy comparable to time-correlated single-photon counting methods.

5.3 EXTRACTING THE DEMODULATION AND PHASE SHIFT VALUES USING A DIGITAL FOURIER TRANSFORM

A very fast data analysis routine, requiring only a single pass over the data, for frequency-domain FLIM is achieved by using a discrete Fourier transform (DFT) to determine the parameters for the different frequency components. The DFT is used to fit the fluorescence emission signal at each pixel, treating each of the frequency components individually and independently. In the case of multifrequency fluorescence lifetime data, which can contain many harmonics, the fast Fourier transform (a fast algorithm for carrying out the DFT; Brigham 1974) can be used to determine all of the frequency components simultaneously. In any case, it is crucial to sample the data at a frequency at least twice the highest frequency desired, in order to satisfy the Nyquist criterion so as to avoid aliasing.

The discrete time/phase data points gathered for the hetero/homodyne signal, $S_{\text{hetero/homodyne},k}$, should be uniformly spaced over the full repetition period (2π) in order to extract the modulation and phase values properly. The DFT for the nth frequency component is then applied at each pixel:

$$F_{\sin,\omega,n} = \sum_{k=0}^{K-1} \left\{ \sin(2\pi nk/K) \cdot S_{\text{hetero/homodyne},k} \right\} \tag{5.21}$$

$$F_{\cos,\omega,n} = \sum_{k=0}^{K-1} \left\{ \cos\left(2\pi nk/K\right) \cdot S_{\text{hetero/homodyne},k} \right\} \tag{5.22}$$

and

$$F_0 = \frac{1}{K} \cdot \sum_{k=0}^{K-1} S_{\text{hetero/homodyne},k} \tag{5.23}$$

where K represents the total number of recorded data points, $S_{\text{hetero/homodyne},k}$ is the intensity of the pixel at the kth time/phase point, and n is the harmonic of interest. The phase and modulation are then calculated by applying the following equalities:

$$\Delta\varphi_n^{EG} - \varphi_n = \tan^{-1}\left(F_{\sin,\omega,n}/F_{\cos,\omega,n}\right) \tag{5.24}$$

and

$$\frac{2 \cdot |G_{\omega,n}| \cdot M_{E,\omega,n} \cdot M_{\omega,n}}{G_0} = \frac{\sqrt{F_{\sin,\omega,n}^2 + F_{\cos,\omega,n}^2}}{F_0} \tag{5.25}$$

In order to compute the demodulation, $M_{\omega,n}$, and phase lag, ϕ_n, the values for $2 \times |G_{\omega,n}|$ $\times M_{E,\omega,n}/G_0$ and $\Delta\varphi_n^{EG}$ are needed. A simple way to measure these parameters is to perform a calibration using a reference of known fluorescence lifetime (e.g., fluorescein in 0.1 N NaOH) or by measuring reflected excitation light. After the calibration procedure the demodulation and phase lag images may be determined. The data set can be fit pixel by pixel to extract the demodulation (M) and phase lag (ϕ) images, or the data can be binned (locally averaged) before the analysis. If it can be assumed that the lifetime information is the same throughout the image (or throughout a chosen locality), the data can be globally analyzed (see Chapter 13). Finally, these M and ϕ images can then be further manipulated to determine the fluorescence lifetimes at each pixel. For instance, the polar plot projection is calculated from these images (see Chapter 11 for details regarding the polar plot projection).

5.4 VIDEO-RATE FLIM

5.4.1 Overview

Full-field homodyne FLIM is a method for imaging fluorescence lifetimes at every pixel of an image, simultaneously employing a gain-modulated image intensifier with a CCD camera detector. Full-field methods have been developed for the frequency domain (Gadella, Jovin, and Clegg 1993; Morgan, Mitchell, and Murray 1990; Redford and Clegg 2005), as well as using time-domain fast-gated image intensifiers (e.g., Elson et al. 2004; Galletly et al. 2008). These techniques are capable of capturing multiple FLIM images per second (Clegg, Holub, and Gohlke 2003; Gadella et al. 1993; Redford 2005; Redford and Clegg 2005), and video-rate image acquisition is possible if a single modulation frequency is used.

Video-rate FLIM is of particular interest for applications requiring rapid acquisition/ analysis, such as recording dynamic processes in live cells. In contrast, laser-scanning FLIM with a single-channel detector typically requires at least several seconds per image for the data acquisition, depending on the signal-to-noise level. Video-rate imaging is possible in the time domain by sampling only a few delay times following the excitation pulse.

For *homodyne FLIM*, as explained earlier and illustrated in Figure 5.2, DC fluorescence images are recorded for several phase offsets, $\Delta\varphi_k^{EG}$, in order to extract the demodulation and phase lag values, which we refer to as *phase-selected fluorescence images*. The phase-selected images are recorded over an exposure time that integrates many cycles of the fluorescence signal. This is convenient for measuring weak signals; that is, the exposure time may be adjusted to increase the signal-to-noise ratio. Typical exposure times for acquiring a single frame are on the order of 100–1,000 μs or until the desired signal-to-noise ratio is achieved. Typically, three to eight phase-selected images are acquired. As already mentioned, the number of points sampled along the signal is a compromise between the time required for the data acquisition and the proper measurement of the phase to avoid the aliasing effects (Van Munster and Gadella 2004).

5.4.2 Instrumentation

5.4.2.1 Illumination Sources and Electro-optics for Modulated Excitation Light

A number of illumination sources are compatible with homodyne FLIM, including: (1) solid-state, gas, and dye lasers; (2) laser diodes (LDs); and (3) light-emitting diodes (LEDs).

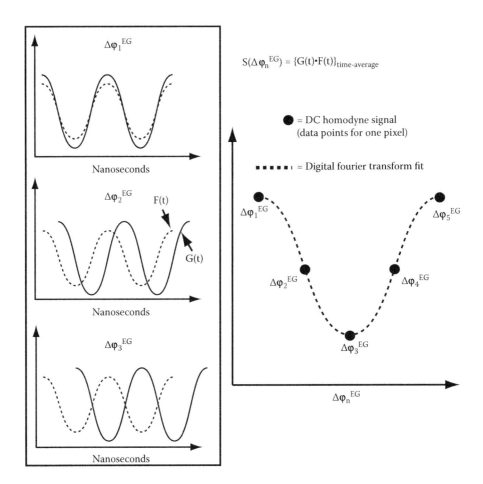

FIGURE 5.2 Illustration of the homodyne FLIM measurement for a single pixel. The box on the left shows three different phase offsets between the image intensifier gain and the excitation light. On the right, the measured signal for homodyne FLIM is shown for a series of phase offsets. The resulting signal, for each pixel, is fit using a single pass of a digital Fourier transform (DFT) to extract the phase and modulation values.

Modulated LED and LD light sources are new to FLIM and are now commercially available (Lambert Instruments, Roden, the Netherlands; ISS, Champaign, Illinois). Modulated LEDs are a welcome alternative to lasers because many different optical wavelengths are available for a fraction of the cost. Moreover, modulated LEDs and LDs are modulated directly by a time-varying voltage signal applied to the diode power supply; thus, the light modulation can be easily controlled without need for external modulation of the light. In addition, high-repetition-rate pulsed lasers naturally contain a rich content of harmonics, so external modulation is not required (see Chapter 11).

Solid-state, gas, and dye continuous-wave lasers (constant amplitude) are modulated externally using either an acousto-optic modulator (AOM) (Gadella et al. 1993) or an electro-optic modulator (EOM) (Redford and Clegg 2005). For instance, the Pockels cell is an EOM constructed from a birefringent crystal, which rotates polarized light when exposed to an electric field, placed between two perpendicular polarizers. Thus, the Pockels cell is a voltage-controlled wave plate. The light intensity output is varied in time by applying a sinusoidal voltage to the Pockels cell.

5.4.2.2 Gain-Modulated Image Intensifiers

Full-field FLIM instruments usually employ an image intensifier coupled to a CCD camera detector for full-field imaging. The data acquisition involves the following steps in the image intensifier (Figure 5.3):

1. Incident photons fall on the photocathode of the image intensifier.

2. The ejected photoelectrons are directed into the MCP by a negative photocathode voltage, where they undergo a large multiplication to amplify the signal (via a 900 V gradient across the MCP).

3. The amplified electron pulses emitted from the MCP bombard a phosphor screen (through a large voltage gradient between the output of the MCP and phosphor screen), which emits photons.

4. The photons emitted by the phosphor screen are focused onto the CCD camera using relay optics (permanent fiber optic coupling of the MCP to the CCD is also used).

In this way, the photoelectrons are amplified and converted back to photons.

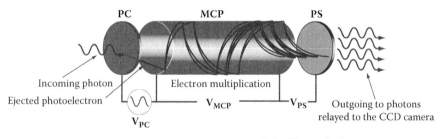

PC = Photocathode
MCP = Microchannel Plate
PS = Phosphor Screen

FIGURE 5.3   Schematic for a single channel of the microchannel plate (MCP) within the image intensifier. An incoming photon is incident on the photocathode (PC) and generates a photoelectron, which is amplified in the MCP. The electron multiplication by the MCP results in many electrons striking the phosphor screen (PS), where the electrons are converted back to photons and then relayed to the CCD. There is substantial noise in the photon gain; that is, the electron multiplication process is noisy (the number of electrons produced from each collision within the MCP has some variance) and results in the so-called exponential pulse height distribution. Thus, the noise characteristics of the ICCD contain both photon noise and substantial detector noise. The voltage applied to the PC (V_{PC}) oscillates to modulate the gain of the detector. The voltage difference from end to end of the MCP (V_{MCP}) is normally several hundred volts to accelerate the electrons along the channel. Finally, tens of thousands of volts are applied between the MCP and PS (V_{PS}) to convert the outgoing electrons into photons.

An HF repetitive voltage signal (often an approximate sinusoid) is applied to the photocathode to modulate the photon gain of the intensifier. For instance, the photoelectrons ejected from the photocathode (by the photoelectric effect) are repelled from the MCP by a positive voltage on the cathode relative to the MCP (zero gain) and are propelled into the MCP for negative cathode voltages (high gain). Finally, FLIM images are acquired by phase shifting the gain of the detector with respect to the excitation light, while modulating both at the same frequency. Every phase image is integrated on the CCD camera, and then the data are transferred to the computer. The FLIM image is constructed and analyzed from the series of phase-delayed images.

5.4.2.3 Optical Setup and Electronics

Figure 5.4 is a schematic of our video-rate FLIM setup. A standard epifluorescence microscope is the base of the optical setup, with a dichroic mirror used to reflect the excitation light and pass the fluorescence emission through the microscope relay optics. A frequency generator outputs a sinusoidal voltage, which is split and used to drive the Pockels cell and

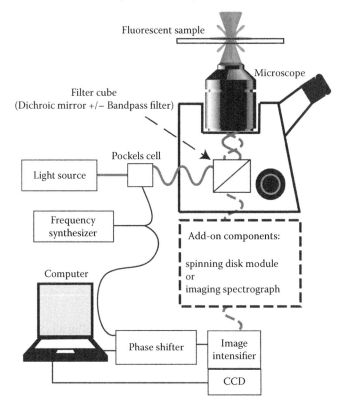

FIGURE 5.4 Rapid, full-field FLIM in the frequency domain with a single frequency (homodyne method) and a top speed of ~30 FLIM images per second (90 frames per second CCD frame rate and a minimum of three phase images per FLIM image). A Pockels cell modulates the laser intensity at high frequency (20–100 MHz). The detection system includes a gain-modulated image intensifier and a CCD camera. The image intensifier is modulated at the same frequency as the laser. A high-frequency digital phase shifter phase shifts the intensifier gain modulation with respect to the laser modulation during the FLIM measurement.

the image intensifier. A digital phase shifter is used to shift the phase of the voltage that reaches the image intensifier relative to the phase sent to the Pockels cell.

5.4.3 Corrections for Random Noise and Systematic Errors

5.4.3.1 Correcting for Laser Fluctuations and Dark Current

Fast fluctuations in the laser intensity and a slow drift in modulation depth during FLIM experiments can introduce random noise and systematic error into the FLIM data. To correct for the random laser fluctuations automatically, reference phase frames are acquired immediately before and after each phase measurement. That is, images collected at a reference phase shift are averaged together to normalize each phase image and therefore take into account any fluctuations in the excitation light intensity during the FLIM image acquisition. For instance, a typical FLIM image stack contains 24 total phase-selected images: eight phase shifts multiplied by three images per phase shift (two reference images plus one image per phase shift). A photomultiplier tube is used to record the modulation depth of the laser beam throughout the FLIM measurements and to correct for drift in the Pockels cell and the laser modulation. For extended FLIM experiments, it is also advisable to perform the calibration periodically (e.g., every half hour) with the fluorescence lifetime standard.

A background image is acquired (a series of dark frames by repeating the data collection with the sample shutter closed to block light from entering the detector, under the same experimental settings and exposure time) for each FLIM experiment. The dark frames (which also contain readout noise of the CCD) are subtracted from the FLIM image to subtract the dark current and the CCD readout noise. This subtraction removes the background from the final image. Dark current is generated by the image intensifier and the CCD. Applying the dark frame subtraction and reference phase image corrections to the data ensures the accuracy and reproducibility of the measurements.

5.4.3.2 Gain-Modulated Image Intensifier Performance

Signal-to-noise ratio. A major source of random noise in FLIM measurements is photon noise and detector noise. The image intensifier adds significant noise by amplifying the photon noise, and the noise associated with electron multiplication within the MCP is added to the photon noise. The signal-to-noise ratio (SNR) of an ICCD (image-intensified CCD) detector has been described previously (Frenkel, Sartor, and Wlodawski 1997). The incident radiation follows Poisson statistics, an inherent source of noise due to the quantum nature of light. The intensifier noise is the dominant noise source of the detection system when using an intensifier (Mitchell et al. 2002a, 2002b).

CCD noise (e.g., readout noise, dark current, and digitization error) is often assumed to be negligible in comparison to the photon noise and intensifier noise. Following the derivation in RCA (1984) and following error propagation for a cascade of sequential events, the SNR of the image intensifier as a whole is expressed as

$$\text{SNR} \approx \sqrt{\eta \cdot \alpha \cdot N_0 \left/ \left(1 + \left(\frac{\sigma_g}{g}\right)^2\right)\right.} \qquad (5.26)$$

where N_0 is the mean rate of photons incident upon the photocathode of the intensifier, g is the effective gain of the MCP, which is the gain of the MCP times the quantum efficiency of the phosphor with variance σ_g^2; η is the quantum efficiency of the photocathode; and α is the fraction of photoelectrons that undergo electron multiplication inside the MCP (i.e., α is modulated by the voltage applied to the photocathode of the intensifier and this controls the gain modulation). The effective intensifier gain is then

$$G(t) = \alpha(t) \cdot \eta \cdot g \tag{5.27}$$

which is the photon gain or, equivalently, the detector gain—see Equation 5.15 (that is, the ratio of the number of photons emitted from the phosphor per incident photon on the photocathode).

Notably, the quantum efficiency of the photocathode, η, for converting incoming photons into photoelectrons degrades the SNR (RCA 1984). (We assume that α is approximately unity at the maximal negative voltage amplitude.) Effectively, the MCP sees a reduction in particle counts due to the low quantum efficiencies of photocathodes (typical values for η are ≤0.4) (RCA 1984). Thus, Equation 5.26 matches intuition that the SNR should decrease following a reduction in the particle counts (in accordance with Poisson statistics). As expected, the signal amplification further degrades the SNR; note the $(\sigma_g/g)^2$ term in the denominator of Equation 5.26.

Noise probability model. Sandel and Broadfoot (1986) derived a probability model for intensified CCD cameras. The probability model accounts for both the photon noise and the intensifier noise. In similar fashion to a photomultiplier tube running in analog mode, the intensifier gain follows an exponential pulse height distribution. The conditional probability to detect d digital levels from the CCD's analog-to-digital converter, given a mean photon signal rate N_0, is

$$C(d|N_0) = \sum_{N=0}^{\infty} P(N|N_0) \cdot S(d|N) \tag{5.28}$$

where

$$P(N|N_0) = \frac{N_0^N \cdot \exp(-N_0)}{N!} \tag{5.29}$$

is the Poisson conditional probability distribution for the actual measured photon signal rate, N. $S(d|N)$ is given by (see Sandel and Broadfoot (1986) for the derivation)

$$S(d|N) = \int_{q_d}^{q_{d+1}} \frac{c^{N-1} \cdot \exp(-c/G)}{(N-1)! \cdot G^N} \cdot dc \tag{5.30}$$

and is the integral of the exponential pulse height distribution over c, the charge accumulated on the CCD pixel during the exposure. G is the effective gain (i.e., the photon gain) of the intensifier. The integration is done over the range of c that falls into the digital level

d (q_d and q_{d+1} are the minimum charges for digitization to levels d and $d + 1$). The sum in Equation 5.28 is necessary to marginalize over the number of photons, N.

For $N_0 > 100$, the integral of Equation 5.30 can be approximated as

$$S\left(d|N\right) \approx \frac{c^{N-1} \cdot \exp\left(-c/G\right)}{(N-1)! \cdot G^N} \cdot \Delta c \tag{5.31}$$

which corresponds to typical homodyne FLIM measurements, as tested by comparing precise distributions calculated by evaluating the integral of Equation 5.30 with approximate distributions $C(d|N_0)$ calculated from Equation 5.31 (not shown). This approximation is very convenient for simulating $C(d|N_0)$ at larger values of N_0, which is otherwise computationally time consuming. The number Δc is the CCD gain (the number of electrons per digital level).

Calculation of the probability distribution for typical values of d reveals that the distribution is substantially broader than a Poisson distribution. In fact, the ratio of the variance to the mean is >1 for many operating conditions. In contrast, the variance is equal to the mean for a Poisson distribution. Thus, the intensifier's pulse height distribution degrades the Poisson statistics of pure photon noise.

We implemented a simplification that further reduces the computations needed to generate the distribution. We find that the entire probability distribution $C(d|N_0)$ can be approximated using a Gaussian approximation for $N_0 > 100$. The Gaussian approximation is convenient for image analysis to remove the signal-dependent photon and intensifier noise (see Chapter 11).

A noteworthy outcome of the noise model is that the photon gain should be as low as possible for homodyne FLIM. Clearly, raising the photon gain increases the amount of uncertainty in the FLIM data. The increase in noise is easily visible on the polar plot at higher photon gain. However, the most rapid acquisition is achieved at maximal photon gain and minimal exposure times. Thus, a trade-off exists between speed of acquisition and level of noise (without image analysis to remove the noise).

5.5 ENHANCED FLIM MODES

5.5.1 Video-Rate Confocal FLIM

Reflected and scattered light can compromise image quality for FLIM, especially because the fluorescence lifetime data are calculated for all signal levels—dim regions as well as high-intensity locations of the image. Confocal imaging enhances the accuracy of FLIM by localizing the fluorescence lifetime values and rejecting out-of-focus light. Confocal imaging is traditionally carried out with a raster-scanned confocal laser beam (confocal laser scanning microscopy; see Chapter 6), which is not compatible with video-rate FLIM applications.

Multibeam, confocal imaging is possible using Nipkow disks (Kawamura et al. 2002; Nakano 2002), which means simultaneous rotation of a microlens and a corresponding pinhole disk (i.e., "spinning disk confocal imaging"). Full-field FLIM with a Nipkow disk has recently been demonstrated for both time-domain (Grant et al. 2005) and homodyne frequency-domain FLIM (Buranachai et al. EPub available online; van Munster et al. 2007). The spinning disk arrangement retains the speed of the normal two-dimensional full-field FLIM while gaining true three-dimensional resolution (Buranachai et al. EPub available online).

Figure 5.5 shows a schematic of the spinning disk setup, for which thousands of excitation beams are scanned in parallel across the sample. The multibeam excitation results in multiple beams of fluorescence emission, which are scanned across the full-field detector to build up the image. The microlens/pinhole pairs are arranged in a spiral pattern such that a full-image is scanned every 30° of rotation. Thus, a rotation rate of 30 Hz results in 360 confocal images per second. The confocal spinning disk arrangement is compatible with FLIM because the residence time of the emission beam on a pixel (~100 μs; Wang, Babbey, and Dunn 2005) is much longer than the nanosecond period of a radio-frequency gain-modulated image intensifier (for frequency-domain FLIM) or the nanosecond "on time" of a fast-gated image intensifier (for time-domain FLIM).

5.5.2 Rapid Spectral FLIM

Spectral FLIM resolves the spectrum of fluorescence at every pixel within the sample, as well as the fluorescence lifetimes. A variety of spectral FLIM instruments have been developed for both frequency-domain and time-domain instruments (Bird et al. 2004; De Beule,

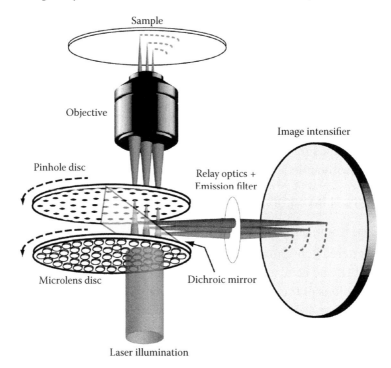

FIGURE 5.5 A schematic of the video-rate, confocal FLIM setup. Approximately 1,000 confocal beams are scanned simultaneously by the microlens and pinhole discs of the spinning disk confocal unit.

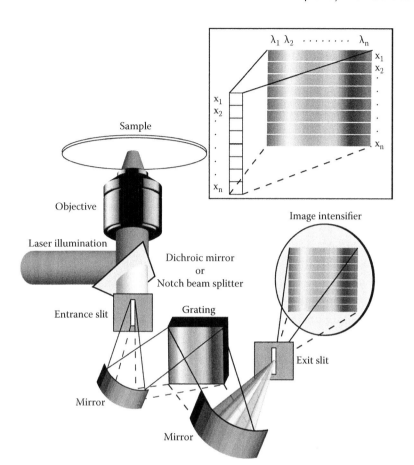

FIGURE 5.6 **(See color insert following page 288.)** A schematic of the spectral FLIM setup. The fluorescence emission is passed through the dichroic mirror and dispersed into separate wavelengths by a diffraction grating. A mirror focuses the spectra from a line of pixels from the sample (see the box inset, top right) onto the image intensifier. Thus, the final image contains the fluorescence lifetime for each wavelength of the fluorescence spectrum from a line of pixels along the sample. Line scanning by a programmable stage can be used to build a complete image of the sample (i.e., hyperspectral FLIM).

Owen, et al. 2007; Forde and Hanley 2006; Hanley, Arndt-Jovin, and Jovin 2002; Pelet, Previte, Kim et al. 2006). The fastest spectral FLIM, just as for normal FLIM, is with full-field detection (De Beule, Owen, et al. 2007). Figure 5.6 is a schematic of our homodyne, spectral FLIM setup. An imaging spectrograph is placed before the detector element and its entrance slit selects a line from the image of the sample. The emission from each point of the line is dispersed into its wavelength components by the diffraction grating. Finally, the full-field detector captures the spectra along the horizontal direction from the individual spatial points, which span the vertical direction (see the inset of Figure 5.6).

Applications of spectral FLIM include imaging tissue autofluorescence (De Beule, Dunsby, et al. 2007) and imaging FRET in live cells (Forde and Hanley 2006). Spectral-based resolution of individual, endogenous fluorophores in tissues increases the throughput for multifluorophore FLIM. That is, the ability to measure multiple wavelengths simultaneously greatly expedites the measurement time relative to sequential measurements with band-pass filters. Thus, spectral FLIM could be a key development for high-throughput characterization of tissues and for recognition of disease. The spectra aid in the separation of several fluorescence species, even if they have different lifetimes.

In regard to FRET imaging, a recent paper by Pelet, Previte, and So (2006) has thoroughly and directly compared the measurement of FRET efficiencies in a microscope using intensity, spectral, and fluorescence lifetime imaging separately. Pelet et al. review the numerous corrections required for determining the FRET efficiency using intensity-based measurements. These corrections include proper consideration of spectral bleed-through, fluorescence quantum yields and extinction coefficients. Pelet et al. conclude that FLIM is the best modality; moreover, FLIM is the only modality that can report on the intensity fraction of the molecules undergoing FRET. This information is invaluable for development of FRET-based assays.

Spectral FLIM furthers the capabilities and advantages of FLIM-based FRET imaging because the FRET donor and acceptor molecules' fluorescence can be completely separated and their fluorescence lifetimes can be measured simultaneously. Both the FRET donor and acceptor fluorescence lifetimes can be used to calculate FRET efficiency. A description of the data analysis and display for spectral FLIM, also in conjunction with the polar plot, can be found in Chapter 11. Spectral FLIM is a powerful approach for fully characterizing a FRET system, and implementing multifrequency spectral FLIM would be an exciting extension. Once a rigorous characterization is performed and the fluorescence lifetime species are identified, it should be possible to interpret dynamic FRET experiments by video-rate FLIM.

5.6 DATA DISPLAY

5.6.1 Dual-Layer FLIM Images

A common way to present FLIM images is to overlay the intensity and FLIM data. This is a lucid way to present the data because the FLIM image contains a fluorescence lifetime at each pixel, regardless of the signal-to-noise ratio of the pixel. The intensity image can be used as a masking layer to block out the dark pixels, which do not convey useful information (if they are too noisy). The intensity masking layer then determines the brightness of the pixels in the dual-layer image. The FLIM data are displayed using a color scale. Thus, the dual-layer image is composed of the intensity layer (brightness) and the fluorescence lifetime image (hue).

A new way to color code FLIM images is by using the polar plot projection (see Chapter 11). Polar-plot-based color coding results in easy visual inspection of the fluorescence lifetime data without need for a model of the number of fluorescence lifetimes or iterative fits. The combination between a polar plot projection and color can be tuned to the application. Usually, shifts in fluorescence lifetimes are the main interest in FLIM. An easy way to display relative shifts in fluorescence lifetimes is by assigning a color to a specific point on the polar plot. Then color can be used to report on the distance of each pixel in the FLIM

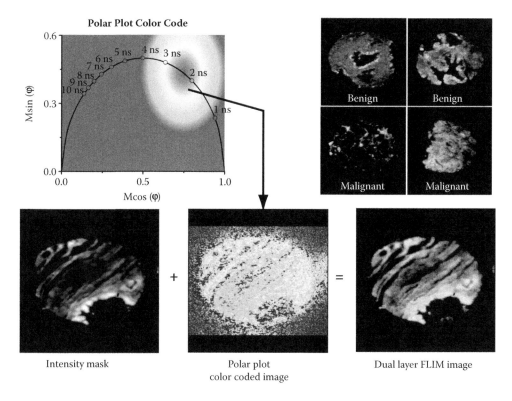

Polar Plot Color Code

Intensity mask   +   Polar plot   =   Dual layer FLIM image
                                                color coded image

FIGURE 5.7 **(See color insert following page 288.)** Dual-layer FLIM images where an intensity image is used to mask the color-coded lifetime image. The color scheme for the color-coded fluorescence lifetime image is derived from the polar plot projection (top left plot). Pixels with values near 2.3 ns are colored red. Pixels located far from 2.3 ns are assigned a blue color. Thus, each pixel of the FLIM image is assigned a color based on where that pixel is located on the polar plot. Blue indicates normal fluorescence lifetimes of prostate tissue; red indicates a significant shift in fluorescence lifetimes that may indicate malignancy according to preliminary studies. The upper right images are dual-layer FLIM images of a few benign and malignant prostate tissue cores from a formalin-fixed tissue microarray.

image from the point, and it is easily and rapidly recognized by the investigator. Figure 5.7 provides an example of the dual-layer FLIM image with a color scheme derived from the polar plot space.

5.6.2 Dual-Layer Fractional Concentration Images

It is very useful to create fractional concentration images to show where specific fluorescence lifetime species are located in an image. Such images can be computed from FLIM data because the contribution of each fluorescent lifetime species to the total modulation and phase is determined by its fractional intensity (see Equation 5.7). This calculation is straightforward from the data produced by multifrequency FLIM (with the assumption that the relative fluorescence quantum yields are approximately the ratio of the fluorescence lifetimes). To create the dual-layer fractional concentration image, each fluorescence lifetime component is assigned a color and the fractional concentrations of the various

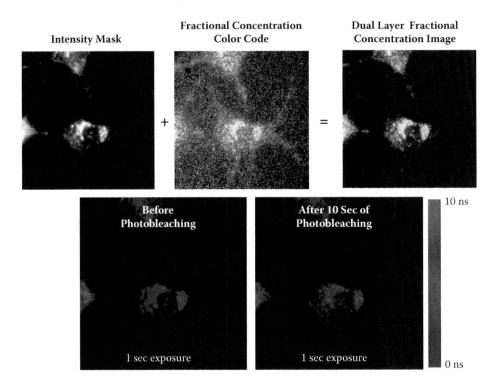

FIGURE 5.8 **(See color insert following page 288.)** Top: overlay of the intracellular locations and concentrations of PpIX (protoporphyrin IX, a common photosensitizing agent for photodynamic therapy) monomer and dimer in live 3T3 cells following treatment with aminolevulinic acid (ALA, a precursor to PpIX). The concentration of monomer is in the red channel of the image; the concentration of dimer is in the green channel. Yellow regions have both monomer and dimer. The image intensity corresponds to the total fluorescence intensity, acting as a mask to show the structure of the cells. Bottom: Dual-layer fluorescence lifetime images of PpIX in live 3T3 cells incubated with ALA. The hue represents the modulation lifetime (i.e., calculating the fluorescence lifetime from the overall measured demodulation value by solving Equation 5.13) from 0 ns (blue) to 10 ns (red), and the intensity is proportional to the fluorescence intensity (autoscaled). The left images are before photobleaching, and the right are after 10 s of photobleaching. Red regions correspond to regions with a higher ratio of monomer; interestingly, the ratio of monomer to dimer PpIX increases with photobleaching. These results can only be obtained from video-rate FLIM because both the monomer and dimer have similar emission spectra and rapid image acquisition is required to resolve the monomer and dimer redistribution dynamics. This unexpected result has important consequences for photodynamic therapy (using ALA—a common therapeutic agent); that is, the PpIX monomer is necessary for producing free radicals to kill tumor cells.

fluorescence lifetime species determine the overall color (e.g., see Figure 5.8). As for the dual-layer FLIM images, an intensity image is used to mask the data.

For single-frequency FLIM, fractional concentrations can only be calculated if the fluorescence lifetimes are known *a priori* or can be measured individually in separate experiments. Redford (2005) has defined the minimal information or assumptions needed to solve the system of equations that determines the fractional concentrations for single-frequency FLIM.

5.7 SUMMARY

We have provided a general description of frequency-domain FLIM and a detailed introduction to video-rate FLIM. The advantages of full-field FLIM are its video-rate imaging speed and model-free, rapid data analysis using the DFT and polar plot projection. Thus, full-field FLIM should be applicable for real-time FRET imaging and for endoscopy and clinical applications of fluorescence lifetime imaging. Recent advances in full-field FLIM include the implementation of video-rate, three-dimensional confocal FLIM with a confocal spinning disk module and fast spectral FLIM with an imaging spectrograph and line scanning.

Future advances and improvements in full-field FLIM will be significantly aided by coupling specific FLIM data analysis, such as that discussed in this chapter, with image analysis methods that have been found valuable and are routinely applied for intensity image analysis, such as those discussed in Chapter 11. In Chapter 11, we demonstrate how image analysis techniques applied in conjunction with frequency-domain FLIM data analysis (to determine the lifetimes and fractional species or, equivalently, the phases and modulation) can greatly enhance the accuracy and separation of the fluorescence lifetime components. For example, there is substantial noise from the image intensifiers, and image analysis can be used to filter out specifically this noise. Better photon efficiency and statistics can be obtained using photon-counting detectors for FLIM. This is advantageous for measuring dim samples.

Until now, multiphoton excitation has been carried out in scanning mode. However, real-time multiphoton excitation with arrays of microlenses (Fujita et al. 1999; Kim et al. 2007) or with a spinning disk (Bewersdorf, Pick, and Hell 1998) is available and could potentially be extended for video-rate FLIM. That is, it may be possible to carry out multiphoton excitation for full-field, homodyne FLIM using a microlens array similar to the present one-photon excitation spinning disk arrangement. Another potentially exciting development for frequency-domain FLIM is the use of modulated LEDs to lower the cost and complexity of the instrumentation and increase flexibility.

REFERENCES

Alcala, J. R., Gratton, E., and Prendergast, F. G. 1987. Resolvability of fluorescence lifetime distributions using phase fluorometry. *Biophysical Journal* 51:587–596.

Bailey, E. A., and Rollefson, G. K. 1953. The determination of the fluorescence lifetimes of dissolved substances by a phase shift method. *Journal of Chemical Physics* 21:1315–1322.

Bewersdorf, J., Pick, R., and Hell, S. W. 1998. Multifocal multiphoton microscopy. *Optics Letters* 23:655–657.

Bird, D. K., Eliceiri, K. W., Fan, C-H., and White, J. 2004. Simultaneous two-photon spectral and lifetime fluorescence microscopy. *Applied Optics* 43:5173–5182.

Birks, J. B. 1970. *Photophysics of aromatic molecules*. London: Wiley.

Birks, J. B., and Dawson, D. J. 1961. Phase and modulation fluorometer. *Journal of Scientific Instruments* 38:282–295.

Birks, J. B., and Little, W. A. 1953. Photo-fluorescence decay times of organic phosphors. *Proceedings of the Physical Society Section A* 66:921–928.

Bracewell, R. N. 1978. *The Fourier transform and its applications*. Tokyo: McGraw–Hill Kogakusha, LTD.

Brigham, E. O. 1974. *The fast Fourier transform*. Englewood Cliffs, NJ: Prentice Hall.

Buranachai, C., Kamiyama, D., Chiba, A., Williams, B., and Clegg, R. M. EPub 2008. Rapid frequency-domain FLIM spinning disk confocal microscope: Lifetime resolution, image improvement and wavelet analysis. *Journal of Fluorescence.* 18:929–942.

Butz, T. 2006. *Fourier transformation for pedestrians.* New York: Springer.

Byron, F. W., and Fuller, R. W. 1969. *Mathematics of classical and quantum physics.* Reading, MA: Addison–Wesley Publishing Company.

Chandler, D. E., Majumdar, Z. K., Heiss, G. J., and Clegg, R. M. 2006. Ruby crystal for demonstrating time- and frequency-domain methods of fluorescence lifetime measurements. *Journal of Fluorescence* 16:793–807.

Clegg, R. M., Holub, O., and Gohlke, C. 2003. Fluorescence lifetime-resolved imaging: Measuring lifetimes in an image. *Methods in Enzymology* 360:509–542.

Clegg, R. M., and Schneider, P. C. 1996. Fluorescence lifetime-resolved imaging microscopy: A general description of lifetime-resolved imaging measurements. In *Fluorescence microscopy and fluorescent probes,* ed. J. Slavik, 15–33. New York: Plenum Press.

Coyler, R. A., Lee, C., and Gratton, E. 2008. A novel fluorescence lifetime imaging system that optimizes photon efficiency. *Microscopy Research and Technique* 71:201–213.

De Beule, P. A. A., Dunsby, C., Galletly, N. P., Stamp, G. W., Chu, A. C., Anand, U., et al. 2007. A hyperspectral fluorescence lifetime probe for skin cancer diagnosis. *Review of Scientific Instruments* 78:123101.

De Beule, P., Owen, D. M., Manning, H. B., Talbot, C. B., Requejo-Isidro, J., Dunsby, C., et al. 2007. Rapid hyperspectral fluorescence lifetime imaging. *Microscopy Research and Technique* 70:481–484.

Dern, H., and Walsh, J. B. 1963. Analysis of complex waveforms. In *Physical techniques in biological research: Electrophysiological methods, part B,* ed. W. L. Nastuk, VI, 99–218. New York: Academic Press.

Elson, D., Requejo-Isidro, J., Munro, I., Reavell, F., Siegel, J., Tadrous, P., Benninger, R., et al. 2004. Time-domain fluorescence lifetime imaging applied to biological tissue. *Photochemical and Photobiological Sciences* 3:795–801.

Feddersen, B. A., Piston, D. W., and Gratton, E. 1989. Digital parallel acquisition in frequency-domain fluorimetry. *Review of Scientific Instruments* 60:2929.

Forde, T. S., and Hanley, Q. S. 2006. Spectrally resolved frequency domain analysis of multifluorophore systems undergoing energy transfer. *Applied Spectroscopy* 60:1442–1452.

Frenkel, A., Sartor, M. A., and Wlodawski, M. S. 1997. Photon-noise-limited operation of intensified CCD cameras. *Applied Optics* 36:5288–5297.

Fujita, K., Nakamura, O., Kaneko, T., Kawata, S., Oyamada, M., and Takamatsu, T. 1999. Real-time imaging of two-photon-induced fluorescence with a microlens-array scanner and a regenerative amplifier. *Journal of Microscopy* 194:528–531.

Gadella, T. W. J., Jovin, T. M., and Clegg, R. M. 1993. Fluorescence lifetime imaging microscopy (FLIM): Spatial resolution of microstructures on the nanosecond time scale. *Biophysical Chemistry* 48:221–239.

Galletly, N. P., McGinty, J., Dunsby, C., Teixeira, F., Requejo-Isidro, J., Munro, I., et al. 2008. Fluorescence lifetime imaging distinguishes basal cell carcinoma from surrounding uninvolved skin. *British Journal of Dermatology* available 159:152–161.

Goldman, S. J. 1948. *Frequency analysis, modulation and noise.* New York: Dover.

Grant, D. M., Elson, D. S., Schimpf, D., Dunsby, C., Requejo-Isidro, J., Auksorius, E., et al. 2005. Optically sectioned fluorescence lifetime imaging using a Nipkow disk microscope and a tunable ultrafast continuum excitation source. *Optics Letters* 30:3353–3355.

Gratton, E., and Limkeman, M. 1983. A continuously variable frequency cross-correlation phase fluorometer with picosecond resolution. *Biophysical Journal* 44:315–324.

Gratton, E., Limkeman, M., Lakowicz, J. R., Maliwal, B. P., Cherek, H., and Laczko, G. 1984. Resolution of mixtures of fluorophores using variable-frequency phase and modulation data. *Biophysical Journal* 46:479-486.

Hanley, Q. S., Arndt-Jovin, D. J., and Jovin, T. M. 2002. Spectrally resolved fluorescence lifetime imaging microscopy. *Applied Spectroscopy* 56:155–166.

Jameson, D., and Gratton, E. 1983. Analysis of heterogeneous emissions by multifrequency phase and modulation fluorometry. In *New directions in molecular luminescence*, ed. D. Eastwood, 67–81. Philadelphia: American Society of Testing and Materials.

Jameson, D. M., Gratton, E., and Hall, R. 1984. The measurement and analysis of heterogeneous emissions by multifrequency phase and modulation fluorometry. *Applied Spectroscopy Review* 20:55–106.

Kawamura, S., Negishi, H., Otsuki, S., and Tomosada, N. 2002. Confocal laser microscope scanner and CCD camera. *Yokogawa Technical Report English Edition* 33:17–33.

Kim, K. H., Ragan, T., Previte, M. J. R., Bahlmann, K., Harley, B. A., Wiktor-Brown, D. M., et al. 2007. Three-dimensional tissue cytometer based on high-speed multiphoton microscopy. *Cytometery Part A* 71A:991–1002.

Lakowicz, J. R., Laczko, G., Cherek, H., Gratton, E., and Limkeman, M. 1984. Analysis of fluorescence decay kinetics from variable-frequency phase shift and modulation data. *Biophysical Journal* 46:463–477.

Maercks, O. 1938. Neuartige fluorometer. *Zeitschrift fur Physik* 109:685–699.

Mitchell, A. C., Wall, J. E., Murray, J. G., and Morgan, C. G. 2002a. Direct modulation of the effective sensitivity of a CCD detector: A new approach to time-resolved fluorescence imaging. *Journal of Microscopy* 206:225–232.

Mitchell, A. C., Wall, J. E., Murray, J. G., and Morgan, C. G. 2002b. Measurement of nanosecond time-resolved fluorescence with a directly gated interline CCD camera. *Journal of Microscopy* 206:233–238.

Morgan, C. G., Mitchell, A. C., and Murray, J. G. 1990. Nanosecond time-resolved fluorescence microscopy: Principles and practice. *Transactions of the Royal Microscopical Society* 1:463–466.

Nakano, A. 2002. Spinning-disk confocal microscopy—A cutting edge tool for imaging of membrane traffic. *Cell Structure and Function* 27:349–355.

Pelet, S., Previte, M. J. R., Kim, D., Kim, K. H., Su, T.-T. J., and So, P. T. C. 2006. Frequency domain lifetime and spectral imaging microscopy. *Microscopy Research and Technique* 69:861–874.

Pelet, S., Previte, M. J. R., and So, P. T. C. 2006. Comparing the quantification of Förster resonance energy transfer measurement accuracies based on intensity, spectral, and lifetime imaging. *Journal of Biomedical Optics* 11:034017.

Piston, D. W., Marriott, G., Radivoyevich, T., Clegg, R. M., Jovin, T. M., and Gratton, E. 1989. Wideband acousto-optic light modulator for frequency domain fluorometry and phosphorimetry. *Review of Scientific Instruments* 60:2596–2600.

RCA. 1984. *RCA photomultiplier handbook* PMT-62. Lancaster, PA: RCA.

Redford, G. I. 2005. Fast fluorescence lifetime imaging using a full-field homodyne system with applications in biology. Physics Department. Urbana, the University of Illinois at Urbana-Champaign: 239.

Redford, G. I., and Clegg, R. M. 2005. Real-Time fluorescence lifetime imaging and FRET using fast gated image intensifiers. In *Molecular imaging: FRET microscopy and spectroscopy*, ed. A. Periasamy and R. N. Day, 193–226. New York: Oxford University Press.

Sandel, B. R., and Broadfoot, A. L. 1986. Statistical performance of the intensified charged coupled device. *Applied Optics* 25:4135–4140.

Schneider, P. C., and Clegg, R. M. 1997. Rapid acquisition, analysis, and display of fluorescence lifetime-resolved images for real-time applications. *Review of Scientific Instruments* 68:4107–4119.

Spencer, R. D., and Weber, G. 1969. Measurement of subnanosecond fluorescence lifetimes with a cross-correlation phase fluorometer. *Annals of the New York Academy of Sciences* 158:361–376.

Squire, A., Verveer, P. J., and Bastiaens, P. I. H. 2000. Multiple frequency fluorescence lifetime imaging microscopy. *Journal of Microscopy* 197:136–149.

Tolstov, G. P. 1962. *Fourier series.* New York: Dover.

Tumerman, L. A. 1941. Law of the quenching of the luminescence of complex molecules. *Journal of Physics (Moscow)* 4:151–166.

Tumerman, L. A., and Sorokin, E. M. 1967. The photosynthetic unit: A physical or statistical model? *Molecular Biology USSR* (Engl. transl.) 1:527–535.

Tumerman, L. A., and Szymanowski, V. 1937. Fluorometer based on the Debye Sears effect. *Comptes Rendus de l'Academie des Sciences USSR* 15:323.

Van Munster, E. B., and Gadella, T. W. J. 2004. φFLIM: A new method to avoid aliasing in frequency-domain fluorescence lifetime imaging microscopy. *Journal of Microscopy* 213:29–38.

Van Munster, E. B., Goedhart, J., Kremers, G. J., Manders, E. M. M., and Gadella, T. W. J. J. 2007. Combination of a spinning disc confocal unit with frequency-domain fluorescence lifetime imaging microscopy. *Cytometry Part A* 71A:207–214.

Wang, E., Babbey, C. M., and Dunn, K. W. 2005. Performance comparison between the high-speed Yokogawa spinning disc confocal system and single-point scanning confocal systems. *Journal of Microscopy* 218:148–159.

Weber, G. 1981a. Resolution of the fluorescence lifetimes in a heterogeneous system by phase and modulation measurements. *Journal of Physical Chemistry* 85:949–953.

CHAPTER 6

Laser Scanning Confocal FLIM Microscopy

Hans C. Gerritsen, Arjen Bader, and Sasha Agronskaia

6.1 INTRODUCTION

When fluorescent molecules are excited by a short light pulse, the excited state is populated and decays back to the ground state of the molecules. For simple molecules, this decay has an exponential shape, $I(t) = I_0\, e^{-t/\tau}$, and the lifetime τ is typically on the order of nanoseconds (10^{-9} s).

The fluorescence lifetime contains information about the photophysical processes of the fluorescent molecules and of the interactions of the fluorescent molecules with their microenvironment. Importantly, the fluorescence lifetime can be affected by the local microenvironment of the fluorescent molecule. For example, ion concentrations (pH, Ca), Förster resonance energy transfer (FRET), quenching, and conformational changes may affect the fluorescence lifetime. Consequently, the fluorescence lifetime can be used to quantify these concentrations and processes (Agronskaia, Tertoolen, and Gerritsen 2004; Bastiaens and Squire 1999; Centonze et al. 2000; Christensen 2000; Clegg 1996; Gadella, Clegg, et al. 1993; Sanders et al. 1994, 1995). Multiple lifetimes may be observed from the same detection volume due to heterogeneities in the local environment of the molecule.

The combination of fluorescence lifetime detection and imaging offers the possibility to employ this photophysical property directly for quantitative imaging. Special, high time resolution, high-throughput detection equipment is required to combine lifetime detection with imaging.

A number of techniques are available to measure the decay curves of fluorescence processes. In general, τ_f measurements are either made in the time or the frequency domain. In the frequency domain, one measures the phase shift of the fluorescent light with respect to the phase of a modulated source (Gadella, Jovin, and Clegg 1993; Lakowicz and Berndt 1991; Morgan, Mitchell, and Murray 1990; see Chapter 5). For excitation, cw (continuous wave) lasers are often employed in combination with an acousto-optical modulator to modulate the laser output. Alternatively, pulsed lasers can be employed. On the detection side, phase sensitive detection is realized by using gain modulated detectors.

The fluorescence lifetime can be derived from the phase difference between the excitation light and the fluorescence emission. The fluorescence lifetime can also be calculated from the demodulation of the fluorescence light with respect to the excitation light. For monoexponential decays, the lifetimes derived from phase and modulation will be identical; however, but in the case of multiexponential decay, they will differ. For nanosecond fluorescence decay times, modulation frequencies of several tens of megahertz are typically used.

Measurements in the time domain are generally performed by measuring the time dependency of the fluorescence intensity with respect to a short excitation pulse (see Figure 6.1). The decay can be analyzed by fitting to a mono- or multiexponential function and yields lifetimes and relative contributions of the lifetime components. Detailed overviews of implementations and applications of time-domain fluorescence lifetime techniques are reported in the literature (Clark and Hester 1989; Lakowicz 1999; Marriott et al. 1991; O'Connor and Phillips 1984).

The most common implementations of fluorescence lifetime imaging are based on laser scanning microscopes (confocal and multiphoton) in combination with time-domain lifetime detection methods. These microscopes are equipped with pulsed lasers, detectors with a high time resolution, and special detection electronics. The fluorescence lifetime is obtained from the fluorescence decay after exciting the fluorescent molecules with a short light pulse. The most common techniques to record the decay are time correlated single-photon counting (TCSPC; see Section 6.2.1) and time gating (TG; see Section 6.2.2).

Lifetime imaging has also been implemented in wide-field microscopy in the time domain and in the frequency domain (Agronskaia, Tertoolen, and Gerritsen 2003; French et al. 1998; Gadella, Jovin, et al. 1993; Lakowicz and Berndt 1991; see Chapters 4 and 5). Wide-field implementations afford direct recording of whole images and are often a bit

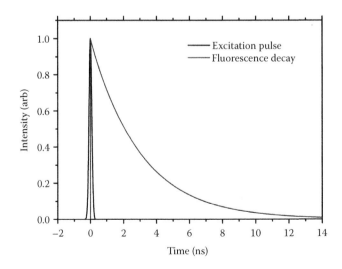

FIGURE 6.1 In time-domain lifetime detection, fluorescence is excited by pulsed excitation and the decay is directly sampled by high time resolution detection methods.

less accurate but considerably faster than scanning microscopy implementations. In this chapter, we will restrict ourselves to combinations of scanning confocal microscopy and fluorescence lifetime imaging. This combination is comparatively straightforward because only single-channel detection is required. Conventional lifetime detection methods used in spectroscopy can be conveniently combined with confocal microscopes.

6.1.1 Historical Background

Before true lifetime imaging was implemented, experiments were carried out that combined electronics intended for fluorescence lifetime spectroscopy with scanning microscopes. Such experiments yielded average lifetimes of regions of interest and required long acquisition times, typically on the order of minutes.

An example of such a measurement is shown in Figure 6.2. Here, an intensity image of Leydig cells stained with coumestrol is illustrated. Coumestrol is an autofluorescent phytoestrogen, mimicking the biological activity of estrogens. In this experiment, the transport and distribution of coumestrol was followed. The nucleus contains significantly less signal than the cytoplasm; this can be caused by a low coumestrol concentration as well as by quenching effects. Dynamic quenching effects would result in a shortening of the fluorescence lifetime (Lakowicz 1999). The decays in the nucleus and cytoplasm have comparable lifetimes (see Figure 6.3). Therefore, dynamic quenching can be excluded.

In this case, conventional TCSPC equipment (Ortec 9307 pico-timing discriminator and Ortec 566 TAC) for recording the decay and a homemade confocal microscope for selecting the ROIs (regions of interest) and imaging were used. Radiation (390 nm) from the Daresbury Synchrotron Radiation Source was used for exciting the coumestrol (Van der Oord et al. 1995). The count rates in the experiment were low—10–25 kHz—and total acquisition time for one ROI amounted to 2 min.

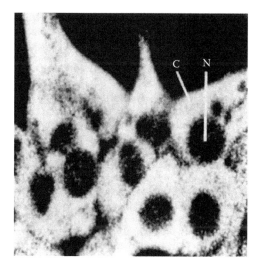

FIGURE 6.2   Intensity image of a Leydig cell loaded with coumestrol. Field of view is $100 \times 100\ \mu m^2$.

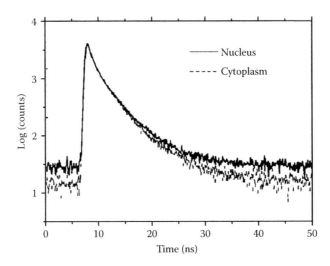

FIGURE 6.3   TCSPC traces recorded from regions in the cytoplasm and nucleus respectively, $\tau_{nucleus} \approx \tau_{cytoplasm}$ (4–8 ns).

The first experiments that combined confocal laser scanning microscopy and life-time imaging were carried out at the end of the 1980s and in the early 1990s using time-domain-based instrumentation (Bugiel, König, and Wabnitz 1989; Buurman et al. 1992; Schneckenburger and Konig 1992; Wang et al. 1991). At the middle and end of the 1990s, optimized detection electronics became available that afforded much faster lifetime imaging (Becker et al. 1999, 2003; Van der Oord et al. 1995). Until this happened, the use of FLIM (fluorescence lifetime imaging microscopy) was restricted to a small number of specialized laboratories that employed homemade setups. An important step in the history of FLIM was the demonstration that FLIM can be used to image FRET between fluorescent molecules (Wouters and Bastiaens 1999). This enables the detection of molecular interactions and extended the use of FLIM with an important application. As a result, the number of FLIM publications increased rapidly after 1999 (see Figure 6.4).

For a long time, the need for a picosecond pulsed laser source made FLIM a comparatively expensive and complicated technique. However, at present, turn-key, affordable picosecond pulsed solid-state lasers are available that can be conveniently combined with commercial FLIM electronics. This and the availability of reasonably priced FLIM electronics make laser scanning confocal FLIM microscopy a very affordable technique.

6.2 LIFETIME DETECTION METHODS IN SCANNING MICROSCOPY

6.2.1 Time-Correlated Single-Photon Counting (TCSPC)

Time-correlated single-photon counting is at present the most commonly used method in fluorescence lifetime imaging. Plug-in boards for PCs that contain all the detection electronics for TCSPC are commercially available, as is software to analyze the lifetimes in TCSPC images. TCSPC is a time-domain base detection method. In confocal fluorescence lifetime imaging, usually picosecond light pulses are used for exciting the fluorescent

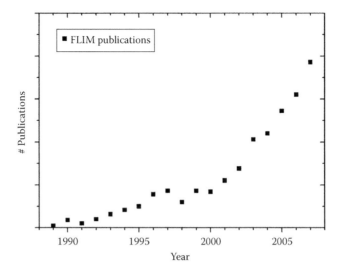

FIGURE 6.4   In the early days of FLIM, the number of FLIM publications per year was compara-tively low. After commercial FLIM hardware became available and the demonstration of its suit-ability to image FRET in 1999, the number of publications per year increased rapidly.

molecules. The fluorescence decay is measured by repeatedly measuring the time difference between the excitation pulse and the detection of a single fluorescence photon emitted by the specimen. This yields a histogram of detection times of individual photons with respect to the excitation pulse, and corresponds to the fluorescence decay curve.

A high overall time resolution of 25–300 ps and a wide dynamic lifetime range charac-terize TSPC. The TCSPC electronics often have a very high timing accuracy (1–200 ps). In most practical cases, however, the time resolution in TCSPC is limited by the timing jitter of the detector (25–300 ps). The finite response time of electronics and detectors affects the shape of the decay curve, especially at short times with respect to the excitation pulse. This "system response" can be taken into account in the analyses of the decays by employing iterative analyses methods and a measured system response. The latter can be obtained, for instance, by recording scattered excitation light from a strongly scattering specimen such as ludox. Alternatively, one can record the decay of a fast reference dye such as rose Bengal ($\tau \sim 70$ ps) (Stiel et al. 1996).

A schematic diagram of a typical TCSPC setup is shown in Figure 6.5. Here, the speci-men is excited by a short light pulse, usually coming from a picosecond pulsed laser. A trigger pulse that is synchronized with the excitation pulse is used to start a high-accuracy timing device such as a time-to-amplitude converter (TAC) or a time-to-digital converter (TDC). The (filtered) fluorescence emitted by the specimen is detected by a high-speed detector (photomultiplier tube, avalanche photodiode) setup to detect single photons (sin-gle-photon counting, SPC). The output pulses of the detector are amplified and fed into a discriminator (see Section 6.3.2). The high-bandwidth amplifier is used to amplify the low signal (millivolts) coming from the detector and the discriminator assures that only pulses above a preset level are employed. The output of the discriminator consists of a short pulse of well defined shape, height, and width. The discriminator serves to suppress electronic

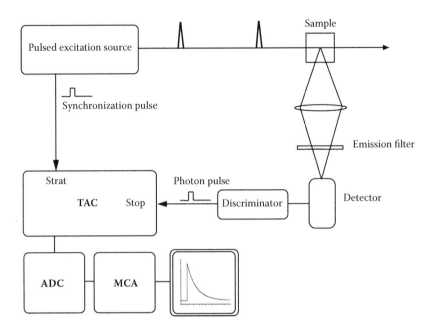

FIGURE 6.5 Schematic diagram of a TCSPC setup (see text for discussion).

background noise of the system, and its properties may have a strong effect on the timing behavior of the system.

The output of the discriminator is used to stop the TAC (or TDC). The TAC produces an analog output signal that is proportional to the time difference between the start and stop pulses. The analog output of the TAC is digitized by an analog-to-digital converter (ADC) and its digital output value is used as a pointer to a channel of a multichannel analyzer (MCA). Next, this specific channel of the MCA is incremented by one. After repeating this process many times, a histogram of arrival times of photons is accumulated in the memory of the MCA.

In fluorescence lifetime spectroscopy, the histogram usually contains 512–2,048 channels and represents the fluorescence decay curve of the specimen. In FLIM, typically fewer channels are used. Note that the recorded decay curve is convoluted with the timing response of the total instrument, the instrument response function (IRF). In TACs, the timing information is derived from the charging of a small capacitor, and a separate ADC is used. Nowadays, TDCs are often employed in TCSPC. These devices directly convert the timing difference between start and stop signals into a digital word. To this end, a series of solid-state delay lines is used. In practice, TDCs have very similar timing accuracies and limitations to those of TACs.

Importantly, the dead time of TACs and TDCs is comparatively long, typically 125–300 ns. When a photon arrives within this time interval after the detection of a photon, it will not be observed. Therefore, care must be taken that the count rate of the experiment is sufficiently low to prevent this pulse pileup. TACs and TDCs usually operate in reversed start–stop geometry. Here, the TAC is started by the fluorescence signal and stopped by the laser trigger. In this way, the TAC is only triggered by usable events, rather than by laser trigger pulses that do not result in a detected fluorescence photon. This mode of operation suffers less from dead-time effects. If two photons arrive within a period equal to the dead-time of the system, pulse pileup occurs and the second photon cannot be detected. In the reversed start–stop geometry, pileup is minimized by reducing the excitation intensity to about one to five detected photons per 100 excitation pulses (O'Connor and Phillips 1984).

Furthermore, in spectroscopy applications, excitation frequencies not exceeding 10 MHz are employed to ensure that the fluorescence decay signal from one excitation pulse is not affected by the tail of the fluorescence decay produced by other excitation pulses. In contrast, in lifetime imaging experiments, higher excitation frequencies are usually employed (40–80 MHz) to increase the duty cycle of the experiments and reduce acquisition time. As a result, the observed decays may be affected by the tails of previously produced decays.

The maximum count rate employed in conventional spectroscopy applications of TCSPC is typically less than 100 kHz. Moreover, the time required to access the histogramming memory and to transfer the decay curve from the histogramming memory to the computer system can be substantial—in particular for TCSPC electronics designed for spectroscopic application. TCSPC boards optimized for imaging do not suffer anymore from a memory transfer bottleneck. Their acquisition speed is determined by the dead time of the TAC or TDC (125–300 ns).

6.2.2 Time Gating

In TG methods, again, picosecond pulsed excitation is used and the fluorescence emission is detected in two or more time gates that are delayed by a different time relative to the excitation pulse (see Figure 6.6). In the case of a system equipped with two time gates, the ratio of the signals acquired in the two time gates is a direct measure of the fluorescence lifetime. For a decay that exhibits only a single exponent, the fluorescence lifetime is given by

$$\tau = \Delta T / \ln(I_1/I_2) \tag{6.1}$$

with ΔT the time offset between the start of the two time gates and I_1 and I_2 the corresponding fluorescence intensities accumulated in the gates. Here, the assumption is made that the two time gates are of equal width.

The two-gate approach affords determination of the fluorescence lifetime in real time. In the case of a multiexponential fluorescence decay, Equation 6.1 yields only an "average" fluorescence lifetime. This limitation can be circumvented by increasing the number of time gates; this enables the recording of multiexponential decays (de Grauw and Gerritsen 2001; French et al. 1998; Scully et al. 1997; Sytsma et al. 1998). Increasing the number of gates requires more sophisticated data analyses approaches, such as fitting the decay to exponential or multiexponential functions. This analysis method is slow and not suitable for real-time visualization of lifetime images.

Time gating is implemented in high-speed electronics like emitter-coupled logic (ECL). In one implementation, a series of high-speed counters are sequentially enabled for a preset time (de Grauw and Gerritsen 2001). This simple but effective scheme has extremely low dead time and allows high-count-rate lifetime detection.

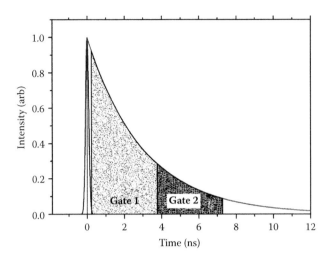

FIGURE 6.6 Schematic diagram of time gating. The specimen is excited with a brief light pulse; next, the fluorescence is detected in a number of time gates. In the diagram, two gates are shown that open after a short delay with respect to the excitation pulse.

Practical implementation of TG requires careful synchronization of the opening of the gates with respect to the laser pulses (see Figure 6.7). In addition, often a small delay is applied between the opening of the excitation pulse and the first gate. This avoids distortion of the measured fluorescence lifetime by the tail of the excitation pulse. In general, a delay of a few hundred picoseconds is sufficient to avoid such artifacts. The optimum value of the delay depends on both the width of the excitation pulse and the rise time of the opening of the gates. In principle, the delay reduces the sensitivity of the time gating method somewhat, in particular at short lifetimes.

Interestingly, the small offset can be employed to suppress background signals that are correlated with the excitation pulse. For instance, direct and multiple scattered excitation light as well as Raman scattering reaches the detector at $t \approx 0$ and can be effectively suppressed by opening the first gate a few hundred picoseconds after $t = 0$. This improves the signal-to-background ratio in the images without a significant loss of signal. TG can also be employed to suppress autofluorescence background in biological specimens. Often, autofluorescence has a comparatively short fluorescence lifetime and the signal-to-background ratio of the images can be improved by offsetting the first gate with respect to the excitation pulse (Vroom et al. 1999).

TG can be implemented using SPC. As in TCSPC, this requires use of a discriminator to separate the photon signal from background noise. One of the advantages of time gating is that it can be very efficient at high fluorescence signals. When time gating is implemented so that all gates open sequentially after each and every excitation pulse, detection efficiency is high and multiple photons can be detected per excitation pulse (Sytsma et al. 1998; Van der Oord, de Grauw, and Gerritsen 2001). Moreover, dead times of time gating electronics can be very low (<0.5 ns), so TG can operate at high count rates. In practice, its performance is limited by that of the detector.

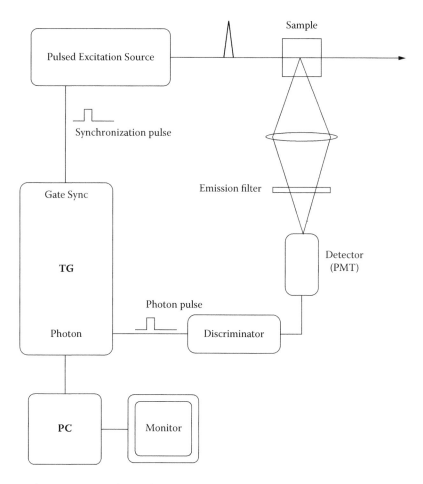

FIGURE 6.7   The time-gating (TG) electronics control the opening and closing of the time gates. This requires synchronization of the electronics with the excitation pulse.

Detector-limited count rates of up to 10 MHz have been reported using such equipment (Gerritsen et al. 2002).

6.3 DETECTORS AND ELECTRONICS

6.3.1 Detectors

The most common type of detector used for detecting photons is the photomultiplier tube (PMT). It consists of a vacuum tube with a transparent entrance window, a photocathode, and several dynodes. An incoming photon strikes the photocathode of the PMT, resulting in the emission of a photoelectron. This electron is then attracted to a dynode by a voltage difference between the photocathode and the dynode. Here, the impact of the electron results in the release of multiple secondary electrons. The secondary electrons are accelerated to a second dynode, which results in the release of more electrons, and so on. Finally, the electrons are collected by an electrode and a detectable electrical pulse is available for further processing.

The quantum efficiency of PMTs depends on the wavelength of the light, and peak quantum efficiencies are usually in the range 10–40%. The PMTs' gain can be adjusted by changing the high voltage across the tube and can be as high as 10^6 or 10^7. For time-domain lifetime detection, PMTs are operated in single-photon detection mode. Here, the PMT is operated at high gain and comparatively low photon fluxes. Each detected photon results in a single pulse that does not overlap with the pulses of other photons at the output of the PMT.

The pulse width at the output of the PMT depends on the type of PMT. Typical values range from 1 to 2 ns full width half maximum (FWHM) for "fast" PMTs to 5–10 ns for "slow" PMTs. There is some delay between the time that a photon hits the photocathode of the PMT and the detection of the pulse at the PMT's output. There is also some time jitter in this delay. This so-called transit time spread (TTS) varies between 150 ps and a few nanoseconds for fast and slow PMTs, respectively. The TTS of a PMT arises from timing variations in the production of photoelectrons by the photocathode and in the multiplication process. Additionally, the electronics may exhibit some TTS. The total TTS of a system has a strong influence on the timing uncertainties in lifetime measurements. In particular, it affects the shortest lifetimes that can be measured with a system. Importantly, standard PMTs used in confocal microscopes are in general slow and exhibit poor timing properties. Therefore, they are usually not suitable for fluorescence lifetime imaging (Gerritsen et al. 2002).

In addition to the TTS, there are variations in the pulse height of the output pulses of the PMT (see Figure 6.8A). In SPC mode, these pulse height variations are minimized by operating the PMT at high gain (voltage). This results in saturation effects; only a maximum amount of charge can be removed from the dynodes of the PMT that reduce the statistical variations. However, even under these circumstances, the pulse height variations can be substantial. An example of a pulse height distribution is shown in Figure 6.8(B). Typically, the output signal of the PMT consists of two contributions: the first from detected photons and the second from noise (dashed curve). The latter consists of comparatively low amplitude pulses. The sum of the noise and photon signals (drawn curve in Figure 6.8B) generally shows a valley (vertical line in Figure 6.8B) that separates the regions dominated by noise and by photon-induced pulses.

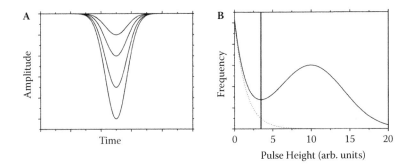

FIGURE 6.8   (A) The output signal from PMTs exhibits pulse height variations. (B) An example of a pulse height distribution of pulses coming out of a PMT (drawn line). It consists of a contribution from noise (dashed line) and detected photons.

The background signal from PMTs can be reduced by cooling the PMT. Without cooling, the background can be as high as several kilohertz. By cooling, this can be reduced to several tens of hertz. In lifetime imaging, pixel dwell times ranging from several microseconds to milliseconds are employed. Therefore, the background signal from the PMT is usually not a serious problem unless very long integration times are being used.

Microchannel plate PMTs (MCP-PMTs) exhibit particularly good timing properties. MCP-PMTs consist of plates with small-diameter channels that act as continuous-dynode electron multipliers. The small thickness of the plates results in narrow output pulses and low TTS (down to 25 ps). These properties make them very well suited for measuring short fluorescence lifetimes. However, MCP-PMTs are also expensive and vulnerable. They should be operated at low count rates to avoid deterioration. For the well known Hamamatsu R3809U-50 MCP, a maximum count rate of only 20 kHz is advised (Hamamatsu). This is too low for lifetime imaging at reasonable acquisition speed. The maximum count rate can be somewhat improved by operating the MCP-PMT at lower gain (voltage).

Cooled avalanche photodiodes are an interesting alternative for PMTs. Single-photon avalanche photodiodes (SPADs) can be used to detect single photons. The solid-state devices are operated close to or slightly above the breakdown voltage (Ekstrom 1981). A photon generates an electron–hole pair that initiates an avalanche breakdown in the diode. Active or passive quenching circuits are used to stop the breakdown. SPADs have high peak quantum efficiencies (~40%) and some of them have excellent timing properties (<100 ps TTS).

6.3.2 Front-End Electronics

A diagram of the PMT and font-end electronics is shown in Figure 6.9. The output of the PMT is usually first amplified by a high-bandwidth preamplifier (typically, 500 MHz–1 GHz). This amplifies the millivolt output pulses coming out of the PMT. The preamplifier should be positioned as close as possible to the PMT to reduce the pickup of noise and RF signals. The output of the preamplifier has low impedance (50 Ω) and is significantly higher than the input signal. This makes the length of the cable between the preamplifier and the next electronic element, the discriminator, less critical.

The discriminator is used to suppress noise; it is triggered by signals that exceed an adjustable threshold level and it outputs standard fast logic pulses. By setting the threshold to a value corresponding to the valley in Figure 6.8(B), noise is suppressed while most of the photon signal is detected. Two types of discriminators exist: constant height discriminators and constant fraction discriminators. The constant height discriminators are triggered when the amplitude of the incoming electrical pulse exceeds a preset value. This

FIGURE 6.9 Diagram of a PMT and front-end electronics.

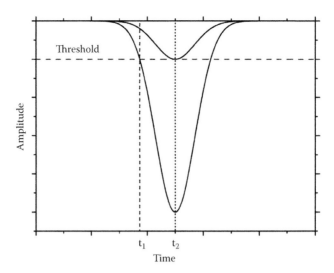

FIGURE 6.10   A constant height discriminator is triggered at a fixed level (dashed horizontal line). The time of triggering depends on the amplitude of the pulses; the high and low pulses are triggered at t_1 and t_2, respectively. Pulse-height variations cause timing jitter.

type of discriminator is less suited for applications that require high timing accuracy. The amplitude of the pulse affects the time of triggering (see Figure 6.10). Low amplitude pulses are detected later than high amplitude pulses. Therefore, the variation in pulse height of the PMT results in significant timing jitter. PMTs with short output pulses are less affected by this effect than slow PMTs are.

The timing behavior of constant fraction discriminators is much less affected by amplitude variations. This type of discriminator first discriminates pulses by their amplitude and next detects the peak position of the pulse. The latter is realized by splitting the input signal into two fractions. The first is only attenuated and the second is delayed and inverted. Next, the two signals are added. This results in a so-called constant fraction signal (see Figure 6.11). By proper choice of the delay, the zero-crossing of the sum signal is at the peak position of the pulse. Constant fraction discriminators produce an output signal that is synchronized with the zero-crossing. The timing properties of a constant fraction discriminator are insensitive to pulse height variations. Typical timing variations on the order of 25 ps are observed.

6.4 COUNT RATE AND ACQUISITION TIME

Laser scanning confocal microscopy is a comparatively slow technique. At high dye concentrations, intensity imaging may be accomplished at one frame per second or even a bit faster. FLIM, however, is a comparatively signal-hungry technique and much higher signal levels are required to obtain quantitative lifetime information than in intensity imaging. Acceptable intensity images may be obtained at 25 detected photons per pixel or even less. In FLIM a signal of at least one more order of magnitude is required for recording monoexponential lifetimes with acceptable accuracy at the single-pixel level. In the case of multiexponential decay, orders of magnitude more signal is required. Work by Kollner et al. showed that about 100 times more fluorescence signal is required to resolve two lifetimes

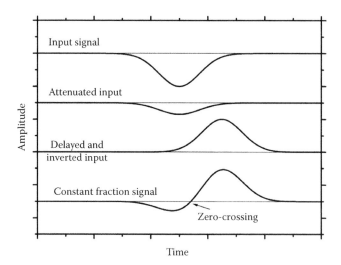

FIGURE 6.11 Principle of a constant fraction discriminator. One part of the input signal is attenuated and the other part is delayed and inverted. The two parts are added and the zero crossing is detected.

differing by a factor of two than to detect a single lifetime with the same accuracy (Kollner and Wolfrum 1992).

The FLIM acquisition speed is determined by the minimum required signal, the efficiency of the lifetime detection method, detector and electronics limitations, and duty cycle of the light source (Gerritsen et al. 2002). Furthermore, the analysis method may also play a role.

Realistic estimates of the total amount of signal available from one detected volume element show that it is difficult to realize high signal levels in a laser scanning confocal microscope. In practice, at most, a few thousand detected photons per pixel can be obtained from a strongly fluorescing specimen. Photobleaching limits the maximum amount of signal that can be detected. This problem may be circumvented by spatial binning; this increases the amount of signal at the price of a loss of spatial resolution. The efficiency of the FLIM system is an important factor for its performance. High quantum efficiency detectors and efficient detection techniques are required for optimum performance.

6.4.1 Detector and Electronics Limitations

Detectors like PMTs have a maximum output current. Above the maximum current, the detector may be damaged. In addition to the maximum current of PMTs, they have a finite dead time. When a photon is detected by the PMT, charge is removed from the dynodes. This results in a drop of the voltages across the dynodes of the PMT and a corresponding drop in gain. The PMT is temporarily dead and will only detect photons again once the dynodes have been recharged. This dead time depends on the type of PMT and the voltage (gain) that has been applied. Typically, standard PMTs are usable at count rates between 1 and 10 MHz without deterioration of their performance. MCP-PMTs, however, should be used at much lower count rates—up to 20 kHz. SPADs usually have count rate limitations similar to those of standard PMTs.

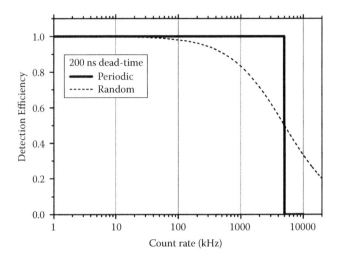

FIGURE 6.12 The dead-time of electronics and detector reduce detection efficiency at high count rates of random signals like fluorescence. This effect is caused by pulse pileup.

Also, detection electronics suffer from dead-time effects. In the case of TACs and TDCs, the dead time can be as high as 125–300 ns. This causes saturation of the detector at a count rate equal to the reciprocal of the dead time. However, well before the saturation count rate is reached, the effects of the dead time have already showed up as a loss in detection efficiency. In time gating, the detection electronics is usually fast and dead times can be ignored (<1 ns). The total dead time of the detection system and PMT limitations are usually the bottleneck in acquisition speed.

In Figure 6.12, the effect of the dead time on the detection efficiency is shown for a dead time of 200 ns. Two cases are considered—the first a periodic signal and the second a random signal. The latter is more representative for a fluorescence signal. For the periodic signal, the efficiency remains constant until the frequency equals the reciprocal of the dead time. At this frequency, the pulses are separated by a time equal to the dead time and as a result they can no longer be detected. In the case of a random signal, the detection efficiency is 100% for low frequencies. In this example, already at 100 kHz the efficiency starts to go down due to the increased probability of pulses arriving at the detector within a period less than or equal to the dead time after a previous pulse. At 1 MHz, the detection efficiency has gone down to 80% and it reaches a value of 50% at a frequency of 5 MHz, the reciprocal of the dead time.

6.4.2 Efficiency of Time-Domain Lifetime Detection Methods

The efficiency of time-domain lifetime detection methods has been studied theoretically (Ballew and Demas 1989; Kollner and Wolfrum 1992) and by computer simulations (Gerritsen et al. 2002, 2006; Periasamy, Sharman, and Demas 1999). Different lifetime detection methods can be compared by using a figure of merit for their performance. Here, we employ a figure of merit, F, that compares the accuracy of the lifetime acquisition method with an intensity measurement with the same number of detected photons,

$F = (\Delta\tau/\tau)/(\Delta I/I)$. Here, τ is the lifetime, $\Delta\tau$ the standard deviation in the lifetime, I the number of detected photons in the intensity measurement, and ΔI the standard deviation of the intensity determination. Assuming Poisson statistics, $\Delta I/I$ equal to $1/\sqrt{I}$ and F can be rewritten as

$$F = \frac{\Delta\tau}{\tau}.\sqrt{I}$$

(6.2)

If we assume that $\Delta\tau/\tau$ is governed by Poisson statistics, F is a direct measure of the performance of the acquisition method. It is independent of the number of counts in the decay, always larger than one, and a low value of F indicates a high sensitivity. The F values can be converted into the number of counts (in the total decay) required to realize a specific accuracy, $I = (F/(\Delta\tau/\tau))^2$. Moreover, the sensitivities of two configurations can be compared by taking the square of the ratio of their figures of merit, $(F_1/F_2)^2$.

The figure of merit of systems with gates of equal width can be calculated analytically (Kollner and Wolfrum 1992). Here, we show an example of calculations on systems with 2, 4, and 64 time channels of equal width. Furthermore, a total collection time per laser pulse of 10 ns is chosen. A 10 ns collection time per laser pulse is typical for FLIM systems equipped with high repetition rate lasers such as Ti:Sa lasers. Note that the 64-time-channel situation is typical for TCSPC FLIM, while two and four channels are commonly used in TG-based FLIM. The results of the figure of merit calculations are summarized in Figure 6.13. The figure of merit of the two gate systems is significantly larger than that of the systems with four or more gates at all lifetimes. At a lifetime of 2 ns, the difference in sensitivity $((F_1/F_2)^2)$ is about 70%. The performance is significantly better than that for 4 and 64 channels.

Interestingly, only very small differences in F are found above 2 ns for systems with four or more gates. The differences between systems with four and more gates only show up below 1.5 ns. They are all well usable down to lifetimes of 0.5 ns or even shorter. The system with 64 gates has a very low figure of merit down to the shortest lifetimes. In practice, the theoretical figures of merit cannot be realized for lifetimes of the order of or shorter than the time response of the detection system. Usually, it is on the order of several hundred picoseconds. Therefore, for short lifetimes, the sensitivity of real FLIM systems will be lower than the theoretical calculations indicate.

At long lifetime, the sensitivity at all gate numbers goes down. Now, total collection time is short compared with the lifetime and counts in the tail of the decay are not detected. Consequently, the sensitivity goes down. The performance for longer lifetimes can be improved by increasing the total detection period and reducing the repetition frequency of the laser. This, however, reduces the duty cycle of the system and acquisition time will go up.

The calculations and simulations presented here concern ideal photon-counting systems. Effects related to noise and timing jitter of the electronics and detectors are not included. Whether these assumptions are reasonable depends on the implementation of the method.

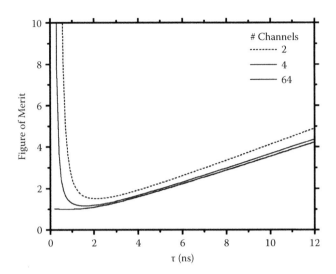

FIGURE 6.13 The results of figure-of-merit calculations for time-domain lifetime detection using 2, 4, and 64 time channels of equal width. Here, the total detection window is fixed to 10 ns.

In methods based on SPC, virtually no noise is present. Timing jitter, in particular of the detector, will affect the sensitivity of the lifetime acquisition method at short lifetimes. In general, such effects will show up below a lifetime of 0.5 ns.

6.5 EXAMPLE

In this example, we focus on demonstrating colocalization of the microdomain markers ganglioside GM1 and glycosylphosphatidylinositol (GPI) anchored GFP by FRET-FLIM. This work is part of a study on EGFR signaling and the possible role of cholesterol-enriched microdomains (rafts) as signaling platforms (Hofman et al. 2008).

Confocal time-gated FRET-FLIM experiments were carried out using a standard confocal microscope (Nikon PCM2000) equipped with a time-gated detection module operating in SPC mode (LIMO). The module contains four time gates with a width of 2 ns each (de Grauw and Gerritsen 2001; Van der Oord et al. 1995). Basically, the LIMO electronics consist of four high-speed counters that are sequentially enabled.

A frequency-doubled picosecond pulsed Ti:Sa laser (Tsunami, Spectra Physics) operating at a repetition frequency of 82 MHz was used for excitation at 460 nm. The excitation pulses were transferred to the confocal microscope using a single-mode optical fiber. For imaging, an NA = 1.20/40x water immersion objective (Plan Apo, Nikon) and a medium-size pinhole were used. The laser power at the specimen was estimated to be 50 μW and typical FLIM acquisition times amounted to 30 s.

The fluorescence emission was detected with a fiber-coupled PMT (Hamamatsu H7422P-40) that was connected to a preamplifier and the LIMO unit. The PMT has a high quantum efficiency (40% max), good TTS (~250 ps), and can be operated at count rates of up to ~3 MHz. The time offset between the excitation pulse and the opening of the first gate was adjusted in such a way that the first gate opened after 90% of the rose Bengal

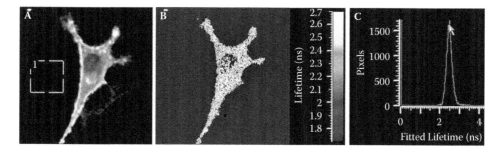

FIGURE 6.14 **(See color insert following page 288.)** Intensity (A) and lifetime (B) images of a GPI-GFP stained Her14 cell. Image size is 100×100 μm^2 (226×226 pixels). The lifetime image is homogeneous and has an average lifetime of 2.47 ns; see lifetime histogram in (C).

fluorescence (lifetime 70 ps) was emitted. The widths of the time gates were calibrated using a continuous wave white light source and a reference dye with known decay time. The four gate intensity decays recorded for each pixel were fitted with a monoexponential decay using the LIMO software.

All the experiments presented here were carried out on fixed Her14 cells at room temperature. Here, GPI and GM1 were used as raft markers. The GPI was coupled to green fluorescent protein (GPI-GFP) and the GM1 was detected by cholera toxin subunit β labeled with Alexa594 (GM1-CTB-Alexa594).

Figure 6.14 shows the intensity (A) and lifetime (B) images of a GPI-GFP (donor only) stained Her14 cell. The brighter areas in the intensity image contain 300–400 counts per pixel. The average background signal was determined from the boxed area in Figure 6.14(A) and found to be 1.8 counts per pixel. After subtracting the background from the original image, the lifetime image was calculated by fitting the four gate intensities to a monoexponential function. The resulting lifetime image is shown in Figure 6.14(B). The corresponding lifetime histogram in Figure 6.14(C) shows that the lifetime is homogeneous with an average value of 2.47 ns. This lifetime value serves as a reference value for the FRET-FLIM experiment.

Figure 6.15(A) shows a lifetime of a HER14 cell labeled with both GPI-GFP and the GM1-CTB-Alexa594. The average lifetime has gone down significantly to an average value of 2.14 ns. The reduction in the lifetime amounts to 0.33 ns and is indicative of FRET between the two raft markers. As a control experiment, cholesterol was extracted from the rafts by the addition of methyl-beta-cyclodextrin. Cholesterol removal causes disruption of the rafts and an increase in the distance between GPI-GFP and the GM1-CTB-Alexa594. As a result, a decrease in FRET is expected. In Figure 6.15(B), the results of the control experiment are shown. Here, the average lifetime has gone up to 2.30 ns, close to the original reference value of 2.47 ns. The incomplete recovery is most likely due to shrinking of the cell caused by the cholesterol removal.

The overall conclusion of these experiments is that GPI-GFP and the GM1-CTB-Alexa594 colocalize in the microdomains and that both markers can be used as lipid raft markers in FRET experiments.

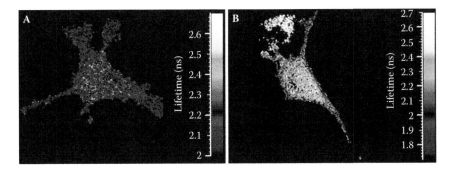

FIGURE 6.15 **(See color insert following page 288.)** (A) Lifetime image of a Her14 cell stained with GPI-GFP and GM1-CTB-Alexa594. Image size is 100 × 100 μm² (226 × 226 pixels). Due to FRET, the lifetime of the GFP is reduced to 2.14 ns. (B) Lifetime image after extraction of cholesterol from the lipid rafts. The average lifetime has gone up to 2.30 ns, close to the original reference value of 2.47 ns.

6.6 SUMMARY

This chapter discussed time-domain lifetime imaging methods in scanning microscopy. Two techniques were described in detail: TCSPC and time gating. Commercial solutions are available for both techniques and both are comparatively simple to implement in scanning microscopes. However, special caution should be taken in the selection of detectors and electronics.

The choice of the detector (PMT, SPAD, MCP-PMT) is critical. Standard scanning microscope detectors are often too slow for FLIM. The choice of the detector is usually a compromise of maximum count rate, timing properties (transit time spread), and quantum efficiency. The choice of the front-end electronics is important; high-bandwidth preamplifiers and high-timing accuracy discriminators (CFDs) are required for optimal performance.

Both TCSPC and time gating are very usable for FLIM and each has its own benefits. TCSPC is capable of recording lifetimes over a broad lifetime range including short lifetimes (<0.5 ns) with good sensitivity. Time gating using four gates offers comparable sensitivity as TCSPC for lifetimes > 1.5 ns. Below 1 ns, sensitivity goes down, but in our experience acceptable lifetime images can be recorded even below 0.5 ns. TCSPC often suffers from dead-time effects that degrade the sensitivity at high count rates. This limits the usable maximum count rate to a few megahertz. The maximum count rate in time gating is determined by the PMT. Proper choice of PMT allows operation at count rates of up to 10 MHz.

6.7 FUTURE PERSPECTIVE

Since its introduction in the late 1980s and early 1990s, FLIM has significantly matured and easily implemented commercial solutions are now available. Current limitations of FLIM systems are caused by both detectors and electronics. Ideally, detectors have high quantum efficiency, small TTS, and high maximum count rate (short dead time). These parameters will certainly improve in the future and further improve FLIM performance.

Solid-state solutions like SPADs are particularly interesting. These detectors have high quantum efficiency and good timing properties, and first steps have been set to integrate SPADs in standard CMOS (complementary metal-oxide semiconductor) technology accompanied by integrated readout technology (Faramazpour et al. 2008; Rae et al. 2008). This will likely result in cheap commercial solutions with, for instance, integrated SPAD and TCSPC electronics. This development would also allow integration of SPADs and time gating electronics with high numbers of time channels. The latter would have the advantage of low dead-time electronics (<0.5 ns), allowing count rates limited only by the SPAD.

Another way of circumventing the count rate limitations in TCSPC would be to integrate multiple TCSPC electronics on the same chip. The use of multiple TCSPC boards has already been successfully applied by Becker and Hickl (see Chapter 7). Here, multiple standard TCSPC boards are integrated in one system and used in parallel in one of their commercial products. Significant progress has been made in multichannel FLIM for simultaneously recording lifetime images in multiple wavelength bands and the combination of FLIM and spectral imaging (see Chapter 9).

Developments in this direction are likely to be continued. Currently, solutions for multichannel and spectral FLIM are available, but they are bulky and/or suffer from count-rate limitations, as in standard commercial systems. Importantly, although hardware solutions are available for multichannel and spectral FLIM, software developments are lagging behind. In particular, global analyses of multichannel and spectral FLIM offer interesting prospects of novel ways of detecting (e.g., FRET with higher sensitivity and accuracy).

REFERENCES

Agronskaia, A. V., Tertoolen, L., and Gerritsen, H. C. 2003. High frame rate fluorescence lifetime imaging. *Journal of Physics D: Applied Physics* 36:1655–1662.

Agronskaia, A. V., Tertoolen, L., and Gerritsen, H. C. 2004. Fast fluorescence lifetime imaging of calcium in living cells. *Journal of Biomedical Optics* 9 (6): 1230–1237.

Ballew, R. M., and Demas, J. N. 1989. An error analysis of the rapid lifetime determination method for the evaluation of single exponential decays. *Analytical Chemistry* 61:30–33.

Bastiaens, P. I. H., and Squire, A. 1999. Fluorescence lifetime imaging microscopy: Spatial resolution of biochemical processes in the cell. *Trends in Cell Biology* 9 (2): 48–52.

Becker, W., Bergmann, A., Biskup, C., et al. 2003. High-resolution TCSPC lifetime imaging. *Proceedings of SPIE* 4963:175–184.

Becker, W., Hickl, H., Zander, C., et al. 1999. Time-resolved detection and identification of single analyte molecules in microcapillaries by time-correlated single photon counting. *Review of Scientific Instruments* 70:1835–1841

Bugiel, I., König, K., and Wabnitz, H. 1989. Investigation of cells by fluorescence laser scanning microscopy with subnanosecond time resolution. *Laser Life Sciences* 3:1–7.

Buurman, E. P., Sanders, R., Draaijer, A., et al. 1992. Fluorescence lifetime imaging using a confocal laser scanning microscope. *Scanning* 14 (3): 155–159.

Centonze, V. E., Takahashi, A., Casanova, E., and Herman, B. 2000. Quantitative fluorescence microscopy. *Journal of Histotechnology* 23 (3): 229–234.

Christensen, K. A. 2000. Measurement of calcium in macrophage vacuolar compartments using ratiometric and fluorescence lifetime imaging microscopy. *Molecular Biology of the Cell* 11:730.

Clark, R. J. H., and Hester, R. E. 1989. *Time-resolved spectroscopy*. Chichester, UK: John Wiley & Sons.

Clegg, R. M. 1996. Fluorescence resonance energy transfer. In *Fluorescence imaging spectroscopy and microscopy,* ed. X. F. Wang and B. Herman, 179–236. New York: John Wiley & Sons, Inc.

de Grauw, C. J., and Gerritsen, H. C. 2001. Multiple time-gate module for fluorescence lifetime imaging. *Applied Spectroscopy* 55 (6): 670–678.

Ekstrom, P. A. 1981. Triggered-avalanche detection of optical photons. *Journal of Applied Physics* 52:6974–6979.

Faramazpour, N., Deen, M. J., Shirani, S., and Fang, Q. 2008. Fully integrated single photon avalanche diode detector in standard CMOS 0.18-μm technology. *IEEE Transactions in Electron Devices* 55:760–767.

French, T., So, P. T. C., Dong, C. Y., Berland, K. M., and Gratton, E. 1998. Fluorescence lifetime imaging techniques for microscopy. *Methods in Cell Biology* 56:277–304.

Gadella, T. W. J., Clegg, R. M., Arndtjovin, D. J., and Jovin, T. M. 1993. Epidermal growth-factor (Egf)-receptor clustering monitored by fluorescence resonance energy-transfer using donor photobleaching and lifetime-resolved fluorescence imaging microscopy. *Biophysical Journal* 64 (2): A130–A130.

Gadella, T. W., Jovin, T. M., and Clegg, R. M. 1993. Fluorescence lifetime imaging microscopy (FLIM)—Spatial resolutions of microstructures on the nanosecond timescale. *Biophysical Chemistry* 48:221–239.

Gerritsen, H. C., Asselbergs, M. A. H., Agronskaia, A. V., and Van Sark, W. 2002. Fluorescence lifetime imaging in scanning microscopes: Acquisition speed, photon economy and lifetime resolution. *Journal of Microscopy–Oxford* 206:218–224.

Gerritsen, H. C., Draaijer, A., van den Heuvel, D. J., Agronskaia, A. V., and Pawley, J. 2006. Fluorescence lifetime imaging in scanning microscopy. In *Handbook of biological confocal microscopy,* ed. J. Pawley, 516–533. New York: Plenum Press.

Hamamatsu. Data sheet Hamamatsu R3809U MCP-PMT, ww.hamamatsu.com

Hofman, E. G., Bader, A. N., Ruonala, M. O., et al. 2008. EGF induces coalescence of different lipid rafts. *Journal of Cell Science* 121:2519–2528.

Kollner, M., and Wolfrum, J. 1992. How many photons are necessary for fluorescence-lifetime measurements? *Chemistry and Physics Letters* 200 (1,2): 199–204.

Lakowicz, J. R. 1999. *Principles of fluorescence spectroscopy.* New York: Plenum Press.

Lakowicz, J. R., and Berndt, K. W. 1991. Lifetime-selective fluorescence imaging using an Rf phase-sensitive camera. *Review of Scientific Instruments* 62 (7): 1727–1734.

Marriott, G., Clegg, R. M., Arndtjovin, D. J., and Jovin, T. M. 1991. Time-resolved imaging microscopy—Phosphorescence and delayed fluorescence imaging. *Biophysical Journal* 60 (6): 1374–1387.

Morgan, C. G., Mitchell, A. C., and Murray, J. G. 1990. Nanosecond time-resolved fluorescence microscopy: Principles and practice. *Transactions of the Royal Microscopical Society* 1:463–466.

O'Connor, D. V., and Phillips, D. 1984. *Time-correlated single photon counting.* London: Academic Press.

Periasamy, A., Sharman, K. K., and Demas, J. N. 1999. Fluorescence lifetime imaging microscopy using rapid lifetime determination method: Theory and applications. *Biophysical Journal* 76 (1): A10–A10.

Rae, B., Griffin, C., Muir, K., et al. 2008. A microsystem for time-resolved fluorescence analysis using CMOS single-photon avalanche diodes and micro-LEDs. In *Conference on International Solid State Circuit,* IEEE, San Francisco, CA, 166–167.

Sanders, R., Draaijer, A., Gerritsen, H. C., Houpt, P. M., and Levine, Y. K. 1995. Quantitative Ph imaging in cells using confocal fluorescence lifetime imaging microscopy. *Analytical Biochemistry* 227 (2): 302–308.

Sanders, R., Gerritsen, H., Draaijer, A., Houpt, P., and Levine, Y. K. 1994. Fluorescence lifetime imaging of free calcium in single cells. *Bioimaging* 2 (3): 131–138.

Schneckenburger, H., and Konig, K. 1992. Fluorescence decay kinetics and imaging of Nad(P)H and flavins as metabolic indicators. *Optical Engineering* 31 (7): 1447–1451.

Scully, A. D., Ostler, R. B., Phillips, D., et al. 1997. Application of fluorescence lifetime imaging microscopy to the investigation of intracellular PDT mechanisms. *Bioimaging* 5 (1): 9–18.

Stiel, H., Teuchner, K., Paul, A., Leupold, D., and Kochevar, I. E. 1996. Quantitative comparison of excited state properties and intensity-dependent photosensitization by rose Bengal. *Journal of Photochemistry and Photobiology B* 33 (3): 245–254.

Sytsma, J., Vroom, J. M., De Grauw, C. J., and Gerritsen, H. C. 1998. Time-gated fluorescence lifetime imaging and microvolume spectroscopy using two-photon excitation. *Journal of Microscopy* 191:39–51.

Van der Oord, C. J., de Grauw, C. J. and Gerritsen, H. C. 2001. Fluorescence lifetime imaging module LIMO for CLSM. *Proceedings of SPIE* 4252:119–123.

Van der Oord, C. J. R., Gerritsen, H. C., Rommerts, F. F. G., Shaw, D. A., Munro, I. H., and Levine, Y. K. 1995. Microvolume time-resolved fluorescence spectroscopy using a confocal synchrotron-radiation microscope. *Applied Spectroscopy* 49 (10): 1469–1473.

Vroom, J. M., De Grauw, K. J., Gerritsen, H. C., et al. 1999. Depth penetration and detection of pH gradients in biofilms by two-photon excitation microscopy. *Applied and Environmental Microbiology* 65 (8): 3502–3511.

Wang, X. F., Uchida, T., Coleman, D. M., and Minami, S. 1991. A two-dimensional fluorescence lifetime imaging-system using a gated image intensifier. *Applied Spectroscopy* 45 (3): 360–366.

Wouters, F. S., and Bastiaens, P. I. H. 1999. Fluorescence lifetime imaging of receptor tyrosine kinase activity in cells. *Current Biology* 9:1127–1130.

CHAPTER 7

Multiphoton Fluorescence Lifetime Imaging at the Dawn of Clinical Application

Karsten König and Aisada Uchugonova

7.1 INTRODUCTION

Fluorescence techniques in life sciences have the advantage of being nondestructive. Combined with time-resolved detection, these techniques offer the possibility to obtain direct information on biochemical processes inside cells and tissues. The first fluorescence lifetime imaging on a microscopic scale (FLIM) was performed in the 1980s at Jena, Germany; it was based on a prototype of a ZEISS confocal picosecond laser scanning microscope in combination with a time-correlated single-photon counting (TCSPC) unit. At that time, the fluorescence decay kinetics of porphyrins were studied inside living cells and even inside the subcutaneous tumor tissue of living mice (Gärtner et al. 1988; König 1989; Bugiel, König, and Wabnitz 1989; König and Wabnitz 1990). Meanwhile, the FLIM technique is used for clinical research on volunteers and patients.

So far, reports of clinical applications of time-resolved fluorescence in vivo imaging has concentrated on the human skin, teeth, and the eye (i.e., parts of the body that can be easily accessed). Imaging of the interior of the human eye can be performed with one-photon excitation sources such as picosecond blue laser pulses. There is the hope that FLIM can help to detect early stages of eye diseases, especially age-related macula degeneration (Schweitzer et al. 2004, 2007).

Fluorescence lifetime imaging using a time-gated camera and UV/blue light excitation has been performed on teeth of volunteers to detect dental plaques and carious lesions. Interestingly, the typical mean fluorescence lifetimes of these lesions are on the order of 10 ns; they arise from the biosynthesis of fluorescent porphyrins by a variety of bacteria (e.g., *Actinomyces odontilyticus*) compared to the shorter mean fluorescence lifetimes of healthy dental hard and soft tissue (König, Flemming, and Hibst 1998; König, Schneckenburger, and Hibst 1999).

Time-resolved fluorescence imaging of skin and other "turbid" tissues is more complicated due to multiple scattering and low light penetration depth. However, in vivo multiphoton excitation of endogenous fluorophores in human skin based on near infrared (NIR) femtosecond laser pulses has been realized (König and Riemann 2003; König et al. 2006; Köhler et al. 2006; König 2008, König et al. 2008, König et al. 2009). For example, the reduced mitochondrial coenzyme nicotinamide adenine dinucleotide (phosphate) (NAD(P)H) with a major absorption band around 340 nm can easily be detected in intratissue cells by nonresonant two-photon absorption of two NIR photons within the wavelength range of 700–800 nm. Additional valuable information on the fluorophores and their in vivo microenvironment, such as the binding status of the fluorescent biomolecules, can be obtained by using time-resolved fluorescence techniques. For example, the fluorescence lifetime of NAD(P)H is significantly increased upon binding to proteins.

Clinical fluorescence applications normally forbid the use of exogenous fluorophores. Apart from the general problem of staining and labeling, only a few fluorescent dyes, such as fluorescein and indocyanine green (ICG), are approved for clinical use. Fortunately, biological tissue contains a wide variety of endogenous fluorophores (e.g., König and Schneckenburger 1994; Schweitzer et al. 2004; Laiho et al. 2005). Of particular interest are the skin pigment melanin and the intracellular coenzymes NAD(P)H and FAD. Melanin is the natural sun blocker that protects the skin against photodamage; NADH and FAD are the primary electron donor and acceptor in the metabolism of the cell.

In the 1970s, Chance et al. showed that the "redox ratio" (i.e., the ratio of the fluorescence intensities of FAD and NADH) is a direct indicator of the oxidation-reduction state of the cell (Chance 1976; Chance et al. 1979). Oxygen supply as well as cell proliferation change the metabolic pathways (e.g., Warburg effect, Pasteur effect; Warburg, Posener, and Negelein 1924; Gulledge et al. 1996; Niesner et al. 2004; Bird et al. 2005; Evans et al. 2005; Schroeder et al. 2005). Interestingly, the extracellular matrix components elastin and collagen also exhibit a weak autofluorescence. More exciting is the fact that collagen structures transfer the incident laser light into light at half the laser wavelength. The process is called second harmonic generation (SHG).

Unfortunately, the in vivo fluorescence spectra of endogenous chromophores are often broad, variable, and poorly defined. It is therefore difficult to disentangle the fluorescence components by their emission spectra alone. Adding fluorescence lifetime information to autofluorescence detection not only provides an additional separation parameter but also yields better information about the metabolic state and the microenvironment of the fluorophores (e.g., Schneckenburger and König 1992; Lakowicz et al. 1992; So et al. 1998; Lakowicz 1999; Vishwasrao et al. 2005; Chorvat and Chorvatova 2006). Fluorescence lifetime detection can also be used to obtain SHG images simultaneously with fluorescence lifetime images (Laiho et al. 2005; Uchugonova and König 2008).

Multiphoton excitation in human skin is typically performed with the flying-spot technology, in which a tiny subfemtoliter excitation volume is scanned through the tissue of interest in three dimensions by x,y galvoscanners and piezodriven focusing optics (z). Optical sections with submicron resolution are obtained. The technique is called "multiphoton tomography" because it can reconstruct the three-dimensional structure of the tissue based on optical sections in different tissue depths. Multiphoton tomography has been applied to patients with skin diseases such as melanoma, psoriasis, and dermatitis. It is also used to track pharmaceutical and cosmetic agents in different skin layers (König 2008). Multiphoton tomographs are being increasingly used to investigate the effect of UV radiation and smoking on skin, to determine the skin aging index, and to trace the accumulation of nanoparticles such as ZnO.

It should be mentioned that there is a billion-dollar market for beauty and antiaging products. Animal experiments are largely banned from testing these products. Consequently, there is a massive demand in the cosmetics industry for noninvasive test technology. Multiphoton tomography has been shown to track the diffusion of cosmetic agents through various skin layers and to prove the stimulation of the biosynthesis of collagen by certain natural products (König 2008).

Multiphoton FLIM tomography (x,y,z,τ) of human skin has in particular been applied to distinguish between different types of endogenous fluorophores (such a melanin types; Ehlers et al. 2007) as well as to distinguish between autofluorescence and cosmetic and pharmaceutical components (König 2008).

There is hope that FLIM can help differentiate cancer cells and normal cells in skin and other tissues (Bird et al. 2005; Skala et al. 2007a; De Beule et al. 2007; Dimitrow et al. 2009a, 2009b). Leppert et al. (2006) and Kantelhardt et al. (2007) recorded two-photon excited lifetime images of glioma and the surrounding brain tissue. They found a significantly increased fluorescence lifetime for the glioma cells compared to the surrounding brain cells.

This chapter describes multiphoton lifetime imaging and its application in clinical high-resolution imaging of human skin and in stem-cell research.

7.2 PRINCIPLE OF MULTIPHOTON IMAGING

Multiphoton tissue imaging should provide information on intratissue and intracellular morphology, metabolism, and the distribution of exogenous fluorophores as well as SHG active materials. It is important that the metabolic functions within the intratissue cells remain unchanged during the laser exposure. Therefore, certain requirements on the laser parameters and the exposure times must be considered. Multiphoton imaging does not require the removal of biopsies in order to obtain images with subcelluar resolution. The nondestructive optical sectioning provides "optical biopsies" under in vivo physiological conditions with a similar resolution as the microscope of the pathologist without any slicing and staining and enables recording over long periods of time. When physically taken, biopsies can be typically imaged with the tomograph 2 hours before significant signs of lysis occur and the autofluorescence pattern changes into a diffuse pattern.

Multiphoton excitation was proposed and theoretically investigated by Maria Göppert-Mayer (Göppert-Mayer 1929, 1931; Masters 2008). The principle of multiphoton excitation is shown in Figure 7.1. A fluorophore excited by one photon at a given wavelength λ in the visible or ultraviolet range can also be excited by two photons at 2λ (two-photon excitation) or at 3λ (three-photon excitation). Of course, multiphoton excitation requires a high power density (100 MW/cm² to 100 GW/cm²) that cannot be provided with conventional light sources.

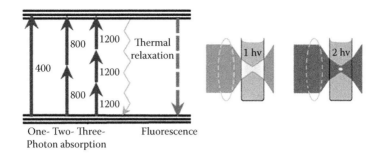

FIGURE 7.1   Principle of multiphoton excitation. h = Planck's constant; ν = frequency.

It was not possible to prove two-photon excitation experimentally at the time of Maria Göppert-Mayer's theoretical work. However, with modern laser sources, excitation with two or even more photons becomes practically applicable. If a picosecond or femtosecond pulsed laser is focused by a microscope objective lens of high numerical aperture, the power density in the focus is high enough to realize multiphoton excitation without damaging photothermal or photodisruptive effects for the short time of the laser excitation. When megahertz laser sources are used, mean NIR powers of less than 5 mW are sufficient for multiphoton imaging of cell monolayers. The application to fluorescence microscopy was proposed by Wilson and Sheppard in 1984 and practically demonstrated by Denk, Strickler, and Webb in 1990.

The most obvious difference between conventional one-photon fluorescence imaging and multiphoton imaging is that the one-photon process excites fluorescence in a full double-cone through the sample, while the two-photon process excites only a small subfemtoliter volume around the focus of the focusing lens. Thus, multiphoton excitation has an inherent optical sectioning capability.

The second advantage of multiphoton imaging is the relatively high depth penetration due to the use of NIR light. At an excitation wavelength in the UV or the visible range, the absorption and scattering coefficients are high. Therefore, the excitation density decreases quickly with increasing depth in the sample. At the NIR wavelength of multiphoton excitation, the one-photon absorption is nearly negligible for most cells. Moreover, the scattering coefficient is lower than in the UV. Therefore, deeper layers of the sample can be excited.

Photobleaching and photodestruction of the sample in multiphoton imaging are strictly confined to the scanned sample plane. Therefore, two-photon excitation has an advantage compared to confocal microscopy when z-stacks of images are recorded.

There is also no need to suppress out-of-focus fluorescence photons as in the case of one-photon three-dimensional microscopy. Consequentially, spatial filters (pinholes) are not required for multiphoton imaging. In principle, more photons from the specimen can be collected using the non-descanned detection method. The photons do not have to pass the galvoscanner, as is the case of a confocal laser scanning microscope. However, an expensive (femtosecond) laser source is required. When multiphoton FLIM is performed, a fast photon detector (PMT, MCP) is necessary (Chapter 6). Today's TCSPC units, such as the SPC830 from Becker & Hickl GmbH, Berlin, are capable of processing the data from femtosecond laser sources with high megahertz repetition rates.

For the sake of completeness, it should be mentioned that there are limitations to the imaging depth in the case of multiphoton high-resolution imaging. The most obvious one is the working distance of the focusing optics on the order of 200 µm when NA 1.2 or 1.3 objectives are used. Moreover, mismatch and inhomogeneity of the refractive index within the sample results in a degradation of the focus quality with increasing depth. A larger focus volume results in decreased excitation power density. Some systems have a feedback control system to enhance the laser power when imaging deep tissue areas (König 2008). Beside the two-photon excitation of fluorophores, second harmonic generation (SHG) may occur when certain non-central symmetric molecules are present. In tissue, collagen structures are the most efficient SHG sources. SHG is a two-photon effect and occurs immediately during intense femtosecond laser exposure at exactly half the laser wavelength ($\lambda/2$).

7.3 CLINICAL MULTIPHOTON TOMOGRAPHY

The only clinically approved multiphoton tomograph on the market is called DermaInspect™, which is designed and produced in Germany (JenLab GmbH, Jena, Germany). So far, the CE-marked system is operational in hospitals and research institutions of cosmetic and pharmaceutical companies as well as in governmental facilities in Europe, Asia, and Australia. It is based on a tunable femtosecond laser source, such as the titanium:sapphire laser MaiTai (Newport/Spectra Physics) or Chameleon (Coherent); an x,y galvoscanner; a piezodriven high NA focusing optics to change the tissue depth of investigation; multiple photon detectors; and a variety of further hardware and software. A detailed description of the femtosecond laser multiphoton tomograph is available in the reference König and Riemann (2003).

For clinical and preclinical applications, one wants to reconstruct the cellular structure within a larger volume of tissue. This requires the sample to be imaged in many focal planes at different depths. The procedure is often called z-scanning, and the results are called z stacks. z-Scanning is performed by moving the focusing lens up and down with a piezoscanner. A particular sample plane is imaged by x,y galvoscanners for a defined period of time, typically 1–12 s. Then the focusing lens is moved to the next z-plane, and the imaging acquisition is repeated. The process is the same as in a conventional confocal or two-photon laser scanning microscope.

The loss of image intensity at increasing depth is compensated by increasing the laser power. Instruments for clinical use need safety provisions, such as laser shutdown in case of excessive excitation power or failure of the scan device.

The clinical system DermaInspect measures multiphoton excited fluorescence and backscattered SHG. SHG radiation signals are originally emitted in a forward direction—that is, deep into the sample. However, multiple scattering occurs that produces backscattered photons. In a confocal system, such photons cannot be detected. However, the multiphoton tomograph is able to detect scattered SHG photons very efficiently. Up to four detectors with single-photon sensitivity enable the simultaneous measurement of autofluorescence images, SHG images, and xenofluorescence images in certain spectral channels defined by band-pass filters. The multichannel SPC830 board (Becker & Hickl GmbH, Berlin) enables multiphoton FLIM in skin with a temporal resolution of better than 300 ps.

7.4 MULTIPHOTON FLIM TECHNIQUE

Multiphoton FLIM of tissue does not excite the sample from above and below the focal plane; that is, multiphoton excitation takes place only in the very small focal volume. All fluorescence and SHG photons including multiple scattered fluorescence/SHG photons can contribute to the signal due to the absence of a (confocal) pinhole. The travel time for scattered photons is for working distances of less than 1 mm negligible. Therefore, multiphoton FLIM sections from different tissue depths can be recorded without significant problems affecting the time resolution.

Autofluorescence of biological material is emitted from a variety of fluorophores. Typically, the fluorescence/SHG of the subfemtoliter two-photon excitation volume is characterized by a multiexponential decay behavior. A typical example of a decay function of an individual "pixel" is shown in Figure 7.2. The FLIM technique should be able to resolve several fluorescence lifetime components, τ_n, present in the same location of the sample with their amplitude factors. The time resolution should be sufficient to resolve even the fastest decay components and distinguish them from SHG and exogenous fluorophores. In order to achieve this, the technique must record full fluorescence decay functions in the individual pixels.

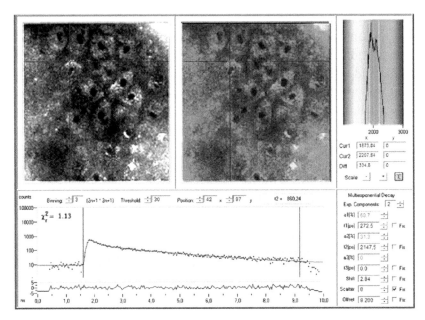

FIGURE 7.2 **(See color insert following page 288.)** Fluorescence intensity image, color-coded FLIM image, and τ_2 distrution of an optical section as well as a particular decay curve of skin autofluorescence. The FLIM image represents color-coded τ_2 values. The fluorescence from a subfemtoliter two-photon excited intracellular focal volume ("pixel") of human stratum spinosum exposed to femtosecond laser pulses at 800 nm is depicted (bottom). The y-scale represents the number of detected photons in each time channel of 0.04 ns. The double-exponential fit delivers the decay components of 272 ps (58.7%) and 2.15 ns (41.3%). The fitting parameter χ^2 of 1.13 is near the optimal value of 1.00.

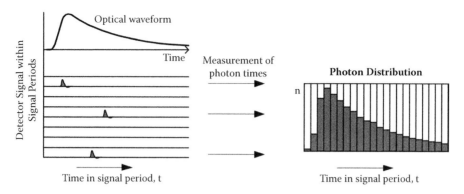

FIGURE 7.3   Principle of classic time-correlated single-photon counting.

Practically, a biexponential fit is typically applied providing the emission lifetime of a fast component τ_1, a slow component τ_2, and a mean lifetime τ_m with $\tau_m = (A_1\tau_1 + A_2\tau_2)$ (Becker 2005, Becker 2008). A color-coded FLIM image is generated by displaying either the τ_1 value, the τ_2 value, or the τ_m value per pixel. An optical section has therefore a third dimension τ depicted as color (Figure 7.2). A stack of FLIM images can be considered as four-dimensional (x,y,z,τ). Nowadays, a fifth dimension can be added. Such a fifth dimension could be the "true" color of the detected photon, which can be obtained by spectral analysis of the emission e.g., based on the PMT arrays. It would be of high value to record FLIM data in a number of spectral channels simultaneously. Simultaneous acquisition of all these parameters is important to avoid artifacts by photobleaching or motion in the sample and to reduce the exposure time for reasons of biosafety.

Time-correlated single-photon counting (O'Connor and Phillips 1984; Gärtner et al. 1988; König 1989; Bugiel et al. 1989; Becker 2005; Becker and Bergmann 2008; Chapter 6) is generally based on

1. detection of single photons emanating from a sample excited by a periodical light signal;
2. measurement of the arrival times of the individual photons using time channels (e.g. 256 channels of minimum 0.8 ps width); and
3. a histogram constructed from the number of photons in each time channel which often reflects the fluorescence decay (multiexponential fluorescence decay curve).

The TCSPC technique makes use of the fact that, for a low-intensity and high-repetition rate of the excitation source, the probability of detecting one signal photon per signal period (distance between excitation pulses) is much smaller than one. The detection of a second photon within one excitation period should be unlikely (Figure 7.3). The TCSPC technique does not use any time gating and therefore, as long as the light intensity is not too high, reaches a counting efficiency close to one.

The drawback of classic TCSPC is that the technique was intrinsically one dimensional (i.e., only the waveform of the light signal in one spot of a sample and at one wavelength is recorded at a time). When applied to laser scanning, classic TCSPC requires placing the laser beam at one pixel, acquiring a decay curve, reading the data into the memory of the TCSPC device, and then advancing to the next pixel. The first FLIM microscopes

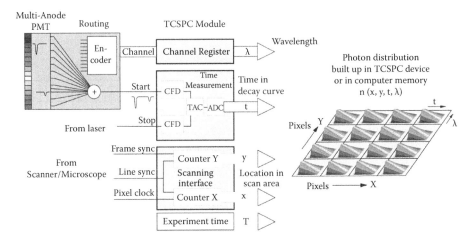

FIGURE 7.4 Principle of multidimensional TCSPC. A photon distribution over the time in the signal period, *t*, and one or several other parameters are built up. PMT = photomultiplier tube; CFD= constant fraction discriminator; TAC = time-to-amplitude converter; ADC = analog-to-digital converter.

built more than 20 years ago were based on such "slow scan" procedures. However, even spectrally resolved FLIM imaging has been performed with a step-motor-driven polychromator (Gärtner et al. 1988; König 1989; Bugiel et al. 1989). Nowadays, fast multidimensional TCSPC with much more advanced compact electronic modules is applied in FLIM microscopy and FLIM tomography. The photon histogram is not only recorded versus the time within the signal period but also as a function of one or several additional parameters, such as spatial coordinates, wavelength, or the time from the start of the experiment (Becker 2005). The principle is illustrated in Figure 7.4. The recording electronics builds up a photon distribution over all these parameters. With the photon stream technology, fast data transport and processing is possible.

The combination of such a multidimensional TCSPC with a laser scanning microscope for spectral FLIM is shown in Figure 7.5. The non-descanned microscope (e.g., TauMap™, JenLab GmbH, Jena, Germany) scans the sample with a focused Ti:sapphire laser. To record spatially resolved data, the TCSPC module has a scanning interface with counters that count the pixels within the lines and the lines within the frames of a single scan. Also, a *frame sync* signal after the particular scan is finished is provided. The counters deliver the coordinates of the excitation laser spot within the scan area at any time within the scan. The recording is continued over as many frames as necessary to obtain the desired number of photons per pixel.

Multiple wavelength FLIM is obtained by using a detector array that detects the spectrally dispersed fluorescence/SHG signal. The array can be a set of fast multipliers in combination with band-pass filters or a PMT array in combination with "colored front windows" or with a polychromator. By adding the wavelength of the photons as an additional dimension to the recording process, a pixel array is produced that contains a number of fluorescence decay functions for different wavelengths in each pixel (Becker 2005; Becker, Bergmann, and Biskup 2007).

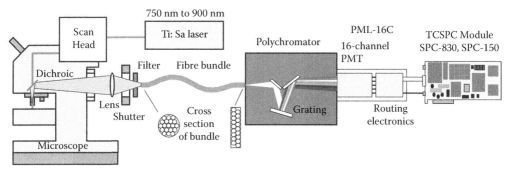

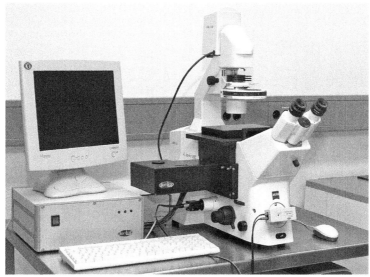

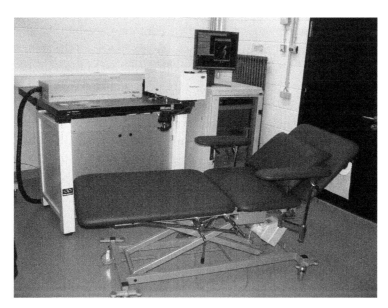

FIGURE 7.5 Scheme of a multiple wavelength TCSPC FLIM system (above). Photograph of the nondescanned microscope "TauMap" for spectrally resolved fluorescence lifetime imaging (middle). Multiphoton tomograph "DermaInspect" for clinical two-photon FLIM of human skin (bottom).

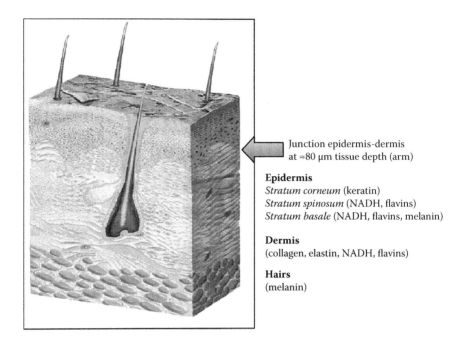

Junction epidermis-dermis
at ≈80 μm tissue depth (arm)

Epidermis
Stratum corneum (keratin)
Stratum spinosum (NADH, flavins)
Stratum basale (NADH, flavins, melanin)

Dermis
(collagen, elastin, NADH, flavins)

Hairs
(melanin)

FIGURE 7.6   Scheme of the skin architecture.

7.5 APPLICATIONS

7.5.1 Multiphoton Skin Imaging

The skin is the largest organ (1.5–2 m^2) of the human and provides a protection shield to the "outer world." In particular, the skin protects the body against successful attack of toxins (chemicals, microorganisms, etc.) against photodamage by UV radiation, as well as prevents the body from significant water loss. Figure 7.6 shows the typical architecture of skin with its major epidermal layers stratum corneum, stratum spinosum, and stratum basale, as well as the upper dermis with its extracellular matrix (ECM) components of elastin and collagen and blood vessels.

Structural and functional imaging of the skin requires subcellular resolution in three spatial dimensions. High-resolution optical sectioning can be performed by two-photon excitation of naturally occurring endogenous fluorescent biomolecules such as flavins, NAD(P)H, metal-free porphyrins (protoporphyrin IX, coproporphyrin), the "age pigment" lipofuscin, melanin, elastin, collagen, and keratin. Melanin has broad and monotonously decreasing absorption spectrum between the UV and the NIR regions and a significant fluorescence in the range from 500 to 650 nm. Three-dimensional fluorescence images can thus be obtained by exclusively using endogenous fluorophores.

As an example, Figure 7.7 shows typical optical sections of normal skin of a Caucasian volunteer obtained with the multiphoton tomograph DermaInspect (Figure 7.5 bottom). An optical section at a resolution of 515 × 512 pixels is obtained in about 8 s acquisition time. The required mean excitation power at the upper skin layers is 2–5 mW. At a depth of 100 μm or deeper, the mean laser power is increased up to 40 mW to compensate for

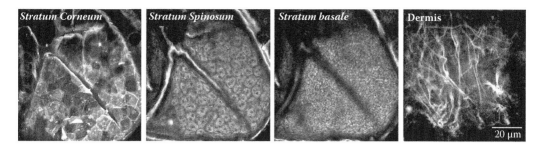

FIGURE 7.7   Typical optical sections of normal Caucasian human skin.

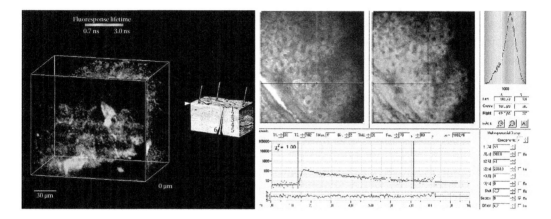

FIGURE 7.8   **(See color insert following page 288.)** Typical autofluorescence decay curves and FLIM images of normal Caucasian human skin.

decreased excitation efficiency due to scattering, refractive index mismatch, and refractive index inhomogeneity. Note the submicron resolution in deep skin areas and the possibility to locate even NAD(P)H in single mitochondria of deep intratissue cells. Also, the elastin and the collagen network can be imaged.

As mentioned before, a variety of endogenous fluorophores are excited simultaneously due to the broad two-photon excitation spectrum. Fluorescence lifetime imaging helps to distinguish between the various autofluorescent species. Figure 7.8 shows typical FLIM images and decay curves in different depths of the tissue. The fluorescence decay kinetics in the outermost layer, the stratum corneum, is mainly attributed to keratin in the "dead" cells that are major parts of the barrier against attacks from the outer world. Deeper tissue layers consist of living cells. The fluorescence decay is dominated by NAD(P)H bound to various proteins. The typical fluorescence lifetime is in the range of 2.2 ns. At a typical depth of 60–90 μm, the mean fluorescence lifetime becomes shorter due to the fluorescence of the pigment melanin in the basal layer. Still deeper, after passing the junction between the epidermis and the dermis, the detected signal is dominated by SHG. The SHG signal comes from collagen, which is present in the dermis, but not in the epidermis.

The investigation of pigmented lesions by multiphoton tomography and three-dimensional multiphoton FLIM provides a possibility of detecting melanoma, the most dangerous skin cancer. In particular, there is a chance that these techniques are able to detect early stages of malignant melanoma. Both normal melanocytic nevi and malignant melanoma are characterized by an enhanced concentration of melanocytes in the basal epidermal layer. However, the presence of melanocytes in the upper epidermal layers is a unique characteristics of malignant melanoma.

The fluorescence of melanin contains an extremely fast decay component, as has been shown by fluorescence decay measurements of the artificial substance DOPA melanin (Ehlers et al. 2007). To resolve the decay profile reliably, the standard detector of the Dermainspect was replaced with an ultrafast Hamamatsu R3809U50 multichannel plate PMT. With this detector, the fluorescence lifetime can be used to distinguish different types of melanin (Ehlers et al. 2007; De Beule et al. 2007). This was demonstrated by recording FLIM data of human hair. Individuals with red hairs (skin type I) possess a high amount of eumelanin, whereas persons with black hairs have mainly pheomelanin. When the fluorescence lifetime distributions were analyzed, significant differences in the fluorescence lifetimes, depending on the type of the hair color (red, blond, black, or gray), were found (Figure 7.9).

The question arises as to whether the fluorescence lifetimes and the fluorescence spectra of melanoma differ in comparison to those of common harmless melanocytic nevi. Differences in the spectra of melanoma and nevi have been discussed (Teuchner et al. 2000; Hoffmann et al. 2001). Also, the absorption process in melanin during femtosecond laser excitation differs from the typical nonresonant two-photon absorption process.

To explore the options of fluorescence lifetime measurements for tumor identification, patients with suspicious pigmented skin lesions were recruited from the Department of Dermatology at the University of Jena, Germany. All patients were extensively informed about the procedure and the risks of multiphoton tomography, and they gave written informed consent. All procedures followed the ethical guidelines of the Helsinki Declaration. The study was approved by the ethical review board.

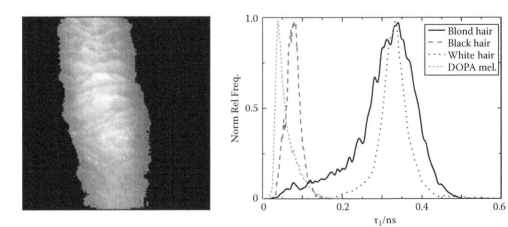

FIGURE 7.9   Lifetime image of a black hair (left) and histograms of fluorescence lifetime distributions for different hair colors (right).

In vivo imaging was performed by the Dermainspect multiphoton tomograph before surgery. After imaging, the tissue was removed and underwent normal histopathological treatment and evaluation (slicing, staining, microscopy). We investigated the pigmented epidermis of 23 cases (13 males, 10 females) and focused on the investigation in melanoma patients of the main epidermal cell populations that are keratinocytes and melanocytes. Ten of the patients were diagnosed with malignant melanoma based on histopathology. Multiphoton sections were taken in different tissue depths up to 200 μm at a typical axial step width of 10 μm.

Scanning was performed with single scans of 4–13 s (512 × 512 pixels covering 320 × 320 μm²) and a laser mean power of 2–50 mW at the target area with respect to the tissue depth. Excitation was performed at 760 and 800 nm. When the excitation wavelength was increased up to 800 nm, the excitation of keratinocytes in the outermost layer and the excitation of NADP(H) in the living cells became less efficient. However, bright and homogeneously fluorescent areas were detected in the basal layer due to luminescent melanin. In the case of malignant melanoma, atypical, pleomorphic, and highly luminescent melanocytes were imaged in upper tissue layers.

Keratinocyte fluorescence decay was found to be dominated by fast fluorescent component of $\tau_1 = 445 \pm 148$ ps (7 6 ± 5%) as well as a minor slow component of $\tau_2 = 2,269 \pm 345$ ps, whereas melanocytes had values of $\tau_1 = 140 \pm 32$ ps (9 3 ± 3%) and $\tau_2 = 1076 \pm 357$ ps. Therefore, multiphoton FLIM sectioning provided evidence of shorter mean fluorescence lifetimes in the epidermal layers of cancer patients with a high amount of melanocytes compared to healthy patients (Figure 7.10). In addition, the ratio of the short to the long component in nonmelanocytes was found to be different, likely due to modifications in the cellular metabolism. A detailed description of the study can be found in Dimitrow et al. (2008a, 2008b).

Multiphoton laser tomography provides optical biopsies for the noninvasive investigation of melanocytic skin lesions with a unique subcellular resolution. However, large clinical trials are required to prove whether multiphoton FLIM tomography is an appropriate tool for the early diagnosis of skin cancer.

7.5.2 Two-Photon FLIM Imaging of Stem Cells

Research on stem cells is one of today's most fascinating fields in biology, pharmacology, biotechnology, tissue engineering, and medicine. What makes these unique cells very attractive is their capability of self-renewal and proliferation and their potential to differentiate into a variety of special cell types. They are attractive candidates to repair damaged or defective tissues of bone, cartilage, liver, heart, and skin (Vassilopoulos, Wang, and Russell 2003; Körbling et al. 2003; Jackson et al. 2007). Stem cells have been used to treat neurological diseases such as Parkinson's and Alzheimer's (Mezey et al. 2003), immunodeficiency (Tsuji et al. 2006; Sato et al. 2007), and diabetes (Dufayet de la Tour et al. 2001), as well as cancer (Joshi et al. 2000; Negrin et al. 2000; Childs et al. 2000).

Current techniques for proving the successful transformation of stem cells into differentiated cell types of interest include polymerase chain reaction (PCR) and immunocytochemistry (ICC). However, often these analyzed cells cannot be used for further in vivo application due to the destructive effects of the examination procedures. There is a need to

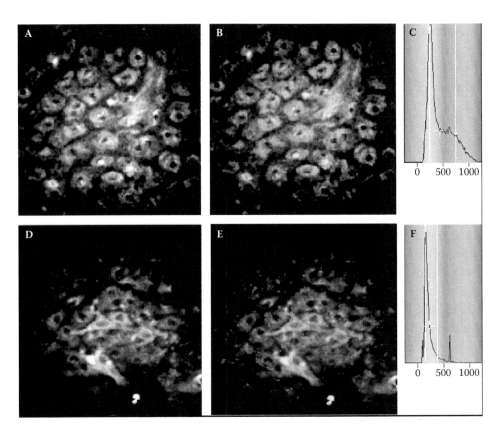

FIGURE 7.10   **(See color insert following page 288.)** False-color-coded FLIM images of nevi (A–C) and melanoma (D–F). Keratinocytes appeared blue/green whereas phleomorphic melanocytes appeared yellow. Color range: 0.25 ns (red) to 1.25 ns (blue) (Dimitrow et al. 2009b).

find a technology to monitor stem cells without an impact on their metabolism and their viability for further therapeutic issues or research. In particular, it would be very helpful to gain information on growth, proliferation, differentiation, and stability of stem cells under physiological conditions. Such technology could be provided by multiphoton imaging.

Currently, adult stem cells are found in most organs. However, the tissue of interest represents a heterogeneous population of stem cells (non-differentiated) and mature cells (differentiated). A significant routine laboratory problem with stem cells is the occurrence of undesired differentiation (Rubio et al. 2005; Kruse et al. 2006). The question arises whether two-photon imaging could distinguish stem cells from mature cells. Another important question is whether the products of biosynthesis of differentiated cells such as collagen in the extracellular three-dimensional environment can be recognized by non-linear imaging.

We performed long-term five-dimensional multiphoton FLIM microscopy with high-spectral, submicron spatial, picosecond temporal resolution. A major advantage of two-photon imaging is the absence of harmful photostress when optimum parameters are used. It is therefore possible to perform long-term studies over several days on the same cells of interest and to monitor the onset of differentiation as well as the further development.

First, information can be gained by imaging the cell's morphology due to autofluorescence intensity signals. Auxiliary information can be acquired by spectral imaging (e.g., using band-pass filters at half the laser wavelength to monitor mainly SHG). A further promising tool is multiphoton FLIM. FLIM can be employed to differentiate between the different types of endogenous fluorophores (in particular NAD(P)H and flavoproteins) as well as to differentiate between autofluorescence and SHG. In addition, FLIM techniques have been employed to monitor the biosynthesis of the ECM protein collagen as well as lipid droplets as a result of stimulated differentiation of animal and human adult stem cells (Uchugonova and König 2008).

Stem-cell lines hSGSC (human salivary gland stem cells), hPSC (human pancreatic stem cells), and rPSC (rat pancreatic stem cells) were used. For imaging, cells were transferred into sterile, miniaturized cell chambers with two 0.17 mm thick glass windows (MiniCeM-Grid, JenLab GmbH, Jena, Germany) and additionally incubated for another day until they were attached to the glass surface. Spheroids of stem cells were prepared by the "pellet culture" method. Approximately 200,000 cells were placed in a 15 ml polypropylene tube and centrifuged to a pellet for 5 min (1,000x/min). The flattened pellet at the bottom of the tube was cultured for 5 or 6 days until a spherical form (spheroid) was formed.

For chondrogenic differentiation, the spheroids were cultured in 1 ml of chondrogenic medium DMEM and supplemented with the transforming growth factor (TGF)-β3. For adipogenic differentiation, the spheroids were cultured in adipogenic medium alpha-MEM supplemented with insulin, isobutylmethylxanthine, and dexamethasone. For microscopy, the pellets were transferred to dishes with 0.17 mm thick glass coverslips.

A ZEISS laser scanning microscope LSM510 Meta NLO equipped with a tunable 80 MHz Ti:sapphire Chameleon laser source was used. For autofluorescence imaging, laser beams at 750, 850, and 900 nm (140 ± 20 fs laser output) and 5 mW mean power were used. The beam splitter HFT SP 650 and the short-pass filter SP 685 were employed. Typically, the cells were scanned in a 512 × 512 pixel field at a 30 s collection time (two scans) for FLIM data. Fluorescence emission was recorded through a Plan-Neofluar 40x NA 1.3 oil objective.

For SHG detection, a PMT with SP 610 filter was mounted to the microscope in the forward direction. SHG signals were collected through a 435 nm band-pass filter (870 nm excitation). Three-dimensional two-photon autofluorescence and SHG images were obtained by optical sectioning with z-intervals of 5–10 μm. FLIM was performed by TCSPS using the SPC 830 board (Becker & Hickl GmbH, Berlin) and PMC-100 detector. Spectral imaging was performed with the 32-channel PMT array (META, Hamamatsu). The grating in the polychromator provided a resolution of 10 nm per channel.

Intrinsic fluorescence from stem cells was detected at different NIR excitation wavelengths in order to separate different endogenous fluorescent molecules. When 750 nm was used, NAD(P)H, as well as flavins and flavoproteins, were excited, whereas 850 or even 900 nm resulted mainly in the fluorescence of flavins and flavoproteins (Uchugonova and König 2008).

Figure 7.11 shows a FLIM image, histogram, and fluorescence decay curve of hSGSC stem cells. The depicted decay curve reflects the autofluorescence of one single intracellular mitochondrium. The biexponential fit (χ^2 = 1.00) reveals a fast component with a

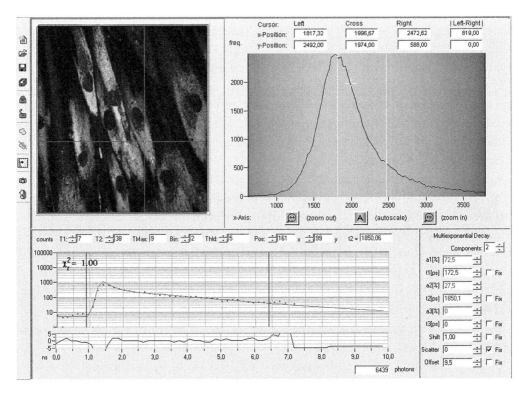

FIGURE 7.11   A FLIM image, histogram, and a fluorescence decay curve of hSGSC stem cells. The decay curve reflects the autofluorescence of one single intracellular mitochondrium.

short lifetime (τ_1) of 0.17 ns and an amplitude of $A_1 = 72\%$ and a second long-lived component with $\tau_2 = 1.8$ ns and $A_2 = 28\%$. When analyzing FLIM images of about 100 cells, we obtained a mean τ_2 value of (1.82 ± 0.02) ns. In contrast, an average value τ_2 of 100 cells of (2.00 ± 0.03) ns was obtained when the excitation wavelength was changed to 900 nm. When the mean lifetime τ_m per pixel was calculated, we obtained an average value of 0.7 ns. The result of a 900 nm excitation wavelength was a longer value of 0.8 ns (Uchugonova and König 2008).

We stimulated the differentiation of stem cells within 0.2–2.2 mm thick three-dimensional spheroids into cells that are able to perform biosynthesis of collagen by adding the growth factors TGF-β3. The spheroid was monitored for a long time period of up to 5 weeks by two-photon imaging (z-steps of 5–10 μm). First, SHG signals were detected after 8 days. As expected, we noticed a highly increased SHG signal from the extracellular matrix after 2 weeks of incubation time. When we used a BP 435/5 filter, we were able to see collagen structures but not living cells and other ECM proteins. When the filter was removed, signals arose from a variety of fluorophores as well as from collagen. In order to differentiate between these different sources of luminescence, we used the FLIM technique. SHG signals have, of course, no "lifetime," whereas the fluorescence lifetime is on the order of hundreds of picoseconds up to 5 ns (flavin mononucleotide, FMN) (König and Schneckenburger 1994; Lakowikz 1999; Skala et al. 2007a, 2007b).

Figure 7.12 shows the differentiation of cells as well as the ECM protein collagen. FLIM images have been used to differentiate between SHG and autofluorescence. If we depict just the very short time interval 0.0–0.3 ns, we observe the SHG signal as well as a fraction of the autofluorescence. Some fluorescence photons always arrive within picoseconds immediately after excitation, even if the fluorescence lifetime is in the range of several nanoseconds. The artifactual "lifetime" of SHG on the order of 0.2 ns is due to the instrument response function. It appears as fast exponential component in the fitting procedure. The image of photons arriving later than 0.3 ns (window of 0.3–5 ns) reflects the autofluorescence only (Uchugonova and König 2008). hSGSCs stem cells can undergo adipogenic differentiation resulting in the formation of mature adipocytes (Rotter 2008). The differentiation process is initiated by the administration of adipogenic differentiation medium.

Adipocyte is a differentiated cell with lipid vacuoles (fat droplets) and specialized in the synthesis and storage of fat. Figure 7.13 shows a transmission as well as a two-photon autofluorescence image of about 10 cells inside a spheroid. The image was taken 7 days

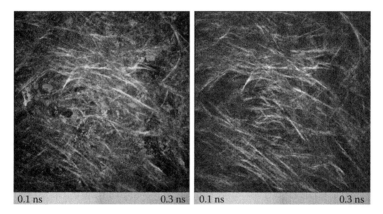

FIGURE 7.12 **(See color insert following page 288.)** False-color-coded FLIM images from cells 4 weeks after onset of differentiation. Left: Autofluorescence and SHG detection in the spectral range of 400–610 nm (870 nm excitation). Fluorescence arises from cells and elastin, SHG from collagen structures. Right: The signal is dominated by SHG of collagen when using a bandpass 435/5 filter.

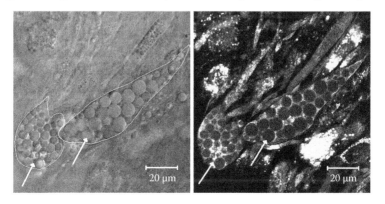

FIGURE 7.13 Transmission image and two-photon autofluorescence image of cells after adipogenic differentiation. (König and Uchugonova, 2009.)

after administration of adipogenic differentiation factors. Two of them are nearly fully developed, with a length of more than 100 μm. The typical diameter of the fat droplets is on the order of 7–10 μm.

A cell can contain more than 20 of these organelles. The nucleus is not longer in the central part of the cell. Some cells at an early stage of differentiation started the formation of these fat organelles. The right cell of the image Figure 7.13 marked with an arrow contains more than 20 organelles with diameters of less than 5 μm. The autofluorescence image shows an intense bright emission of three cells without these organelles. The fluorescence intensity is higher than the normal autofluorescence level of stem cells without any administration of differentiation factors. Very likely these highly fluorescent cells are in very early stages of differentiation and possess a high metabolic activity.

After formation of the fat droplets, the total fluorescence intensity decreases. In fact, no visible autofluorescence was found inside the organelles. We were able to monitor the onset of the formation of the non-fluorescent organelles 3 days after administration. If we perform time-resolved single-photon counting, we obtain information on the fluorescence lifetime and the occurrence of SHG (König and Uchugonova 2008).

In addition, mean parameters per cell could be determined by defining a region of interest (ROI) that covers the cell of interest. The ROI parameters were determined for a highly fluorescent cell without any signs of nonfluorescent organelles (fat droplets) in the right lower corner of the image Figure 7.14. The results show a fast fluorescence decay with a mean τ-value of 0.59 ns, a short 0.22 ns component with an amplitude of 78%, and a slow 1.93 ns component (22%). Interestingly, differentiated cells with large organelles demonstrate a completely different fluorescence decay behavior. The overall mean lifetime is in general much longer (false blue color) and has values of more than 0.85 ns.

A typical fluorescence decay of an area in the periphery exhibited a short lifetime component of (0.26 ± 0.04) ns with an amplitude of $(67 \pm 3)\%$, as well as a strong long-lived component with lifetimes of (2.33 ± 0.16) ns. Because the product of amplitude and fluorescence lifetime describes the contribution to the total fluorescence intensity, the long component determines the steady-state fluorescence behavior. However, many cells possess also a small area where the mean fluorescence decay time becomes shorter. Typically, this area has lifetime parameters $\tau_m = (0.60 \pm 0.04)$ ns, $\tau_1 = (0.17 \pm 0.04)$ ns, $A_1 = (78.0 \pm 1.4)\%$, and τ_2 of (2.00 ± 0.26) ns. This region is likely the area of high accumulation of lysosomes.

Figure 7.14(A) was taken with an excitation wavelength of 750 nm. This wavelength is excellent to excite NAD(P)H. In order to get information on the contribution of flavins, we change the wavelength to 900 nm (Figure 7.14B). The autofluorescence image is much weaker. Only some fluorescent structures are seen. The cell without fat droplets showed a long fluorescence lifetime of 1.99 ns (21%), whereas the short one received values of 0.26 ns (79%). The mean lifetime was determined to be 0.68 ns (König and Uchugonova 2009).

Our first experimental results show that FLIM imaging provides the ability to detect changes in cellular metabolism of endogenous fluorophores from single cells during the differentiation as well as to monitor the results of the differentiation procedure, such as the biosynthesis of collagen and lipids.

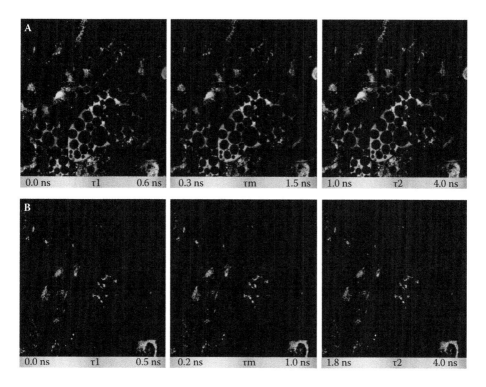

FIGURE 7.14 **(See color insert following page 288.)** False-color-coded FLIM images of cells after adipogenic differentiation at 750 nm excitation (A) and at 900 nm excitation (B). (König and Uchugonova, 2009.)

7.6 CONCLUSION

Multiphoton FLIM has found its way into medical high-resolution tissue imaging. The multiphoton tomograph DermaInspect™ with a time-resolved single-photon counting module has been employed for ex vivo and clinical in vivo examination of human skin to differentiate between different fluorescent endogenous fluorophores, for separation of xenofluorescence from autofluorescence, and for selection and discrimination of SHG signals from a fluorescence background. It can provide information on the intracellular and extracellular microenvironment. In the near future, two-photon microendoscopes that allow high-resolution imaging of intrabody tissues will be available.

Further multicenter clinical studies must be carried out to prove whether multiphoton FLIM can be used to diagnose cancer at a very early stage, to track pharmaceutical and cosmetic compounds, and to improve the efficacy of treatments in dermatology.

Stem cells are becoming novel biotech tools for the treatment of a variety of diseases as well as for tissue engineering. Two-photon imaging can be used to analyze and to trace stem cells in heterogeneous cell populations and to monitor the biosynthesis of fat droplets, collagen, and other biomaterials during the differentiation process. In particular, multiphoton multispectral FLIM has the potential to become a valuable technology in stem cell research (Uchugonova and König 2008; König and Uchugonova 2009).

ACKNOWLEDGMENT

We would like to thank Dr. Wolfgang Becker from the company Becker & Hickl GmbH, Berlin, for his helpful discussions during the preparation of this chapter.

REFERENCES

Becker, W. 2005. *Advanced time-correlated single-photon counting techniques.* Berlin: Springer.

Becker, W. 2008. *The bh TCSPC handbook,* 3rd ed. Becker & Hickl GmbH.

Becker, W., and Bergmann, A. 2008. Lifetime-resolved imaging in nonlinear microscopy. In *Handbook of nonlinear optical microscopy,* ed. B. R. Masters and P. T. C. So. New York: Oxford University Press.

Becker, W., Bergmann, A., and Biskup, C. 2007. Multispectral fluorescence lifetime imaging by TCSPC. *Microscopy Research and Techniques* 70:403–409.

Bird, D. K., Yan, L., Vrotsos, K. M., et al. 2005. Metabolic mapping of MCF10A human breast cells via multiphoton fluorescence lifetime imaging of the coenzyme NADH. *Cancer Research* 65:8766–8773.

Bugiel, I., König, K., and Wabnitz, H. 1989. Investigations of cells by fluorescence laser scanning microscopy with subnanosecond time resolution. *Lasers in the Life Sciences* 3 (1): 47–53.

Chance, B. 1976. Pyridine nucleotide as an indicator of the oxygen requirements for energy-linked functions of mitochondria. *Circulation Research* 38:I31–I38.

Chance, B., Schoener, B., Oshino, R., Itshak, F., and Nakase, Y. 1979. Oxidation–reduction ratio studies of mitochondria in freeze-trapped samples. NADH and flavoprotein fluorescence signals. *Journal of Biological Chemistry* 254:4764–4771.

Childs, R., Chernoff, A., Contentin, N., et al. 2000. Regression of metastatic renal-cell carcinoma after nonmyeloablative allogenic peripheral-blood stem cell transplantation. *New England Journal of Medicine* 343:750–758.

Chorvat, D., and Chorvatova, A. 2006. Spectrally resolved time-correlated single photon counting: A novel approach for characterization of endogenous fluorescence in isolated cardiac myocytes. *European Biophysics Journal* 36:73–83.

De Beule, P. A. A., Dunsby, C., Galletly, N. P., et al. 2007. A hyperspectral fluorescence lifetime probe for skin cancer diagnosis. *Review of Scientific Instruments* 78:123101.

Denk, W., Strickler, J. H., and Webb, W. W. 1990. Two-photon laser scanning fluorescence microscopy. *Science* 248:73–76.

Dimitrow, E., Ziemer, M., Köhler, M. J., et al. 2009a. Sensitivity and specificity of multiphoton laser tomography for in vivo and ex vivo diagnosis of the malignant melanoma. *Journal of Investigative Dermatology,* doi:10.1038/jid.2008.439

Dimitrow, E., Ziemer, M., Köhler, M. J., et al. 2009b. Spectral fluorescence lifetime detection and selective melanin imaging by multiphoton laser tomography for melanoma diagnosis. *Experimental Dermatology,* doi:10.1111/j.1600-0625.2008.00815.x

Dufayet de la Tour, D., Halvorsen, T., Demeterco, C., et al. 2001. B-cell differentiation from a human pancreatic cell line in vitro and in vivo. *Molecular Endocrinology* 15:476–483.

Ehlers, A., Riemann, I., Stark, M., and König, K. 2007. Multiphoton fluorescence lifetime imaging of human hair. *Microscopy Research and Technique* 70:154–161.

Evans, N. D., Gnudi, L., Rolinski, O. J., Birch, D. J., and Pickup, J. C. 2005. Glucose-dependent changes in NAD(P)H-related fluorescence lifetime of adipocytes and fibroblasts in vitro: Potential for noninvasive glucose sensing in diabetes mellitus. *Journal of Photochemistry and Photobiology B Biology* 80:122–129.

Gärtner, W., Gröbler, B., Schubert, D., Wabnitz, H., and Wilhelmi, B. 1988. Fluorescence scanning microscopy combined with subnanosecond time resolution. *Experimental Techniques in Physics* 36:443–451.

Göppert-Mayer, M. 1929. Über die Wahrscheinlichkeit des Zusammenwirkens zweier Lichtquanten in einem Elementarakt. *Die Naturwissenschaften* 17:932.

Göppert-Mayer, M. 1931. Über Elementarakte mit zwei Quantensprüngen. *Annalen der Physik* 9:273–294.

Gulledge, C. J., and Dewhirst, M. W. 1996. Tumor oxygenation: A matter of supply and demand. *Anticancer Research* 16:741–749.

Hoffmann, K., Stucker, M., Altmeyer, P., Teuchner, K., and Leupold, D. 2001. Selective femtosecond pulse-excitation of melanin fluorescence in tissue. *Journal of Investigative Dermatology* 116:629–630.

Jackson, K., Jones, D. R., Scotting, P., Sottile, V., and Postgrad, J. 2007. Adult mesenchymal stem cells: Differentiation potential and therapeutic applications. *Journal of Postgraduate Medicine* 53:121–127.

Joshi, S. S., Tarantolo, S. R., Kuszynski, C. A., and Kessinger, A. 2000. Antitumor therapeutic potential of activated human umbilical cord blood cells against leukemia and breast cancer. *Clinical Cancer Research* 6:4351–4358.

Kantelhardt, S. R., Leppert, J., Krajewski, J., et al. 2007. Imaging of brain and brain tumor specimens by time-resolved multiphoton excitation microscopy ex vivo. *Neuro-Oncology* 9 (2): 103–112.

Köhler, M. J., König, K., Elsner, P., Bückle, R., and Kaatz, M. 2006. In vivo assessment of human skin aging by multiphoton laser scanning tomography. *Optics Letters* 31:2879–2881.

König, K. 1989. PhD thesis. Beiträge zur Photochemtherapie und optischen Diagnostik von Tumoren. Archive of the Friedrich Schiller University, Jena, Germany.

König, K. 2008. Clinical multiphoton tomography. *Journal of Biophotonics* 1:13–23.

König, K., Ehlers, A., Stracke, F., and Riemann, I. 2006. In vivo drug screening in human skin using femtosecond laser multiphoton tomography. *Skin Pharmacology and Physiology* 19:78–88.

König, K. Flemming, G., and Hibst, R. 1998. Laser-induced autofluorescence spectroscopy of dental caries. *Cellular and Molecular Biology* 44:1293–1300.

König, K., and Riemann, I. 2003. High-resolution multiphoton tomography of human skin with subcellular spatial resolution and picosecond time resolution. *Journal of Biomedical Optics* 8:432–439.

König, K., and Schneckenburger, H. 1994. Laser-induced autofluorescence for medical diagnosis. *Journal of Fluorescence* 4 (1): 17–40.

König, K., Schneckenburger, H., and Hibst, R. 1999. Time-gated in vivo imaging of dental caries. *Cellular and Molecular Biology* 45:233–239.

König, K., and Uchugonova, A. 2009. Multiphoton autofluorescence lifetime imaging in stem cell research. Submitted.

König, K., Weinigel, M., Hoppert, D., Schubert, H., Köhler, M.J., Kaatz, M., and Elsner, P. 2008. Multiphoton tissue imaging using high-NA microendoscopes and flexible scan heads for clinical studies and small animal research. *Journal of Biophotonics.* 1:506–513.

König, K., Speicher, M., Bückle, R., Reckfort, J., McKenzie, G., Welzel, J., Koehler, M.J., Elsner, P., and Kaatz, M. 2009. Clinical optical coherence tomography combined with multiphoton tomography of patients with skin diseases. *Journal of Biophotonics.* In press.

König, K., and Wabnitz, H. 1990. Fluoreszenzuntersuchungen mit hoher zeitlicher, spektraler und räumlicher Auflösung. *Labortechnik* 23:26-31.

Körbling, M., Estrov, Z., and Champlin, R. 2003. Adult stem cells and tissue repair. *Bone Marrow Transplant* 32:23–24.

Kruse, C., Kajahn, J., Petschnik, A. E., et al. 2006. Adult pancreatic stem/progenitor cells spontaneously differentiate in vitro into multiple cell lineages and form teratoma-like structures. *Annals of Anatomy* 18:503–517.

Laiho, L. H., Pelet, S., Hancewicz, T. M., Kaplan, P. D., and So, P. T. 2005. Two-photon 3-D mapping of ex vivo human skin endogenous fluorescence species based on fluorescence emission spectra. *Journal of Biomedical Optics* 10:024016 (1–10).

Lakowicz, J. R. 1999. *Principles of fluorescence spectroscopy,* 2nd ed. New York: Springer Science/Business Media.

Lakowicz, J. R., Szmacinski, H., Nowaczyk, K., and Johnson, M. L. 1992. Fluorescence lifetime imaging of free and protein-bound NADH. *Proceedings of the National Academy of Sciences* 89:1271–1275.

Leppert, J., Krajewski, J., Kantelhardt, S. R., et al. 2006. Multiphoton excitation of autofluorescence for microscopy of glioma tissue. *Neurosurgery* 58:759–767.

Masters, B. R. 2008. English translation of and translators notes on Maria Göppert-Mayers's theory of two-quantum processes. In *Handbook of nonlinear optical microscopy*, ed. B. R. Masters sand P. T. C. So. New York: Oxford University Press.

Mezey, E., Key, S., Vogelsang, G., Szalayova, I., Lange, G. D., and Crain, B. 2003. Transplanted bone marrow generates new neurons in human brains. *Proceedings of the National Academy of Sciences USA* 100:1364–1369.

Negrin, R..S., Atkinson, K., Leemhuis, T., et al. 2000. Transplantation of highly purified CD34+ Thy-1+ hematopoietic stem cells in patients with metastatic breast cancer. *Biology of Blood and Marrow Transplantation* 6:262–271.

Niesner, R., Peker, B., Schlusche, P., and Gericke, K. H. 2004. Noniterative biexponential fluorescence lifetime imaging in the investigation of cellular metabolism by means of NAD(P)H autofluorescence. *ChemPhysChem* 5:1141–1149.

O'Connor, D. V., and Phillips, D. 1984. *Time-correlated single photon counting.* London: Academic Press.

Rotter, N., Oder, L., Schlenke, P., et al. 2008. Isolation and characterization of adult stem cells from human salivary glands. *Stem Cells and Development.* 17 (3): 509–518.

Rubio, D., Garcia-Castro, J., Martin, M. C., et al. 2005. Spontaneous human adult stem cell transformation. *Cancer Research* 65:3035–3039.

Sato, T., Kobayashi, R., Toita, N., Kaneda, M., et al. 2007. Stem cell transplantation in primary immunodeficiency disease patients. *Pediatrics International* 49 (6): 795–800.

Schneckenburger, H., and König, K. 1992. Fluorescence decay kinetics and imaging of NAD(P)H and flavins as metabolic indicators. *Optical Engineering* 31:1447–1451

Schroeder, T., Yuan, H., Viglianti, B. L., et al. 2005. Spatial heterogeneity and oxygen dependence of glucose consumption in R3230Ac and fibrosarcomas of the Fischer 344 rat. *Cancer Research* 65:5163–5171.

Schweitzer, D., Hammer, M., Schweitzer, F., et al. 2004. In vivo measurement of time-resolved autofluorescence at the human fundus. *Journal of Biomedical Optics* 9:1214–1222.

Schweitzer, D., Schenke, S., Hammer, M., et al. 2007. Towards metabolic mapping of the human retina. *Microscopy Research and Technique* 70:403–409.

Skala, M. C., Riching, K. M., Bird, D. K., et al. 2007a. In vivo multiphoton fluorescence lifetime imaging of protein-bound and free nicotinamide adenine dinucleotide in normal and precancerous epithelia. *Journal of Biomedical Optics* 12 (2): 024014(1 10).

Skala, M. C., Riching, K. M., Gendron-Fitzpatrick, A., et al. 2007b. In vivo multiphoton microscopy of NADH and FAD redox states, fluorescence lifetimes, and cellular morphology in precancerous epithelia. *Proceedings of the National Academy of Sciences* 104:19494–19499.

So, P., König, K., Berland, K., et al. 1998. New time-resolved techniques in two-photon microscopy. *Cellular and Molecular Biology* 44:771–794.

Teuchner, K., Ehlert, J., Freyer, W., et al. 2000. Fluorescence studies of melanin by stepwise two-photon femtosecond laser excitation. *Journal of Fluorescence* 10:275–281.

Tsuji, Y., Imai, K., Kajiwara, M., et al. 2006. Hematopoietic stem cell transplantation for 30 patients with primary immunodeficiency diseases: 20 years experience of a single team. *Bone Marrow Transplant* 37 (5): 469–477.

Uchugonova, A., and König, K. 2008. Two-photon autofluorescence and second harmonic imaging of adult stem cells. *Journal of Biomedical Optics* 13:054068.

Vassilopoulos, G., Wang, P. R., and Russell, D. W. 2003. Transplanted bone marrow regenerates liver by cell fusion. *Nature* 422:901–904.

Vishwasrao, H. D., Heikal, A. A., Kasischke, K. A., and Webb, W. W. 2005. Conformational dependence of intracellular NADH on metabolic state revealed by associated fluorescence anisotropy. *Journal of Biological Chemistry* 280:25119–25126.

Warburg, O., Posener, K., and Negelein, E. 1924. Ueber den Stoffwechsel der Tumoren. *Biochemische Zeitschrift* 152:319–344

Wilson, T., and Sheppard, C. 1984. *Theory and practice of scanning optical microscopy.* San Diego, CA: Academic Press.

CHAPTER 8

FLIM Microscopy with a Streak Camera

Monitoring Metabolic Processes in Living Cells and Tissues

V. Krishnan Ramanujan, Javier A. Jo,
Ravi Ranjan, and Brian A. Herman

8.1 INTRODUCTION

A critical understanding of subcellular phenomena—from the perspectives of various metabolic processes—stems from a strategic alliance between technological innovations and elegant model systems. Intravital fluorescence imaging modalities rely on exciting specific fluorophores in cells and tissues and interpreting the measured fluorescence emission signals in relation to the cellular processes that can modulate these signals (Richards-Kortum and Sevick-Muraca 1996). A recent upsurge in multiple modes, including spectral, lifetime, polarization etc., has shed valuable information on the various aspects of molecular–cellular and physiological complexity (Lyons 2005; Pawley 2006; Periasamy 2001).

Fluorescence imaging modalities are now beginning to find their niche in disease diagnosis owing to their supreme specificity and sensitivity. One of the emerging trends in understanding various disease pathologies is to view them as metabolic disorders: They either begin as metabolic malfunctions at their early stages (e.g., altered glucose metabolism in tumors or in diabetes) or end up as metabolic dysfunctions (e.g., altered free radical/redox metabolism in age-dependent neurodegeneration).

This implies that our current trends of steady-state fluorescence imaging methods of imaging multiple fluorescent labels in live cells or fixed tissues must evolve into a more "dynamic" imaging platform for monitoring not merely fixed cells or tissues but also metabolic "processes" in live animals. Although reasonably successful, steady-state intensity imaging methods have their own limitations (Hoffman 2002, 2007, 2008; Yang et al. 2000,

2005; Yang, Jiang, and Hoffman 2007). For instance, currently available intensity-based methods for noninvasive optical imaging of solid tumors in small animals are severely affected by spectral artifacts and ubiquitous autofluorescence background (Troy et al. 2004). Thus, these approaches serve merely as visualization tools and are unable to quantify the size and shape of tumors in vivo precisely.

In order to address these vital issues and to test the efficacy of fluorescence lifetime imaging methodology in monitoring metabolic processes, we recently developed a multiphoton fluorescence lifetime imaging microscopy system using a streak camera (StreakFLIM) with excellent spatial resolution (≤200 nm) and time resolution (≤50 ps) (Krishnan, Masuda, et al. 2003; Krishnan, Saitoh, et al. 2003; Chapter 9). By applying this technique in a variety of cell biological situations, we have demonstrated high sensitivity in monitoring subtle fluorescence dynamics of proteins as well in studies of protein–protein interactions in single living cells (Biener et al. 2005; Biener-Ramanujan et al 2006; Gertler et al. 2005; Krishnan, Masuda, et al. 2003; Ramanujan and Herman 2008; Ramanujan et al. 2008).

We have further demonstrated the utility of this MP-FLIM (multiphoton fluorescence lifetime-resolved imaging microscopy) methodology in discriminating background autofluorescence from the fluorescent proteins in thick tissues, thereby achieving a 10-fold increase in signal-to-background ratio over conventional intensity-based approaches (Ramanujan et al. 2005). In this chapter, we will demonstrate a few critical applications of the StreakFLIM system and will discuss future perspectives of the system for advanced clinical applications as well.

8.2 STREAKFLIM: SYSTEM INTEGRATION

A streak camera operates by transforming the temporal profile of a light pulse into a spatial profile on a detector by causing a time-varying deflection of the light across the width of the detector (Krishnan, Masuda, et al. 2003; Krishnan, Saitoh, et al. 2003). The resulting image forms a streak of light, from which the duration of the light pulse can be inferred. In conventional time-domain FLIM methods such as multigate detection, the lifetime is extracted by measuring the fluorescence signal in at least two different time-gated windows.

In streak-camera-based imaging, an optical two-dimensional image with spatial axes (x,y) is converted into a streak image with temporal information and with the axes (x,t) (Ramanujan et al. 2005). The streakscope consists of a photocathode surface, a pair of sweep electrodes, a microchannel plate (MCP) to amplify photoelectrons coming off the photocathode, and a phosphor screen to detect this amplified output of MCP. Two beam scanner mirrors are employed for scanning along the x- and y-directions in the effective field of view (~40 × 40 μm). The x-beam scanner scans a single excitation spot along a single line (658 × 494 points for 1 × 1 binning) across the x-axis. Individual optical pulses (fluorescence emission) are collected from every point along a single line as a beam scanner scans across the x-axis.

When these optical pulses strike the photocathode surface, photoelectrons are emitted. The photoelectrons pass between the pair of sweep electrodes when a high voltage, synchronized with the excitation pulse, is applied to these sweep electrodes. This sweep

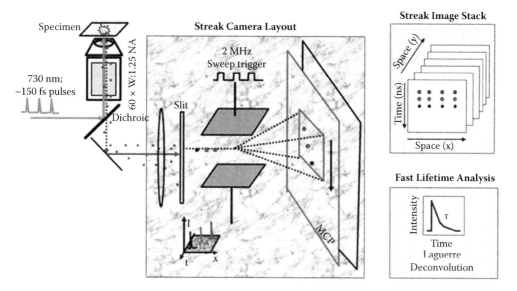

FIGURE 8.1 Schematic of multiphoton StreakFLIM system. Near infrared excitation with femto-second laser pulses elicits fluorescence photon emission from the specimen. The photocathode and the slit assembly allow the corresponding photoelectrons to pass through high-voltage sweep trigger plates, which, in turn, deflect the photoelectrons onto the microchannel plate (MCP). The deflection is in accordance with the time of arrival of photoelectrons at the sweep trigger plates, where early arriving photoelectrons are deflected to the top and late arriving photoelectrons are deflected down-ward. This renders the two-dimensional representation of the line scan along the x-axis as a streak image, with space as the x-axis and time as the y-axis. By a synchronous y-scanning, the entire x,y,t stack, from which fluorescence lifetime decays can be analyzed on a pixel-by-pixel basis, is obtained. More information on the data acquisition and analysis can be found in the main text.

voltage steers the electron paths away from the horizontal direction at different angles, depending on their arrival time at the electrodes (Figure 8.1). The photoelectrons then get amplified in MCP and reach the phosphor screen, forming an image of optical pulses arranged in a vertical direction according to the time of their arrival at the sweep elec-trodes. The earliest pulse is arranged in the uppermost position and the latest pulse is in the bottom-most portion of the phosphor image. The resulting streak image has space as the x-axis and time as the y-axis.

When a synchronous y-scanning is carried out on the region of interest, the streak imaging process gives a complete stack of (x,y,t) streak images. This stack contains the complete information of optical intensity as well as the spatial and temporal information from the optical image. Numerical processing of all these streak images (i.e., the exponen-tial decay profiles at every pixel of the raw streak image) gives the final FLIM image. Every pixel in the FLIM image now contains the lifetime information (in contrast to intensity information in optical image). In this way, complete information on fluorescence decays is obtained on a pixel-per-pixel basis.

The laser system (Model Mira 900, Coherent Inc.) consists of a Ti:sapphire gain medium providing mode-locked, ultrafast femtosecond pulses with a fundamental frequency of 76 MHz. To accomplish the lower repetition rates required for Streak FLIM imaging, we employed a pulse picker (Model 9200, Coherent Inc.), which uses the input signal from the Mira's fast photodiode output to produce variable repetition rates ranging from 146 kHz to 4.75 MHz. The pulse picker employs a high-speed acousto-optic modulator (tellurium dioxide crystal) to extract a single pulse from the input Mira pulse train (see Box 8.1). It derives its synchronization signal from the 76 MHz photodiode output of the Mira laser.

BOX 8.1: PULSE PICKER OR NEUTRAL DENSITY FILTER?

It is pertinent to mention here that the pulse picker is preferred to a simple neutral density filter to attenuate power in the context of multiphoton imaging. This can be rationalized from the following equation for the number of photons absorbed per fluorophore per excitation pulse:

$$N_a \sim [(p_o^2 \times \delta)/(\tau_p\, f_p^2)] \times [(NA)^2/2hc\lambda]^2$$

where

p_o = the average laser power
δ = the fluorophore's two-photon absorption cross section at wavelength λ
τ_p = the pulse duration (~200 fs)
f_p = the laser repetition rate
NA = the numerical aperture of the focusing objective
h = the Planck's constant
c = the velocity of light

As can be seen from this equation, for a constant pulse duration, the two-photon excitation probability increases by increasing the average laser power or by decreasing the laser repetition rate. A neutral density filter attenuates the laser power regardless of the repetition rate and thereby decreases the two-photon excitation probability. On the other hand, a pulse picker can increase the two-photon excitation probability by decreasing the laser repetition rate while keeping the peak laser power the same. In addition to the power requirements, another major advantage of the pulse picker is in modulating the "trigger rate" for the streakscope (for measuring fluorescence lifetimes of various probes) without compromising the peak power of excitation. Operating the Ti:sapphire laser at the full repetition rate (76 MHz) implies a pulse period of ~13.2 ns. This is suitable for measuring lifetimes in the range of 0.5–6 ns without the problems of excitation pulse overlaps.

However, for probes that have longer lifetimes, it is imperative to employ a modulation mechanism where the repetition rate is changed without compromising the two-photon excitation probability. A neutral density filter decreases both the average and pulse power with an additional problem of pulse broadening. However, pulse picker does not introduce pulse broadening and keeps the peak power of the laser pulses. This allows the present system to be able to measure fluorescence lifetimes over a wide dynamic range of ~0.5 to ~2.1 μs (corresponding to a maximum pulse picker repetition rate of 4.75 MHz). This justifies the use of pulse picker instead of a neutral density filter in both the multiphoton and FLIM paths in the present system for optimizing laser power at the sample.

A 60X (1.2 NA) water immersion objective—optimized for NIR excitation—was used in all the experiments reported in this chapter. Nonlinear effects of photobleaching in two-photon excitation mode can be severe in high-resolution biological imaging (Patterson and Piston 2000). However, in our experiments, the average power at the entrance of the microscope was maintained in the range of 8–12 mW. We observed no appreciable photo-bleaching with this imaging condition.

8.2.1 Step-by-Step Demonstration of StreakFLIM System Application

In this section we will demonstrate a unique application of the MP-StreakFLIM system where we measured spatially resolved, nanosecond-scale fluorescence dynamics associated with binding and unbinding kinetics of endogenous metabolic cofactor NADH with enzymes in intact living cells (Ramanujan et al. 2008). Mitochondria serve multiple roles in cellular metabolism. They are the metabolic control centers and bioenergetic hubs (Wallace et al. 1992, 1995, 1998). An emerging picture of a mitochondrial role in the cellular decision- making process is that when they are challenged with exogenous or endogenous perturbations, mitochondria synergistically integrate their apparently dissimilar functions (energy metabolism and programmed cell death) to respond to these perturbations.

Traditionally, enzyme activities in cells and individual organelles have been monitored in cell-free systems in cuvettes. Precise biochemical data obtained from these earlier studies set the so-called "gold standards" in enzymology. However, these biochemical data have only partial physiological significance because enzyme activities in cuvettes need not bear any resemblance to what is really happening in a living cell or even in vivo. It is therefore imperative that we develop tools to monitor enzyme activity on a pixel-per-pixel basis within a living cell.

Subcellular processes and various enzyme activities occur at a wide range of timescales. A real-time monitoring of mitochondrial dynamics (and other subcellular activities) will be a valuable addition to the repertoire of high-resolution optical imaging. It is possible that spatially resolved enzyme kinetics imaging in living cells will shed light on the complex energy landscapes that govern cellular metabolism and eventually tissue metabolism.

8.2.1.1 Data Acquisition

1. Primary hepatocytes were isolated from C57BL/6 mice (young and aged wildtype obtained from the National Institute of Aging) by perfusing the liver as described in Zhang et al. (2002). Mice were anesthetized with Ketamine/S.A. Rompun cocktail and liver was perfused through the hepatic portal vein with 0.5 mM EGTA in Ca^{2+}-free Earle's balanced salt solution (EBSS) for 5 min. This was followed sequentially by wash buffer (0.5% bovine serum albumin and 1.8 mM $CaCl_2$ in Ca^{2+}-free EBSS) and by collagenase buffer (0.06% collagenase, 0.5% bovine serum albumin, and 5 mM $CaCl_2$ in Ca^{2+}-free EBSS) for 10–15 min. The liver was then

further perfused with wash buffer for 5 min and hepatocytes were collected. After being washed three times with cold wash buffer, the hepatocytes were cultured in William's E media supplemented with 5% FBS, 4 μg/mL insulin, 1,000 units/mL penicillin G sodium, 100 μg/mL streptomycin sulfate, and 250 ng/mL amphotericin B in collagen-coated chambered cover glasses (Nalge Nunc International, Naperville, Illinois).

2. Cells were maintained at 37°C in an atmosphere of 5% of CO_2 and 95% air. In order to minimize the dedifferentiation effects due to cell culture, all the experiments were performed within 9–14 h after isolation and the results reported correspond closely to in vivo conditions. All animal surgeries were performed according to the institutional animal protocol guidelines. Before imaging, medium with serum was replaced with Hanks balanced salt solution with low glucose.

3. Fluorescence intensity imaging was performed with a home-built multiphoton imaging platform on an Olympus FVX scanning unit. For data acquisition and analysis software, we employed FluoView 2.1 version (Olympus Inc., New York). Fluorescence lifetime imaging was controlled by the AquaCosmos 2.5 software version (Hamamatsu Photonics, Japan). All experiments were performed with the following imaging conditions: 2p-excitation wavelength 730 nm; 440/90 nm emission bandwidth; Olympus 60X water; 1.2 NA. Typical time required for intensity and lifetime imaging respectively was ~5 s/scan (512 × 512) and 11.1 s/scan (612 × 498). All measurements described in this chapter were carried out at room temperature. More details on the specific conditions for data acquisition and analysis are given in later sections.

4. The lifetime imaging system was previously calibrated with standard solutions and the lifetime resolution was found to be better than 0.16 ns (Krishnan, Saitoh, et al. 2003). After acquiring the raw data (measuring fluorescence decay in every pixel of the field of view in a 10 ns time window), roughly 100 pixels were selected randomly in the time-integrated image (summed over all 64 time channels) and fluorescence decay curves were computed in these pixels.

5. Fluorescence lifetime kinetics in living hepatocytes were obtained in the same configuration described previously. We used mitochondrial complex I inhibition as a metabolic perturbation model in our experiments. For real-time kinetics experiments, complex I inhibitor (10 μM rotenone) was added at the indicated time point during the time-lapse recording of lifetime data.

6. An average decay curve calculated from these pixel decays was analyzed for single- or multicomponent lifetime analysis using Origin 5.0 software (Microcal Inc.). This program calculates lifetime by the best fit by minimizing variance between the proposed model and the experimental data.

8.2.1.2 Data Analysis

7. The raw files from Fluoview were analyzed by ImageJ (NIH) software (http://rsbweb.nih.gov/ij/). A specific set of plug-ins installed in this software allowed us to import FluoView raw image files and convert them to a readable TIFF format for further quantitative analysis. This program can perform quantitative image analysis, including intensity histogram. Lifetime images were analyzed by AquaCosmos software. This software can also convert the raw StreakFLIM images into TIFF files that can further be exported for the Laguerre deconvolution analysis.

8. In addition to the conventional, single-exponential model-fitting procedure (Ramanujan et al. 2005), we have taken advantage of a recently developed Laguerre deconvolution algorithm (Jo et al. 2004, 2006a, 2006b, 2007). In this method, the intrinsic fluorescence decay at a given pixel is expanded as a linear combination of Laguerre functions. Due to their built-in exponential term, these functions are suitable for modeling physical systems with asymptotically exponential dynamics, such as the intrinsic fluorescence decays. In addition, because the Laguerre functions form a complete orthonormal basis, they can always provide a unique and complete expansion of any intrinsic fluorescence decay of arbitrary complexity. A detailed description of the method is given in Box 8.2. Some advantages of this method over standard nonlinear least-squares iterative reconvolution methods are also summarized in Box 8.4.

8.3 CRITICAL DISCUSSION

Figure 8.2(A, B) shows in vitro validation of the system for real-time spatial mapping of the binding and unbinding kinetics of enzyme–substrate complexes by measuring fluorescence lifetimes of NADH in the presence of the classical mitochondrial enzyme, malic dehydrogenase. For in vitro solution measurements, ~100–500 µM NADH was imaged in glass coverslip-bottom chambers and malic dehydrogenase was added at different concentrations (Figure 8.2B). As found in earlier studies, free NADH has a lifetime ~ 0.4 ns, whereas enzyme binding leads to rapid fivefold increase in lifetime.

Over time, a dynamic redistribution of free versus bound NADH species was observed as evidenced by the decrease in lifetime. As can be seen, the equilibrium value of the fluorescence lifetime has a sensitive dependence on enzyme and substrate stoichiometry. A detailed Laguerre analysis and global analysis of these images can be found in Ramanujan et al. (2008).

Complex I (NADH: ubiquinone oxidoreductase) of the mitochondrial respiratory chain is the entry point of substrate metabolism; this catalyzes the electron transfer from reduced NADH to mobile ubiquinones, thereby resulting in conversion of NADH to NAD⁺. Efficiency of electron transport at the site of complex I therefore depends on a dynamic balance between free NADH, which acts as electron carrier, and the enzyme-bound NADH that participates in enzymatic reactions responsible for overall metabolism. During the mammalian life span, there is a chronic progression of mitochondrial dysfunctions accentuated in respiratory chain complexes and ATP synthesis, which results in mitochondrial instability.

BOX 8.2: LAGUERRE FLIM DECONVOLUTION METHOD

The measured FLIM data consist of a set of fluorescence intensity maps as a function of space (x,y) and time (t), forming a three-dimensional cube $H(x,y,t)$ of intensity values. The FLIM data $H(x,y,t)$ represent the convolution of the instrument response $x(t)$ and the undistorted intrinsic fluorescence decay $h(x,y,t)$ that would result under the ideal condition of an instrument response being a delta pulse function. Thus, mathematically, the relation between $H(x,y,t)$ and $h(x,y,t)$ is given by

$$H(x,y,t) = T \cdot \sum_{m=0}^{M-1} h(x,y,m)x(t-m); \quad t = 0,\ldots,N-1 \tag{8.1}$$

Here, T is the sampling time interval at which the intensity maps were acquired, and N is the total number of intensity maps. In order to extract the intrinsic fluorescence decay $h(x,y,t)$ and the lifetime map $\tau(x,y)$, the instrument response must be deconvolved from $H(x,y,t)$. The Laguerre FLIM deconvolution method expands the intrinsic fluorescence decay $h(x,y,t)$ in a set of orthonormal Laguerre functions $b_j(t)$:

$$h(x,y,t) = \sum_{j=0}^{L-1} c_j(x,y).b_j(t) \tag{8.2}$$

Here, $c_j(x,y)$ are the Laguerre expansion coefficients to be estimated, and L is the number of Laguerre functions included in the expansion.

Inserting the Laguerre expansion Equation 8.2 into the convolution Equation 8.1, we obtain the following system of linear equations:

$$H(x,y,t) = \sum_{j=0}^{L-1} c_j(x,y)v_j(t); \quad t = 0,\ldots,N-1 \tag{8.3}$$

Here, $v(t)$ is the convolution of the instrument response with each Laguerre function:

$$v_j(t) = T \cdot \sum_{m=0}^{M-1} b_j(m)x(t-m); \quad t = 0,\ldots,N-1 \tag{8.4}$$

The system of linear Equation 8.3 can be expressed in a matrix notation as follows:

$$\underbrace{\begin{bmatrix} H(x,y,0) \\ H(x,y,1) \\ \vdots \\ H(x,y,N-1) \end{bmatrix}}_{b_{N\times1}} = \underbrace{\begin{bmatrix} v_0(0) & v_1(0) & \cdots & v_{L-1}(0) \\ v_0(1) & v_1(1) & \cdots & v_{L-1}(1) \\ \vdots & \vdots & \ddots & \vdots \\ v_0(N-1) & v_1(N-1) & \cdots & v_{L-1}(N-1) \end{bmatrix}}_{A_{N\times L}} \bullet \underbrace{\begin{bmatrix} c_0(x,y) \\ c_1(x,y) \\ \vdots \\ c_{L-1}(x,y) \end{bmatrix}}_{c_{L\times1}} \tag{8.5}$$

The least-squares analytical solution for Equation 8.5 is given as

$$c = (A^T A)^{-1} A^T b \tag{8.6}$$

Equation 8.6 can be expressed in a term-by-term form as

$$c_j(x,y) = \sum_{k=1}^{L} a_{j+1,k} \sum_{t=0}^{N-1} v_j(t)H(x,y,t); \quad j=0,\ldots,L-1$$

(8.7)

$$(A^T A)^{-1} = [a_{j,k}]_{L \times L}$$

Because the estimation of $c_j(x,y)$ involves only matrix additions, the full deconvolution process can be performed extremely fast. We have demonstrated that our deconvolution method performs at least two orders of magnitude faster than the traditional multiexponential iterative reconvolution.

Once the coefficient maps $c_j(x,y)$ have been calculated, the real fluorescence decays $h(x,y,t)$ can be computed using Equation 8.2. The Laguerre coefficient fully quantifies the fluorescence decay dynamics of the sample under investigation, thus representing a new domain for representing time-resolved fluorescence information. In addition, other, more standards quantities representing the fluorescence decay can also be computed from the estimated intrinsic decay. For instance, the average lifetime map $\tau(x,y)$ can be estimated by averaging the time over the intrinsic fluorescence decay $h(x,y,t)$:

$$\tau(x,y) = \frac{\displaystyle\sum_{t=0}^{M-1} t.h(x,y,t)}{\displaystyle\sum_{t=0}^{M-1} h(x,y,t)}$$

Moreover, the standard multiexponential parameters can also be derived directly from $h(x,y,t)$ by nonlinear least-squares curve fitting, which represents a less computationally intensive task than nonlinear least-squares iterative reconvolution.

Representative steady-state images (Figure 8.2C) of control and complex-I inhibited conditions show significant modifications in median lifetime as well as in LEC coefficients in primary hepatocytes isolated from young (5-month-old) and aged (28-month-old) mice livers. Upon comparing young and aged hepatocytes, it is clear that the effect of rotenone is more acute in young cells than in aged cells. This concurs with the earlier observations in isolated mitochondria that reduced sensitivity to rotenone can be a reliable indicator of age-related mitochondrial dysfunction. Unlike earlier spectroscopic studies, our data clearly show the spatial heterogeneity of free-bound NADH populations at the level of single cells.

We also carried out a two-component lifetime analysis to monitor the real-time modifications in the free and bound NADH populations during complex I inhibition. As seen in Figure 8.2(D), there is a dramatic difference between the kinetics of free and bound populations in young and aged cells. At equilibrium, young cells have almost equal fractions (50%) of these two components and complex I inhibition systematically increases the enzyme-bound fractions while decreasing the free NADH fractions. However, aged cells have a larger (~70%) fraction of enzyme-bound NADH species that does not change

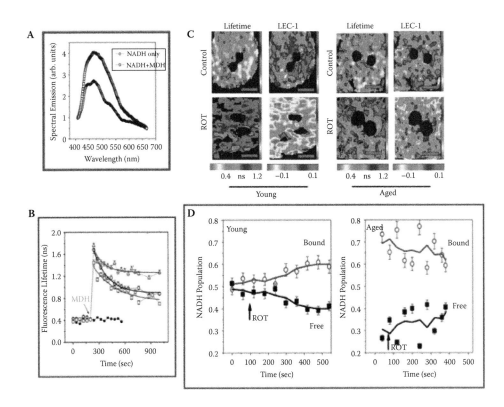

FIGURE 8.2 **(See color insert following page 288.)** Imaging dynamic compartmentalization of free and bound fractions of NADH. (A) Emission spectra of free NADH solution and NADH + malic dehydrogenase (MDH) solution. (B) Temporal kinetics of fluorescence lifetime of free NADH (■); NADH + 0.75 units of MDH (□); NADH + 1.9 units of MDH (○) and NADH + 3.8 units of MDH (△) (one unit will oxidize 1 μm of NADH per minute). (C) Fluorescence lifetime images and representative Laguerre maps in primary hepatocytes isolated from young (5-month-old) and aged (28-month-old) mice under control and complex I inhibited conditions. Scale bars = 10 μm. Reduced rotenone sensitivity is one of the biochemical indicators of age-related modifications in mitochondrial complex I function; the normalized Laguerre expansion coefficients (LECs) reflect how much power the corresponding Laguerre functions contribute to the expansion of the fluorescence decay. Thus, fast fluorescence decays would show larger LEC-1, while negative LEC-1 would reflect a longer tail (delayed longer lifetime) in the fluorescence decay. Similarly, positive and large LEC-2 would reflect a longer tail in the fluorescence decays. Upon rotenone addition, all the normalized LECs show modifications that signify dynamic redistribution of free versus bound NADH species. Addition of rotenone changes this dynamic distribution in such a way that more bound NADH correspondingly yields smaller LEC-0 values. This implies that Laguerre expansion is able to render a realistic picture of dynamic compartmentalization of multiple lifetime components in a spatially resolved manner. An Olympus objective (60X, 1.2 NA, water immersion) was used in this experiment. (D) Real-time two-population lifetime analysis (free and enzyme-bound NADH) in young and aged hepatocytes during complex I inhibition by rotenone.

significantly with exogenous rotenone addition. This can be reasoned by understanding that the aged cells have acquired complex I dysfunction (structural defects in NADH binding sites and/or functional defects in NADH oxidation) during their life span, thereby leading to different equilibrium kinetics as compared with young cells.

The preceding example illustrates the efficacy of the StreakFLIM system in monitoring single-cell responses to metabolic perturbations in real time. Here the lifetime difference between free and protein-bound NAD(P)H is exploited to provide discrimination between mitochondrial activity in young and aged hepatocytes. It should be mentioned here that these measurements did not give any specific information on the NAD(P)H binding partners and/or binding affinity in living cells. To complement the lifetime data, one must carry out detailed biochemical assays to get specific information on any given metabolic pathways.

Furthermore, the two-component lifetime analysis relies strongly on the signal-to-noise ratio (SNR) in the images because the photon utilization and time resolution of the StreakFLIM system are still limited by the detector performance. Typically, 100 photons or pixels give reproducible results for single-exponential decay analysis. In our system, we could achieve this condition easily by an image acquisition time (CCD exposure time × image size) of about 11 s for an image size of 612 pixels × 498 pixels. Time-lapse imaging (kinetics) did not affect the accuracy and reproducibility of the calculated lifetimes. However, for two-component analysis, in order to obtain reproducible results without significant photobleaching, we limited the image size by 2 × 2 binning (306 pixels × 249 pixels) and/or increasing the data acquisition time up to 44 s/image.

As explained in Boxes 8.2 and 8.4, the Laguerre deconvolution approach could yield a dynamic range of two-component lifetimes from 0.4 to 4 ns without significant errors (<10%) in the estimated lifetimes. Although the streak camera has good radiant sensitivity over a wide bandwidth of wavelengths, one cannot avoid background noise arising from multiple scattering events and absorption in multilayer cell structures (e.g., tissues).

Different approximations in multicomponent lifetime analysis are applicable in limited situations and one must take care to validate the assumptions in each model in the biological situation before interpreting the results. For instance, assigning certain parameters to be invariant (i.e., globally shared between the data sets) invokes an a priori assumption of a physical model on the system. As Ladokhin and White (2001) point out, among the several models of good fit with exponentials, physical or biological relevance should be the chief criterion for accepting a model.

In this context, the Laguerre expansion technique employed in our study offers distinct advantages over standard deconvolution algorithms. First, intrinsic fluorescence intensity decays of any form can be estimated at every pixel of the image as an expansion on a Laguerre basis without a priori assumption of its functional form, thus providing a robust and versatile method for FLIM analysis. Furthermore, because the fluorescence IRF at every pixel is expanded in parallel using a common Laguerre basis, the computation speed is extremely high, performing at least two orders of magnitude faster than standard deconvolution algorithms. More details are given in Boxes 8.2, 8.3, and 8.4.

BOX 8.3: GLOBAL ANALYSIS OF LIFETIME DATA

In some instances, we can assume that the overall sample fluorescence is a linear combination of the fluorescence signal from a few specific fluorophores present in the field of view. Let us assume that the sample fluorescence decay can be expanded as

$$h(x,y,t) = \sum_{j=0}^{L-1} c_j(x,y).b_j(t) \qquad (8.8)$$

Let us also assume that M fluorophores are present in the sample, each with intrinsic fluorescence decays:

$$h_k(x,y,t) = \sum_{j=0}^{L-1} a_{k,j}(x,y)b_j(t); \quad k = 1,...,M \qquad (8.9)$$

Here, $a_{k,j}(x,y)$ are the expansion coefficients of the kth fluorophore decay.

Finally, let us assume that the sample fluorescence decay can also be modeled as the linear combination of the M components' decays as

$$h(x,y,t) = \sum_{k=1}^{M} A_k(x,y)h_k(x,y,t) \qquad (8.10)$$

Here, $A_k(x,y)$ is the spatial distribution of the relative concentration of the individual fluorophores excited in the sample. By rearranging these equations, we obtain the following relation:

$$c_j(x,y) = \sum_{k=1}^{M} A_k(x,y)a_{k,j}(x,y) \qquad (8.11)$$

In practice, $c_j(x,y)$ and $a_{k,j}(x,y)$ can be estimated from the FLIM measurements, and the spatial distributions of the relative concentration $A_k(x,y)$ can be computed by solving the system of linear Equation 8.11.

8.4 FURTHER APPLICATIONS

8.4.1 Imaging Cancer Cells in Three-Dimensional Architecture

A major issue that confounds tumor imaging in small animal models is the autofluorescence background of the tissues (Ramanujan et al. 2005; Troy et al. 2004). A simple situation of imaging subcutaneously generated tumor can be complicated by the multiple scattering and absorption due to the intervening skin tissues as well as the normal tissue surrounding the tumor. We needed to optimize the system performance in the two emission windows of interest (EGFP: 525/50 nm and NADPH: 440/90 nm) and to extend the scope of lifetime imaging to tissue depths more than 100 μm. Therefore, we measured intensity- and lifetime-depth profiles using a long working-distance objective lens (LUMFL, 60X, 1.1 NA, Olympus

BOX 8.4: ADVANTAGES OF LAGUERRE EXPANSION METHOD COMPARED WITH NONLINEAR LEAST-SQUARES FITTING METHODS

This approach for fitting fluorescence lifetime decays is an effective method for deconvolving the instrument response function, and it provides a number of advantages over forward convolution and nonlinear least-squares fitting methods for calculating ideal fluorescence decays (Jo et al. 2004a, 2006a). The Laguerre polynomials need only be convolved once each with the recorded instrument response function before they are used in the expansion of the recorded fluorescence decay. This is a linear process and is computationally much less intensive than fitting methods—a property that is especially important in FLIM, where a large number of parallel lifetime measurements must be analyzed.

Furthermore, the Laguerre functions form a complete and orthogonal basis, allowing an unambiguous expansion of the fluorescence decay. This is a distinct advantage over multiple exponential decay approaches, which often produce ambiguous results due to the presence of noise and the well-known correlation between the lifetimes and initial intensities.

After the expansion is calculated, the initial intensity and average fluorescence lifetime can be calculated along with the Laguerre coefficient maps, which have previously been shown to produce additional useful contrast in biological specimens. The normalized Laguerre expansion coefficients (LECs) reflect how much power the corresponding Laguerre functions contribute to the expansion of the fluorescence decay. Thus, fast fluorescence decays would show larger LEC-1, while negative LEC-1 would reflect a longer tail (delayed longer lifetime) in the fluorescence decay. Similarly, positive and large LEC-2 would reflect a longer tail in the fluorescence decays. With the advent of novel imaging strategies to interrogate living cells in real time, it is equally important to develop image analysis methods that can extract maximum information from the acquired data quickly and reliably. In this chapter, we have reported fast fluorescence lifetime data acquisition using a streak camera, as well as the fast data lifetime decay analysis method based on Laguerre deconvolution. This method presents a number of advantages over conventional lifetime data analysis methods:

1. The intrinsic fluorescence intensity decays of any form can be estimated at every pixel of the image as an expansion on a Laguerre basis, without a priori assumption of its functional form, thus providing a robust and versatile method for FLIM analysis.
2. Because the fluorescence decay at every pixel is expanded in parallel using a common Laguerre basis, the computation speed is extremely fast, performing at least two orders of magnitude faster than standard FLIM deconvolution algorithms (Jo et al. 2004b, 2006b).
3. The estimated maps of Laguerre expansion coefficients represent a new domain of representing the spatial distribution of the time-resolved characteristics of the fluorescence system being imaged by FLIM.
4. Only a few images (~5–10 images) are required for deconvolution, resulting in short acquisition time (typically 6–10 s for an image size of 150 × 150 pixels).
5. Because actual time deconvolution of the instrument response is performed at every pixel of the images, ultrafast light sources are no longer required, allowing for development of less expensive FLIM systems.

Inc., New York) in a gel phantom containing fluorescein and NADH. Figure 8.3(A) and 8.3(B) compare the intensity and lifetime profiles in these gel phantoms down to ~1,000 μm. As one can see, lifetime measurements yield robust data with a good dynamic range (fluorescein: pH 7.5: $\tau = 3.5 \pm 0.15$ ns and NADH: pH 7.5: $\tau = 0.48 \pm 0.10$ ns).

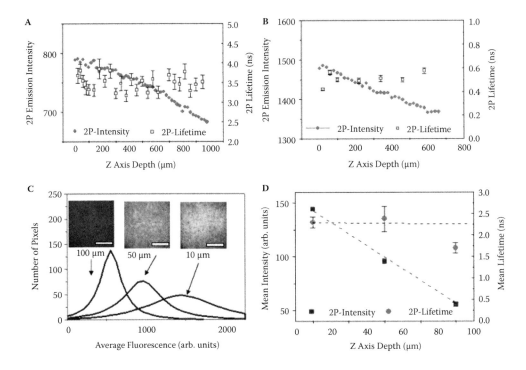

FIGURE 8.3 Lifetime imaging in ex vivo tissues and gel phantoms. (A) and (B) Intensity and lifetime values at different depths (maximum depth of 1,000 μm) measured in fluorescein and NADH solidified in 3% agarose gel, respectively. Note that lifetime stability is good over a large dynamic range (0.48 to ~4 ns). (C) 2p-Fluorescence intensity images obtained in a prostate tumor tissue section at representative depths (10, 50, and 100 μm) within the tissue. Mean fluorescence histograms are also shown that indicate a clear reduction in signal (note that the intensity is in the x-axis). Scale bars = 50 μm. An Olympus objective (60X, 1.2 NA, water immersion) was used in this experiment. (D) Comparison of intensity- and lifetime-based methods for the dependence of SNR on tissue depths. Note that lifetime values are fairly constant (<20%), whereas intensity values are significantly reduced (~66%).

Recently, we hypothesized that in addition to EGFP imaging in tumor tissues, lifetime measurements can shed more light on metabolic differences in tumor tissue as compared with their normal counterparts. In order to obtain a high SNR, either the fluorescence markers of the tumor have to be increased or the autofluorescence background has to be discriminated reliably well. An established practice in generating tumor xenografts is to inject cancer cells stably expressing one of the visible fluorescent proteins and then allow the tumors to reach a certain size to be imaged. Because the autofluorescence background has a relatively broad fluorescence emission spectrum, a reasonable SNR is achieved only by allowing the tumor to grow sufficiently large (beyond the early detection limits). Near-infrared fluorescent proteins are currently being investigated to circumvent this problem. However, it would be useful to discriminate autofluorescence background from the fluorescent protein signal so that even smaller tumors can be detected.

The major goal of this section is to lay the foundation for extending the scope of the current MP-FLIM methodology to imaging intrinsic autofluorescence signals and transgenic fluorescent signals in live cells and in thick tissues. We demonstrated reasonably rapid (~20–40 s) lifetime acquisition for imaging EGFP and NAD(P)H in living cells without compromising on spatial–temporal resolution. We compared the depth dependence of SNR in both MP intensity and lifetime detection mode and found that lifetime measurements give a significant improvement in detection sensitivity over the intensity measurements.

For ex vivo tissue imaging, we used a small section (~1 mm³) from a prostate tumor tissue originally produced by orthotopic injection of EGFP-expressing tumor cells in prostate gland. After 6–8 weeks of injection, the tumor formed in prostate gland was isolated and frozen at −80°C. Data acquisition and analysis were carried out as described in an earlier section. The mean lifetime of EGFP in cells is $\tau_{mean} = 2.65 \pm 0.08$ ns, which is in accordance with literature reports. When measured in a frozen prostate cancer tissue, the mean lifetime of EGFP was slightly lower—$\tau_{mean} = 2.25 \pm 0.20$ ns, which may be due to the combined effects of tissue scattering and absorption and autofluorescence contributions. We have recently shown that a multicomponent lifetime analysis strategy faithfully extracts the EGFP and autofluorescence lifetime contributions (Ramanujan et al. 2005). However for the present discussion, we confine ourselves to single-component analysis because the main purpose is to contrast the intensity and lifetime-based methods in deep tissue depth profiling.

Figure 8.3(C) shows the mean fluorescence intensity histogram from the tissue section obtained using 2p-intensity imaging mode. As one can see clearly, there is a considerable reduction in signal as one moves from 10 to 100 μm (plotted in Figure 8.3D). On the other hand, the lifetime values extracted at these depths are fairly constant (also plotted in Figure 8.3D), indicating that lifetime-based detection can offer a significant improvement in SNR over 2p-intensity methods. This feature is very crucial in reconstructing the three-dimensional tumor in deeper layers of small animal tissues. Figure 8.3(D) is only a proof-of-concept demonstration of the inherent difference between intensity- and lifetime-based tumor imaging. As described in the earlier section, spatially resolved redox imaging in tumor tissues is a novel direction in detecting early signatures of tumorigenesis.

8.4.2 Kinetic Imaging of pH Transients during Glucose Metabolism

Spatially resolved imaging of glucose metabolism and its subtle alterations might provide valuable diagnostic information in vivo. A classical example is positron emission tomography that exploits this feature in obtaining preferential accumulation of fluorescent analog of glucose in tumors, thereby achieving an imaging contrast. Here we demonstrate the efficacy of the StreakFLIM system in real-time monitoring of glucose-induced pH transients in neuronal cell cultures. Earlier studies employed ratiometric pH probes such as BCECF to measure intracellular pH in living cells. Another elegant study used this same probe for measuring pH gradients in skin tissues by frequency-domain fluorescence lifetime imaging microscopy (Hanson et al. 2002). We used the same probe in our measurements to monitor the effect of glucose metabolism on pH, as depicted in Figure 8.4(A).

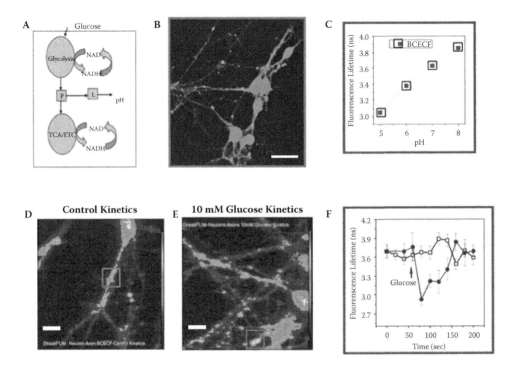

FIGURE 8.4 **(See color insert following page 288.)** (A) Schematic of bioenergetic pathway involving glycolysis, mitochondrial tricarboxylic acid cycle (TCA). Glycolytic by-product pyruvate (P) has multiple fates: It can enter the mitochondrial TCA cycle or can be converted to lactic acid. Acute glucose metabolism can lead to increased conversion of pyruvate to lactate, thereby leading to transient acidosis of the cytosol. (B) Representative fluorescence intensity image of primary neurons labeled with BCECF; scale bar = 50 μm. An Olympus objective (60X, 1.2 NA, water immersion) was used in this experiment. (C) pH calibration curve of BCECF measured by StreakFLIM system. (D) and (E) Representative time-resolved (FLIM) image of neurons labeled with BCECF; scale bar = 20 μm. (F) Fluorescence lifetime kinetic profiles of BCECF-labeled neurons in the marked regions in (D) and (E). Blue symbols represent control kinetics (D) and red symbols represent lifetime kinetics during glucose metabolism (E).

Primary neuronal cultures from a 5-month-old rat brain were labeled with 1μM BCECF at 37°C for 30 min before imaging at room temperature. Figure 8.4(B) shows a representative BCECF fluorescence intensity image of live neuronal cultures. Systematic pH calibration was carried out by maintaining the cells in high K^+ buffers at varying pH (5, 6, 7, and 8) at 37°C for 30 min before imaging at room temperature. All the experiments at fixed pH were performed by clamping pH with 5 μM nigericin in the corresponding pH buffers. BCECF lifetime displayed a robust pH-sensitive response of $\Delta\tau = \pm 10\%$ above and below neutral pH. Figure 8.4(D, E) shows a representative time-resolved intensity image of the live neuronal cells before and during glucose metabolism. The kinetic profiles of fluorescence lifetime (Figure 8.4F) show distinctly the extreme sensitivity of the StreakFLIM system to report pH transients in real time.

We recently reported a novel scaling analysis of glucose metabolism in Cerulean (cyan fluorescent protein variant) fluorescence (Ramanujan and Herman 2008). Because cancer cells usually have altered glucose metabolism, we hypothesized that normal and cancer cells could be discriminated based on their differential glucose metabolism, as depicted in Figure 8.4(A). We monitored pH-dependent fluorescence lifetime of Cerulean in normal and cancer cells during glucose metabolism and attempted to correlate the FLIM data with the time-series data.

Fluorescence fluctuations of Cerulean (time series) were found to be indicative of dynamic pH changes associated with glucose metabolism. Acute dependence of cancer cells on glycolysis as compared with normal cells is exploited to yield a statistically significant difference in the scaling exponent, thereby providing discrimination between normal and cancer cells in vitro. By careful experiments in vivo, the proposed scaling approach might even have diagnostic potential for early detection of cancer lesions in small-animal models.

8.4.3 FLIM-Based Enzyme Activity Assays In Vivo

As a final application of FLIM-based metabolic imaging, we present an interesting case of monitoring real-time enzymatic activity in the Malpighian tubule system of a live fly (*Drosophila melanogaster*). The Malpighian tubule system is a type of excretory system and osmoregulatory system found in many insects, including *Drosophila melanogaster*. The system consists of branching tubules extending from an alimentary canal that absorbs solutes, water, and wastes from the surrounding hemolymph. Insect Malpighian (renal) tubules perform jobs analogous to the human kidney. They purify the blood of waste materials, excrete and adjust a primary urine, and thus play a major role in ion and water homeostasis. They are also major immune tissues, and they detoxify many compounds, like a human liver does.

We expected the *Drosophila* tubule model to be useful at many levels: as a model epithelium to study the basic biology of transport and signaling, as a potential target tissue for novel insecticides, and as a model system to study human genes and human renal diseases. We attempted to image the NAD(P)H activity in the Malpighian tubes in real time while a living fly was positioned underneath the glass cover slip (Figure 8.5). Figure 8.5(B) shows a two-photon excited NAD(P)H intensity image of a live fly obtained with a 20X objective. Figure 8.5(C) shows the high-magnification steady-state intensity image of the Malpighian tubules of the same fly. Figure 8.5(D) shows the time-resolved image obtained with the StreakFLIM system.

As can be seen from the steady-state intensity image (Figure 8.5D), the two segments marked 1 and 2 have identical intensity values and one cannot discriminate these two segments based only on the intensity values. However, the time-resolved image (Figure 8.5E) clearly shows differences in decay profiles from these two segments, indicating differential NAD(P)H activity in the two segments of the Malpighian tubes. Free NADH has a very short lifetime (0.4 ns), whereas protein-bound NADH has longer lifetimes (1.5–2 ns). By differential NAD(P)H activity, we mean that there is a distribution of free and protein-bound NADH populations (metabolic activity) in real time.

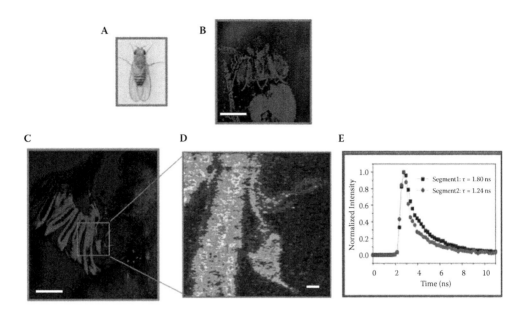

FIGURE 8.5 **(See color insert following page 288.)** (A) A male *Drosophila melanogaster.* (B) Multiphoton fluorescence NAD(P)H intensity image of a partial field of view of a live *Drosophila;* scale bar = 100 μm. An Olympus objective (20X, 0.7 NA) was used to collect this image. (C) Multiphoton intensity image of the Malpighian tubule system of the live *Drosophila;* scale bar = 50 μm. (D) Time-resolved intensity image of a partial section of the Malpighian tubes; scale bar = 10 μm. An Olympus objective (60X, 1.2 NA, water immersion) was used for obtaining (C) and (D). (E) Representative fluorescence decay profiles from segments 1 and 2 in (D); free NADH has a very short lifetime (0.4 ns), whereas protein-bound NADH has longer lifetimes (1.5–2 ns). This figure illustrates a unique niche for fluorescence lifetime imaging for monitoring spatially resolved, differential metabolic activity (in the present case, different distribution of free and protein-bound NADH population) in live animals. As mentioned in the text, this feature is very valuable in employing endogenous fluorescence contrast agents (autofluorescence) that can also report intrinsic metabolic activity of normal and disease tissues (e.g., cancer).

Many aspects of *Drosophila* biology and physiology are waiting to be explored by genetics and the new genomic approaches. We expect the future of *Drosophila* research to turn increasingly to questions beyond the cellular level—to questions of physiology, maintenance, and regeneration of whole organs, and to systemic processes. For example, some human renal disorders are associated with defects in genes involved in fluid and electrolyte transport. *Drosophila* orthologs of these genes found in the genomic sequence should spur studies of the physiology and function of Malpighian tubules, which serve as the *Drosophila* "kidney." This possibility opens an interesting avenue to implement functional imaging of neuromuscular activity in the various organs. By correlating with specific neuronal markers, it is even possible to extend this imaging capability to map a larger network of neuronal activity in live brain. We anticipate new collaborations between researchers of vertebrates and flies to study behavior, neurodegeneration, aging, and drugs and that important new biological principles and pathways will emerge.

8.5 SUMMARY AND FUTURE PERSPECTIVE

Multiphoton excitation methods are finding increasing applications in deep-tissue imaging owing to near-infrared wavelengths (700–1,000 nm) and hence increased penetrability as well as reduced phototoxicity. Fluorescence lifetime imaging microscopy (FLIM) has gained attention in recent times owing to its detection sensitivity, temporal resolution, and ability to probe subtle fluorescence dynamics and intrinsic cellular environment (pH changes, ion concentration, etc.). A synergetic combination of multiphoton excitation and fluorescence lifetime detection is a valuable tool to exploit the combined advantages of deep tissue imaging and high-sensitivity lifetime detection of tissue microenvironments.

As exemplified by the aforementioned applications, we believe that the MP-StreakFLIM system will find its unique niche in metabolic imaging of vital processes in animal models. We demonstrated three salient features of FLIM methodology: namely, (1) autofluorescence discrimination, (2) pH sensing, and (3) metabolic imaging.

Traditional cuvette-based enzyme studies lack spatial information and do not allow real-time monitoring of the effects of modulating enzyme functions in vivo. In order to probe the realistic time scales of steric modifications in enzyme/substrate complexes and functional binding/unbinding kinetics in living cells without losing spatial information, it is imperative to develop sensitive imaging strategies that can report enzyme kinetics in real time over a wide dynamic range of timescales. In this context, our results on monitoring endogenous metabolic markers (NAD(P)H) obviate the necessity of using exogenous fluorescent labels. Kinetic measurements of LEC parameters and rapid lifetime analysis of multiple components can be valuable diagnostic tools in monitoring metabolic dysfunctions in vivo.

Although NADH imaging in both single- and two-photon modalities has been reported earlier, to the best of our knowledge, our results (Figure 8.2) are the first report of real-time NADH-enzyme kinetics in living cells using a multiphoton FLIM approach. Owing to the fast acquisition and analysis, our approach opens up avenues in clinical diagnostic applications where on-site rapid measurements of enzyme modifications can be valuable.

Noninvasive monitoring of metabolic dysfunctions in intact animal models is an attractive strategy for gaining insight into the dynamics of tissue metabolism in health and in various metabolic syndromes such as cancer, diabetes, and aging-induced metabolic dysfunctions. In addition to the example demonstrated in Figure 8.5, we envisage that the proposed method can find applications in a variety of other situations where intensity-based approaches fall short owing to spectroscopic artifacts. For clinical applications, it will be useful to consider employing hybrid modalities (radiology imaging + optical imaging) to maximize the benefits of multiple imaging platforms.

REFERENCES

Biener, E., Charlier, M., Ramanujan, V. K., et al. 2005. Quantitative FRET imaging of leptin receptor oligomerization kinetics in single cells. *Biology of the Cell* 97:905–919.

Biener-Ramanujan, E., Ramanujan, V. K., Herman, B., and Gertler, A. 2006. Spatio-temporal kinetics of growth hormone receptor signaling in single cells using FRET microscopy. *Growth Hormone & Igf Research* 16:247–257.

Gertler, A., Biener, E., Ramanujan, V. K., Djiane, J., and Herman, B. 2005. Fluorescence resonance energy transfer (FRET) microscopy in living cells as a novel tool for the study of cytokine action. *Journal of Dairy Research* 72 (spec. no.): 14–19.

Hanson, K. M., Behne, M. J., Barry, N. P., Mauro, T. M., Gratton, E., and Clegg, R. M. 2002. Two-photon fluorescence lifetime imaging of the skin stratum corneum pH gradient. *Biophysical Journal* 83:1682–1690.

Hoffman, R. M. 2002. Whole-body fluorescence imaging with green fluorescence protein. *Methods in Molecular Biology* 183:135–148.

Hoffman, R. M. 2007. Subcellular imaging of cancer cells in live mice. *Methods in Molecular Biology* 411:121–129.

Hoffman, R. M. 2008. Recent advances on in vivo imaging with fluorescent proteins. *Methods in Cell Biology* 85:485–495.

Jo, J. A., Fang, Q., Papaioannou, T., and Marcu, L. 2004. Fast model-free deconvolution of fluorescence decay for analysis of biological systems. *Journal of Biomedical Optics* 9:743–752.

Jo, J. A., Fang, Q., Papaioannou, T., et al. 2006a. Laguerre-based method for analysis of time-resolved fluorescence data: Application to in vivo characterization and diagnosis of atherosclerotic lesions. *Journal of Biomedical Optics* 11:021004.

Jo, J. A., Fang, Q., Papaioannou, T., et al. 2006b. Diagnosis of vulnerable atherosclerotic plaques by time-resolved fluorescence spectroscopy and ultrasound imaging. *Conference Proceedings: Annual International Conference of the IEEE Engineering in Medicine & Biology Society* 1:2663–2666.

Jo, J. A., Marcu, L., Fang, Q., et al. 2007. New methods for time-resolved fluorescence spectroscopy data analysis based on the Laguerre expansion technique—Applications in tissue diagnosis. *Methods of Information in Medicine* 46:206–211.

Krishnan, R. V., Masuda, A., Centonze, V. E., and Herman, B. 2003. Quantitative imaging of protein–protein interactions by multiphoton fluorescence lifetime imaging microscopy using a streak camera. *Journal of Biomedical Optics* 8:362–367.

Krishnan, R. V., Saitoh, H., Terada, H., Centonze, V. E., and Herman, B. 2003. Development of a multiphoton fluorescence lifetime imaging microscopy (FLIM) system using a streak camera. *Review of Scientific Instruments* 74:2714–2721.

Ladokhin, A. S., and White, S. H. 2001. Alphas and taus of tryptophan fluorescence in membranes. *Biophysical Journal* 81:1825–1827.

Lyons, S. K. 2005. Advances in imaging mouse tumor models in vivo. *Journal of Pathology* 205:194–205.

Patterson, G. H., and Piston, D. W. 2000. Photobleaching in two-photon excitation microscopy. *Biophysical Journal* 78:2159–2162.

Pawley, J. B. 2006. *Handbook of biological confocal microscopy.* New York: Plenum.

Periasamy, A. 2001. *Methods in cellular imaging.* New York: Oxford University Press.

Ramanujan, V. K., and Herman, B. A. 2008. Nonlinear scaling analysis of glucose metabolism in normal and cancer cells. *Journal of Biomedical Optics* 13:031219.

Ramanujan, V. K., Jo, J. A., Cantu, G., and Herman, B. A. 2008. Spatially resolved fluorescence lifetime mapping of enzyme kinetics in living cells. *Journal of Microscopy* 230:329–338.

Ramanujan, V. K., Zhang, J. H., Biener, E., and Herman, B. 2005. Multiphoton fluorescence lifetime contrast in deep tissue imaging: Prospects in redox imaging and disease diagnosis. *Journal of Biomedical Optics* 10:051407.

Richards-Kortum, R., and Sevick-Muraca, E. 1996. Quantitative optical spectroscopy for tissue diagnosis. *Annual Review of Physical Chemistry* 47:555–606.

Troy, T., Jekic-McMullen, D., Sambucetti, L., and Rice, B. 2004. Quantitative comparison of the sensitivity of detection of fluorescent and bioluminescent reporters in animal models. *Molecular Imaging* 3:9–23.

Wallace, D. C., Brown, M. D., Melov, S., Graham, B., and Lott, M. 1998. Mitochondrial biology, degenerative diseases and aging. *BioFactors* (Oxford, England) 7:187–190.

Wallace, D. C., Shoffner, J. M., Trounce, I., et al. 1995. Mitochondrial DNA mutations in human degenerative diseases and aging. *Biochimica et Biophysica Acta* 1271:141–151.

Wallace, D. C., Shoffner, J. M., Watts, R. L., Juncos, J. L., and Torroni, A. 1992. Mitochondrial oxidative phosphorylation defects in Parkinson's disease. *Annals of Neurology* 32:113–114.

Yang, M., Baranov, E., Jiang, P., et al. 2000. Whole-body optical imaging of green fluorescent protein-expressing tumors and metastases. *Proceedings of the National Academy of Sciences USA* 97:1206–1211.

Yang, M., Jiang, P., and Hoffman, R. M. 2007. Whole-body subcellular multicolor imaging of tumor–host interaction and drug response in real time. *Cancer Research* 67: 5195–5200.

Yang, M., Jiang, P., Yamamoto, N., et al. 2005. Real-time whole-body imaging of an orthotopic metastatic prostate cancer model expressing red fluorescent protein. *Prostate* 62:374–379.

Zhang, Y., Chong, E., and Herman, B. 2002. Age-associated increases in the activity of multiple caspases in Fisher 344 rat organs. *Experimental Gerontology* 37:777–789.

CHAPTER 9

Spectrally Resolved Fluorescence Lifetime Imaging Microscopy

SLIM/mwFLIM

Christoph Biskup, Birgit Hoffmann,
Klaus Benndorf, and Angelika Rück

9.1 INTRODUCTION

Fluorescence imaging methods can provide a wealth of information about living cells. Scientists now have a versatile toolkit of fluorescence probes and techniques at their hands that can be used to monitor cellular function. Fluorescent probes can be used to monitor noninvasively the concentration of ions and biomolecules (Tsien 1989). Fluorescent analogues of biomolecules provide the possibility to study the binding of ligands to their receptor (Biskup, Kusch, et al. 2007). Fluorescence labeling techniques can be used to reveal and follow up the spatial distribution of proteins inside a cell (Zhang et al. 2002).

This chemical toolkit has been nicely complemented by GFP (green fluorescent protein) and other visible fluorescent proteins that can be fused by simple cloning techniques selectively to the protein of interest and develop fluorescence without additional cofactors (Tsien 1998; Chapter 3). In this way, proteins can be labeled selectively, as opposed to classical dye labeling techniques. By exploiting physical phenomena such as Förster resonance energy transfer, it is even possible to test whether molecules are in close vicinity (< 10 nm) to each other, thereby closing the gap between the limited optical resolution and molecular dimensions. Moreover, a variety of biological specimens exhibit autofluorescence when they are irradiated. Autofluorescence can provide additional information about the sample, which in contrast to some expensive fluorescent probes, is available free of charge to the researcher.

Together with new developments in the field of imaging techniques these tools have contributed significantly to our understanding of biological systems. Confocal and multiphoton laser scanning microscopes have become standard techniques of biomedical imaging on the cellular level (Pawley 2006). Scanning (single-photon confocal and two-photon) techniques have a number of advantages over wide-field imaging techniques. The suppression of out-of focus light results in high-contrast images of the selected sample plane (White, Amos, and Fordham 1987). By recording a stack of images in subsequent focal planes, the three-dimensional structure of the sample can be reconstructed. Three-dimensional imaging becomes especially interesting in conjunction with two-photon excitation (Denk, Strickler, and Webb 1990).

However, in standard confocal techniques, only fluorescence intensities are recorded and not all the information contained in the fluorescence light emitted by the sample is exploited. Other characteristics exhibited by a fluorophore, such as its fluorescence lifetime, its anisotropy, and its spectroscopic properties, are ignored. But how can we exploit all the information nature gives us?

In previous years, more dimensions have been added to laser scanning microscopy. In a new generation of confocal microscopes, the fluorescent light emitted by the sample can be dispersed by a grating and projected onto a multichannel detector consisting of 32 photomultiplier elements (Dickinson et al. 2001). In this way, 32 channels per pixel can be recorded simultaneously. This technique is extremely powerful for separating fluorescence signals of different fluorescence species. Also, the kinetics of the fluorescence decay bear a lot of useful information because the fluorescence lifetime τ of a fluorophore depends not only on the intrinsic characteristics of the fluorophore but also on its local environment. In general, the viscosity, refractive index, and pH, as well as interactions with other molecules (e.g., by collision or energy transfer), can affect the fluorescence lifetime (Suhling, French, and Phillips 2005; Lakowicz 2006). In turn, measurements of the fluorescence lifetime can be used to probe the surroundings of a fluorophore; consequently, images of the fluorescence lifetime can directly visualize local changes in the fluorescence lifetime of the fluorophore.

Fluorescence lifetimes can be obtained in the time domain or in the frequency domain. Time-domain techniques are based on pulsed excitation of the fluorophore and the direct measurement of the fluorescence decay by gated detection (Chapter 4), streak cameras (Chapter 8), or time-correlated single-photon counting (TCSPC; Chapters 6 and 7). Wide-field techniques have the advantage that the fluorescence decay can be acquired in all pixels of an image in parallel (Dowling et al. 1997; Chapter 4). In laser scanning microscopes, lifetimes are recorded sequentially by scanning the sample and recording the photon distribution in each pixel. This approach has the benefit that the resolution is enhanced by introducing a pinhole or taking advantage of the intrinsic sectioning capabilities of two-photon excitation. Another advantage is that the fluorescence decay in each pixel of the sample can be easily recorded in a spectrally resolved manner. For this, only a spectrograph is needed to disperse the fluorescence light emitted by the sample. The resulting spectrum has then to be recorded in a time-resolved manner. In this chapter, we describe two techniques that are able to do this. One is based on a streak camera; the other technique is based on TCSPC.

To our knowledge, the first setup consisting of a fluorescence microscope and a streak camera was introduced by Kusumi et al. (1991). This setup allowed measurement of the time-resolved fluorescence spectra in a small spot of a microscopic sample. Excitation light from a pulsed cavity-dumped dye laser was guided into the optical axis of the excitation path of an epifluorescence microscope. The beam was focused by a 40x or 100x objective onto the sample and the subcellular structure of interest was transported to the focused laser spot by mechanically moving the microscope stage. The fluorescence light emitted by the sample was collected with the objective lens and focused onto a pinhole placed at the image plane of the microscope. The fluorescence light was then dispersed by a grating and guided to a synchroscan streak camera to record the fluorescence signal. The acquisition time in this setup was between one second and a minute. The spectral resolution was ~10 nm, and the temporal resolution was ~100 ps.

In our setup, which we constructed almost a decade later, we took advantage of the capabilities of the available commercially confocal laser scanning microscopes (Biskup, Zimmer, and Benndorf 2001, 2004). Here, the beam can be moved to the spot of interest by using scanning mirrors instead of moving the specimen (see Section 9.3.2 for details). The setup benefits from the excellent temporal and spectral resolution of streak camera technology. However, one disadvantage is the long acquisition time necessary to obtain a temporally *and* spectrally resolved image (see Section 9.4).

However, fluorescence lifetime imaging (FLIM) based on streak cameras does not necessarily need to be slow. A combination of a microscope and a streak camera optimized to acquire fast fluorescence lifetime images has been introduced by Krishnan et al. (2003, Chapter 8). In this setup, one dimension of the streak image represents the x-axis of the specimen. In this way, the lifetimes of all pixels along one line in the specimen are recorded simultaneously. This saves acquisition time, but is done at the expense of the spectral information, which is lost. The reader interested in this setup is referred to Chapter 8.

A fascinating compromise between scan speed and spectral resolution has been presented by Qu et al. (2006). In this setup, the sample was excited by a 4 × 4 microlens array. To record the fluorescence originating from the 16 foci simultaneously, the streak camera entrance slit was replaced by a 4 × 4 pinhole array. Fluorescence was then dispersed with a prism. In this way, the active area of the streak camera was divided into 16 squares, in which small, spectrally and temporally resolved streak images could be recorded simultaneously. This speeds up acquisition time, but is done at the expense of spectral and temporal resolution.

The TCSPC approach was first extended to spectrally resolved measurements by Becker et al. (2003), who used several individual photomultiplier tubes (PMTs) or a multichannel PMT (Becker et al. 2002; Bird et al. 2004). Here, the fluorescence obtained from a spot of the sample is dispersed by a spectrograph and focused onto a multianode 16-channel PMT array. For each detected photon, it delivers a timing pulse and a 4 bit number that indicates in which channel the photon was detected. Together with the frame, line, and pixel scan signals, a four-dimensional histogram of the photon distribution over time (τ), wavelength (λ), and spatial coordinates (x,y) can be acquired in the memory of the TCSPC module. This technology has been applied to image autofluorescence (Becker, Bergmann, and

Biskup 2007) and photosensitizers (Kinzler et al. 2007, Rück et al. 2007) or to determine FRET in biological samples (Biskup, Zimmer, et al. 2007; Rück et al. 2008) (see Section 9.3.3 for details).

Another very elegant approach to resolve images temporally and spectrally has been recently realized using a line-scanning microscope to which a spectrograph and a time-gated EM-CCD (electron multiplying charge-coupled device) camera have been attached (De Beule et al. 2007). In this setup, the frequency-doubled beam of a titanium:sapphire laser is focused via a cylindrical lens through a 5 μm wide slit and imaged onto the sample. The resulting line of fluorescence is then guided to an imaging spectrograph that disperses the fluorescence along the vertical axis. The resulting output (x,λ) is imaged onto the photocathode of a gated optical intensifier and imaged with a cooled EM-CCD camera that records a (x,λ,τ) series of time-gated images at different delays after excitation. To obtain a full (x,y,λ,τ) data set, the sample is then scanned along the y-axis. Because a line scanning microscope with a slit detection provides confocal optical sectioning, the sample can be also scanned in the z-direction.

All these techniques allow one to exploit the information contained in the emission spectrum and in the fluorescence lifetime. They add the temporal dimension to spectrally resolved confocal measurements and, vice versa, the spectral dimension to classical fluorescence lifetime measurements (Figure 9.1). Accordingly, these techniques have been referred to as time-resolved spectroscopy or as spectrally resolved fluorescence lifetime measurements. Other commonly used names are hyperspectral FLIM (De Beule et al. 2007), multiwavelength TCSPC (mwTCSPC; Becker et al. 2002), multiwavelength FLIM (mwFLIM), or spectrally resolved fluorescence lifetime imaging (SLIM; Rück et al. 2007).

9.2 BACKGROUND

9.2.1 The Spectral Axis of the Fluorescence Decay Surface

The phenomenon of fluorescence can be best understood by a Jablonski diagram (Figure 9.2). The ground and the first excited electronic state are depicted by S_0 and S_1, respectively. At each of these electronic energy levels, the fluorophores can exist at a number of vibrational and rotational energy levels. The energy difference between rotational levels is much smaller than the difference between the vibrational levels. Therefore, they are not shown in Figure 9.2. When a fluorophore absorbs light, the molecule can be excited to a higher electronic, vibrational, and rotational state. With a few rare exceptions, excited molecules in condensed phases relax within picoseconds to the lowest vibrational level of S_1. From this level, the excited molecule can return to the ground state S_0 either by emitting a photon or by a variety of other radiationless processes, such as interaction with solvent molecules, quenching by solutes, or excited-state reactions.

According to the Franck–Condon principle, electronic transitions are most likely to occur between vibrational states in which the nuclei are most probably at the same positions of the ground and upper electronic state, (i.e., between vibrational states whose wave functions show the greatest spatial overlap). Hence, the shape of the absorption and fluorescence spectrum is determined by the energy differences between the energy levels of the

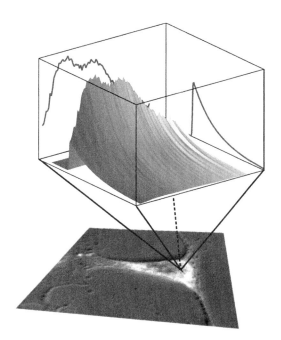

FIGURE 9.1 Principle of spectrally resolved fluorescence lifetime measurements. For each pixel of the specimen, spectral and temporal information of the fluorescence decay are recorded. Thus, the wavelength (λ) and the "microtime" on the timescale of the fluorescence decay (t_{micro}) are added to the three spatial dimensions (x,y,z) and the "macrotime" on the timescale of the experiment (t_{macro}), as additional dimensions to confocal microscopy (six-dimensional microscopy). (Adapted from Biskup, C. et al. 2004. *Nature Biotechnology* 22:220–224.)

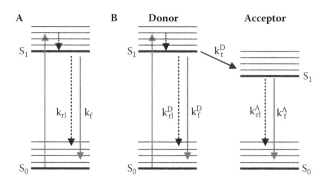

FIGURE 9.2 (A) Jablonski diagram of a fluorescent molecule. When a fluorophore absorbs light, the molecule is excited to a higher electronic, vibrational, and rotational state. With a few rare exceptions, excited molecules in condensed phases relax within picoseconds to the lowest vibrational level of S_1. From this level, the excited molecule can return to the ground state S_0 either by emitting a photon (k_f) or by a variety of other radiationless processes (k_{rl}). (B) Jablonski diagram of FRET. In the presence of an acceptor, an additional pathway is opened through which the excited donor molecule can return to the ground state. Donor fluorescence intensity and donor fluorescence lifetime decrease, whereas acceptor fluorescence intensity increases.

ground state and the excited state, as well as by the transition probabilities between these states. At room temperature, predominantly the lowest vibrational state is occupied; thus, the absorption spectrum is determined by the vibrational states of the excited electronic state and the respective transition probabilities. The fluorescence spectrum has a shape characteristic of the vibrational states of the electronic ground state.

9.2.2 The Time Axis of the Fluorescence Decay Surface

The time during which a molecule resides in the excited state is not only a function of the transition probabilities between the excited and ground state of an isolated molecule. Also, interactions with other molecules can affect the fluorescence lifetime. In general, the rates of all pathways that depopulate the excited state determine the lifetime τ_D of the excited state:

$$\tau = \frac{1}{k_f + k_{rl}} \tag{9.1}$$

Here, k_f is the rate constant of the fluorescence pathway and k_{rl} is the sum of the rate constants of all radiationless transitions, including *intramolecular* transitions like internal conversion (IC), as well as *intermolecular* interactions with solvent molecules or quenching by solutes.

A short pulse of light will excite a certain number of fluorophores, N_0^*. According to the Beer–Lambert law, the actual number depends on the intensity of the light pulse, the absorption coefficient of the fluorophore at the respective wavelength, the concentration of the fluorophore, and the thickness of the specimen.

The number dN^* of molecules that return from the excited state to the ground state during a time interval dt follows first-order kinetics (provided that the light pulse is the only way to excite the molecule) and is determined by the sum of the rate constants and the number of molecules N^* in the excited state:

$$\frac{dN^*(t)}{dt} = -\left(k_f + k_{rl}\right)N^*(t) - -\frac{1}{\tau}N^*(t) \tag{9.2}$$

Integration from $t = 0$ to t and N_0^* to $N^*(t)$ yields

$$N^*(t) = N_0^* \exp{-\left(k_f + k_{rl}\right)t} = N_0^* \exp{-\left(\frac{t}{\tau}\right)} \tag{9.3}$$

The fluorescence intensity $I(t)$ emitted by the sample is proportional to the number $N^*(t)$ of molecules in the excited state and can be calculated by multiplying $N^*(t)$ with k_f. This yields

$$I(t) = I_0 \exp{-\left(k_f + k_{rl}\right)t} = I_0 \exp{-\left(\frac{t}{\tau}\right)} \tag{9.4}$$

where $I(t)$ is the intensity at time t and I_0 is the intensity at time 0. I_0 is also called the amplitude of the fluorescence decay. The fluorescence lifetime can be determined from the time course of the fluorescence decay by plotting ln $I(t)$ versus t or by fitting the data to a monoexponential decay model (see Section 9.3.6 for details).

Ideally, an isolated fluorophore in a homogeneous solution exhibits a monoexponential fluorescence decay. However, in a biological sample, the fluorescence lifetime of a fluorophore can vary depending on its environment (because k_{rl} varies). In this case, a bi- or multiexponential decay is observed:

$$I(t) = \sum_i A_i \exp-\left(\frac{t}{\tau_i}\right) \qquad (9.5)$$

where τ_i is the individual lifetime of the fluorophore in environment i. The amplitudes A_i represent the fractional contributions of the respective fluorophores to the overall intensity decay. Moreover, in biological samples not only one, but many different fluorophores can contribute to the fluorescence emitted by a sample. In these and other cases, the fluorescence decay can be fitted only by a multiexponential function, and it might be difficult to identify the individual components, especially when the fluorescence signals are weak. In these cases, it is helpful to exploit all properties of the fluorescence light emitted by the sample to disentangle the individual contributions.

9.2.3 Global Analysis

When a mixture of n fluorophores, where each component i exhibits a monoexponential fluorescence decay, is measured at m different wavelength bands j, one can describe the resulting fluorescence decay surface by a set of m multiexponential equations:

$$I(t,\lambda(j)) = \sum_i^n A_{i,j} \exp-\left(\frac{t}{\tau_{i,j}}\right) \qquad (9.6)$$

As a result, one needs $n \bullet m$ parameters to describe the shape of the decay surface. However, as discussed in Section 9.2.1, excited molecules in condensed phases relax within picoseconds to the lowest vibrational level of S_1. Because fluorescence emission occurs from this level, provided that the environmental interactions of the molecules are not strong enough to change the electronic structure of the S_1 state, the fluorescence lifetime is independent of the emission wavelength and the lifetime of the respective components i observed in each wavelength band j should be equal. Thus, Equation 9.6 can be reduced to

$$I(t,\lambda(j)) = \sum_i^j A_{i,j} \exp-\left(\frac{t}{\tau_i}\right) \qquad (9.7)$$

In this example, the lifetimes are the same in all data sets. They are called global parameters. The amplitudes, however, are nonglobal. They differ for each data set and indicate the contribution of the fluorophore i to the fluorescence intensity decay at the respective wavelength band. Accordingly, only $n + m$ parameters are necessary to describe the shape of the decay surface. This reduced set of parameters can then be determined more reliably in a fitting routine because it imposes more constraints to the fit than a set of n • m parameters (Knutson, Beechem, and Brand 1983; Beechem 1992).

9.2.4 A Special Case: Global Analysis of FRET Measurements

A molecule in an excited electronic state can also return to the ground state by nonradiatively transferring its energy via dipole–dipole interaction to a suitable acceptor molecule that is only few nanometers away. This process is referred to as "resonance energy transfer" (RET) or "Förster resonance energy transfer" (FRET). The phenomenon was first described by Perrin in 1927. The underlying theory was developed by Förster, who first published a classical and later a quantum mechanical derivation of his theory (Förster 1949a, 1949b, 1959).

According to Förster, the rate of energy transfer (k_t) between a pair of donor and acceptor molecules is given by

$$k_t = \frac{1}{\tau_D} \left(\frac{R_0}{r} \right)^6 \tag{9.8}$$

where τ_D is the lifetime of the donor in the absence of an acceptor, and r is the distance between the centers of the donor and acceptor chromophores. R_0 is the so-called Förster distance, at which energy transfer is 50%. R_0 for a given donor and acceptor pair is a function of the spectral properties of donor and acceptor (donor fluorescence quantum yield in the absence of the acceptor, the extent of overlap of the emission spectrum of the donor with the absorption spectrum of the acceptor), the relative orientation of the donor and acceptor transition dipoles, and the refractive index of the medium (see Clegg 1996 and Periasamy and Day 2005 for a more detailed introduction into the theory of FRET).

The distance dependence of FRET was experimentally verified by studies of donor–acceptor pairs separated by a known distance (Latt, Cheung, and Blout 1965; Stryer and Haugland 1967). These studies demonstrated nicely that FRET can be exploited to determine inter- and intramolecular distances in the range of a few nanometers. However, in most cases, FRET is only used in a more qualitative way to prove close vicinity of donor and acceptor molecules. According to Equation 9.8, the efficiency of FRET depends on the inverse sixth power of the distance between the donor and the acceptor molecule. It is mainly this steep dependence of the energy transfer efficiency on the distance that results in considerable potential for applications in biological research.

When donor and acceptor are separated by the distance $r = \frac{1}{2} R_0$, FRET efficiency is over 98%; at a distance $r = R_0$, FRET efficiency is 50%, and, if the distance is doubled to $2 R_0$, FRET efficiency is decreased by a factor of $2^6 = 64$ to less than 2%. Thus, between molecules that are separated by a distance greater than $2 R_0$, almost no FRET occurs. Hence, by performing FRET measurements in the fluorescence microscope, it is possible to distinguish proteins that are merely colocalized in the same compartment from proteins that are associated on the molecular level.

In the absence of an acceptor, the lifetime τ_D of the donor is given by Equation 9.1:

$$\tau_D = \frac{1}{k_f + k_{rl}} \tag{9.9}$$

k_f is the rate constant of the fluorescence pathway and k_{rl} is the sum of the rate constants of all radiationless transitions. In the presence of an acceptor, an additional pathway contributes to the depopulation of the excited donor state. The lifetime of the donor in presence of an acceptor decreases accordingly:

$$\tau_{DA} = \frac{1}{k_f + k_{rl} + k_t} \tag{9.10}$$

The rate of energy transfer k_t can be determined experimentally by comparing the lifetime of the donor in presence of an acceptor (Equation 9.10) with the lifetime of the donor in the absence of an acceptor (Equation 9.9):

$$k_t = \frac{1}{\tau_{DA}} - \frac{1}{\tau_D} \tag{9.11}$$

When k_t and the other rate constants are known, the so-called FRET efficiency (E) can be calculated as the first equality in Equation 9.12. The FRET efficiency is defined as the fraction of excited donor molecules that relax to the ground state by transferring a part of their excitation energy to the acceptor. The FRET efficiency is the ratio of the rate constant of energy transfer k_t to the total decay rate of the donor $k_{rl} + k_f + k_t$:

$$E = \frac{k_t}{k_{rl} + k_f + k_t} = \frac{1/\tau_{DA} - 1/\tau_D}{1/\tau_{DA}} = 1 - \frac{\tau_{DA}}{\tau_D} \tag{9.12}$$

Hence, in order to determine FRET efficiency by lifetime measurements, it is only necessary to measure τ_{DA} and τ_D. Because fluorescence lifetimes are independent of the actual number of fluorophores, FRET and control measurements can be carried out in different cells.

The total fluorescence intensity I_X recorded from a sample X is equal to the area under the fluorescence decay curve:

$$I_X = \int_0^\infty I_X^0 \exp\left(-\frac{1}{\tau}t\right) dt = I_X^0 \tau \tag{9.13}$$

Thus, I is directly proportional to the fluorescence lifetime and Equation 9.12 can be transformed to

$$E = 1 - \frac{I_{DA}/I_{DA}^0}{I_D/I_D^0} \qquad (9.14)$$

In fluorescence intensity measurements, the amplitudes I_D^0 and I_{DA}^0 of the respective fluorescence decays are not accessible. However, when fluorescence intensities are measured in the same sample, the numbers of donor molecules would be equal; consequently, both amplitudes I_D^0 and I_{DA}^0 are equal. In this case, Equation 9.14 can be simplified to

$$E = 1 - \frac{I_{DA}}{I_D} \qquad (9.15)$$

Hence, to calculate FRET efficiency, either the donor fluorescence lifetime or the donor fluorescence intensity in the presence of an acceptor has to be compared with the respective value in the absence of an acceptor. Although Equations 9.12 and 9.15 look similar at first sight, there is one important difference: In contrast to fluorescence lifetimes, fluorescence intensities depend on the actual concentration of fluorophores. Therefore, in intensity-based FRET estimates, fluorescence intensities of the donor in the presence and in the absence of the acceptor have to be measured in the same sample. This is where most problems of intensity-based FRET measurements arise. Whereas fluorescence intensities of the donor in the presence of the acceptor (I_{DA}) can be measured easily, fluorescence intensities of the donor in the absence of the acceptor (I_D) are not directly accessible.

I_D can be determined by measuring the donor fluorescence after photobleaching the acceptor or by estimating the loss of donor fluorescence intensity due to FRET by measuring the acceptor fluorescence intensity. The first approach has the disadvantage that photobleaching can only be performed once, so, consequently, the FRET efficiency can also be determined only once. Dynamic protein–protein interactions cannot be monitored in this way. The success of the second approach depends on how well fluorescence excitation and fluorescence emission spectra of the donor and the acceptor can be separated.

Moreover, in intensity measurements, only the integral over the time of the fluorescence decay is measured; any additional information that can be derived from the kinetics of the fluorescence decay is not accessible. The examples shown in Figure 9.3 demonstrate this. In the first case (upper left panel), it is assumed that every donor molecule is associated with an acceptor molecule. In this case, only one species of excited donor molecules exists and, consequently, the fluorescence decay is monoexponential. However, in a case where only a fraction of the donor molecules is associated (lower left panel), two populations of donor molecules—bound and unbound—are present, giving rise to a biexponential fluorescence decay. In this case, the fast-decay component is due to associated donor molecules whose excited-state lifetimes are shortened by FRET. The slow-decay component is caused by unbound donor molecules whose lifetime is not affected by energy transfer.

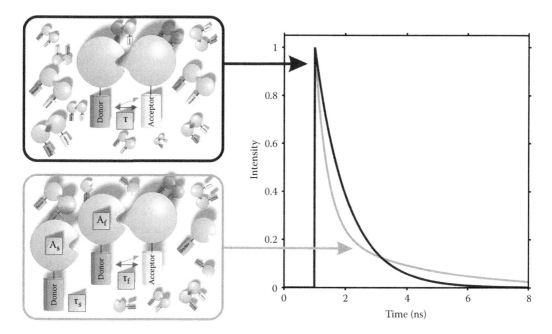

FIGURE 9.3 Models of interacting proteins and the respective fluorescence decays. In the case that all donor molecules are associated with an acceptor molecule (upper left panel) only one class of donor molecules exists and the resulting fluorescence decay is monoexponential. In a case where only a part of the donor molecules is associated (lower left panel), two populations of donor molecules—bound and unbound—are present, giving rise to a biexponential decay. By measuring and analyzing the fluorescence decay one can distinguish between the two cases. Moreover, the parameters recovered in the biexponential fit have a meaning: The slow lifetime constant (τ_s) is the lifetime of unbound donor molecules. From the lifetime of the fast component (τ_f), the distance between interacting donor and acceptor pairs can be calculated. From the relative amplitudes (A_s, A_f), one can get an estimate for the fraction of bound and unbound molecules. (Adapted from Biskup, C. et al. 2007. *Proceedings of SPIE* 6442:64420V.)

However, for intensity measurements, only the time integral of the fluorescence decay curve is measured. Thus, it is not possible to distinguish whether all donor molecules are separated from acceptor molecules by a single distance or whether only a fraction of donor molecules is bound close to an acceptor, and the remaining donor molecules are separated from the acceptor by a distance exceeding $2\,R_0$. In many fluorescence intensity-based FRET efficiency calculations, it is assumed that all donor molecules are bound to acceptor molecules—without any proof. All calculations of intensity-based FRET efficiency are implicitly based on this assumption, unless the fraction of the associated and unassociated donor and acceptor species can be estimated in an independent way.

However, in fluorescence lifetime-based FRET measurements, associated and free donor populations can be distinguished and it is particularly this facility that makes these measurements superior to approaches based only on fluorescence intensity measurements. Moreover, in fluorescence lifetime-based FRET measurements, it is possible to check whether assumed models are consistent with the data. By fitting a model function to the data it can be tested whether this function and the underlying model are able to describe the fluorescence decay. For example, a monoexponential function would describe perfectly the fluorescence decay in the first case (where all donors can transfer energy to an acceptor) and the corresponding model would be considered to be consistent with the data. However, in the second case (where only a fraction of donors can transfer energy to an acceptor), a monoexponential function would not account for the fluorescence decay, provided that the signal-to-noise ratio of the data is sufficient to distinguish a single- from a double-exponential decay. Thus, in the second case, a model assuming that all donor molecules are associated with an acceptor molecule is not consistent with the data and must be rejected. It has to be replaced by a model better suited to describing the data. In this example, a biexponential function could account for sufficiently accurate data, showing that a model consisting of at least two donor populations is consistent with the measured data.

In addition, the parameters recovered by the correct model that fits the data have a meaning. In this example, the slow lifetime constant corresponds to the lifetime of unbound donor molecules. The lifetime of the fast component is determined by the distance between interacting donor and acceptor pairs. From the relative amplitudes of the lifetime components, one can get an estimate of the fraction of bound and unbound molecules.

The analysis of the data can be improved if the fluorescence decay of the acceptor is included in the analysis. Donor and acceptor fluorescence are linked by a set of differential equations.

In analogy to Equation 9.2, the decrease of donor molecules D^* in the excited state is given by

$$\frac{d[D^*]}{dt} = -\left(k_f^D + k_{rl}^D + k_t^D\right)\left[D^*(t)\right] \tag{9.16}$$

The change in the number of acceptor molecules A^* in the excited state depends on the rate constant of FRET and all pathways that depopulate the excited acceptor state:

$$\frac{d[A^*]}{dt} = k_t^D\left[D^*(t)\right] - \left(k_f^A + k_{rl}^A\right)\left[A^*(t)\right] \tag{9.17}$$

Thus, the information about the FRET rate constant k_T^D can be derived from both the donor and the acceptor fluorescence decay. k_T^D determines the steepness of the decrease in donor fluorescence and the steepness of the rise of the acceptor fluorescence.

Figure 9.4 gives an example. For the simulation, it was assumed that the donor has a fluorescence lifetime of $\tau_D = 3$ ns in the absence of the acceptor and exhibits monoexponential fluorescence decay. Furthermore, it was assumed that all donor molecules are covalently linked to the acceptor molecules in a single fixed distance such that energy transfer efficiency is 50%.

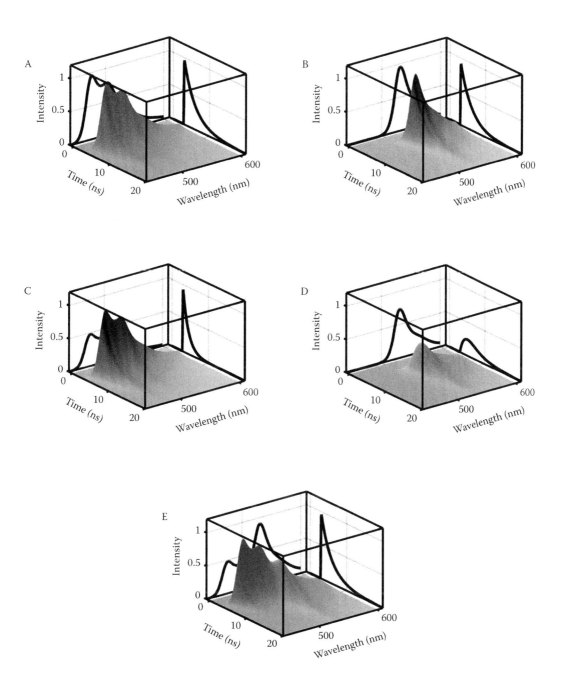

FIGURE 9.4 Simulated donor and acceptor fluorescence decay curves. Fluorescence decay surface of (A)—the donor in the absence of an acceptor; (B)—the acceptor after direct excitation; (C)—the donor in the presence of an acceptor; (D)—the acceptor when excited via FRET; and (E)—the total fluorescence emitted from the sample. For the simulation, the following assumptions are made: (1) Donor and acceptor alone exhibit a monoexponential fluorescence decay with a fluorescence lifetime $\tau_D = \tau_A = 3$ ns. (2) Donor and acceptor molecules are separated by a single fixed distance so that energy transfer efficiency is 50%. (3) Quantum yield of donor and acceptor is 0.6.

9.3 METHODS

9.3.1 The Setup

The following list gives an overview of the components that are necessary for spectrally resolved fluorescence lifetime measurements. A scheme of the setup is shown in Figure 9.5. It consists of the following components:

1. To record fluorescence decays in the time domain, the sample must be excited with a *pulsed laser* whose pulses are short compared to the timescale on which the fluorescence decay occurs in order to avoid the convolution of the excitation pulse with the fluorescence response. Thus, with lifetimes in the range of few nanoseconds, the length of the laser pulse should be in the subnanosecond range. With the use of titanium:sapphire lasers for two-photon excitation (Denk et al. 1990), pulses with a pulse length around 150 fs full width at half maximum (FWHM) are routine. The repetition rate of the available titanium:sapphire lasers is between 75 and 80 MHz, which is almost ideally suited to excite most fluorophores. The pulse spacing is approximately 13.3 ns with a repetition rate of 75 MHz. This gives most fluorophores, whose lifetime is in the range of 1–2 ns, enough time to return to the ground state. In our experiments, we used a combination of a 5 W, frequency-doubled Nd:YVO$_4$ (Neodymium:Yttrium Vanadate) laser (5 W Verdi, Coherent GmbH, Dieburg, Germany) and a mode-locked titanium:sapphire laser (Mira 900, Coherent). In this setup, the pulse length is 150 fs (FWHM) and repetition rate is 76 MHz. Alternatively, pulsed diode lasers, which are available at various wavelengths (see www.becker-hickl.de), can be used for pulsed one-photon excitation. Depending on the diode type, repetition rates in the 20–80 MHz range can be chosen. However, the pulses of laser diodes are considerably broader compared to the pulses of a titanium: sapphire laser. Their pulse width is usually greater than 50 ps (FWHM).

2. A *pulse picker* can be used to reduce the pulse repetition rate of the excitation source. This might be necessary, for example, to measure the luminescence decay of chromophores with longer fluorescence or phosphorescence lifetimes. Decreasing the repetition rate and increasing the pulse-to-pulse distance assures that the fluorescence decays elicited by sequential pulses do not overlap. Reducing the repetition rate of the laser is also necessary if the repetition rate of the detection device is limited. This helps to avoid unnecessary overexposure and photobleaching of the sample. In our setup, for example, we use a pulse picker (Model 9200, Coherent) to decrease the repetition rate of the titanium:sapphire laser to conform to the maximum repetition rate (2 MHz) of the single-sweep module of our streak camera (see Section 3.2).

3. For one-photon excitation, the laser beam can be frequency doubled by a *second harmonic generator* (SHG). An SHG consists of a nonlinear crystal (i.e., β-barium borate [β-BaB$_2$O$_4$, BBO] or lithium borate [LiB$_3$O$_5$, LBO]) and a lens or curved mirror to

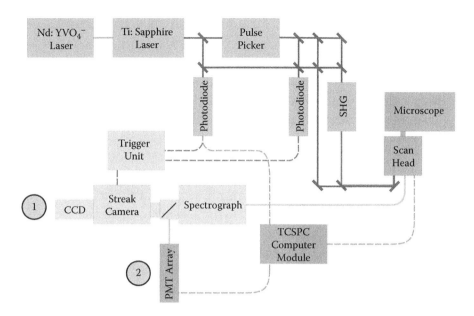

FIGURE 9.5 Setup for multiple wavelength lifetime measurements. The sample is excited by ultra-short (130 fs FWHM) light pulses generated by a combination of a mode-locked titanium: sapphire laser, a pulse picker, and a SHG crystal. Emitted fluorescence is guided through an optical glass fiber to a polychromator, which disperses the fluorescence light according to its wavelength along the horizontal axis and focuses it either on the entrance slit of the streak camera (1) or on the PML-16 detector module (2). (Adapted from Biskup, C. et al. 2007. *Proceedings of SPIE* 6442:64420V.)

focus the laser beam onto the crystal and a second lens or curved mirror to parallelize the frequency-doubled beam. To block the fundamental wavelength a Schott BG39 filter (ITOS, Mainz, Germany) is inserted into the beam path after the SHG unit.

4. A *laser scanning microscope* (LSM) is used to scan the sample. In our studies, we use a LSM 410 or LSM-510 from Zeiss (Carl Zeiss GmbH, Jena, Germany) equipped with a C-Apochromat 63x (NA 1.20) water immersion objective (Carl Zeiss GmbH) to scan the sample and to direct the pulsed laser beam to the regions of interest. To guide emitted fluorescence to the polychromator, an optical fiber needs to be coupled to one of the confocal channels. When the sample is excited by two-photon excitation, fluorescence can be collected via the non-descanned (NDD) port. In this case, a special fiber bundle has to be used; it transforms the circular cross section of the beam at the NDD port into a linear shape required for the input slit of the polychromator (Becker et al. 2007).

5. A *polychromator* is used to disperse the light along the horizontal axis. In principle, dispersive elements such as prisms or diffraction gratings can be used. In our labs, we use commercial polychromators (250is, Chromex, Albuquerque, New Mexico, or MS125, LOT-Oriel, Darmstadt, Germany), which select the wavelength range of interest automatically by rotating the grating.

6. Several devices can be used to record the fluorescence decay in a time-resolved manner. In this chapter, we focus on a streak camera (C5680 with M5677 sweep unit, Hamamatsu Photonics, Herrsching, Germany) and a 16-channel TCSPC detector head (PML-16, Becker & Hickl, Berlin, Germany). The operation principle of both devices is explained in the following paragraphs. Another interesting approach to record spectral and temporal information simultaneously has been implemented in a line-scanning microscope to which a time-gated EM-CCD camera has been attached (De Beule et al. 2007).

9.3.2 Operation Principle of the Streak Camera

The operation principle of a streak camera is explained in more detail in Chapter 8. However, the setup used for the experiments presented in this chapter is slightly different. Here, the x-axis is used to encode the spectral information; in the setup presented in Chapter 8, the pixels along the x-axis correspond to the spatial position along the scanned line.

Figure 9.6 explains the operation principle of the setup we use for our experiments (Biskup et al. 2004). In brief, photons dispersed by the spectrograph hit the photocathode, where they are converted into photoelectrons. The photoelectrons are accelerated by the electrode potential, pass through a pair of deflection plates, are multiplied in a multichannel plate (MCP), and hit a phosphor screen, where they are reconverted into an optical image, the so-called streak image. At the instant at which the photoelectrons pass through the deflection electrodes, a voltage ramp is applied so that the electrons are "swept" from top to bottom. Electrons leaving the photocathode at earlier times arrive at the phosphor screen in a position close to the top, whereas those electrons that leave the photocathode at later times arrive at a position closer to the bottom. In this way, a time axis is imposed onto the spectrum.

The streak images are read out with a set frequency by a CCD camera and transferred to a computer. In the so-called photon counting mode, a threshold value is used with the streak image to remove noise. Only signals exceeding the threshold value are counted as one single-photon event. The events are added up in the computer memory, and as a result, integrated photon-counting streak images of the fluorescence signal are obtained. If the integration time is long enough, a high signal-to-noise ratio can be obtained. Depending on the brightness of the sample, it may take a few seconds up to several minutes to acquire a streak image that warrants a reliable fit.

9.3.3 Operation Principle of the mwFLIM/SLIM Setup

The configuration for the multiple wavelength TCSPC setup in conjunction with a laser scanning microscope is shown in Figure 9.7. The setup is similar to the streak camera setup shown in Figure 9.5, except that the spectrum is now focused on a PMT array (PML-16, Becker & Hickl). The PML-16 detector module has 16 parallel PMT channels in a linear arrangement. For each detected photon, it delivers a timing pulse and a 4-bit number that indicates the channel in which the photon was detected. These signals are fed into the

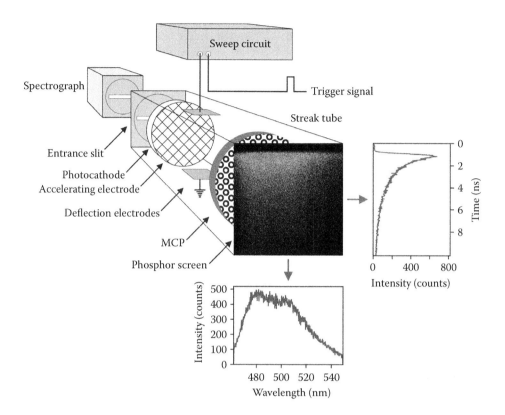

FIGURE 9.6 Operation principle of a streak camera. The light pulse to be measured is focused onto the photocathode of the streak tube, where photons are converted into electrons. The photoelectrons are accelerated by the accelerating electrode, pass through a pair of deflection plates, are multiplied in a microchannel plate (MCP), and hit the phosphor screen of the streak tube, where they are converted into an optical image, the so-called streak image. At the instant at which the photoelectrons pass through the deflection electrodes, a voltage ramp is applied so that the electrons are swept from top to bottom. Electrons generated at earlier times arrive at the phosphor screen in a position close to the top of the screen, while those electrons generated at later times arrive at a position close to the bottom of the screen. Hence, the time at which a photon left the sample can be determined by the vertical position of the photoelectron in the streak image. The horizontal position of the photoelectron depends on the wavelength of the incident light pulse because a spectrograph was used to focus the spectrum onto the photocathode. The streak image is read out by a CCD camera and transferred to a computer. From the streak image, the fluorescence spectrum can be obtained by summing up fluorescence intensities along the time axis (vertical axis) and plotting the resulting intensities versus the wavelength (horizontal axis). Accordingly, the fluorescence decay curve can be obtained by summing up the fluorescence intensities in the wavelength region of interest and plotting the resulting intensities versus the time axis. (Figure reproduced from Biskup, C. et al. 2004. *Nature Biotechnology* 22:220–224.)

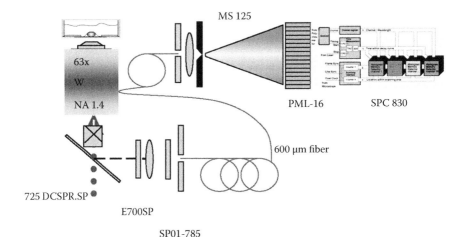

FIGURE 9.7   Operation principle of multichannel TCSPC imaging. For each detected photon, the spectral detector delivers a timing pulse and a 4-bit number that indicates in which channel the photon was detected. These signals are fed into the TCSPC imaging module, where they are used to build up a four-dimensional histogram of the photon distribution over time, wavelength, and spatial coordinates. The imaging process is synchronized with the scanning process via the frame sync, line sync, and pixel sync signal of the laser scanning microscope.

SPC-730 or SPC-830 TCSPC imaging module (Becker & Hickl), where they are used to build up a four-dimensional histogram of the photon distribution over time, wavelength, and spatial coordinates. The imaging process is synchronized with the scanning process via the frame sync, line sync, and pixel sync signal of the laser scanning microscope (Becker et al. 2002; Bird et al. 2004).

With a special fiber adapter, the TCSPC setup can also be used in conjunction with two-photon microscopy, thereby exploiting the advantages of two-photon excitation and maximizing fluorescence collection efficiency (Becker et al. 2007). Depending on the brightness of the sample, it takes up to several minutes to collect a lifetime image with 16 spectral channels and 128 × 128 pixels.

9.3.4 Benefits of the Techniques

The streak camera setup and the multiple wavelength TCSPC setup benefit from the high spatial resolution of the laser-scanning microscope and are able to record the fluorescence decay as a function of time *and* wavelength (Figure 9.8). Accordingly, the plot of the data yields a decay *surface* instead of a simple decay *curve*. The decay curve obtained by the conventional lifetime techniques is just the sum of the intensities recorded in the respective wavelength intervals. Accordingly, the spectrum obtained by spectral detection techniques is just the sum of the intensities recorded at the respective wavelengths along the time axis. Both techniques exploit the information contained in the emission spectrum and in the fluorescence lifetime. They add the spectral dimension to classical fluorescence lifetime measurements and the temporal dimension to spectrally resolved confocal measurements. Section 9.3.6 will show how this additional information can be used.

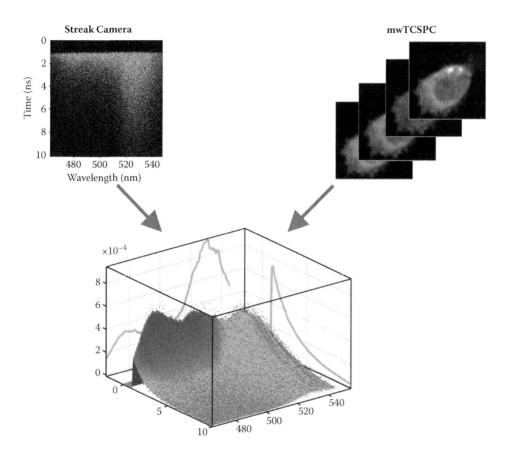

FIGURE 9.8   Reconstruction of fluorescence decay surfaces from streak images or TCSPC data. Both techniques—the streak camera setup and the multiple wavelength TCSPC setup—are able to record the fluorescence decay as a function of time *and* wavelength. For each spot of the sample, a decay surface is recorded. (Adapted from Biskup, C. et al. 2007. *Proceedings of SPIE* 6442: 64420V.)

9.3.5 Calibration

As discussed in the preceding sections, the data recorded by the streak camera setup and the multiple wavelength TCSPC setup contain all information about the spectral and temporal intensity distribution of the photons emitted by the sample. However, to make this information useful, a scale has to be assigned to all axes, the wavelength axis, the time axis, and the intensity axis. This can be done by the calibration procedures described in the following subsections.

9.3.5.1 Calibration of the Spectral Axis

To calibrate the spectral axis of the streak images, the internal lasers of the confocal microscope can be used. Two attenuated laser lines, whose wavelengths are close to the short- and the long-wavelength edges of the recorded spectrum, are reflected by a mirror specimen or a cover slip to the spectrograph and the streak camera. By correlating the centers of the signals to the pixels and calculating the dispersion (nanometers) per pixel, a wavelength can be assigned to each pixel. The calibration can be verified by evaluating the streak measurement of a third laser wavelength. The calibration procedure for the PMT array is similar. In this case, a wavelength band is assigned to each channel.

9.3.5.2 Calibration of the Time Axis

The time axis is usually calibrated by the manufacturer. This applies for both the streak camera and the TCSPC setup. In the streak camera setup, the scaling of the time axis depends on the slope of the voltage ramp that is applied to the pair of deflecting electrodes. However, the slope is not exactly linear, so the time axis needs to be scaled. For this, scaling tables are usually provided by the manufacturer for each sweep speed so that nonlinearities can be corrected. It is advisable to save the scaling information together with the streak images. For instance, the ITEX file format, which is used to store streak data recorded with Hamamatsu systems, allows saving the scaling information in the status string of a streak image.

In the TCSPC setup, the scaling of the time axis is determined by the time-to-amplitude converter (TAC). The TAC core generates a linear voltage ramp that starts (in the reversed start–stop setup) with the detector pulse and is stopped with the next reference pulse, which is generated by the subsequent light pulse of the laser. The voltage is then amplified and fed to an analog-to-digital converter (ADC), which delivers the digitized photon detection time (Becker 2005). All these components must work with extremely high precision to ensure proper scaling of the time axis.

The TAC parameters must be chosen by the user so that only the linear part of the TAC curve is used. For small and large offsets, the time-to-amplitude conversion is not linear and can cause peaks in the recorded photon distribution. The amplifier and the ADC are responsible for amplifying the TAC signal and dividing it into up to more than 1,000 equally wide time channels. Any variation in the width of the time channels will bias the number of counted photons and distort the recorded fluorescence decay curve. Usually, the user can rely on the proper calibration of the time axis by the manufacturer. However, in sophisticated TCSPC setups, some care has to be taken to adjust the relevant parameters properly.

TCSPC in the reversed start–stop setup relies not only on perfect electronics but also on a stable laser source. Because the time from the detected photon to the subsequent laser pulse is measured, pulse-to-pulse stability is vital. Pulse-to-pulse jitter of the laser broadens the instrument response function (see Section 9.3.6.1) and can—in the worst case—deteriorate the measurement.

In the streak camera and in the TCSPC setup, the linearity of the time axis can be verified by sending a laser pulse through an etalon and recording the signal. Another option is to measure the fluorescence decay of fluorescence lifetime standards like POPOP (1,4-Bis(5-phenyloxazol-2-yl)benzene) or coumarin derivatives and to compare the lifetimes with reference values documented in the literature (Lakowicz 2006; Boens et al. 2007)

9.3.5.3 Calibration of the Intensity Axis

The transmission efficiency of a spectrograph depends on the wavelength and the grating. It is highest at the blaze wavelength, but drops at higher or lower wavelengths. Thus, intensities recorded at different wavelengths are not comparable. If the data are only used to compare data obtained in the same setup or to calculate the lifetimes in a given wavelength range, this does not matter. But it does matter when intensities have to be compared (e.g., in the FRET experiments described in Section 9.3.7.3). In these cases, it is necessary to know the wavelength-dependent efficiency of the detection system. One simple way to obtain these correction factors is to measure the emission spectrum of fluorescence standards and to compare it with published spectra. Such spectra have been published for a variety of substances (see Lakowicz, 2006, for a survey). In our laboratory, we usually use solutions of quinine sulfate in 1 N H_2SO_4 ($\lambda_{em,max}$ = 457.2 nm), harmaline in 0.1 N H_2SO_4 ($\lambda_{em,max}$ = 498 nm), or fluorescein in 0.1 N NaOH ($\lambda_{em,max}$ = 512 nm) depending on the wavelength interval to be calibrated.

9.3.6 Data Analysis

9.3.6.1 The Instrument Response Function

On the timescale of fluorescence decays, the width of a light pulse used for excitation of the fluorophore, and the response time of the detector, the trigger circuit and the timing electronics are not negligible. The response of the entire system, also called "instrument response function" (IRF), can be measured by reflecting a part of the laser pulse to the detector. In this case, the observed "decay" is caused by the measurement system alone and represents the shortest time profile that can be measured by the setup.

In general, the response time $\Delta\tau_{sr}$ of a system depends on the pulse width $\Delta\tau_p$ of the excitation source, the response time $\Delta\tau_{dr}$ of the detector, and, if signals are integrated, on the trigger jitter $\Delta\tau_j$. All of these factors can be limiting to the time resolution of the system. If one assumes that all sources of error have the shape of a Gaussian distribution, then the system response time $\Delta\tau_{sr}$ depends on the square root of the sum of the squares of all limiting factors (Nordlund 1991):

$$\Delta\tau_{sr}^2 = \Delta\tau_p^2 + \Delta\tau_{dr}^2 + \Delta\tau_j^2 \tag{9.18}$$

The pulses of a mode-locked titanium:sapphire laser usually have a pulse width (FWHM) of about 150 fs. Thus, their contribution to $\Delta\tau_{sr}$ is negligible. This, however, does not apply to other excitation sources. For example, the pulses of laser diodes are considerably broader. Their pulse width is usually greater than 50 ps (FWHM).

The detector response time $\Delta\tau_{dr}$ of a streak camera depends upon several factors (Campillo and Shapiro 1983; Nordlund 1991). The first is the so-called technical time resolution. It is determined by the speed at which electrons are swept across the phosphor screen and the spatial resolution of the streak tube. The second factor is the transit-time spread (TTS), which is due to the difference in velocities that photoelectrons acquire when they leave the photocathode. Detector response times of the best currently available streak tubes can reach 200 fs. The streak camera module we use (C5680 + M5677, Hamamatsu) has a temporal resolution of approximately 70 ps (Figure 9.9A).

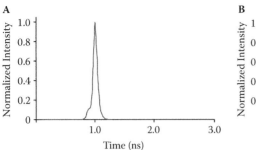
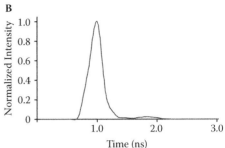

FIGURE 9.9 Instrument response function (IRF) of (A) the streak camera and (B) the PMT array. To determine the IRF, the beam of the titanium:sapphire laser was tuned to 860 nm, frequency doubled by a BBO crystal to 430 nm, and coupled into the scan head of the laser scanning microscope. The beam was reflected by a mirror specimen to one of the confocal channels and guided to the spectrograph and the streak camera or to the PMT array. Because the light pulse used is comparably short (150 fs), the recorded time courses reflect the response characteristics of the detector.

The TTS is also the main factor that determines the IRF of a PMT. MCP-PMTs are very fast. Their TTS can be as low as 30 ps (Becker 2005). The IRF of the PMT array used in this setup, however, is in the range of 250–300 ps, depending on the wavelength of the incident photons (Figure 9.9B). As can be seen from Figure 9.9, PMTs often have a tail and afterpulses, whereas streak cameras have a nearly Gaussian IRF.

The trigger jitter $\Delta\tau_j$ depends on the jitter of all components in the trigger circuit, including the laser pulse jitter, the jitter of the trigger photodiode, the jitter of the delay unit, and the jitter of the electronics. In our system, the overall trigger jitter is less than 10 ps. Thus, according to Equation 9.18, the system response of our system is mainly determined by the detector response time $\Delta\tau_{dr}$.

9.3.6.2 Deconvolution and Data Fitting

If the system response time is in the same order of magnitude or longer than the fluorescence lifetime of a fluorophore, then not only the characteristics of the fluorophore and its environment but also the characteristics of the setup determine the shape of the observed fluorescence decay. In this case, the observed fluorescence decay profile $I(t)$ of the fluorophore is a convolution of the true fluorescence decay function $F(t)$ of the sample and the IRF of the entire measurement system, including the excitation and detection parts (McKinnon, Szabo, and Miller 1977):

$$I(t) = \int_0^t IRF(t')\, F(t-t')dt' = IRF(t) \otimes F(t) \qquad (9.19)$$

The IRF can be measured by reflecting a part of the laser pulse to the detector, as described previously, or by measuring the fluorescence decay of a fluorophore with a short lifetime. However, one must take into account that the time response of the detector might vary with the wavelength of the incident photons. Thus, for the measurement of the IRF, the laser should be tuned to the range where fluorescence emission is measured.

When the fluorescence decay $I(t)$ and the IRF are measured, the true fluorescence decay $F(t)$ can be determined (i.e., deconvolved) by a variety of techniques (McKinnon et al. 1977; O'Connor, Ware, and Andre 1979). Difficulties in deconvolving the fluorescence decay are mostly due to the fact that $I(t)$ and the IRF are only known at a limited number of discrete points with a certain numerical error. Especially under these conditions, the "iterative reconvolution method" has proven to be reliable (Grinvald and Steinberg 1974; O'Connor et al. 1979).

The analysis starts with a model that is assumed to describe the data adequately. The goal is then to optimize the parameters of the model such that the calculated decay $I_c(t_k)$ matches the measured decay $I(t_k)$. As a parameter for the goodness of the fit, the sum of the weighted squared deviations between the measured decay $I(t_k)$ and the calculated decay $I_c(t_k)$ is calculated:

$$\chi^2 = \sum_{k=1}^{n} \frac{1}{\sigma_k^2} \left[I(t_k) - I_c(t_k) \right]^2 \tag{9.20}$$

The deviations are weighted by the variance of each data point, σ_k^2. The distribution of photons follows Poisson statistics and the standard deviation, σ_k, is equal to the square root of the number of detected photons. Thus, Equation 9.20 can be transformed to

$$\chi^2 = \sum_{k=1}^{n} \frac{\left[I(t_k) - I_c(t_k) \right]^2}{I(t_k)} \tag{9.21}$$

Several mathematical methods, such as the Gauss–Newton, the Levenberg–Marquardt, or the Nelder–Mead algorithms, can be used to approximate the parameter of the model, such that χ^2 is minimum (Straume, Frasier-Cadoret, and Johnson 1991; Press et al. 1992; Bevington and Robinson 2003).

The χ^2 value depends on the number of data points. To facilitate the interpretation of the χ^2 value obtained by a fit, a reduced value χ_v^2 can be calculated:

$$\chi_v^2 = \frac{\chi^2}{v} = \frac{\chi^2}{n-p} \tag{9.22}$$

where n is the number of data points, p is the number of floating parameters, and $v = n - p$ is the number of degrees of freedom. If deviations of the data from the calculated curve are due only to random errors, each data point k contributes σ_k^2/σ_k^2 to the sum. Thus, for a large number of data points, χ_v^2 should be close to unity. If a chosen model cannot describe the data sufficiently, deviations of the data from the calculated curve are not random and χ_v^2 will be greater than 1.

When least-squares parameter estimation procedures are used, one has to be aware that one of the inherent assumptions to all algorithms is that the dependent variables follow a Gaussian distribution. A photon (i.e., Poisson) distribution, however, is only similar to a Gaussian distribution when the number of events is large enough. Thus, for low photon counts, a least-squares analysis might not be justified.

9.3.7 Applications

9.3.7.1 Functional Staining of Cell Structures

Figure 9.10 shows human myoblasts (C2C12) that have been co-incubated with the nuclear staining dye 4′,6-diamidino-2-phenylindol dihydrochloride (DAPI) and the mitochondrial stain rhodamine 123. The sample was excited by two-photon excitation at 800 nm. The spectrometer with 600 lines/millimeter diffraction grating was adjusted such that fluorescence could be recorded by the PMT array within a wavelength of 450–650 nm. The bandwidth of one channel was approximately 12 nm.

For all spectral channels, the same color coding was used. The lifetime τ was calculated in each channel by fitting a monoexponential function to the photon distribution in consecutive time channels. To obtain reasonable lifetime accuracy for the given number of photons, the lifetimes were calculated by binning a 3×3 pixel area. Overlapping binning was used so that the spatial resolution of the intensity data remained unchanged (Rück et al. 2007).

The lifetime histograms in Figure 9.10 show a sharp peak for the lifetime of rhodamine 123, whereas a broad distribution was found for DAPI. The reason could be a statistical broadening due to a lower photon count rate in the case of DAPI and a low signal-to-noise ratio. In addition, a more detailed analysis of the fluorescence lifetime of DAPI by SLIM showed two lifetimes in the spectral range between 478 and 490 nm, which could be correlated to different binding sites of DAPI in the cell nucleus (Rück et al. 2007).

9.3.7.2 Photodynamic Therapy (PDT)

An interesting application of SLIM is the functional characterization of photosensitizers in photodynamic therapy (PDT). PDT is based on the absorption of light by a photosensitizer and subsequent energy transfer to molecular oxygen, leading to the production of cytotoxic reactive oxygen species (ROS) (Fischer, Murphree, and Gomer 1995). Thus far, time-resolved methods have been used to distinguish different species of tumor-localizing porphyrins, such as monomers and aggregates as well as ionic species located at different cellular sites (Schneckenburger et al. 1993, 1995; Strauss et al. 1997). Time-gated video microscopy with highly intensifying camera systems has been used for imaging of tumor-localizing dyes (Scully et al. 1996).

A picosecond laser line-scanning microscope combined with a sub-nanosecond-gated image intensifier and a CCD camera was used to study FLIM of disulphonated aluminium phthalocyanine (AlPcS$_2$), pyridinium zinc(II)phthalocyanine (ZnPPC) and meta-tetra(hydroxyphenyl)chlorine (m-THPC) (Connelly et al. 2001). Using a time-gated video camera system, FLIM of hematoporphyrin derivative (HpD) in tumor-bearing mice revealed that the fluorescence decay was appreciably slower in the tumor than in

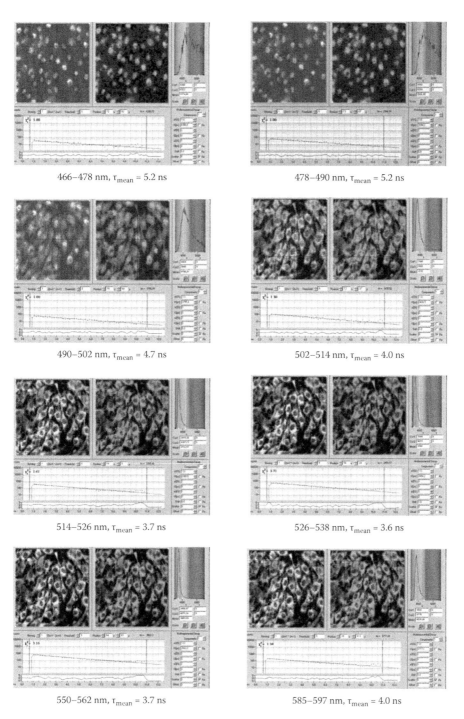

466–478 nm, τ_{mean} = 5.2 ns

478–490 nm, τ_{mean} = 5.2 ns

490–502 nm, τ_{mean} = 4.7 ns

502–514 nm, τ_{mean} = 4.0 ns

514–526 nm, τ_{mean} = 3.7 ns

526–538 nm, τ_{mean} = 3.6 ns

550–562 nm, τ_{mean} = 3.7 ns

585–597 nm, τ_{mean} = 4.0 ns

FIGURE 9.10 **(See color insert following page 288.)** SLIM of C2C12 myoblast cells incubated with DAPI and rhodamine 123. DAPI stained the cell nucleus. Its fluorescence could be detected in the spectral range between 466 and 490 nm. The mean lifetime of DAPI was 5.2 ns. Rhodamine 123 stained the mitochondria. Its fluorescence could be detected in the spectral range between 514 and 562 nm. Its mean lifetime was 3.6–3.7 ns. (Adapted from Rück, A. et al. 2007. *Microscopy Research and Technique* 70:485–492.)

the healthy tissues nearby (Cubeddu et al. 1997). The uptake of photosensitizer aggregates and their monomerization inside tumor cells could also be demonstrated by using TCSPC (Kelbauskas and Dietel 2002).

Different short-pulsed diode lasers were coupled to a confocal laser scanning microscope to measure FLIM of photosensitizers (Kress et al. 2003). With this system, the time-resolved characteristics of the mitochondrial marker rhodamine 123 and 5-ALA (5-aminolevulinic-acid), as well as HAL (5-aminolevulinic-acid-hexylester)-induced protoporphyrine IX (PPIX), were investigated (Rück et al. 2003). A precursor of photosensitizing porphyrins, 5-ALA is approved for the treatment of actinic keratosis and basal cell carcinomas, as well as fluorescence-guided diagnosis of bladder cancer (Bissonette, Bergeron, and Liu 2004; Ickowicz Schwartz et al. 2004). Although it is in widespread use, the specificity of fluorescence diagnosis by 5-ALA and other photosensitizers is still not sufficient.

As discussed recently, SLIM can improve fluorescence diagnosis by analyzing simultaneously different metabolites synthesized from the precursor 5-ALA or other photosensitizers (Kinzler et al. 2007). In addition, with SLIM different compounds involved in autofluorescent signals can be separated. Figure 9.11 shows human hepatoblastoma cells incubated with Photofrin and UDCA (ursodeoxycholic acid) and analyzed by SLIM. Photofrin was the first photosensitizer approved by the U.S. Food and Drug Administration (FDA) for

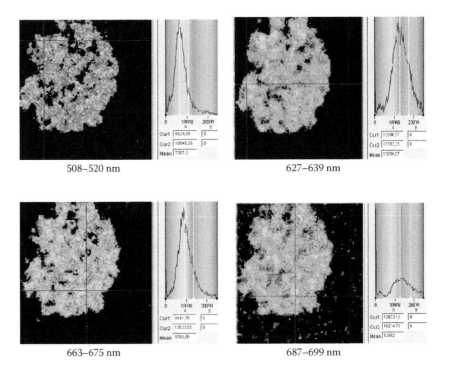

508–520 nm   627–639 nm

663–675 nm   687–699 nm

FIGURE 9.11 **(See color insert following page 288.)** SLIM of HepG2 cells incubated 24 h with 2 µg/mL Photofrin and 150 µM UDCA, demonstrated within different spectral ranges. See the text for explanation. (Adapted from Kinzler, I. et al. 2007. *Photochemical & Photobiological Sciences* 6:1332–1340.)

PDT of bladder cancer, palliative treatment of lung and esophageal cancer, and for other disorders including Barrett's esophagus (Wolfsen 2005; Litle et al. 2003; Schenkman and Lamm 2004; Ortner 2004; Jones et al. 2001). UDCA is a bile acid, which was discussed to enhance PDT efficiency (Kessel, Caruso, and Reiners 2000). Photofrin is a complex mixture of monomeric porphyrins and aggregates and the efficiency of PDT depends on the relative amounts of these species. In order to distinguish between those compounds, pure spectral unmixing is not sufficient because spectral overlap is high. However, a clear separation is possible when spectral and time-resolved analyses are performed simultaneously.

For the measurements, a short pulsed diode laser emitting at 398 nm (PicoQuant GmbH, Berlin, Germany) was coupled to the laser scanning microscope. The PML-16 detector (Becker & Hickl) was used and the spectrograph MS125 was calibrated between 500 and 700 nm, with a spectral bandwidth of 12 nm for one channel. The fluorescence lifetime within the different spectral channels was calculated by fitting a biexponential function to the data using the SPCImage version 2.8.3 software (Becker & Hickl), which uses a Levenberg–Marquardt algorithm to optimize the parameters.

The fluorescence lifetime images of selected channels in Figure 9.11 are represented in pseudocolor. Different compounds of Photofrin could be distinguished. The lifetimes in the two channels of 687–699 nm and 627–639 nm were in the range of 12.7–12.9 ns. The lifetime distribution in the channel of 663–675 nm was significantly different and the maximum lifetime was shifted toward shorter values (around 8 ns). The first two channels could be correlated with the second and first fluorescence bands of monomeric protoporphyrin IX and hematoporphyrin, whereas the spectral range between 663 and 675 nm belongs to aggregates and photoproducts of Photofrin (Kinzler et al. 2007). Lifetimes from aggregates and photoproducts are generally shifted toward shorter values (Schneckenburger et al. 1993, 1995). Figure 9.11 also demonstrates the spectral range between 508 and 520 nm, which coincides with autofluorescent molecules of the cells, mainly flavin molecules. The lifetime was significantly shorter; a biexponential fit delivered a value of about 6 ns.

9.3.7.3 Förster Resonance Energy Transfer

To demonstrate how the FRET efficiency can be determined from multiple wavelength FLIM data, we constructed an EYFP–Cerulean hybrid protein in which the donor fluorophore (Cerulean) and the acceptor fluorophore (EYFP) were closely linked together by a short sequence of 10 amino acids. In this study, Cerulean was chosen as donor fluorophore because of its superior fluorescent properties. Compared to ECFP, which is commonly used as donor fluorophore, Cerulean has a higher quantum yield and a higher extinction coefficient (Rizzo et al. 2004).

To calculate the FRET efficiency of the EYFP–Cerulean hybrid protein, its fluorescence lifetime has to be compared with the fluorescence lifetime of Cerulean in the absence of an acceptor as outlined in Section 9.2.4. To this aim, a streak image was obtained from a HEK293 cell expressing Cerulean as a cytosolic protein (Figure 9.12A). From this streak image, both the fluorescence spectrum (Figure 9.12B) and the fluorescence decay curve (Figure 9.12C) were derived. The fluorescence spectrum was obtained by summing up fluorescence intensities along the time axis (vertical axis) and plotting the resulting intensities

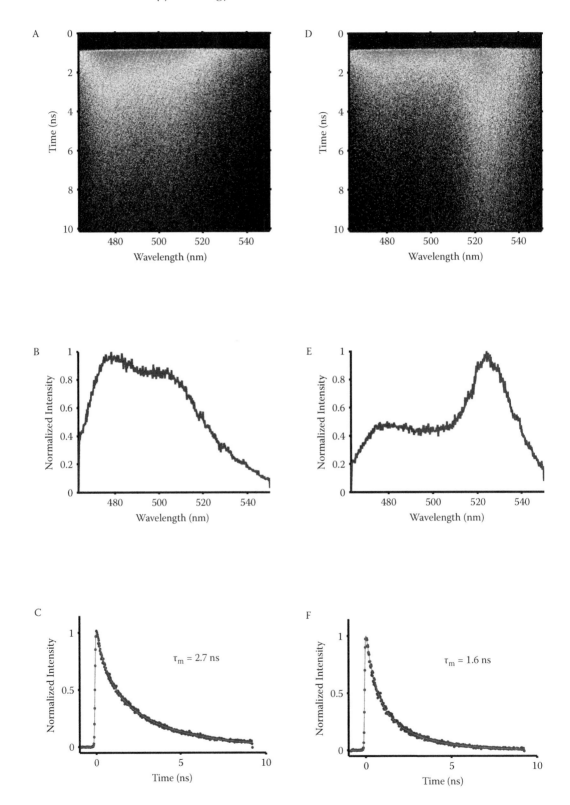

FIGURE 9.12 See caption on opposite page and see color insert following page 288.

FIGURE 9.12 **(See color insert following page 288.)** Analysis of streak images. All streak images were acquired with the setup described in Section 9.3.1. Acquisition time was 5 min. (A) Streak image obtained from a HEK293 cell expressing Cerulean. (B) Fluorescence spectrum obtained from the streak image (A) by adding up fluorescence intensities along the vertical axis. (C) Fluorescence decay curve obtained from the streak image (A) by summing up fluorescence intensities in the wavelength band of 465–495 nm. In this case, the fit of the fluorescence decay of Cerulean yielded a fluorescence lifetime of 2.7 ns. (D) Streak image obtained from a HEK293 cell expressing an EYFP–Cerulean hybrid protein. The comparison with the streak image of Cerulean (A) shows that the donor (Cerulean) fluorescence signal in the wavelength range between 465 and 495 nm is decreased, whereas the acceptor (EYFP) fluorescence in the wavelength range between 510 and 540 nm is increased considerably. (E) Fluorescence spectrum extracted from the streak image. (D) Compared to the fluorescence spectrum of HEK293 cells expressing Cerulean alone, the peak of the fluorescence spectrum is shifted toward acceptor emission wavelengths. (F) Fluorescence decay in the wavelength range from 465 to 495 nm extracted from the streak image. (D) Compared to the fluorescence decay of Cerulean alone, the fluorescence decay of EYFP–Cerulean hybrid protein is considerably accelerated.

versus the wavelength (horizontal axis). The fluorescence decay curve was extracted by summing up the fluorescence intensities in the wavelength band from 465 to 495 nm, which is devoid of acceptor fluorescence. The resulting intensities were plotted against the time axis and the fluorescence lifetime was determined by fitting an exponential function to this histogram using a Levenberg–Marquardt algorithm. In the case of Cerulean, the fit yielded a lifetime $\tau = 2.7$ ns.

An example of a streak image obtained from a HEK293 cell expressing the EYFP–Cerulean hybrid protein is shown in Figure 9.12(D). Comparison with the Cerulean (= donor only) streak image shows that the donor (Cerulean) fluorescence signal in the wavelength range between 465 and 495 nm is decreased, whereas the acceptor (EYFP) fluorescence between 510 and 540 nm is increased. This is also demonstrated by the spectrum (Figure 9.12E). Compared to the spectrum of Cerulean, the emission peak of the fluorescence spectrum of the EYFP–Cerulean hybrid protein was shifted toward acceptor emission wavelengths. Furthermore, the fluorescence decay of the Cerulean moiety (Figure 9.12F) was accelerated in the hybrid protein. The fluorescence lifetime τ decreased to 1.6 ns. From these data, a FRET efficiency of 41% can be calculated.

9.4 CRITICAL DISCUSSION

The techniques presented in this chapter are able to record the fluorescence decay as a function of time *and* wavelength. Accordingly, the plot of the fluorescence decay yields a three-dimensional decay *surface* instead of a two-dimensional decay *curve*. The data contain more information than conventional fluorescence lifetime measurements or spectral imaging techniques can provide. In conventional fluorescence lifetime measurements, a filter is used to select the wavelength band of interest and the rest of the spectrum is discarded. By using the spectrally resolved lifetime techniques, one can exploit the information contained in this part of the spectrum as well. This is especially advantageous when many fluorophores are excited simultaneously in one sample because then these techniques have the power to disentangle the individual contributions.

TABLE 9.1 Comparison of Streak Camera and mwTCSPC/FLIM Techniques

	Streak camera	mwTCSPC
Spectral resolution	512 Channels	16 Channels
Time resolution	512 Channels	256 Channels
IRF (profile)	Almost Gaussian	After-bumps
IRF (FWHM)	~70 ps	~250 ps
xy-Resolution	Determined by resolution of the xy-stage	Determined by scanner and computer memory
Synchronization with the LSM scan	Not possible	Via LSM user interface and router
	Determined by resolution of the xy-stage	Determined by scanner and computer memory
Time needed for a 128 × 128 pixel image	~15–30 min!	Few minutes

However, one must be aware that in the spectrally resolved setups an additional optical element—the spectrograph—is inserted into the fluorescence emission path, which can decrease the overall transmission efficiency of the setup. A considerable part of the fluorescence signal may be lost due to inefficient coupling of the spectrograph to the non-descanned or descanned ports of the confocal microscope. In the first case, specially designed fiber bundles can help to increase collection and coupling efficiency (Becker et al. 2007).

Another factor that might contribute significantly to a loss in the fluorescence signal is the spectrograph itself. Even at the blaze wavelength, transmission efficiency is often only around 70–80%. At other wavelengths, transmission efficiency might decrease considerably. In summary, the advantage to collecting fluorescence from the entire spectrum goes along with a loss of overall efficiency. This is no problem if fluorescence signals are bright, but in dim samples it might be advantageous to revert to conventional lifetime techniques.

When the streak camera and the mwTCSPC approaches are compared (see Table 9.1), the streak camera has far better resolution on both the spectral and the temporal scales. Streak images usually have a size of more than 500 × 500 pixels. Thus, there are 500 parallel wavelength channels in the streak camera setup compared to 16 channels in the mwTCSPC/SLIM setup. The resolution on the temporal axis depends on the sweep time and the number of pixels. Depending on the streak camera type, even measurements in the subpicosecond time range are possible. As discussed in Section 9.3.6.1, the IRF of a streak camera has an almost Gaussian shape. This makes data analysis simple because the center of the laser pulse and the FWHM of the system response function can be included as additional free parameters in the fitting routine.

Also, the IRF of a streak camera does not depend on the wavelength of the incident light. As a result, the IRF does not need to be recalibrated when recording fluorescence emission at different wavelengths. In contrast to this, PMTs and PMT arrays exhibit after-pulses and the IRF of the multichannel PMT is rather broad. It has an FWHM of more than 250 ps and the time-resolution is limited accordingly. Moreover, in PMTs and multichannel PMTs, the IRF depends on the wavelength of the incident light. Thus, the IRF should be determined for each channel separately.

Despite the numerous arguments in favor of the streak camera for spectrally resolved measurements, there are two strong arguments against its use in routine measurements.

One is the high price and the other one is the long acquisition time, which is mainly determined by the readout time of the CCD camera. Typical readout times are around 50 ms; depending on the numbers of pixels to be measured, the total acquisition time for a spectrally resolved fluorescence lifetime image can reach several minutes up to 1 hour for a single hyperspectral image. In our lab, we use the streak camera only to measure the fluorescence decay surface at certain *spots* of interest inside a cell. To obtain spectrally resolved fluorescence lifetime *images,* we use the mwTCSPC/SLIM technique, despite its lower spectral and temporal resolution.

9.5 SUMMARY

This chapter described two techniques that can be used for spectrally resolved fluorescence lifetime measurements in the time domain in conjunction with a laser scanning microscope and a pulsed excitation source. One technique is based on a streak camera and the other is based on a TCSPC approach.

With these techniques, it is possible to acquire fluorescence decays in several wavelength bands simultaneously. The fluorescence emitted by the sample can be recorded in one single measurement. No filters are used to separate the contributions of different fluorophores to the overall fluorescence signal. When applied to FRET measurements, the techniques separate the decay components of the donor and acceptor fluorescence. In this way, it is possible to determine FRET efficiencies between acceptor and donor fluorophores reliably in given subcellular structures.

REFERENCES

Becker, W. 2005. *Advanced time-correlated single photon counting techniques.* Berlin: Springer.

Becker, W., Bergmann, A., and Biskup, C. 2007. Multispectral fluorescence lifetime imaging by TCSPC. *Microscopy Research and Technique* 70:403–409.

Becker, W., Bergmann, A., Biskup, C., et al. 2003. High-resolution TCSPC lifetime imaging. *Proceedings of the SPIE* 4963:175–184.

Becker, W., Bergmann, A., Biskup, C., Zimmer, T., Klöcker, N., and Benndorf, K. 2002. Multiwavelength TCSPC lifetime imaging. *Proceedings of SPIE* 4620:79–84.

Beechem, J. M. 1992. Global analysis of biochemical and biophysical data. *Methods in Enzymology* 210:37–54.

Bird, D. K., Eliceiri, K. W., Fan, C. H., and White, J. G. 2004. Simultaneous two-photon spectral and lifetime fluorescence microscopy. *Applied Optics* 43:5173–5182.

Biskup, C., Hoffmann, B., Kelbauskas, L. et al. 2007. Multidimensional microscopy in the biomedical sciences VII. *Proceedings of SPIE* 6442, 64420V.

Biskup, C., Kusch, J., Schulz, E., et al. 2007. Relating ligand binding to activation gating in CNGA2 channels. *Nature* 446:440–443.

Biskup, C., Zimmer, T., and Benndorf, K. 2001. Fluorescence lifetime measurements in living cells with a confocal laser scanning microscope and a streak camera. *Biophysical Journal* 80:165A.

Biskup, C., Zimmer, T., and Benndorf, K. 2004. FRET between cardiac Na+ channel subunits measured with a confocal microscope and a streak camera. *Nature Biotechnology* 22:220–224.

Biskup, C., Zimmer, T., Kelbauskas, L., et al. 2007. Multi-dimensional fluorescence lifetime and FRET measurements. *Microscopy Research and Technique* 70:442–451.

Bissonette, R., Bergeron, A., and Liu, Y. 2004. Large surface photodynamic therapy with aminolevulinic acid: Treatment of actinic keratoses and beyond. *Journal of Drugs in Dermatology* 3:26–31.

Bevington, P. R., and Robinson, D. K. 2003. *Data reduction and error analysis for the physical sciences,* 3rd ed. Boston: McGraw–Hill.

Boens, N., Qin, W. W., Basaric, N., et al. 2007. Fluorescence lifetime standards for time and frequency domain fluorescence spectroscopy. *Analytical Chemistry* 79:2137–2149.

Campillo, A. J., and Shapiro, S. L. 1983. Picosecond streak camera fluorometry—A review. *IEEE Journal of Quantum Electronics* 19:585–603.

Clegg, R. M. 1996. Fluorescence resonance energy transfer. In *Fluorescence imaging spectroscopy and microscopy,* Chemical Analysis Series, vol. 137, ed. X. F. Wang, and B. Herman, 179–251. New York: John Wiley & Sons.

Connelly, J. P., Botchway, S. W., Kunz, L., Pattison, D., Parker, A. W., and MacRobert, A. J. 2001. Time-resolved fluorescence imaging of photosensitiser distributions in mammalian cells using a picosecond laser line-scanning microscope. *Journal of Photochemistry and Photobiology A* 142:169–175.

Cubeddu, R., Canti, G., Pifferi, A., Taroni, P., and Valentini, G. 1997. Fluorescence lifetime imaging of experimental tumors in hematoporphyrin derivative-sensitized mice. *Photochemistry and Photobiology* 66:229–236.

De Beule, P., Owen, D. M., Manning, H. B., et al. 2007. Rapid hyperspectral fluorescence lifetime imaging. *Microscopy Research and Technique* 70:481–484.

Denk, W., Strickler, J. H., and Webb, W. W. 1990. Two-photon laser scanning fluorescence microscopy. *Science* 248:73–76.

Dickinson, M. E., Bearman, G., Tille, S., Lansford, R., and Fraser, S. E. 2001. Multispectral imaging and linear unmixing add a whole new dimension to laser scanning fluorescence microscopy. *Biotechniques* 31:1272–1278.

Dowling, K., Hyde, S. C. W., Dainty, J. C., French, P. M. W., and Hares, J. D. 1997. 2-D fluorescence lifetime imaging using a time-gated image intensifier. *Optics Communications* 135:27–31.

Fischer, A. M. R., Murphree, A. L., and Gomer, C. J. 1995. Clinical and preclinical photodynamic therapy. *Lasers in Surgery and Medicine* 17:2–31.

Förster, T. 1949a. Versuche zum zwischenmolekularen Übergang von Elektronenanregungsenergie. *Zeitschrift für Elektrochemie* 53:93–99.

Förster, T. 1949b. Experimentelle und theoretische Untersuchung des zwischenmolekularen Überganges von Elektronenanregungsenergie. *Zeitschrift für Naturforschung* 4A:321–327.

Förster, T. 1959. Transfer mechanisms of electronic excitation. *Discussions of the Faraday Society* 27:7–17.

Grinvald, A., and Steinberg, I. Z. 1974. On the analysis of fluorescence decay kinetics by the method of least-squares. *Analytical Biochemistry* 59:583–598.

Ickowicz Schwartz, D., Gozlan, Y., Greenbaum, L., Babushkina, T., Katcoff, D. J., and Malik, Z. 2004. Differentiation-dependent photodynamic therapy regulated by porphobilinogen deaminase in B16 melanoma. *British Journal of Cancer* 90:1833–1841.

Jones, B. U., Helmy, M., Brenner, M., et al. 2001. Photodynamic therapy for patients with advanced non-small-cell carcinoma of the lung. *Clinics in Lung Cancer* 3:37–41.

Kelbauskas, L., and Dietel, W. 2002. Internalization of aggregated photosensitizers by tumor cells: Subcellular time-resolved fluorescence spectroscopy on derivatives of pyropheophorbide-a ethers and chlorin e6 under femtosecond one- and two-photon excitations. *Photochemistry and Photobiology* 76:686–694.

Kessel, D., Caruso, J. A., and Reiners, J. J. 2000. Potentiation of photodynamic therapy by ursodeoxycholic acid. *Cancer Research* 60:6985–88.

Kinzler, I., Haseroth, E., Hauser, C., and Rück, A. 2007. Role of mitochondria in cell death of hepatic cancer cells induced by Photofrin PDT and ursodeoxycholic acid. *Photochemical & Photobiological Sciences* 6:1332–1340.

Knutson, J. R., Beechem, J. M., and Brand, L. 1983. Simultaneous analysis of multiple fluorescence decay curves: A global approach. *Chemical Physics Letters* 102:501–507.

Kress, M., Meier, T., Steiner, R., et al. 2003. Time-resolved microspectrofluorometry and fluorescence lifetime imaging of photosensitizers using ps pulsed diode lasers in laser scanning microscopes. *Journal of Biomedical Optics* 8:26–32.

Krishnan, R. V., Saitoh, H., Terada, H., Centonze, V. E., and Herman, B. 2003. Development of a multiphoton fluorescence lifetime imaging microscopy system using a streak camera. *Review of Scientific Instruments* 74:2714–2721.

Kusumi, A., Tsuji, A., Murata, M., et al. 1991. Development of a streak-camera based time resolved microscope fluorometer and its application to studies of membrane fusion in single cells. *Biochemistry* 30:6517–6527.

Lakowicz, J. R. 2006. *Principles of fluorescence spectroscopy,* 3rd ed. Singapore: Springer.

Latt, S., Cheung, H. T., and Blout, E. R. 1965. Energy transfer. A system with relatively fixed donor–acceptor separation. *Journal of the American Chemical Society* 87:995–1003.

Litle, V. R., Luketich, J. D., Christie, N. A., et al. 2003. Photodynamic therapy as palliation for esophageal cancer: Experience in 215 patients. *Annals of Thoracic Surgery* 76:1687–1692.

McKinnon, A. E., Szabo, A. G., and Miller, D. R. 1977. The deconvolution of photoluminescence data. *Journal of Physical Chemistry* 81:1564–1570.

Nordlund, T. M. 1991. Streak cameras for time-domain fluorescence. In *Topics in fluorescence spectroscopy, volume 1: Techniques,* ed. J. R. Lakowicz, 183–260. New York: Plenum Press.

O'Connor, D. V., Ware, W. R., and Andre, J. C. 1979. Deconvolution of fluorescence decay curves. A critical comparison of techniques. *Journal of Physical Chemistry* 83:1333–1343.

Ortner, M. A. 2004. Photodynamic therapy in cholangiocarcinomas. *Best Practice and Research Clinical Gastroenterology* 18:147–54.

Pawley, J. B. 2006. *Handbook of biological confocal microscopy,* 3rd ed. Berlin: Springer.

Periasamy, A., and Day, R. N. 2005. *Molecular imaging: FRET microscopy and spectroscopy.* Oxford, England: Oxford University Press.

Perrin, J. 1927. Fluorescence et induction moléculaire par résonance. *Comptes Rendus Hebdomadaires des Seances de l'Academie des Sciences* 184:1097–1100.

Press, W. H., Teukolsky, S. A., Vetterling, W. T., and Flannery, B. P. 1992. *Numerical recipes in C. The art of scientific computing,* 2nd ed. Cambridge, England: Cambridge University Press.

Qu, J., Liu, L., Chen, D., et al. 2006. Temporally and spatially resolved sampling imaging with a specially designed streak camera. *Optics Letters* 31:368–370.

Rizzo, M. A., Sprinter, G. H., Granata, B., and Piston, D. W. 2004. An improved cyan fluorescent protein variant useful for FRET. *Nature Biotechnology* 22:445–449.

Rück, A., Dolp, F., Scalfi-Happ, C., Steiner, R., and Beil, M. 2003. Time-resolved microspectrofluorometry and fluorescence lifetime imaging using ps pulsed diode lasers and TCSPC techniques in laser scanning microscopes. *Proceedings of the SPIE* 5139:166–172.

Rück, A., Dolp, F., Steiner, R., Steinmetz, C., von Einem, B., and von Arnim, C. A. F. 2008. SLIM for multispectral FRET imaging. *Proceedings of the SPIE* 6860:68601F.

Rück, A., Hülshoff, C. H., Kinzler, I., Becker, W., and Steiner, R. 2007. SLIM: A new method for molecular imaging. *Microscopy Research and Technique* 70:485–492.

Schenkman, E., and Lamm, D. L. 2004. Superficial bladder cancer therapy. *ScientificWorldJournal* 28:387–399.

Schneckenburger, H., Gschwend, M. H., Sailer, R., Rück, A., and Strauß, W. S. L. 1995. Time-resolved pH dependent fluorescence of hydrophilic porphyrins in solution and in cultivated cells. *Journal of Photochemistry and Photobiology B* 27:251–255.

Schneckenburger, H., König, K., Kunzi-Rapp, K., Westphal-Frösch, C., and Rück, A. 1993. Time-resolved in vivo fluorescence of photosensitizing porphyrins. *Journal of Photochemistry and Photobiology B* 21:143–147.

Scully, A. D., MacRobert, A. J., Botchway, S., et al. 1996. Development of a laser-based fluorescence microscope with a subnanosecond time resolution. *Journal of Fluorescence* 6:119–125.

Straume, M., Frasier-Cadoret, S. G., and Johnson, M. L. 1991. Least-squares analysis of fluorescence data. In *Topics in fluorescence spectroscopy, volume 2: Principles*, ed. J. R. Lakowicz, 177–240. New York: Plenum Press.

Strauss, W. S. L., Sailer, R., Schneckenburger, H., et al. 1997. Photodynamic efficacy of naturally occurring porphyrins in endothelial cells in vitro and microvasculature in vivo. *Journal of Photochemistry and Photobiology B* 39:176–184.

Stryer, L., and Haugland, R. P. 1967. Energy transfer. A spectroscopic ruler. *Proceedings of the National Academy of Sciences USA* 58:719–726.

Suhling, K., French, P. M. W., and Phillips, D. 2005. Time-resolved fluorescence microscopy. *Photochemical & Photobiological Sciences* 4:13–22.

Tsien, R. Y. 1989. Fluorescent probes of cell signaling. *Annual Review of Neuroscience* 12:227–253.

Tsien, R. Y. 1998. The green fluorescent protein. *Annual Review of Biochemistry* 67:509–544.

White, J. G., Amos, W. B., and Fordham, M. 1987. An evaluation of confocal versus conventional imaging of biological structures by fluorescence light microscopy. *Journal of Cell Biology* 105:41–48.

Wolfsen, H. C. 2005. Present status of photodynamic therapy for high-grade dysplasia in Barrett's esophagus. *Journal of Clinical Gastroenterology* 39:189–202.

Zhang, J., Campbell, R. E., Ting, A. Y., and Tsien, R. Y. 2002. Creating new fluorescent probes for cell biology. *Nature Reviews. Molecular Cell Biology* 3:906–918.

CHAPTER 10

Time-Resolved Fluorescence Anisotropy

Steven S. Vogel, Christopher Thaler,
Paul S. Blank, and Srinagesh V. Koushik

10.1 INTRODUCTION

In this chapter, we will discuss how molecular rotation and protein–protein interactions can be measured using time-resolved fluorescence anisotropy—a variation of fluorescence lifetime imaging (FLIM) (Bastiaens and Squire 1999; Gadella, Jovin, and Clegg 1993; Lakowicz, Szmacinski, 1992; Wang, Periasamy, and Herman 1992). We wish to state that this chapter is not intended to be a review of the fluorescence polarization and anisotropy literature; rather, our goal is to complement the limited but growing list of recent studies where polarization and anisotropy microscopy has been applied to biological questions (Bader et al. 2007; Blackman et al. 1996; Blackman, Piston, and Beth 1998; Clayton et al. 2002; Clegg, Murchie, and Lilley 1994; Gautier et al. 2001; Heikal et al. 2000; Hess et al. 2003; Jameson and Mocz 2005; Piston and Rizzo 2008; Rao and Mayor 2005; Rizzo and Piston 2005; Runnels and Scarlata 1995; Sharma et al. 2004; Suhling et al. 2004; Thaler et al. 2009; Varma and Mayor 1998; Volkmer et al. 2000; Yan and Marriott 2003; Yeow and Clayton 2007) by conveying an intuitive appreciation of the methods used for measuring anisotropy and the strengths and weaknesses of the approach, as well as identifying some of the current technical limitations.

Often in textbooks, complex methods such as fluorescence anisotropy and polarization are described using mathematical formulae. Although a formula can be a precise and concise means of conveying complex ideas between experts well trained in mathematics, the use of formulae can actually impede comprehension for a lay audience. Similarly, the use of technical jargon can aid communication between experts with a common appreciation of underlying concepts, but its use can hinder communication with the uninitiated. Alternatively, attempts to present complex methods in an intuitive fashion often require making simplifying assumptions, which can lead to confusion. Here, we attempt to avoid these communication pitfalls by first presenting the fundamental photophysical concepts underlying fluorescence polarization and anisotropy. This we hope will provide some foundation for the more detailed explanations, occasionally using formulae, found in subsequent sections and serve as a starting point from which a thorough understanding of the theoretical principles governing the technique can be attained.

At its most fundamental level, *time-resolved fluorescence anisotropy monitors changes in the orientation of a fluorophore.* These measurements are typically monitored over a time course spanning picoseconds to nanoseconds using the time-correlated single-photon counting technology (TCSPC; Becker 2005) more commonly used for fluorescence lifetime imaging. Changes in fluorophore orientation can reflect the rotation of a macromolecule to which a fluorophore is attached. Rapid changes in fluorescence orientation (anisotropy) can also reveal resonance energy transfer between fluorophores in close proximity and, by inference, indicate that the macromolecules to which the fluorophores are attached are also in close proximity. The immediate challenge is to understand how the orientation of a fluorophore can be measured using an optical microscope.

10.2 UNDERLYING CONCEPTS

Two key concepts of photophysics are critical to gaining a clear understanding of fluorescence anisotropy imaging. The first concerns the nature of the electric field of light. This is important because the orientation of the electric field of the light used to excite fluorophores in an anisotropy experiment is used as a fiduciary orientation against which all changes in orientation are measured. The second concept involves absorption and emission dipoles of fluorophores and the role that they play in the absorption and emission of photons. The orientation of the absorption dipole of a fluorophore is important because it is a key factor that determines the probability that a fluorophore will absorb a photon to reach an excited state. Similarly, the orientation of the emission dipole of a fluorophore is important because it dictates the orientation of the electric field of the photons emitted by the fluorophore. The orientation of the electric field of these emitted photons will subsequently be used to deduce the orientation of the fluorophore itself relative to the electric field orientation of the excitation light. We begin with the nature of the electric field of light.

The electric field of light: Light is a form of electromagnetic radiation with wave-like behavior and composed of elementary particles called *photons.* As photons travel through space, they generate both electric and magnetic fields. The electric field of light oscillates, and the vector of these oscillations defines the orientation of its electric field. This orientation is always perpendicular to the direction in which the light is traveling and orthogonal to the vector orientation of its magnetic field. If the electric field orientation does not change as light travels through space, it is called *linearly polarized light.* This light is primarily used as the excitation light in fluorescence anisotropy experiments, and the invariant electric field orientation of the linearly polarized light source serves as a fiduciary orientation ($\theta = 0°$) against which other orientations are measured. We will show shortly that analyzing the orientation of the electric fields of light emitted from a sample yields information regarding the speed of molecular rotation.

Absorption and emission dipoles: Imaging fluorescence involves excitation by the absorption of a photon at one wavelength and the emission and detection of a photon at a different wavelength. With one-photon excitation, one high-energy photon is absorbed by the fluorophore, and a single lower energy (longer wavelength) photon is subsequently emitted. With multiphoton excitation, two or more than two lower energy photons are absorbed simultaneously by the fluorophore (Denk, Strickler, and Webb 1990).

The absorption of a photon by a fluorophore involves the excitation of an electron from a ground state to an excited state. In general, a *dipole* is formed by the separation of a positive and a negative charge. With fluorescence excitation, an *absorption dipole* defines the preferred orientation of the chromophore to absorb a photon. The orientation of this dipole is specific to the chemical structure of the chromophore (often oriented along the axis of alternating single and double bonds that define the extended p-orbital).

The probability that a fluorophore will be excited is a direct function of the relative orientations of the electric field of the excitation light and the absorption dipole of the chromophore. When the electric field orientation is parallel to the absorption dipole, a fluorophore has the greatest chance to absorb a photon. When the electric field orientation is perpendicular, there is little chance that the fluorophore will be excited. In a population of randomly oriented fluorophores, those with absorption dipole orientations similar to the electric field's orientation will be preferentially "selected" by the excitation light.

Between the absorption and the emission of a photon, the shape of a fluorophore can change. If this happens, the orientation of the absorption dipole might be different from the fluorophore's dipole orientation immediately prior to emission (the *emission dipole*). Under these conditions, it is the orientation of the emission dipole, rather than the absorption dipole, that determines the orientation of the electric field of emitted light. Obviously, if the orientations of the absorption and emission dipoles are the same (termed *collinear*), the orientation of the electric field of the emitted light can be inferred from the orientation of either dipole.

Having conveyed these basic underlying concepts of photophysics, we can now proceed to the following sections. There we will attempt to build progressively an intuitive understanding of the theory behind fluorescence polarization and anisotropy measurements, beginning with a thorough treatment of the polarization of light.

10.3 LIGHT HAS AN ORIENTATION

Time-resolved fluorescence anisotropy measurements are based on detecting a property of light resulting from the orientation of the electric field—that is, its *polarization*. Unlike other aspects of light, such as intensity and frequency, that are readily perceived by the human eye, polarization is poorly detected by humans. The intensity of light that we perceive is proportional to the number of photons detected by our eyes. The human eye can perceive changes in the intensity of light over a range of a few photons per trial to light intensities that are greater than 10 orders of magnitude brighter when gazing upon the sun on a clear day (Hecht, Shlaer, and Pirenne 1941; Inoue and Spring 1997). The human eye can also detect and discriminate between light with wavelengths ranging from about 350 to 750 nm (violet to red; Inoue and Spring 1997).

Thus, it is no surprise that we can readily and intuitively understand the meaning of a change in light intensity or color as encountered in fluorescence microscopy. As mentioned earlier, in addition to intensity and color, light also has a direction of propagation, and the electric field has an orientation in a plane perpendicular to this direction. This attribute of light is called *polarization,* which can be thought of as the orientation of the electronic oscillations of a light wave as it propagates through space. If this orientation does not change with propagation, it is called *linearly polarized light;* if the orientation does change, it can vary in a circular or an elliptical pattern. Many insects, cephalopods, and several other aquatic organisms have the ability to perceive the polarization state of light.

In contrast, human eyes have very low sensitivity to the polarization state of light and accordingly most people do not perceive differences in polarization. Humans can perceive polarization by viewing the world through a linear polarizer—a type of filter that dramatically attenuates linearly polarized light as a function of its orientation relative to the axis of the polarizer. In essence, the use of these polarizers transforms the directional orientation of light into changes in light intensity—an attribute that our eyes can readily distinguish. The use of a linear polarizer to visualize the orientation of light is illustrated in Figure 10.1. Linear polarizers can also be used to parameterize the polarization of emitted light and form the basis for building microscopes that can image changes in polarization. Before we describe how polarization is parameterized and the construction of microscopes for imaging the polarization state of light in biological samples, we first explore what happens when a population of randomly oriented fluorophores is excited by linearly polarized light.

10.4 PHOTOSELECTION

The typical light sources used in most conventional fluorescence light microscopes (mercury arc lamps) emit light with random orientation; these sources are said to emit *natural light.* Individual photons still have polarization, but all possible orientations of photons are present in natural light.* As a consequence, fluorophores whose absorption dipoles are in a plane perpendicular to the direction of the excitation light will have the same chance of absorbing a photon when excited using a mercury arc lamp. Thus, the emission from this population of fluorophores will also lack orientation bias if the light is observed in the direction parallel to the direction of travel of the excitation light.

This picture is very different when linearly polarized light is used to excite a population of fluorophores. Linearly polarized light can be generated by passing light with random polarization through a linear polarizer (to filter out most of the photons whose orientation does not match the orientation of the filter axis). Linearly polarized light is also produced by most of the lasers used in laser scanning microscopy. When this light is used to excite a population of randomly oriented fluorophores, not all molecules are excited with equal probability. Molecules whose absorption dipole is oriented parallel to the electric field of the light source will be preferentially excited, while molecules with absorption dipoles

---

* These orientations are, however, constrained to the plane of polarization, which is perpendicular to the direction in which the light is traveling.

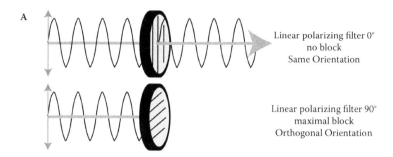

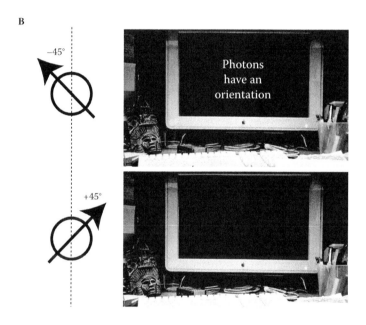

FIGURE 10.1 **(See color insert following page 288).** Detecting polarization. (A) The electric field of linearly polarized light is depicted as a wave orthogonal to the wave vector (green arrow). The electric field vector (blue double arrow) is also orthogonal to the wave vector and resides in a single plane for linearly polarized light. The light is transmitted if the electric field vector is parallel to the orientation of a polarizing filter (top; 0°), but is attenuated if it is perpendicular (bottom; 90°). (B) A digital camera was used to photograph a desktop with LCD computer monitor through a polarizing filter oriented either at −45 or +45° relative to the monitor's height axis. Notice how the writing displayed on the monitor is attenuated by rotating the polarizing filter by 90°. Also notice that the small white pilot light on the monitor casing (lower right) is not attenuated, nor are the other objects on the desk. The LCD monitor emits polarized light and therefore the intensity of its emission can be attenuated by rotation of the filter. Presumably, at −45° the orientation of the polarizing filter is parallel to the electric vector of the light being emitted by the LCD monitor. In contrast, the pilot light emits natural light, and the objects on the desk primarily reflect natural light, so their image is not attenuated.

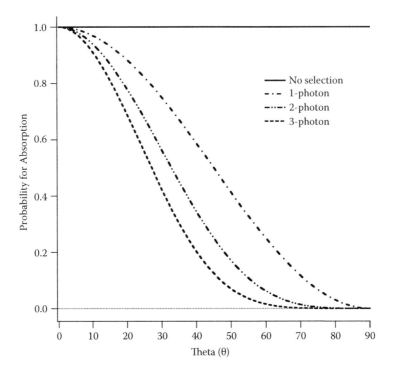

FIGURE 10.2 Fluorescence excitation with linearly polarized light. The probability of exciting a fluorophore with linearly polarized light is plotted as a function of the angle formed between the electronic vector of the excitation and the dipole orientation of the fluorophore (θ). This probability is also a function of the type of excitation as shown in the figure. In contrast, all fluorophores are excited with equal probability, regardless of orientation, when excited with natural unpolarized light (solid line). Note that the actual probabilities for excitation are also attenuated by the fluorophores' absorption coefficient (or multiphoton cross section and by the intensity of the excitation; not shown).

oriented exactly 90° to the electric field will never be excited. This excitation biased by orientation is called *photoselection* and is depicted in Figure 10.2.

A more quantitative appreciation for photoselection can be attained by considering a single fluorophore excited by linearly polarized light. For now we will assume that this fluorophore's orientation is rigid and fixed relative to the electric field of the light source. If theta (θ) is the angle formed between the electric field of the linearly polarized light source and the absorption dipole of a fluorophore, then the probability that a fluorophore will absorb a photon will be proportional to $\cos^2 \theta$. For two-photon excitation, two low-energy photons need to be absorbed almost simultaneously. Because the probability for two independent events occurring is the product of their individual probabilities, the probability for two-photon absorption will be proportional to $\cos^4 \theta$, and for three-photon excitation it is proportional to $\cos^6 \theta$ (Lakowicz, Gryczynski, et al. 1992). This angular dependence of excitation is depicted in Figure 10.2. There is a sharper transition in the θ dependence with multiphoton excitation with preferential excitation at smaller angles.

We will now consider what happens when a population of fluorophores is excited by linearly polarized light. With natural light, most of the fluorophores in a sample can be excited.* Our objective is to understand both intuitively and quantitatively what fraction of a population of fluorophores can be excited by linearly polarized light. We shall start with a population of fluorophores that have the same absorption dipole orientation (as might be encountered in a crystal or for some hydrophobic fluorophores that partition into a membrane bilayer). If θ is $0°$, then the relative probability of excitation will be 1 ($cos^x \theta = 1$) for one-, two-, or three-photon excitation. All fluorophores can be excited. Obviously, the actual percentage of fluorophores that *will* get excited is also a function of their absorption coefficient (or multiphoton cross section), the excitation light intensity, and wavelength. Conversely, if θ is exactly $90°$, then none of the fluorophores can absorb a photon.

If, however, the population of uniformly oriented fluorophores has a θ value between 0 and $90°$, we will get different amounts of excitation for one-, two-, or three-photon excitation. For example, if θ is $45°$, then for one-photon excitation we expect a 50% probability that a fluorophore will absorb a photon ($cos^2(45°) = 0.5$). With two-photon excitation, we expect that no more than 25% ($cos^4(45°) = 0.25$) of the fluorophores will be excited, and with three-photon excitation the probability for excitation drops to only 12.5% ($cos^6(45°) = 0.125$).

10.4.1 Photoselection of a Randomly Oriented Population of Fluorophores

We now wish to consider what happens when a randomly oriented population of fluorophores is excited by linearly polarized light. What do we mean by "randomly oriented" and how do we model it both mathematically and conceptually? In the context of the defined orientation of the electric field of our polarized excitation light source, we can imagine a sphere where the electric field orientation is represented by a double arrow running through the center of the sphere and emerging out of each pole at 0 and $180°$ (see Figure 10.3). In this framework, the absorption dipole orientation of a fluorophore (relative to the electric field of the light source) is represented by a single arrow beginning at the center of the sphere and pointing to a location on the sphere's surface (black arrow with radius r).

All possible arrow orientations define the sphere and must have the same length in this visualization because the length of the arrow represents the dipole strength and we are modeling all orientations of the *same* dipole. In the example depicted in Figure 10.3, the angle formed between the electric field of the linearly polarized light source and the absorption dipole of a fluorophore, θ, is $45°$. Clearly, many other dipole orientations exist that would also have a θ value of $45°$; together, this subpopulation of fluorophore orientations can be thought of as defining a *circle of latitude* (red circle in Figure 10.3).

As mentioned previously, all fluorophores that share the same θ value (same circle of latitude) would have the same probability for excitation. Similarly, other subpopulations of fluorophores will define other circles of latitude (e.g., at $15°$, $30°$, etc.) with their own unique probabilities for excitation defined by their θ value (ranging from 0 to $180°$). In this

---

* As previously noted, when a sample is excited with "natural light" on a microscope, the orientations of photons available for excitation is constrained to a plane that is perpendicular to the direction in which the excitation light is traveling. Thus, due to the geometry of the excitation path of light microscopes, a small subset of fluorophores with absorption dipoles oriented orthogonal to this plane will not be excited, even with "unpolarized" light.

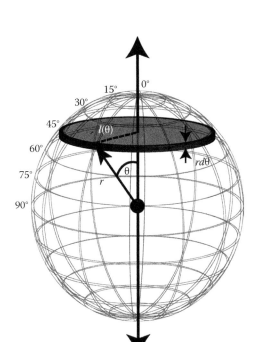

FIGURE 10.3 **(See color insert following page 288).** Random dipole orientation distributions. Electric field orientation is represented by a double arrow running through the center of the sphere and emerging out of each pole at 0 and 180°. The absorption dipole orientation of a fluorophore is represented by a single arrow beginning at the center of the sphere and pointing to a location on the surface of the sphere. In the example shown, the angle formed between the electric field of the linearly polarized light source and the absorption dipole of a fluorophore, θ, is 45°. Many other dipole orientations exist that would also have a θ value of 45° and together this subpopulation of fluorophore orientations defines a circle of latitude (red circle). All fluorophores that share the same θ value (same circle of latitude) would have the same probability for excitation. Other subpopulations of fluorophores will define other circles of latitude (e.g., at 15°, 30°, etc.) with their own unique probabilities for excitation that are defined and parameterized by their θ value (ranging from 0 to 90°). In this scheme, the population of all fluorophore orientations is represented by the collection of all circles of latitude, which is equivalent to the total surface area of the sphere. Accordingly, a random distribution of dipole orientations would have the same density of arrowhead distributed evenly over the entire surface of the sphere. The relative abundance of fluorophores with a specific θ value will therefore be proportional to the surface area of its circle of latitude. This area is the product of its circumference ($2\pi\,l$) times its width ($rd\theta$), where r is the radius of our sphere and l is the radius of a circle of latitude. Note that the radius of a circle of latitude (l) is itself a function of θ ($l = r{\cdot}\sin\theta$), the circumference of a circle of latitude is $2\pi r{\cdot}\sin\theta$, and its surface area will be $2\pi r^2{\cdot}\sin\theta d\theta$. Thus, the surface area of a circle of latitude will be proportional to $\sin\theta d\theta$.

scheme, the population of all fluorophore orientations is represented by the collection of all circles of latitude, which is equivalent to the total surface area of the sphere. Accordingly, *a random distribution of dipole orientations would have the same density of arrowheads distributed evenly over the entire surface of the sphere.*

Next, we explore how to model this random distribution of fluorophore orientations mathematically in anticipation of calculating the distribution of fluorophores expected to be excited by linearly polarized light. Specifically, we would like to know how the number of fluorophores in a random distribution changes as a function of θ. We might expect that, in a population of randomly oriented fluorophores, we will find the same number of fluorophores residing in each circle of latitude. That is, we will find approximately the same number of molecules with a θ value of 0° as we would at 90°. This intuitive guess is wrong! Furthermore, the intellectual conflict generated by the failure of this hunch, we believe, is one of the main stumbling blocks toward achieving a quantitative understanding of polarization.

To understand why this distribution must be wrong we calculate the surface area defined by the area between two neighboring circles of latitude (Dill and Bromberg 2003). The surface area of two neighboring circles of latitude is simply the circumference of the circle ($2\pi \ell$) times its width ($r d\theta$), where r is the radius of our sphere and ℓ is the radius of a circle of latitude (see Figure 10.3). Because the radius of a circle of latitude (ℓ) is itself a function of θ ($\ell = r \cdot \sin \theta$), the circumference of a circle of latitude is $2\pi r \cdot \sin \theta$, and its surface area will be $2\pi r^2 \cdot \sin \theta d\theta$. Because $2\pi r^2$ is a constant (the value of r does not change), we see that the surface area of a circle of latitude will be proportional to $\sin \theta d\theta$. Thus, the surface area will be small when θ has values close to 0° and it will be largest when θ is 90°.

If, as proposed, the *same* number of fluorophores were to be distributed evenly over the *surface area of each differential circular area of latitude* (as the value of θ changes), the surface density would have to decrease as the value of θ increased from 0 to 90°. As mentioned earlier, the hallmark of a random distribution of fluorophore orientations is that the surface density of fluorophores will be constant over the entire surface of the sphere. To offset the influence of changing surface area, a random distribution of fluorophore orientations will occur only when the number of fluorophores found at a specific θ value changes proportionally to the change in surface area. Specifically, for a random distribution of fluorophore orientations, the *abundance* of fluorophores having a specific θ value must be proportional to $\sin \theta$. Thus, in a population of *randomly* oriented fluorophores, many more fluorophores will be oriented with θ angles close to 90° than at angles close to 0°. A random distribution of orientations is called *isotropic,* from the Greek words *iso* (equal) and *tropos* (direction).

Starting with an isotropic distribution of fluorophores, we are now ready to calculate the distribution of fluorophores that are expected to be excited by linearly polarized light. As mentioned previously, the probability of exciting a fluorophore will be proportional to $\cos^x \theta$, where x is zero for excitation with natural "unpolarized" light (no photoselection), two for one-photon excitation, four for two-photon excitation, and six for three-photon excitation. With polarized light, it will be easier to excite molecules at low θ values than at larger angles. In contrast, the probability for finding a fluorophore with a particular θ value is proportional to $\sin \theta$. It will be easier to find molecules with θ values near 90° than those with values closer to 0°. Thus, the distribution of fluorophores that are expected to be excited by linearly polarized light (as a function of θ) is determined by two opposing factors and will be proportional to the product of these two factors:

$$p \propto \cos^x \theta \cdot \sin \theta \qquad (10.1)$$

In Figure 10.4, the value of p is plotted as a function of θ for excitation with natural light, and for one-, two-, and three-photon excitation with linearly polarized light. Natural light (no photoselection) has no predominant electric field orientation to serve as a fiduciary orientation (θ = 0) against which the orientation of excited fluorophores can be compared. Nonetheless, the distribution of fluorophore orientations excited with natural light can be considered relative to the electric field orientation of a linearly polarized light source (as would be used for one-, two-, and three-photon photoselection). Under these conditions,

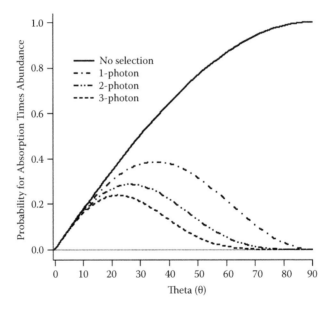

FIGURE 10.4 Photoselection with linearly polarized light. The probability of exciting a fluorophore from an isotropic population with linearly polarized light is plotted as a function of the angle formed between the electronic vector of the excitation and the dipole orientation of the fluorophore (θ). This probability is a function of the type of excitation as shown in the figure. Note that the shape of these curves is determined by two opposing factors, both functions of θ; the probability that a fluorophore will absorb a photon—$\cos^x \theta$, where x is zero for excitation with natural light, two for excitation by one-photon excitation, four for two-photon excitation, and six for three-photon excitation; and the probability that a fluorophore will have a specific θ value (sin θ). All fluorophores are excited with equal probability, regardless of orientation, when excited with natural unpolarized light (solid line trace); thus, in this instance, the curve simply represents the relative abundance of fluorophores at different θ value (sin θ).

the distribution of fluorophore orientations excited with natural light will simply be proportional to the random orientation of the molecules in solution; most excited molecules will have an orientation with a θ value close to 90° because, without photoselection, Equation 10.1 reduces to $p \propto \sin \theta$.

In contrast, with photoselection, the distribution of fluorophores that are excited will be skewed to lower θ values and have modes centered at 35, 27, and 22° for one-, two-, and three-photon excitation, respectively. Photoselection with linearly polarized light transforms an isotropic distribution of ground-state fluorophore orientations into an *anisotropic* distribution of excited fluorophores.

10.5 HOW DO WE DETECT POLARIZED EMISSIONS?

With the absorption of a photon, a fluorophore will be excited; from the excited state, the fluorophore will ultimately decay by a radiative or a nonradiative pathway. For simplicity, we will assume that fluorophores are immobile (they do not rotate while in the excited state) and that their absorption and emission dipole are collinear (the impact of molecular rotation and having absorption and emission dipoles that are not collinear will be discussed later). Under these conditions, the orientation of an emitted photon will be correlated with the orientation of the linearly polarized light as a result of photoselection.

Ultimately, we would like to use microscopy to follow changes in the orientation of isotropic populations of molecules to learn about their behavior and proximity to other molecules in living cells. This requires a way to parameterize and measure the orientation of emitted photons relative to the orientation of the electric vector of our linearly polarized light source. In Figure 10.5, we see a diagram depicting a transition dipole (blue double arrow) of a fluorophore (from an isotropic solution) that was excited by a linearly polarized light source (L.S.) whose electric field is shown as a black double arrow. The three-dimensional orientation of the dipole can be characterized by two angles, θ and ϕ, where θ is the angle formed between the dipole and the X-axis in the XY-plane* and ϕ is the angle formed between the projection of the dipole on the YZ-plane (green disk) and the Z-axis.

The X-axis is parallel to the electric vector of our light source and is therefore an axis of symmetry. The Y- and Z-axes are perpendicular to the electric vector of our light source and are therefore not axes of symmetry. The light intensity emitted from our sample dipole will be proportional to the square of the dipole length, and the dipole vector can be thought of as being composed of three directional components: x, y, and z. A signal proportional to the total intensity of light emitted by our fluorophore can be measured by placing photomultiplier (P.M.) detectors on each of the three axes. The light emitted will be proportional to the sum of the three signals ($I_{\text{total}} = I_x + I_y + I_z$). P.M.x will only detect light related to the yz-vector components of the dipole; x-component information is absent in this direction. Similarly, P.M.y will only detect light related to the xz-vector components, and P.M.z will only detect light related to the xy-vector components.

---

* Note that although the assignment of axes' names is arbitrary, in most published treatments θ is the angle formed between the dipole and the Z-axis. Because this chapter is describing anisotropy measurements on a light microscope, we chose to maintain the commonly accepted axes where the Z-axis projects out of the objective, and the X- and Y-axes are perpendicular to the Z-axis in the image plane.

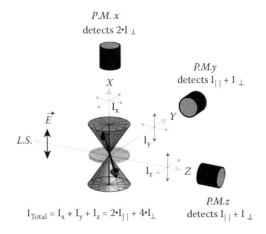

FIGURE 10.5 **(See color insert following page 288).** Detecting polarization on a microscope. The transition dipole (blue double arrow) of a fluorophore excited by a linearly polarized light source (L.S.) is shown. The electric field vector of the light source is shown as a black double arrow. The three-dimensional orientation of the dipole can be characterized by two angles, θ and ϕ, where θ is the angle formed between the dipole and the X-axis (pink cone) and ϕ is the angle formed between the projection of the dipole on the YZ-plane (green disk) and the Z-axis. The light intensity emitted from this fluorophore will be proportional to the square of the dipole strength, and the dipole vector can be thought of as being composed of three directional components: x, y, and z. A signal proportional to the total intensity of light emitted by our fluorophore can be measured summing the signals detected by photomultiplier detectors (P.M.) positioned on each of the three axes. The intensity information encoded in the x-vector component by convention is called $I_{\parallel}$; the intensity information encoded in the y- and z-vectors components is called $I_{\perp}$. As a result of photoselection and the distribution symmetry of excited molecules formed around the X-axis for an isotropic solution, the y-vector component is equal to the z-vector component. For an isotropic solution of fluorophores excited with linearly polarized light whose electric vector is parallel to the X-axis, the P.M.x detector will measure a signal whose intensity is proportional to $2 \cdot I_{\perp}$ (see the crossed double-headed green arrows); P.M.y and P.M.z will each measure light signals whose intensity is proportional to $I_{\parallel} + I_{\perp}$ (see the crossed double-headed red and green arrows). The total emission intensity will therefore be proportional to $2 \cdot I_{\parallel} + 4 \cdot I_{\perp}$ or more simply $I_{\parallel} + 2 \cdot I_{\perp}$. Note that the xy-plane depicted here corresponds to the sample plane on a microscope, and the P.M.z detector corresponds to a photomultiplier placed after the microscope condenser (or at an equivalent position on the epifluorescence path).

The intensity information encoded in the x-vector component by convention is called $I_{\parallel}$ and the intensity information encoded in the y- and z-vector components is called $I_{\perp}$. As a result of photoselection and the distribution symmetry of excited molecules formed around the X-axis for an isotropic solution, the y-vector component is equal to the z-vector component. Thus, for an isotropic solution of fluorophores excited with linearly polarized light whose electric vector is parallel to the X-axis, the P.M.x detector will measure an unpolarized light signal whose intensity is proportional to $2 \cdot I_{\perp}$. P.M.y and P.M.z will each measure a polarized light signal whose intensity is proportional to $I_\parallel + I_\perp$. The total emission intensity will therefore be proportional to $2 \cdot I_\parallel + 4 \cdot I_\perp$ or, more simply, $I_\parallel + 2 \cdot I_\perp$.

The most accessible axis available on a light microscope for measuring the polarization of emitted light from a sample is the Z-axis. As mentioned before, a photomultiplier placed along the Z-axis will collect light proportional to $I_\parallel + I_\perp$. Two general schemes are used to separate the $I_\parallel$ signal and the $I_\perp$ signal (Figure 10.6). The first arrangement measures $I_\parallel + I_\perp$ sequentially (panels A1 and A2 in Figure 10.6), and the second arrangement measures $I_\parallel + I_\perp$ in parallel (panel B). In the first scheme a linear polarizer (L.Pol.) is placed between the sample (S) and the light detector. A photomultiplier would typically be used for laser scanning microscopy such as confocal microscopy or two-photon microscopy and is depicted here.

The use of photomultipliers in conjunction with a pulsed laser light source allows time-resolved polarization measurements using TCSPC (Becker 2005). Essentially, two fluorescent lifetime decay curves are generated; one represents the decay of $I_\parallel(t)$ and the other for $I_\perp(t)$. Alternatively, an EMCCD (electron multiplying charge coupled device) camera would typically be used for wide-field imaging of *steady-state* polarization and for polarization imaging in TIRF (total internal reflection fluorescence) mode. When the linear polarizer is oriented at 0° relative to the electric field of the excitation source, the light detector will measure a signal proportional to $I_\parallel$ (panel A1). When the polarizer is oriented at 90°, the detector will measure a signal proportional to $I_\perp$ (panel A2). The orientation of the polarizing filter can be changed manually between acquisition of $I_\parallel$ and $I_\perp$ images; alternatively, the rotation of the polarizing filter can be mechanized using a motorized rotation stage.

Proper alignment of polarizing filters at 0 and 90° is key to measuring the polarization of emitted light accurately. In our laboratory, this is achieved by removing fluorescence emission filters (not depicted in Figure 10.6) from the light path (such that the linearly polarized light source projects onto the light detector directly) and then finding the rotational orientation of the polarizing filter that yields the weakest signal (by definition, 90°). Once the 90° orientation is found, the 0° orientation is a simple 90° offset. With this calibration, precautions must be taken to prevent accidental damage to the light detector, particularly when using photomultipliers or EMCCD cameras. This can be prevented by the use of neutral density filters, low laser power settings, and low gain settings on the detector. It is worth noting that this alignment procedure assumes that the detectors are insensitive to the polarization of the light. This is not always the case; for example, side-on photomultipliers are often very sensitive to polarization while end-on tubes typically are not.

In the second scheme, $I_\parallel$ and $I_\perp$ signals are separated using a beam splitter and then measured in parallel (Figure 10.6, panel B). A polarizing beam splitter (Pol.B.S.) is placed after the sample to separate the $I_\parallel$ fluorescent signal from the $I_\perp$ signal. Typically, the $I_\perp$ signal is reflected orthogonally while the $I_\parallel$ signal is transmitted through the cube. It is important to note that polarizing beam splitters are wavelength dependent. Thus, it is important to choose a beam splitter that has a flat response over a broad wavelength range that is matched to the emission spectrum of the fluorophore of interest. It is also important to realize that although the *contrast ratio* (the intensity ratio of the transmitted polarization state vs. the

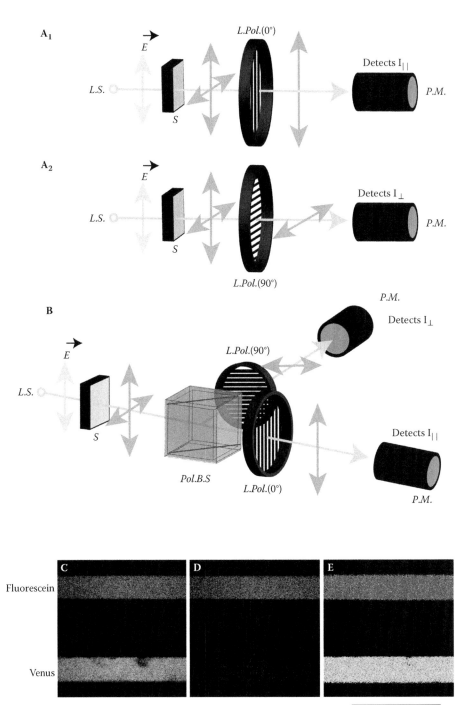

FIGURE 10.6   See figure caption on opposite page.

attenuated state) of polarizing beam splitters is quite reasonable in the transmitted pathway (typically, ≥500:1), their performance in the reflected pathway can be poor (20:1).

For this reason, in our laboratory we augment a broadband polarizing beam splitter with two linear polarizing filters whose orientation is matched to the output of the beam splitter. Linear polarizing filters typically have contrast ratios that are at least 500:1. Higher selectivity is rarely needed because the laser outputs of most lasers used for photoselection in biological imaging applications are rarely polarized greater than 500:1. Finally, in this imaging scheme, each $I_{\parallel}$ and $I_{\perp}$ signal pathway has a dedicated photomultiplier.

The imaging scheme portrayed in panels A and B can also be adapted for steady-state polarization imaging (i.e., *not* time resolved). We show an example of steady-state polarization imaging (and anisotropy analysis) using a two-photon microscope configured for sequential acquisition (as depicted in panel A) in Figure 10.6(C–E). It is worth noting that the data acquisition arrangement depicted in panel B is particularly well suited for steady-state polarization imaging using cameras. Andor Technology (Belfast, Northern Ireland) manufactures a dual port camera adapter that allows two EMCCD cameras to be aligned to image $I_{\parallel}$ and $I_{\perp}$ in parallel. Alternatively, Cairn Research Ltd. (Faversham, UK) and MAG Biosystems (Pleasanton, California) both manufacture devices that can split an emission image into $I_{\parallel}$ and $I_{\perp}$ images and project them side by side onto a single EMCCD camera.

Both the sequential and parallel imaging approaches outlined in Figure 10.6 can effectively measure $I_{\parallel}$ and $I_{\perp}$, but it is worth considering the pros and cons of each method. The sequential approach is simple to implement and requires only a single photodetector. This is not a very photon-efficient approach because, when $I_{\parallel}$ data are collected, $I_{\perp}$ data are discarded and vice versa. Furthermore, any motion occurring between acquiring $I_{\parallel}$ and $I_{\perp}$ images will result in pixel registration artifacts. The parallel approach is more photon efficient than the sequential approach, and it is less susceptible

FIGURE 10.6 **(See color insert following page 288.)** Separating $I_{\parallel}$ and $I_{\perp}$. When a sample (S) is excited on a microscope by a linearly polarized light source (L.S.) with electronic vector E, the fluorescent emission along the z-axis (yellow arrow) will comprise $I_{\parallel}$ and $I_{\perp}$. The magnitudes of these two intensity components can be measures either sequentially (panels A1 and A2) or in parallel (panel B). The sequential configuration uses a single photomultiplier (P.M.) and a linear polarizing filter (L.Pol.) that is first positioned parallel (panel A1; 0°) to the electronic vector of the light source to measure $I_{\parallel}$ and then perpendicular (panel A2; 90°) to measure $I_{\perp}$. In the parallel detection configuration (panel B), a polarizing beam splitter (Pol.B.S.) is used in conjunction with two linear polarizing filters and two photomultipliers to measure $I_{\parallel}$ and $I_{\perp}$ simultaneously. Panel C: An $I_{\parallel}$ intensity image of two capillaries filled with a solution of fluorescein (top) or Venus (bottom) acquired as in panel A1 with 950 nm two-photon excitation. Panel D: An $I_{\perp}$ intensity image of two capillaries filled with a solution of fluorescein (top) or Venus (bottom) acquired as in panel A2 with 950 nm two-photon excitation. Panel E: A steady-state anisotropy image calculated from the $I_{\parallel}$ and $I_{\perp}$ images depicted in panels C and D using Equation 10.2. Note that the fluorescein capillary had a steady-state anisotropy value of 0 (blue) while the Venus capillary had a value of ~0.3 (green). In this instance, we are using anisotropy to "image" the difference in the molecular rotation of these molecules.

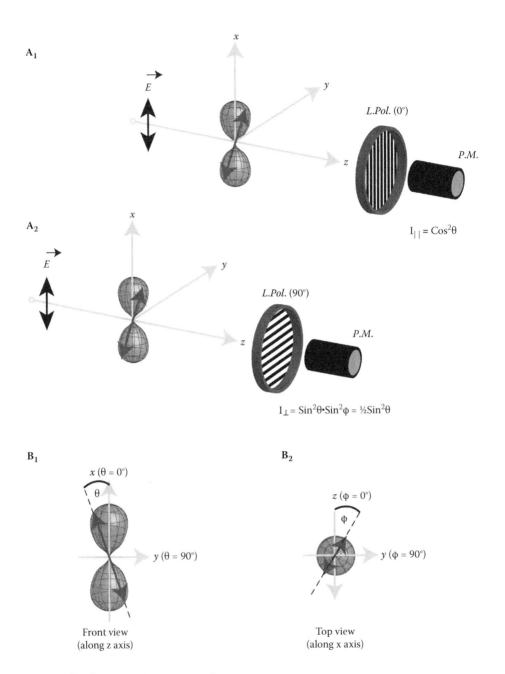

FIGURE 10.7   See figure caption on opposite page.

to motion and photobleaching artifacts. This is advantageous for live cell imaging. It does, however, require two photodetectors and is therefore more expensive to implement and more difficult to align. Furthermore, the detectors might have different efficiencies for detecting photons and different instrument response functions requiring correction factors.

FIGURE 10.7   **(See color insert following page 288.)** The probability of detection through a polarizing filter. The probability that a photon emitted by a fluorophore will pass through a linear polarizing filter (L.Pol.) and be detected by a photomultiplier tube (P.M.) is a function of the orientation of the fluorophore's emission dipole (green double arrow) and the orientation of the filter. When the filter is situated along the z-axis and is oriented at 0° relative to the electric field (E) of the light source, the photomultiplier will detect $I_\parallel$ (A1). When the filter is rotated to 90° relative to the electric field, the photomultiplier will detect $I_\perp$ (A2). $I_\parallel$ will be proportional to $\cos 2\theta$, where θ is the angle formed between electric field of the light source and the emission dipole of the fluorophore (see panel B1). $I_\perp$ will be proportional to $\sin^2\theta \cdot \sin^2\phi$, where ϕ is the angle formed between the emission dipole of the fluorophore and the z-axis (see panel B2). For an isotropic distribution of fluorophores excited with linearly polarized light, the distribution of excited-state dipole orientations (pink hour-glass-shaped cloud) will have a symmetrical distribution of ϕ values around the x-axis (B2). Due to this symmetry, the value of $\sin^2\phi = 1/2$. Thus, for an isotropic distribution of fluorophores $I_\perp$ will be proportional to $1/2 \sin^2\theta$. Notice that for an isotropic population of fluorophores, the values of $I_\parallel$ and $I_\perp$ are functions of θ alone.

10.6 HOW DO WE QUANTIFY POLARIZED EMISSIONS?

When an excited fluorophore emits a photon, the orientation of that photon's polarization will be correlated with the orientation of the fluorophore's emission dipole (Weber 1952). As mentioned previously, for a randomly oriented population of static fluorophores excited by linearly polarized light whose absorption and emission dipoles are collinear, the orientation of emitted photons will be strongly correlated with the electric field orientation of the polarized light source. Before we can understand this correlation quantitatively and apply it to biological questions we must first cover two more concepts: (1) how the orientation of a fluorophore's emission dipole affects the probability of detecting the emitted photon through either parallel or perpendicularly oriented linear polarizers, and (2) how we can use measured $I_\parallel$ and $I_\perp$ values to parameterize the orientation of the emission from an isotropic population of fluorophores.

In Figure 10.7, we illustrate how the orientation of an individual fluorophore's emission dipole (double green arrow) from an isotropic population of fluorophores excited with polarized light will influence the signal intensity measured through a filter polarizer oriented either parallel (A1) or perpendicular (A2) to the electric field of the light source. The three-dimensional excited-state distribution as calculated using Equation 10.1 is depicted in pink. The orientation of any single fluorophore from this excited-state population can be described by two angles: θ (B1) and ϕ (B2). When the filter polarizer is oriented to 0° (that is, parallel to the electric field polarization), the light intensity measured through the filter will be proportional to $\cos^2\theta$ (where θ is the polar angle of the emitting molecule relative to the electric field polarization).

For the population of fluorophores, the measured $I_\parallel$ intensity will be proportional to an average of all the $\cos^2\theta$ values weighted by their abundance. When the filter polarizer is rotated to 90° (so that it is perpendicular to the orientation of the excitation electric field polarization), the intensity measured will be proportional to $\sin^2 \theta \cdot \sin^2 \phi$. Because the excited-state distribution is symmetrical around the X-axis, $\sin^2 \phi = 1/2$.* Thus, for the population of fluorophores, $I_\perp$ will be proportional to an abundance weighted average of all $1/2\sin^2 \theta$ values. This equation is important because it indicates that when populations of randomly oriented fluorophores are excited by linearly polarized light, the values of both $I_\parallel$ and $I_\perp$ will be determined by the value of θ, the polar angle of the emitting molecules relative to the electric field polarization alone.

Finally, we must discuss how $I_\parallel$ and $I_\perp$ values are used to parameterize the orientation of populations of fluorophores. Two main conventions have been used in the literature: the *polarization ratio* (p) and *emission anisotropy* (r). The polarization ratio is simply the intensity difference between $I_\parallel$ and $I_\perp$ divided by the intensity observed by a photodetector placed along either the Y- or Z-axis ($I_\parallel + I_\perp$; see Figure 10.5):

$$p = (I_\parallel - I_\perp)/(I_\parallel + I_\perp)$$

When $I_\perp$ or $I_\parallel$ is 0, the polarization ratio will have values of –1 or 1, respectively. This represents the full range of polarization ratio values possible with a value of 1 indicating a perfect alignment of emission dipoles with the orientation of the light source electric field; a value of –1 indicates an orthogonal orientation. The emission anisotropy is the intensity difference between $I_\parallel$ and $I_\perp$ divided by an emission intensity with parallel and perpendicular components proportional to the *total* intensity ($I_\parallel + 2 \cdot I_\perp$; see Figure 10.5):

$$r = (I_\parallel - I_\perp)/(I_\parallel + 2 \cdot I_\perp) \tag{10.2}$$

Now, when $I_\perp$ or $I_\parallel$ is 0, the emission anisotropy will have values of 1 to –0.5, respectively. This represents the full range of anisotropy values possible; a value of 1 indicates a perfect alignment of emission dipoles with the orientation of the light source and a value of –0.5 indicates an orthogonal orientation of emission dipoles. It is important to realize that the polarization ratio and anisotropy are just different expressions used to parameterize the same phenomenon: the orientation of light emitted relative to the orientation of the linearly polarized light source electric field. The relationship between p and r is simply:

$$r = 2 \cdot p/(3 - p)$$

---

* In an isotropic population of fluorophores, there is the same number of molecules with ϕ values falling between 0 and 180° as between 180 and 360°. The sine function yields values between +1 and –1. When the value of ϕ falls between 0 and 180°, $\sin \phi$ has positive values, and $\sin \phi$ has negative values when ϕ has values between 180 and 360°. Thus, the average value of $\sin \phi$ will be 0 (positive values cancel negatives). In contrast, $\sin^2 \phi$ has values that fall between 0 and 1 (positive) and therefore the average value of $\sin^2 \phi$ is 1/2.

For the remainder of this chapter, we will use *emission anisotropy* because in many biological applications it is more amenable to analysis. Two important examples to illustrate this follow. First, the average anisotropy of a population of fluorophores is (Lakowicz 1999)

$$\langle r \rangle = \sum_i f_i r_i \tag{10.3}$$

where f_i is the fractional intensity and r_i is the anisotropy of a single fluorophore. This equation indicates that the *anisotropy* of a population of fluorophores is simply the intensity weighted sum of the anisotropy values of the individual fluorophores in the population.

An example that illustrates how Equation 10.3 might be useful for interpreting a biological experiment is the use of anisotropy to monitor the transition of monomers into dimers of fluorescent protein-tagged proteins upon stimulation. Prior to stimulation, the anisotropy of a population of monomers should be high because green fluorescent proteins (GFPs) rotate slowly and because Förster resonance energy transfer (FRET) does not occur with isolated fluorophores. In contrast, FRET between tagged proteins in close proximity, as might be encountered for a dimer, can result in a large decrease in anisotropy. The impact of molecular rotation and FRET on anisotropy will be discussed in detail shortly. In this type of experiment, the use of Equation 10.3 allows interpretation of intermediate anisotropy values in terms of a population comprising a mixture of monomers and dimers with different r values, whose relative abundance changes with time. Interpretation of ensemble anisotropy values from populations with more than two species of fluorophores (each having unique anisotropy values) is more problematic. Similarly, for a spherical molecule that is free to rotate and excited repeatedly with short pulses of polarized light, anisotropy will decay with time (Lakowicz 1999; Valeur 2002):

$$r(t) = r_0 \cdot e^{-t/\tau_{rot}} \tag{10.4}$$

where r_0 is the *limiting anisotropy*—that is, the anisotropy measured at the instant of photo selection—and τ_{rot} is the *rotational correlation time* of the molecule, an indicator of how rapidly a molecule rotates.

Equation 10.4 states that for an isotropic population of spherical molecules that are free to rotate, the initial anisotropy measured immediately following photoselection (at $t = 0$) will be r_0, the average anisotropy value for all of the excited molecules in the population. With time, the anisotropy of this population will decay as a result of stochastic rotation until it reaches an average value of 0. If stochastic rotations occur in every axis of rotation with equal probability, the anisotropy will decay as a single-exponential function with a decay constant τ_{rot}. Thus, by plotting the change of anisotropy as a function of time following excitation, Equation 10.4 can be used to reveal the value of the limiting anisotropy as well as the rotational correlation time.

10.7 THE ANISOTROPY OF RANDOMLY ORIENTED POPULATIONS OF FLUOROPHORES

What anisotropy values are expected immediately following excitation when an isotropic population of fluorophores is excited with one-, two-, or three-photon excitation using linearly polarized light? These *theoretical* values are called the *fundamental anisotropies* (r_f; Valeur 2002). We will first explore what these values are and then discuss reasons why the *limiting anisotropy* measured in experiments is almost always less than the *fundamental anisotropy*. Figure 10.8 shows the results of a Monte Carlo simulation of an initial population of randomly oriented fluorophores (proportional to sin θ). Next, fluorophores are stochastically *activated* as a function of $\cos^x θ$, with $x = 0$ for no photoselection, $x = 2$ for one-photon excitation, $x = 4$ for two-photon selection, and $x = 6$ for three-photon selection (panel A). With photoselection, only a fraction of the fluorophores in the population will

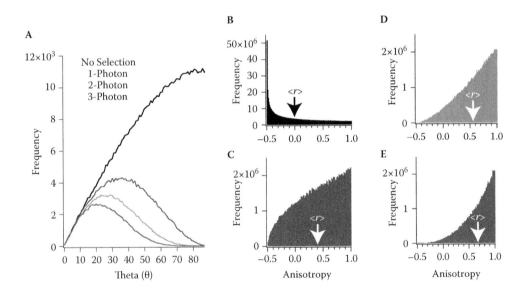

FIGURE 10.8 **(See color insert following page 288.)** The fundamental anisotropy value of an isotropic population. Monte Carlo simulations can be used to predict the fundamental anisotropy value of an isotropic population of fluorophores excited with natural light (panels A and B, black trace and histogram), or with one-photon (panel A and C, blue trace and histogram), two-photon (panels A and D, green trace and histogram), or three-photon (panels A and E, red trace and histogram) excitation using linearly polarized light. The distribution of theta values (A) generated by the stochastic simulation was identical to the theoretical distribution shown in Figure 10.4. Notice that the histogram of anisotropy values generated for the different excitation conditions (panels B–E) contain essentially all possible anisotropy values possible (–0.5 to 1). However, the relative abundance of these values and therefore the mean anisotropy value of the population, ⟨r⟩, are different with a value of 0 for excitation with natural light, 0.4 for one-photon excitation with polarized light, 0.57 for two-photon excitation, and 0.67 for three-photon excitation.

get excited; for example, compare the black trace (no selection, which is all molecules in all orientations excited) with the blue trace (one-photon photoselection).

For each excited fluorophore in the population, we next determine what proportion of its emitted photons will be detected by a $I_\parallel$ detector, and by a $I_\perp$ detector based on the individual fluorophore's θ value. Recall that when the filter polarizer is oriented at $0°$, the intensity measured will be proportional to $\cos^2 \theta$ (the $I_\parallel$ detector) and that when the polarizer is rotated to $90°$, the intensity measured will be proportional to $1/2\sin^2 \theta$ (the $I_\perp$ detector). Once individual $I_\parallel$ and $I_\perp$ values are calculated for each activated fluorophore in a population, we next can use the $I_\parallel$ and $I_\perp$ values to calculate an anisotropy value for each excited fluorophore in the population using Equation 10.2. We can plot anisotropy histograms for randomly oriented populations of excited fluorophores (no selection; panel B), and those excited with one-photon photoselection (panel C), two-photon photoselection (panel D), and three-photon photoselection (panel E). These anisotropy distributions are useful for conceptualizing how the dipole orientation (θ value) of individual excited fluorophores changes with photoselection and how this change impacts the anisotropy values measured from populations of fluorophores.

Notice that, in all four distributions, individual fluorophores with every possible anisotropy value (ranging from –0.5 to 1.0) are present. Remember that the average anisotropy value is simply the intensity weighted sum of the individual anisotropy values (Equation 10.3). For excited molecules oriented randomly (no photoselection), the fundamental anisotropy (the average theoretical anisotropy value, r_f) is 0. For molecules excited by one-photon excitation, r_f is 0.40; by two-photon excitation it is 0.57, and by three-photon excitation it is 0.67. These values are the *fundamental anisotropy* values expected for one-, two-, and three-photon photoselection of randomly oriented populations of fluorophores (Callis 1997; Lakowicz 1999; Lakowicz, Gryczynski, et al. 1992; McClain 1972; Scott, Haber, and Albrecht 1983).

10.8 DEPOLARIZATION FACTORS AND SOLEILLET'S RULE

As mentioned previously, the limiting anisotropy values measured experimentally by analyzing the time-resolved decay of anisotropy using Equation 10.4 are often lower than the fundamental anisotropy values predicted by theory. Factors responsible for a decrease in the measured anisotropy are called *depolarization factors* (*d*). One of the most important reasons for using anisotropy rather than polarization ratios is that the measured anisotropy is simply the fundamental anisotropy times the product of all depolarization factors (Lakowicz 1999; Valeur 2002):

$$r = r_f \cdot \prod_i d_i \tag{10.5}$$

This equation is called Soleillet's rule (Soleillet 1929). For biological experiments, typically only four depolarization factors are considered to account for a discrepancy between theory and a measured anisotropy value:

- depolarization due to the instrumentation used to measure anisotropy;

- depolarization due to noncollinear absorption and emission dipoles;

- depolarization due to molecular rotation occurring between fluorophore excitation and emission; and

- depolarization occurring as a result of FRET.

Although these depolarization factors can complicate the interpretation of anisotropy measurements, they also represent the basic reason why fluorescence anisotropy measurements are used for investigating biological processes; that is, the anisotropy measurements deliver a wealth of information about the molecular system. In the following sections, we will first illustrate how anisotropy is measured and then discuss each of these factors in greater detail.

As mentioned previously, high anisotropy values ($I_{\parallel} > I_{\perp}$) indicate a strong correspondence between the orientation of the electric field vector of the excitation light and the polarization of the emitted photons. This could be observed from a population of photoselected fluorophores whose absorption and emission dipoles are approximately collinear, have slow molecular rotation, and do not transfer energy by FRET efficiently. An anisotropy value of 0 ($I_{\parallel} = I_{\perp}$) indicates that no correspondence takes place between the orientation of the electric field vector of the excitation light and the polarization of the emitted photons. This could be observed in a randomly oriented population of excited fluorophores (as a result of rapid stochastic molecular rotation or by efficient energy transfer to acceptor fluorophores with random orientations).

Negative anisotropy values ($I_{\perp} > I_{\parallel}$) indicate an inverse correspondence between the orientation of the electric field vector of the excitation light and the polarization of the emitted photons as would arise if a fluorophore's absorption dipole was approximately orthogonal to its emission dipole or if efficient FRET occurs between a donor whose absorbance dipole is approximately orthogonal to the FRET acceptor's emission dipole. We have indicated that the highest anisotropy values that can be measured for a population of randomly oriented fluorophores, the *fundamental anisotropy*, is 0.4 with excitation by one-photon linearly polarized light, 0.57 with two-photon excitation, and 0.67 for three-photon excitation. In our laboratory we primarily use two-photon excitation using a mode-locked linearly polarized Ti:sapphire laser. Accordingly, the examples included here all have a fundamental anisotropy of 0.57.

In the first example, we show the decay of fluorescence from the protein Venus, a yellow spectral variant of GFP (Nagai et al. 2002), when observed by polarizers and photomultipliers positioned to observe either $I_{\parallel}$ or $I_{\perp}$ (Figure 10.9A, B). This pair of fluorescence lifetime decay curves was collected by TCSPC (Becker 2005) using a laser scanning microscope configured, as described in Figure 10.6(B). The sample was a HEK293 cell expressing Venus. The curves in Figure 10.9(A) depict $I_{\parallel}(t)$ (∆, which are proportional to the probability of Venus emitting a photon that can pass through a linear polarizer oriented at 0°) (parallel to the electric field polarization) as a function of time

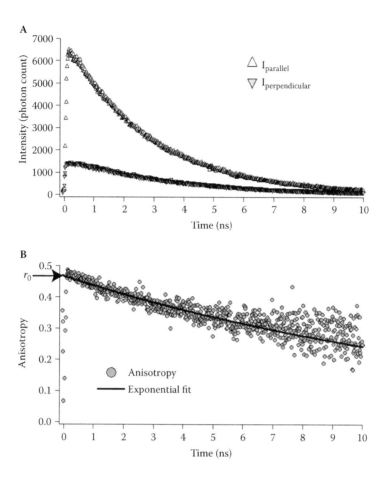

FIGURE 10.9 Anisotropy decay of Venus, a fluorescent protein. HEK cells expressing monomeric Venus were imaged using a 20X 0.5 NA water objective on a Zeiss 510 META/NLO two-photon microscope using ultrafast, 950 nm excitation pulses (at 90 MHz) from a mode-locked linearly polarized Ti:sapphire laser (Coherent). The fluorescent emission was collected through a low NA air condenser, a BG39 filter to attenuate the laser, and a 535 ± 15 nm band-pass filter. $I_{\parallel}$ and $I_{\perp}$ were measured on a pair of MCP photomultipliers (R3809U-52; Hamamatsu) using a parallel detector anisotropy arrangement as depicted in Figure 10.6(B). Time-correlated single-photon counting was used to acquire $I_{\parallel}(t)$ and $I_{\perp}(t)$ lifetime decay curves (panel A; Δ and ∇, respectively). A Becker & Hickl SPC-830 card was used as an interface to measure timing between photodetector pulses and laser pulses. "Parallel" and "perpendicular" traces were collected in 1,024 channel histograms. Background subtraction was performed on both traces and anisotropy values calculated using Equation 10.6 yielding an anisotropy decay curve (gray circles; panel B). Solid line indicates a single-exponential fit of the anisotropy decay data using IGOR Pro (Wavemetrics). Arrow indicates the value of the limiting anisotropy, r_0.

following an ultrashort, <200 fs excitation pulse) and $I_\perp(t)$ ∇ (proportional to the probability of Venus emitting a photon that can pass through a linear polarizer oriented at 90°). In Figure 10.9(B), we plot the fluorescence anisotropy decay curve, $r(t)$, of Venus calculated from $I_\parallel(t)$ and $I_\perp(t)$ using a variation of Equation 10.2 that includes an experimentally measured constant, G,* to account for differences in the sensitivity between the two photomultipliers:

$$r(t) = [I_\parallel(t) - G{\cdot}I_\perp(t)]/[(I_\parallel(t) + 2{\cdot}G{\cdot}I_\perp(t)] \tag{10.6}$$

Two features of the Venus anisotropy decay curve should be noted: the value of $r(t)$ when $t = 0$ and how the value $r(t)$ changes with time. In this example, the value of $r(t)$ when $t = 0$ was 0.47 (see Figure 10.9B arrow). This *measured t = 0* anisotropy is the *limiting anisotropy* (r_0). Clearly, in this example r_0 is greater than the fundamental anisotropy value expected for one-photon excitation (0.4), indicating that this sample was excited by a multiphoton absorption process. The discrepancy between r_0 (0.47) and the fundamental anisotropy expected for two-photon excitation (0.57) also suggests that some other process is responsible for the depolarization of our sample at $t = 0$. This will be explored momentarily.

Also note that during a 10 ns period following excitation and photoselection, the anisotropy value of Venus did not remain constant; that is, it did not remain at a value of 0.47. Rather, it decayed with time, and this decay can be modeled as a single exponential with a decay constant of ~15 ns (see solid line in Figure 10.9, panel B). This indicates that some additional process is responsible for the further depolarization of our sample. The depolarization factor or factors responsible for the discrepancy between our measured r_0 value and the limiting anisotropy must occur on a timescale significantly faster than the time resolution of our photomultipliers (~38 ps) because they appear to occur instantaneously. In contrast, the depolarization factors responsible for the decay in anisotropy are occurring on a much slower timescale (nanoseconds).

10.8.1 Instrumental Depolarization

One assumption of anisotropy measurements is that the light rays forming the beam of polarized light used to excite a population of fluorophores are aligned parallel to each other. Furthermore, when the emission from a population of fluorophores is conveyed to the photodetector by way of a polarizing filter set at either 0 or 90°, the paths that the emission light rays travel should also be aligned parallel to each other and orthogonal to the surface of the polarizing filter. Although these assumptions are reasonably met when anisotropy is measured in a spectrofluorimeter, this rarely is the case when anisotropy is measured using a microscope. This is because when parallel light beams are focused by high numerical aperture (NA) lenses (such as those found in microscope objectives), the light rays no longer travel on parallel paths; rather, they converge at a focal spot (Axelrod 1979, 1989). Essentially, when anisotropy is measured using high NA optics, the lens curvature

---

* The G factor for this microscope setup was measured by tail fitting of fluorescein (as described in Hess et al. 2003) and was found to be 1.26.

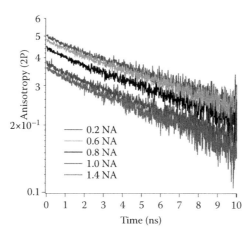

FIGURE 10.10 **(See color insert following page 288.)** The influence of numerical aperture on anisotropy. Purified Venus was imaged using a 10X 0.3 NA air objective on a Zeiss 510 META/NLO two-photon microscope using ultrafast, 950 nm excitation pulses (at 90 MHz) from a linearly polarized Ti:sapphire laser. The fluorescent emission of Venus was collected through an oil condenser, a BG39 filter, and a 535 ± 15 nm band-pass filter. The NA of the condenser was varied from 0.2 to 1.4 by opening up the condenser aperture, and anisotropy was measured as described for Figure 10.9. As the NA of the condenser was increased, the value of r_0 decreased (y-axis intercept), but the decay rate was not significantly altered. Note that the data are plotted on a semilog scale.

transforms polarized light into light composed of a mixture of electric field vector angles (elliptical polarization).

A similar process can also occur when the trajectories of polarized rays of emitted light are redirected by high NA lenses relative to the angle of a polarizing filter. The impact of this NA depolarization factor (d_{NA}) is illustrated in Figure 10.10. In this example, linearly polarized two-photon excitation conveyed to our sample (purified Venus) through a 10X 0.3 NA objective, and the polarized emission was collected through a condenser with numerical aperture settings adjusted from 0.2 to 1.4. We see that the r_0 value decreases from 0.5 to 0.36 as the NA of the condenser lens increases. Also note that despite the dramatic change in r_0, the decay rate was not significantly altered.

To maximize the dynamic range of anisotropy measurements (as well as to maximize signal-to-noise ratios), it is important to minimize depolarization due to lens curvature. Obviously, for the most accurate anisotropy measurements, the lowest NA objectives should be used. Unfortunately, low NA objectives are also the least efficient for collecting emitted photons. For live-cell imaging, photon efficiency is often an overriding concern because photons are almost always in short supply. In our laboratory, we typically use microscope optics with numerical apertures between 0.8 and 0.9 for live-cell anisotropy measurements because under these conditions, $d_{NA} > 0.95$ (with 1 = no depolarization), the anisotropy decay kinetics are not significantly altered, and photon efficiency is not drastically compromised. It should also be noted that it is possible to use postprocessing to correct anisotropy measurement values computationally, based on the NA of the imaging system (Axelrod 1979).

10.8.2 Depolarization Caused by Absorption and Emission Dipole Orientation

In our previous examples, we assumed that the absorption dipole of a fluorophore was collinear with its emission dipole. We designate the angle difference between the absorption and emission dipole of a fluorophore as β. β is thought to be an intrinsic property of a specific fluorophore and should not change during the course of a biological experiment. $\beta = 0°$ for collinear fluorophores. This assumption of collinearity, however, is not always valid. Often, the absorption of a photon can cause molecular rearrangements that subtly alter the structure of the fluorophore. The depolarization factor that accounts for the relative orientation of the absorption and emission dipoles, d_β (Lakowicz 1999), is

$$d_\beta = \tfrac{3}{2}\cos^2\beta - \tfrac{1}{2} \tag{10.7}$$

If $\beta \neq 0°$, there will be a randomized offset in the orientation of emitted photons relative to the orientation of the absorption dipole. Accordingly, r_0 of an isotropic solution of fluorophores will be less than the fundamental anisotropy. Assuming that the only other depolarization factor operant is d_{NA}, the predicted value of r_0 using Soleillet's rule (Equation 10.5) is

$$r_0 = r_f \cdot d_{NA} \cdot d_\beta = r_f \cdot d_{NA} \cdot \left(\tfrac{3}{2}\cos^2\beta - \tfrac{1}{2} \right) \tag{10.8}$$

Note that r_f is 2/5 (0.4) for one-photon excitation, 4/7 (0.57) for two-photon excitation, and 6/9 (0.67) for three-photon excitation. If $d_{NA} = 1$ (no instrumental depolarization), then the measured Venus r_0 value of 0.5 with two-photon excitation using low NA optics (see Figure 10.10) suggests that Venus's absorption and emission dipoles are not collinear (but see Volkmer et al. 2000) and are consistent with Venus having a β value of ~16°. This number, however, is only an upper estimate because d_{NA} might have a value less than 1 (e.g., β for Venus would be approximately 13° if d_{NA} is 0.95). Furthermore, β values might be different for the same fluorophore, with one-photon and multiphoton excitation as a result of different selection rules (Callis 1997; McClain 1972).

10.8.3 Timescale of Depolarization

Thus far we have covered two possible sources of depolarization in experiments. The first factor, d_{NA}, is a function of the instrumentation used to measure anisotropy; although it clearly affects anisotropy measurements, it tells us little about our fluorescent biological samples per se. The second depolarization factor, d_β, can reveal information on the structure of the fluorophore used in a biological experiment (that is, the angle between the absorption and emission dipoles), and this structural trait could potentially change as a function of the environment of the fluorophore (e.g., in different solvents). Both of these factors operate on a timescale that is significantly faster than the time resolution of most imaging systems used to measure fluorescence lifetimes or anisotropy decay (typically tens to hundreds of picoseconds). Practically speaking, they can be thought of as acting

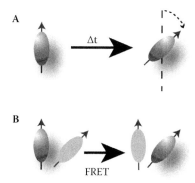

FIGURE 10.11 Two causes of depolarization on a nanosecond scale. Venus has a fluorescent lifetime of 3.4 ns and fluorescein (at pH 10) has a lifetime of 4.1 ns as measured by TCSPC-FLIM. Effectively, these fluorophores can remain in the excited state for ~16–22 ns. During this period, two types of events can result in depolarization of anisotropy measurements, molecular rotation (panel A), and FRET (panel B).

instantaneously. The same cannot be said for the next two depolarization factors that we will discuss: molecular rotation and FRET.

The fluorescence lifetime of a fluorophore is the average time that a fluorophore remains in the excited state. Ultimately, the fluorophore will decay to the ground state, often by emitting a photon. It is this photon that we detect and characterize in an anisotropy experiment. Most fluorophores used in biological imaging have lifetimes ranging from a few hundred picoseconds to tens of nanoseconds. For example, Venus has an average lifetime of about 3 ns, and fluorescein has an average lifetime of a little over 4 ns. This means that, following excitation by a short pulse of light, virtually every excited Venus or fluorescein molecule will decay to the ground state by five lifetimes (15–20 ns). It is useful to contemplate what events can happen within this time span, how they will affect the orientation of the fluorophore's emission dipole, and how this can be exploited for biological research. In Figure 10.11, we show the two main events (molecular rotation and FRET) that can occur within this time frame, which have the potential of altering the orientation of a fluorophore's emission dipole.

10.8.4 Depolarization Caused by Molecular Rotation

If a fluorophore can freely rotate between excitation and emission (Figure 10.11A), the anisotropy signal detected for that molecule will change. With time, the biased orientation of excited fluorophores resulting from the action of photoselection with linearly polarized light will be negated by random molecular rotations. Fluorophore excitation to an excited state occurs on a femtosecond timescale (1×10^{-15} s), and subsequent molecular vibration rapidly allows for the dissipation of energy as the molecule assumes the lowest vibrational energy of the S_1 excited state. This process occurs on a picosecond time scale (1×10^{-12} s). Molecular rotation typically is a much slower process, ranging from 10s of picoseconds to milliseconds or longer! Therefore, the longer that a fluorophore remains in the excited state, the higher the probability is that it will rotate before emitting a photon. In essence, if molecular rotation is occurring, we expect that the anisotropy value will decay from an initial value (r_0) to a randomized state ($r_\infty = 0$).

This concept is well illustrated by observing the anisotropy decay curve of a solution of fluorescein in water in Figure 10.12. In panel A, we show $I_\parallel(t)$ and $I_\perp(t)$ for a solution of fluorescein. Notice that, by 1 ns after photoselection, these two intensities are identical. This means that, by 1 ns, the probability of detecting an emitted photon through a parallel oriented polarizing filter is the same as detecting it through a perpendicularly oriented filter. In panel B, we show the anisotropy decay curve calculated from $I_\parallel(t)$ and $I_\perp(t)$. By 1 ns after photoselection, the anisotropy value has decayed to 0, indicating that the orientation of fluorescein's emission dipole has randomized as expected. We can model this time-dependent decrease in fluorescein's anisotropy values as a single-exponential decay (panel B, dashed line) using Equation 10.4.

Notice that the anisotropy of fluorescein's emission decays from its initial r_0 value (~0.3) to a final value of 0. With such a sharp decay, it is difficult to determine the value of r_0 accurately. Also note that the value of r_0 is itself a function of r_f, d_{NA}, and d_β (see Equation 10.8). The decay constant for this fit is called the *rotational correlation time*, τ_{rot}, and for a fluorescein solution in water has a value of 140 ps. The rotational correlation time, τ_{rot}, is related to D_r, the *rotational diffusion coefficient*, and is a measure of how rapidly a molecule can rotate; the smaller the value of τ_{rot} is, the faster the fluorophore can rotate. The relationship between τ_{rot} and D_r for a small molecule in solution is (Cantor and Schimmel 1980; Lakowicz 1999):

$$\tau_{rot} = \frac{1}{6D_r} \tag{10.9}$$

Several factors can modulate the rotational correlation time because D_r is itself a function of the absolute temperature (T), the viscosity (η), and the molar volume of the fluorophore (V) as described by the Stokes–Einstein relationship (Lakowicz 1999):

$$D_r = \frac{RT}{6V\eta} \tag{10.10}$$

where R is the gas constant. Thus,

$$\tau_{rot} = \frac{\eta V}{RT} \tag{10.11}$$

In Figure 10.12(C), we plot how the anisotropy decay of fluorescein's emission changes as the viscosity of the solution is increased by the addition of glycerol. As the solution becomes more viscous, fluorescein rotates slower and therefore it takes longer for its anisotropy to decay to 0. Notice that all five curves were well fit to a single-exponential model, but the value of τ_{rot} increased with higher viscosity. Also notice that all five curves now appear

FIGURE 10.12 Depolarization due to rotation: the anisotropy decay of fluorescein. $I_{\parallel}(t)$ (panel A, circles), $I_{\perp}(t)$ (panel A, diamonds), and the anisotropy decay (panel B, gray circles) of a 2.5 µM solution of fluorescein in water was measured as described for Venus in Figure 10.9. Notice how rapidly the anisotropy decays to a value of 0. Fitting of the anisotropy decay curve to a single-exponential decay model revealed a decay constant of 140 ps (panel B, dashed line). To confirm that this value represents the rotational correlation time of fluorescein in water, we measured its decay as a function of added glycerol (to increase viscosity) while maintaining the fluorescein concentration at 2.5 µM (panel C). Equation 10.11 predicts that as viscosity increases so does the rotational correlation time (τ_{rot}). Curve fitting to a single-exponential decay model (solid black lines) revealed that the decay constant became longer as the viscosity increased, consistent with an increase in the rotational correlation time.

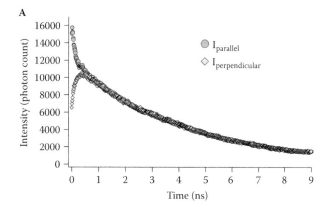

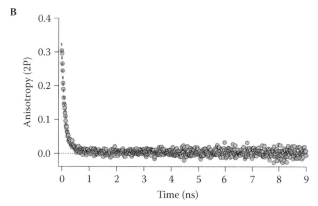

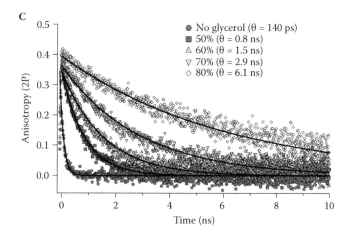

to decay from a common origin at an anisotropy value of ~0.4, r_0. It is also worth comparing the anisotropy decay curves for fluorescein in Figure 10.12(C) with the anisotropy decay curve for Venus in Figure 10.9(B). Clearly, Venus has a rotational correlation time that is much slower than that for fluorescein, even when fluorescein was placed in 80% glycerol. We have found that the rotational correlation time of Venus when expressed in HEK cells is similar to the rotational correlation time of purified Venus in 5% glycerol (data not shown).

Thus, it is unlikely that viscosity alone can explain this difference between Venus and fluorescein. The most likely explanation is that the molecular volumes of these two fluorophores are dramatically different. This is expected because fluorescein is a small molecule with a molecular weight of only 332 g/mol, and Venus is a large, 28,000 molecular weight, can-shaped protein with its fluorophore rigidly fixed inside. From a biologist's perspective, Equations 10.4 and 10.9–10.11 indicate that any factors that can change how fast a fluorophore will rotate, such as viscosity, molecular weight, hydration, molecular shape, association with other molecules, etc., can be detected using anisotropy decay analysis.

Strictly speaking, only spherically shaped fluorophores will have anisotropy values that decay as a single exponential, as described by Equation 10.4 (though this type of behavior is typically seen for small fluorophores). This is because nonspherical fluorophores will rotate at different rates along different axes. The anisotropy decay of nonspherical fluorophores is often modeled as a sum of exponentials (Lakowicz 1999):

$$r(t) = r_0 \cdot \sum_i a_i \cdot e^{-t/\tau_{rot_i}} \tag{10.12}$$

where a_i is the amplitude of each decay component.

Furthermore, in some instances (such as for some membrane dyes), fluorophores are free to rotate in some directions but not in others, thus requiring more complicated equations to model their anisotropy decay (Lakowicz 1999). It is also important to realize that anisotropy decay curves can become very complicated if multiple fluorophores are present in a sample.

It is worth noting that although anisotropy measurements are a powerful tool for measuring molecular rotation, even inside living cells, there are some obstacles to interpreting these types of experiments accurately. We will highlight three of these obstacles. First, in many biological experiments, a fluorophore is attached to a biological molecule of interest with the intention of monitoring the molecular rotation of that molecule. If the fluorophore is rigidly attached to the molecule of interest, then the orientation of that fluorophore's dipole will reflect the orientation of the molecule to which it is attached. All too often, however, the linker used to attach a fluorophore to another molecule is not rigid. Under these circumstances, the rotational component of the anisotropy decay of the fluorophore will reflect the rotational behavior of the free fluorophore as constrained by the flexibility of the linker and by the size and shape of the molecule to which the linker is attached.

Another limit to using anisotropy to measure molecular rotation results from the lifetime of the fluorophore and the time resolution of the anisotropy instrumentation. As previously mentioned, the fluorescent lifetime of a fluorophore (τ) is the average time that it remains in the excited state. By $t = 5\tau$, virtually all fluorophores have already decayed to the ground state. Furthermore, TCSPC imaging systems rarely have time resolutions that are better than 35 ps (full width at half maximum), and the $I_{\parallel}$ and $I_{\perp}$ lifetime decay curves will be a convolution of this instrument response function (IRF). Accordingly, anisotropy decay curves will only have data covering a time window spanning $t = 0$ to 5τ, with a time resolution of a few 10s of picoseconds. It is therefore difficult to measure rotational correlation times greater than 10τ or less than 10 ps *accurately*. For fluorophores like Venus and fluorescein, it will be difficult to measure rotational correlation times much larger than 30–40 ns accurately. One solution to this limitation is to use fluorophores with longer lifetimes or phosphorescence (Austin, Chan, and Jovin 1979; Dixit et al. 1982; Eads, Thomas, and Austin 1984).

A third obstacle worth considering for using anisotropy to measure molecular rotation is based on the accuracy of nonlinear fitting of multiexponential decay curves. A single-exponential decay will have at least two fitting parameters, a biexponential fit will have at least four parameters, etc. With more free fitting parameters, there is a greater possibility that there will not be a unique solution to a fit. The answer to this problem is to find ways to constrain the number of free fitting parameters. Global fitting is one such solution (Beechem, Knutson, and Brand 1986; Knutson, Beechem, and Brand 1983).

10.8.5 Depolarization Caused by FRET

Another source of depolarization shown in Figure 10.11 is FRET (Jablonski 1970). Förster resonance energy transfer is a physical phenomenon where the excited-state energy of a fluorophore is transferred to another molecule by a nonradiative mechanism (Förster 1948). For FRET to occur, several criteria must be met. First, the donor fluorophore and the acceptor must be in close proximity—typically less than 10 nm separating the two molecules. Second, for FRET to occur, the acceptor dipole must not be oriented perpendicular to the electric field of the donor. To transfer energy, FRET uses a dipole–dipole coupling mechanism in which the acceptor's dipole resonates with the electric field oscillations of the donor. When the acceptor dipole is perpendicular to the electric field, it cannot "sense" the oscillations and therefore FRET does not occur.

The electric field of a donor fluorophore can be envisioned as a three dimensional *curved* wave emanating out from its oscillating dipole. Concentric field lines connecting the two poles of the dipole define equipotential surface contours of this field, and the tangent at any point on these field lines represents the direction of the electric field at that specific location. At positions close to the dipole, the curved nature of these equipotential surfaces can change dramatically, even with small changes in position. Because FRET only occurs when donor and acceptor dipoles are close to each other (within a fraction of a wavelength), the relative positions of the donor and acceptor dipoles must be accounted for. Thus, FRET will be dependent on the spatial orientation of the emission dipole of the donor and the orientation of the absorbance dipole of the acceptors, as well as on the relative positions of these dipoles in space.

This relationship between the orientation of the donor and acceptor dipoles and of their relative positions in space is called the dipole orientation factor, κ^2 (Förster 1948). A final requirement for FRET is that there must be significant overlap between emission spectrum of a donor fluorophore and the absorption spectrum of the acceptor (Clegg 1996; Jares-Erijman and Jovin 2003; Periasamy and Day 2005; Vogel, Thaler, and Koushik 2006; Wallrabe and Periasamy 2005).

The two general types of FRET reactions, hetero-FRET and homo-FRET, differ based on spectral overlap. When the donor fluorophore and the acceptor have different absorption and emission spectra, the transfer is called hetero-FRET. In this case, the forward FRET transfer rate (from donor to acceptor) is typically much faster than the backward rate (acceptor to donor). The basis of this difference is that there will typically be a much larger spectral overlap between the emission spectrum of the donor and the absorption spectrum of the acceptor than of the emission spectrum of the acceptor and the absorption spectrum of the donor. Therefore, in most cases, hetero-FRET can be thought of as a unidirectional transfer.

In contrast, homo-FRET (Clayton et al. 2002; Jameson, Croney, and Moens 2003) is FRET occurring between fluorophores having identical spectra. Thus, with homo-FRET the forward and backward FRET transfer rates must be the same. Unlike hetero-FRET, homo-FRET from a donor to an acceptor can immediately be transferred back again to the donor. With homo-FRET, a dynamic situation is created where energy can readily migrate back and forth between donors and acceptors; therefore, another, more descriptive name for homo-FRET is *energy migration FRET* (emFRET). Although hetero-FRET is typically measured by monitoring changes in the fluorescence intensities or lifetimes of donors and/or acceptors, theoretically there should be no net change in the intensity or lifetime of a fluorophore with emFRET (Koushik and Vogel 2008; Rizzo et al. 2004). Therefore, these classical approaches for measuring FRET normally cannot be used to monitor emFRET.

Energy migration FRET can be routinely measured using fluorescence anisotropy. This approach is based on detecting a decrease in emission polarization resulting from the transfer of excitation energy from photoselected donors to typically more randomly oriented acceptors. This approach will be described in much more detail shortly. At this point, one might naively think that using fluorescence anisotropy to monitor FRET cannot work because the dipole orientation rules dictating photoselection will be the same as the dipole orientation rules required for FRET. This is not the case. Although the efficiency of FRET is in part a function of the angle formed between absorption and emission dipoles as encoded by κ^2, the angular selectivity for FRET transfer is much more permissive than the angular selectivity required for photoselection itself.

Now let us consider how the fluorescence anisotropy of a static sample will be affected by emFRET. For simplicity, we assume that molecular rotation is not occurring. Photons emitted from a population of fluorophores not undergoing emFRET will all originate from directly excited fluorophores. Thus, the measured anisotropy should be equal to the limiting anisotropy. Now consider a sample that has clusters of several fluorophores in close proximity. Let N equal the number of fluorophores in a cluster that participate in emFRET. Photons emitted from a cluster of fluorophores undergoing emFRET can originate from the directly excited fluorophore that was photoselected, or they can be emitted by fluorophores residing in the same cluster that were indirectly excited by nonradiative energy migration (Berberan-Santos and Valeur 1991; Gautier et al. 2001; Tanaka and Martaga 1982; Valeur 2002). Photons emitted by the directly excited fluorophore have dipole orientations highly correlated with the orientation of the electric field vector of the excitation light source as a result of photoselection.

In contrast, the orientation of the emission dipoles of fluorophores that were indirectly excited by emFRET should have little if any correlation with the orientation of the electric field vector of the light source. Accordingly, the anisotropy measured from fluorophore clusters should reflect the fraction of *emitters* that were directly excited and the fraction that were indirectly excited by emFRET. For example, if all of the molecules that were photoselected end up emitting, the anisotropy should be the limiting anisotropy of the fluorophore. In contrast, if all of the photons emitted were from fluorophores excited indirectly by emFRET, the anisotropy should be close to zero. Because of the additivity of anisotropy values (Equation 10.3), we expect that a population of fluorophores where half of the photons emitted come from directly excited fluorophores and half from indirectly excited fluorophores will have an anisotropy of approximately $r_0/2$.

It should be noted that when a fluorophore in a cluster emits, the series of energy transfer events that occurred prior to the emission are not usually known. For example, even when a fluorophore that was originally photoexcited emits, we cannot simply assume that it was excited and then emitted. An alternative possibility is that the fluorophore was photoexcited and then transferred energy to a neighbor. Next, that neighbor transferred the energy back to the original fluorophore prior to emission. In fact, much more complicated energy migration pathways are also possible involving multiple neighbors and multiple energy transfer steps before an emission event occurs. The complexity of an energy migration pathway, as well as the number of different energy transfer pathways possible (that can occur when a specific fluorophore is excited and a specific fluorophore emits) dramatically increases with the number of fluorophores in a cluster, and it can be influenced by the spatial arrangement of the fluorophores in the cluster (e.g., consider the different energy migration pathways possible between a pair of fluorophores or among six fluorophores when arranged in a row, a ring, or in a branched structure).

Due to this complexity, anisotropy measurements are not as a rule used to deduce the pathways that occur prior to an emission (though the complexity of an energy migration pathway could influence the kinetics of an anisotropy decay curve and will be mentioned shortly). More typically, anisotropy measurements are used to deduce the fraction of emission events occurring from fluorophores originally photoselected and the fraction occurring from fluorophores indirectly excited by emFRET without regard to the energy migration pathways that occurred prior to emission.

If we assume that the single-step energy transfer rate (ω) in a cluster of fluorophores is much greater than the emission rate of the fluorophore (Γ, where Γ is the reciprocal of the fluorescent lifetime τ), then as excitation energy jumps from fluorophore to fluorophore in the cluster, the probability that the fluorophore that was directly excited will emit a photon decreases with the number of fluorophores participating in emFRET in a cluster (Jameson et al. 2003; Runnels and Scarlata 1995). As a result, the fraction of directly excited emitters will decrease, and the anisotropy will drop toward zero. This provides the basis through which anisotropy can reveal the number of fluorophores engaging in energy migration.

Accordingly, the anisotropy observed for a pair of fluorophores undergoing emFRET will have an anisotropy value of approximately $r_0/2$, a cluster of three fluorophores undergoing emFRET will have an anisotropy value of $\sim r_0/3$, and a cluster of four will have an anisotropy value of $\sim r_0/4$. In general, if N is the number of fluorophores in a cluster participating in emFRET and $\omega \gg \Gamma$, then $r \approx r_0/N$. If ω is not much greater than Γ, the anisotropy of a complex with multiple fluorophores will have a value $\geq r_0/N$ and can be estimated using the following equation of Runnels and Scarlata (1995) (Jameson et al. 2003):

$$r_N = r_0 \cdot \frac{1+\omega \cdot \tau}{1+N \cdot \omega \cdot \tau} + r_{et} \cdot \frac{(N-1) \cdot \omega \cdot \tau}{1+N \cdot \omega \cdot \tau} \quad (10.13)$$

where

$$\omega = \frac{1}{\tau} \cdot \left(\frac{R_0}{R} \right)^6 \quad (10.14)$$

R_0 is the Förster distance, assuming a κ^2 value of 2/3 (for example, 4.95 nm for Venus-to-Venus transfer); τ is the fluorescence lifetime (for Venus $\tau = 3.4 \pm 0.1$ ns, mean $\pm$ SD, $n = 6$; data not shown); and R is the separation distance. Essentially, when ω is not much greater than Γ, a fraction of the directly excited fluorophores never transfers energy to neighbors by FRET and thus the population has a higher anisotropy value. Note that because r_{et} is small (0.016 for one-photon excitation; Berberan-Santos and Valeur 1991), the second term of Equation 10.13 can often be ignored.

As mentioned before, the molecular rotation of a fluorophore when attached to a protein can be measured by anisotropy decay analysis as a slow decay component (with a rotational correlation time τ_{rot}). Fluorophore rotation will also attenuate a steady-state anisotropy measurement. A variation of the Perrin equation (Perrin 1926) can be used to calculate a depolarization factor to account for this (Lakowicz 1999):

$$d_\theta = \frac{1}{1+\dfrac{\tau}{\tau_{rot}}} \quad (10.15)$$

Soleillet's rule (Lakowicz 1999; Soleillet 1929) for the multiplication of depolarization factors can then be used to combine the equation of Runnels and Scarlata (Equation 10.13) with the Perrin equation:

$$r = \left(r_0 \cdot \frac{1+\omega \cdot \tau}{1+N \cdot \omega \cdot \tau} + r_{et} \cdot \frac{(N-1) \cdot \omega \cdot \tau}{1+N \cdot \omega \cdot \tau} \right) \cdot \frac{1}{1+\dfrac{\tau}{\tau_{rot}}} \quad (10.16)$$

This equation can be used to predict steady-state anisotropy as a function of the number of fluorophores in a cluster and the rotational time constant of those fluorophores. Note that any depolarization occurring as a result of the optical design of the microscope used to measure anisotropy, as well as due to noncollinear absorption and emission dipoles, will be accounted for by measuring r_0 under the same imaging conditions.

How does emFRET alter fluorescence anisotropy decay curves? *In the absence of rotation*, fluorophores directly excited will emit with anisotropy values similar to r_0. In contrast, fluorophores excited indirectly by emFRET will typically emit with anisotropy values less than r_0. This value is r_{et}. The anisotropy value of r_{et} is a function of the dipole–dipole angle between the emFRET donor and acceptor with a value ranging between r_0 and 0. When an isotropic population of fluorophores is excited by one-photon linearly polarized light, the value of r_{et} will be approximately 0.016. Upon excitation, the anisotropy of this population will decay from a value of r_0 to a value of $I \cdot r_0 + j \cdot r_{et}$, where i and j are the fraction of directly and indirectly excited fluorophores emitting, respectively. For clusters of two fluorophores, if $\omega \gg \Gamma$, the value of $i = j = 0.5$. Thus, for dimmers, the anisotropy will decay from r_0 to a value of $\sim r_0/2$. Similarly, for trimers, $i = 1/3$ and $j = 2/3$; thus, the anisotropy will decay from r_0 to a value approaching $r_0/3$, etc.

The kinetics of this emFRET-related anisotropy decay reflects the *net* rate of energy transfer from photoselected fluorophores to the other fluorophores in the cluster. It is important to realize that this *ensemble* transfer rate must account for back transfer to the originally excited fluorophore, but it is still proportional to the single-step emFRET transfer rates occurring within the cluster. Equation 10.14 defines the single-step energy transfer rate, ω, from a donor to an acceptor as a function of fluorophore separation distance (again assuming that κ^2 is 2/3). For a dimer ($N = 2$), the anisotropy decay constant related to emFRET, ϕ, is thought to be the following (Berberan-Santos and Valeur 1991; Gautier et al. 2001; Tanaka and Martaga 1979):

$$\phi = \frac{1}{2\omega} \tag{10.17}$$

Thus, for dimers, the anisotropy is expected to decay two times faster than the single-step emFRET transfer rate predicted by Equation 10.14. Clearly, the closer two fluorophores are, the faster the anisotropy of a dimer should decay to a value approaching $r_0/2$. This can be seen for constructs composed of two Venus molecules separated by 5, 17, and 32 amino acid linkers in Figure 10.13. Notice that these anisotropy decay curves are well fit with a double-exponential decay model. The slow decay components (ranged between 15.9 and 23.5 ns) were similar to the rotational correlation time observed for a single Venus molecule shown in Figure 10.9 (15 ns), and they represent the molecular rotation of the fluorescent protein when tethered to another fluorescent protein.

An interesting aspect of the photophysics of fluorescent proteins that can be exploited in anisotropy studies is their slow rotation compared to more classical fluorophores (compare the anisotropy decay curves of Venus in Figure 10.9 and fluorescein in Figure 10.12). The comparatively slow rotation of Venus results from its large size and molecular weight, and because its fluorophore is rigidly anchored within the proteins' β-barrel structure. Accordingly, anisotropy decay curves of fluorescent proteins when attached to other proteins should never decay faster than the rotational correlation time of the free fluorophore (~15 ns) unless emFRET is occurring.

In Figure 10.13, the fast decay components (ϕ = 0.60, 0.81, and 1.5 ns for V5V, V17V, and V32V, respectively) most prominent between 0 and 2 ns are therefore interpreted as changes in the emFRET transfer rate as the separation distance between the two Venus

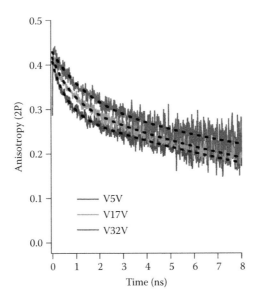

FIGURE 10.13 (**See color insert following page 288.**) Depolarization due to FRET: the anisotropy decay of Venus dimers. Three constructs consisting of two Venus molecules tethered to each other by 5 (V5V), 17 (V17V), and 32 (V32V) amino-acid linkers were expressed in HEK cells to demonstrate the effect of increasing the separation distance between two fluorescent proteins on anisotropy decay curves. Because these constructs are dimeric with a cluster size (N) of two, their fast anisotropy decay components should asymptote to a value that, at most, should be half that of their limiting anisotropy. This is observed. Curve fitting with a biexponential decay model (black dashed lines) yielded a fast anisotropy decay constant (ϕ) of 0.60 ns and a slow rotational correlation time (τ_{rot}) of 16.3 ns for V5V; for V17V, ϕ was 0.81 and τ_{rot} was 15.9 ns and, for V32V, ϕ was 1.5 and τ_{rot} was 23.5 ns. Notice how the value of ϕ increases with increased separation distance between the fluorophores.

molecules increases with linker size. Comparing this set of curves with the decay of free Venus in Figure 10.9, it should be obvious that, for fluorescent proteins with large Förster distance (R_0; 4.95 nm for Venus-to-Venus transfer), the presence of emFRET is determined by an anisotropy decay component faster than the slow rotational component of the free fluorophore. If we now assume that κ^2 has a value of 2/3, using a Venus fluorescence lifetime (τ) of 3.4 ns, we can use Equations 10.14 and 10.17 to estimate the separation distances for V5V (4.2 nm), V17V (4.4 nm), and V32V (4.9 nm). Interestingly, a hetero-FRET study of a related set of constructs (C5V, C17V, and C32V) composed of a single Cerulean acting as donor and a Venus acting as an acceptor separated by 5, 17, or 32 amino-acid linkers suggested that the separation distances between the Cerulean and Venus in those constructs were 5.7, 5.9, and 6.2 nm, respectively.

 Although both sets of experiments qualitatively displayed the expected increase in FRET as the linker size was decreased, it is surprising that the separation distance measured for the Venus–Venus constructs were ~1.4 nm shorter than the equivalent Cerulean–Venus constructs. This difference might perhaps reflect a true difference in the separation distance or dipole–dipole angle between the Cerulean–Venus and Venus–Venus constructs.

Alternatively, this discrepancy might simply reflect the difficulty of accurately measuring anisotropy decay constants by fitting multiexponential decays (having four or more free fitting parameters). Regardless, it highlights the difficulty of analyzing anisotropy decay curves *quantitatively*, and it indicates aspects of anisotropy decay analysis that need further study.

It is important to mention that the kinetics of the emFRET-related anisotropy decay components become much more complicated when the fluorophore cluster size is greater than two. Under these circumstances, the emFRET component of the anisotropy may no longer decay as a single exponential and will be dependent not only on transfer rates between fluorophores in a cluster, but also on the spatial arrangement of the fluorophores (e.g., in a row, in a ring, tetrahedron, branched, etc.). Kinetic models of the migration of energy for many of these distributions are so intractable mathematically that they are best approached using Monte Carlo simulations (Blackman et al. 1996, 1998; Marushchak and Johansson 2005). The absence of closed-form mathematical models for energy migration for various cluster arrangements is another important problem limiting our ability to use curve fitting to analyze the decay of anisotropy *quantitatively*.

Qualitatively, the amplitude of the emFRET anisotropy decay component should increase dramatically with the number of fluorophores in a cluster. This is illustrated in Figure 10.14, where the anisotropy decay curves of three fluorescent protein constructs, AAV, AVV, and VVV, are compared. In this nomenclature, V stands for Venus and A stands for Amber, a point mutation in Venus that prevents the formation of its fluorophore (Koushik et al. 2006). All three constructs should have essentially the same structure and therefore should have similar molecular rotation. AAV should have no emFRET, AVV should have an emFRET cluster size of two, and VVV should have a cluster size of three. Notice the dramatic change in the fast decay component of these anisotropy curves when comparing cluster sizes of one, two, or three Venus molecules. This change in the fast anisotropy decay component related to cluster size is much more dramatic than the subtle change observed with changes in transfer rate (Figure 10.13), and it supports the idea that the amplitude of the fast anisotropy decay component, particularly when its decay constant is significantly faster than the rotational correlation time, encodes information about cluster size.

Notice that the fast decay component of the AVV anisotropy decay curve has an asymptote that appears to be significantly less than $r_0/2$ but that decayed faster than the rotational correlation time of AAV. The VVV curve has an asymptote that appears to be equal to $r_0/2$ and was much less than $r_0/3$. This can be accounted for by Equation 10.13 if one considers that r_{et} is not equal to 0 and that ω may not be much greater than τ. In general, after accounting for molecular rotation, if the anisotropy drops below r_0/N, the cluster size must be greater than N.

Note, however, that the converse of this rule is not necessarily true. For example, the amplitude of the emFRET decay component of a cluster of three fluorophores might be less than $r_0/2$ if the energy transfer rate (ω) is less than the emission rate of the fluorophore (Γ). Under these conditions, most fluorophores will never transfer energy to a neighbor, so the vast majority of photons emitted from these clusters will be from fluorophores that were

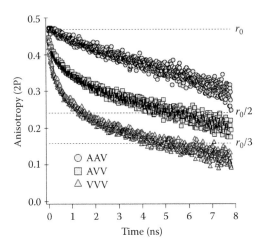

FIGURE 10.14 Depolarization due to energy migration: the impact of cluster size. Three constructs consisting of one (AAV), two (AVV), or three (VVV) Venus molecules tethered to each other and to either two (AAV), one (AVV), or zero (VVV) Amber molecules (A) were expressed in HEK cells to demonstrate the effect of cluster size (N) on anisotropy decay curves. Amber is a single-point mutant of Venus that lacks its fluorophore and does not act as a dark absorber. AAV has a cluster size of one, AVV has a cluster size of two, and VVV has a cluster size of three. All three constructs should have approximately the same mass and shape and therefore should have similar slow rotational correlation times. Notice that all three anisotropy decay curves between 4 and 8 ns are parallel. Also notice that AAV (N = 1) has no fast anisotropy decay component, while both AVV (N = 2) and VVV (N = 3) have prominent fast emFRET-related anisotropy decay components between 0 and 4 ns. The amplitude of the AVV fast decay component was significantly less than $r_0/2$, and the amplitude of the VVV fast component was approximately equal to $r_0/2$.

directly photoselected. It should be pointed out that because its amplitude would be small and its correlation time would be slow, under these circumstances, it would be very difficult to observe a "fast" emFRET anisotropy decay component, even for fluorophores like Venus that have slow rotational correlation times.

10.9 FLUORESCENCE ANISOTROPY APPLICATIONS

By this point it should be clear that fluorescence anisotropy measurements detect changes between the orientation of the absorption dipole of a fluorophore, as it is elevated to its excited state, and the orientation of the emission dipole of the fluorophore as it emits a photon. In the absence of energy migration, differences in anisotropy are primarily attributed to the rotation of a fluorophore while in the excited state. If energy migration can occur, depolarization can be attributed to both rotation and emFRET.

Furthermore, anisotropy can be used to measure FRET and the magnitude of the anisotropy drop can be used to determine the number of fluorophores participating in energy migration. Steady-state anisotropy measurements cannot directly differentiate between the various causes of depolarization, but time-resolved anisotropy decay analysis often can. Clearly, fluorescence anisotropy decay analysis can be used to measure molecular rotation and emFRET to reveal the rotational correlation time (τ_{rot}) of a fluorophore, emFRET transfer rate (ω) for dimers, and cluster size (N). In this section, we will outline three other applications of fluorescence anisotropy to illustrate other uses for these measurements in biomedical research.

10.9.1 Phosphorylation Assay

Anisotropy has been used to measure the binding of a phospho-specific antibody to a kinase substrate peptide (Figure 10.15; Jameson and Mocz 2005). Kinase-specific substrate peptides can be synthesized—one in the phosphorylated form and two in the dephosphorylated forms—with one having a fluorophore rigidly attached. The phosphorylated peptide is used as an antigen to generate phospho-specific antibodies. Binding specificity is determined by screening against both the phosphorylated and dephosphorylated forms of the peptide. Samples are incubated with the dephosphorylated fluorophore-tagged form of the substrate peptide.

Following incubation, samples are inactivated, and an anisotropy decay curve is acquired before and after incubation with excess antibody. Free unphosphorylated peptide should have a fast rotational correlation time (see green curve). The antibody should bind to the phosphorylated peptide during the sample incubation period. Antibody should dramatically increase the fluorophore's effective mass and therefore it should have a much slower rotational correlation time (see red decay curve in Figure 10.15). A mixture of free and bound peptide should generate a biexponential anisotropy decay curve where the amplitude of the fast rotational component indicates the fraction of free peptide (primarily unphosphorylated), and the amplitude of the slow component indicates the fraction of bound phosphorylated peptide (see blue decay curve in Figure 10.15).

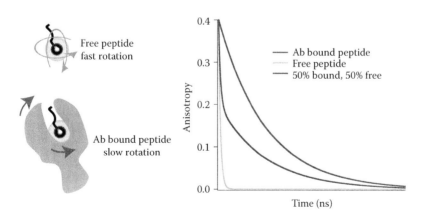

FIGURE 10.15 **(See color insert following page 288).** Using anisotropy to measure phosphorylation. An illustration depicting the expected anisotropy decay curve of a free fluorophore tagged short kinase substrate peptide (green trace) that can undergo fast rotations, and the curve generated by the same peptide when bound to a large antibody (upon phosphorylation; red curve). The blue decay curve depicts the biexponential decay expected from a population where 50% of the peptides are phosphorylated and bound to antibody, and 50% remain unphosphorylated and free.

10.9.2 Putting Limits on the Value of κ^2

As mentioned previously, the interpretation of FRET experiments in terms of separation distances requires knowledge of the value of the dipole–dipole orientation factor, κ^2. This information is almost never known in biological experiments and is often assumed to have a value of 2/3 (van der Meer 2002). This assumption is appropriate if both donor and acceptor fluorophores are free to rotate in any direction at a rate that is significantly faster than their lifetimes (van der Meer 2002).

Unfortunately, this assumption may not be correct because tethering a fluorophore to a protein might hinder its rotation to some degree. Furthermore, for fluorescent proteins, this assumption is never valid (though often made) because their rotational correlation times (typically 15 ns or slower) are never faster than their lifetimes (typically ranging from 1.5 to 4 ns). Although anisotropy measurements will not reveal the actual value of κ^2, a comparison of the limiting anisotropy of donors and acceptors (r_{0donor}, $r_{0acceptor}$, respectively) with the steady-state anisotropy values of the donors and acceptors (r_{donor}, $r_{acceptor}$) can be used to show whether these fluorophores have rotational mobility when attached to other moieties (i.e., they are not rigidly attached).

The bigger the difference is between the limiting anisotropy of a fluorophore used as a FRET donor or acceptor and the steady state anisotropy value of that fluorophore when attached to a construct, the greater their rotational freedom will be. This in turn can be used to constrain the possible values of κ^2 in that experiment and to set ranges of probabilities to certain values in the absence of any other information. This formalism was developed by Robert Dale and colleagues (Eisinger 1976; Dale, Eisinger, and Blumberg 1979; Eisinger and Dale 1974) and is outlined as applied by Lakowicz (1999):

$$\kappa_{min}^2 = \frac{2}{3} \cdot \left[1 - \left(0.5 \cdot \left(\sqrt{\frac{r_{donor}}{r_{0donor}}} + \sqrt{\frac{r_{acceptor}}{r_{0acceptor}}} \right) \right) \right] \tag{10.18}$$

$$\kappa_{max}^2 = \frac{2}{3} \cdot \left[1 + \sqrt{\frac{r_{donor}}{r_{0donor}}} + \sqrt{\frac{r_{acceptor}}{r_{0acceptor}}} + \left(3 \cdot \sqrt{\frac{r_{donor}}{r_{0donor}}} \cdot \sqrt{\frac{r_{acceptor}}{r_{0acceptor}}} \right) \right] \tag{10.19}$$

The limiting anisotropy values of the donor (r_{0donor}) and acceptor ($r_{0acceptor}$) are measured as the y-intercept (at $t = 0$) of time-resolved anisotropy decay measurements. The steady-state anisotropy values of donor- and acceptor-tagged constructs can be calculated by pooling all photons counted by the parallel and perpendicular detectors and then using Equation 10.2. Once these limits for the value of κ^2 have been calculated, a range of separation distances (min and max) can be calculated using the following equations:

$$R_{min} = \sqrt[6]{\frac{\kappa_{min}^2}{\frac{2}{3}}} \cdot \sqrt[6]{\frac{R_0^6}{E} - R_0^6} \tag{10.20}$$

$$R_{max} = \sqrt[6]{\frac{\kappa_{max}^2}{2/3}} \cdot \sqrt[6]{\frac{R_0^6}{E} - R_0^6} \qquad (10.21)$$

where E is the FRET efficiency measured for a sample and R_0 is the Förster distance for a specific donor–acceptor pair, assuming a κ^2 value of 2/3. It is important to note that the range of possible separation distances generated by this formalism is only valid if the assembly transferring energy by FRET consists of a single donor and single acceptor.

10.9.3 Differentiating between Directly Excited Acceptors and FRET

Hetero-FRET measurements are often acquired by exciting a sample with a wavelength optimized for the donor fluorophore while observing emission through a filter specific for the acceptor. One common problem encountered in these types of measurements is that, in addition to FRET, often a fraction of acceptors is also directly excited. The fluorescence signals from the acceptors that become excited via FRET are often weak (due to a low FRET efficiency and/or a low acceptor quantum yield); therefore, the emission of these directly excited acceptors might be interpreted as FRET, thus resulting in erroneous measurements.

Anisotropy measurements have been used to differentiate between directly excited acceptors and those excited by FRET (Piston and Rizzo 2008; Rizzo and Piston 2005). This approach is based on the idea that directly excited acceptors will have high anisotropy values and that those excited by FRET will have low anisotropy values. A sample is excited with linearly polarized light at wavelengths spanning a range covering those thought to be specific for exciting the donor and those that preferentially excite the acceptor. At each wavelength, the steady-state anisotropy is measured through a filter specific for the acceptor. The steady-state anisotropy values are plotted against excitation wavelength. If FRET is not occurring, the anisotropy will be high and will not change with excitation wavelength. If FRET is occurring, at shorter wavelengths, the anisotropy will be low; at longer excitation wavelengths, the anisotropy will be high. Intermediate anisotropy values result from a mixture of directly excited acceptors and acceptors excited by FRET.

10.10 CONCLUSION

Fluorescence anisotropy decay is a powerful tool for investigating molecular rotation, binding reactions, protein–protein interactions and the assembly of multimeric complexes in living cells. The inability of humans to perceive polarization and a complex theory has been a barrier to the general application of polarization-based biological imaging—specifically, anisotropy imaging. Here we have provided a simplified explanation for the theory behind this approach.

This methodology is particularly well suited for analyzing proteins tagged with spectral variants of green fluorescent protein because anisotropy decay analysis can readily differentiate between depolarization caused by rotational diffusion of this large fluorophore and energy migration FRET. We fully expect that many studies in the near future will adopt this approach to understand in vivo molecular assemblies because fluorescence anisotropy decay remains one of the few methods that can differentiate among monomers, dimers, trimers, and higher order assemblies in living cells. Additional work remains, particularly in modeling anisotropy decays from more complicated fluorophore cluster geometries and from nonisotropic distributions.

ACKNOWLEDGMENTS

This work was supported by the intramural program of the National Institutes of Health, National Institute on Alcohol Abuse and Alcoholism, Bethesda, Maryland.

REFERENCES

Austin, R. H., Chan, S. S., and Jovin, T. M. 1979. Rotational diffusion of cell surface components by time-resolved phosphorescence anisotropy. *Proceedings of the National Academy of Sciences USA* 76:5650–5654.

Axelrod, D. 1979. Carbocyanine dye orientation in red cell membrane studied by microscopic fluorescence polarization. *Biophysical Journal* 26:557–573.

Axelrod, D. 1989. Fluorescence polarization microscopy. *Methods in Cell Biology* 30:333–352.

Bader, A. N., Hofman, E. G., van Bergen en Henegouwen, P. M. P., and Gerritsen, H. C. 2007. Imaging of protein cluster size by means of confocal time-gated fluorescence anisotropy microscopy. *Optics Express* 15:6934–6945.

Bastiaens, P. I., and Squire, A. 1999. Fluorescence lifetime imaging microscopy: Spatial resolution of biochemical processes in the cell. *Trends in Cell Biology* 9:48–52.

Becker, W. 2005. *Advanced time-correlated single photon counting techniques.* Berlin: Springer.

Beechem, J. M., Knutson, J. R., and Brand, L. 1986. Global analysis of multiple dye fluorescence anisotropy experiments on proteins. *Biochemical Society Transactions* 14:832–835.

Berberan-Santos, M. N., and Valeur, B. 1991. Fluorescence depolarization by electronic energy transfer in donor–acceptor pairs of like and unlike chromophores. *Journal of Chemical Physics* 95:8048–8054.

Blackman, S. M., Cobb, C. E., Beth, A. H., and Piston, D. W. 1996. The orientation of eosin-5-maleimide on human erythrocyte band 3 measured by fluorescence polarization microscopy. *Biophysical Journal* 71:194–208.

Blackman, S. M., Piston, D. W., and Beth, A. H. 1998. Oligomeric state of human erythrocyte band 3 measured by fluorescence resonance energy homotransfer. *Biophysical Journal* 75:1117–1130.

Callis, P. R. 1997. Two-photon-induced fluorescence. *Annual Review of Physical Chemistry* 48:271–297.

Cantor, C. R., and Schimmel, P. R. 1980. *Biophysical chemistry part II: Techniques for the study of biological structure and function.* San Francisco: W. H. Freeman & Co.

Clayton, A. H., Hanley, Q. S., Arndt-Jovin, D. J., Subramaniam, V., and Jovin, T. M. 2002. Dynamic fluorescence anisotropy imaging microscopy in the frequency domain (rFLIM). *Biophysical Journal* 83:1631–1649.

Clegg, R. M. 1996. Fluorescence resonance energy transfer. In *Fluorescence imaging spectroscopy and microscopy*, ed. X. F. Wang and B. Herman, 137, 179–252. Chichester, England: John Wiley & Sons, Inc.

Clegg, R. M., Murchie, A. I., and Lilley, D. M. 1994. The solution structure of the four-way DNA junction at low-salt conditions: A fluorescence resonance energy transfer analysis. *Biophysical Journal* 66:99–109.

Dale, R. E., and Eisinger, J. 1976. Intramolecular energy transfer and molecular conformation. *Proceedings of the National Academy of Sciences USA* 73:271–273.

Dale, R. E., Eisinger, J., and Blumberg, W. E. 1979. The orientational freedom of molecular probes. The orientation factor in intramolecular energy transfer. *Biophysical Journal* 26:161–193.

Denk, W., Strickler, J. H., and Webb, W. W. 1990. Two-photon laser scanning fluorescence microscopy. *Science* 248:73–76.

Dill, K. A., and Bromberg, S. 2003. *Molecular driving forces.* New York: Garland Science.

Dixit, B. P., Waring, A. J., Wells, K. O., III, Wong, P. S., Woodrow, G. V., III, and Vanderkooi, J. M. 1982. Rotational motion of cytochrome c derivatives bound to membranes measured by fluorescence and phosphorescence anisotropy. *European Journal of Biochemistry* 126:1–9.

Eads, T. M., Thomas, D. D., and Austin, R. H. 1984. Microsecond rotational motions of eosin-labeled myosin measured by time-resolved anisotropy of absorption and phosphorescence. *Journal of Molecular Biology* 179:55–81.

Eisinger, J., and Dale, R. E. 1974. Letter: Interpretation of intramolecular energy transfer experiments. *Journal of Molecular Biology* 84:643–647.

Förster, T. 1948. Intermolecular energy migration and fluorescence. *Annalen der Physik* 2:55–75.

Gadella, T. W. J., Jovin, T. M., and Clegg, R. M. 1993. Fluorescence lifetime imaging microscopy (FLIM): Spatial resolution of microstructures on the nanosecond time scale. *Biophysical Chemistry* 48:221–239.

Gautier, I., Tramier, M., Durieux, C., et al. 2001. Homo-FRET microscopy in living cells to measure monomer–dimer transition of GFP-tagged proteins. *Biophysical Journal* 80:3000–3008.

Hecht, S., Shlaer, S., and Pirenne, M. H. 1941. Energy at the threshold of vision. *Science* 93:585–587.

Heikal, A. A., Hess, S. T., Baird, G. S., Tsien, R. Y., and Webb, W. W. 2000. Molecular spectroscopy and dynamics of intrinsically fluorescent proteins: coral red (dsRed) and yellow (Citrine). *Proceedings of the National Academy of Sciences USA* 97:11996–20001.

Hess, S. T., Sheets, E. D., Wagenknecht-Wiesner, A., and Heikal, A. A. 2003. Quantitative analysis of the fluorescence properties of intrinsically fluorescent proteins in living cells. *Biophysical Journal* 85:2566–2580.

Inoue, S., and Spring, K. R. 1997. *Video microscopy: The fundamentals.* New York: Plenum Press.

Jablonski, A. 1970. Anisotropy of fluorescence of molecules excited by excitation transfer. *Acta Physiologica Pol A* 38:453–458.

Jameson, D. M., Croney, J. C., and Moens, P. D. 2003. Fluorescence: Basic concepts, practical aspects, and some anecdotes. *Methods in Enzymology* 360:1–43.

Jameson, D. M., and Mocz, G. 2005. Fluorescence polarization/anisotropy approaches to study protein–ligand interactions: Effects of errors and uncertainties. *Methods in Molecular Biology* 305:301–322.

Jares-Erijman, E. A., and Jovin, T. M. 2003. FRET imaging. *Nature Biotechnology* 21:1387–1395.

Knutson, J. R., Beechem, J. M., and Brand, L. 1983. Simultaneous analysis of multiple fluorescence decay curves: A global approach. *Chemistry and Physics Letters* 102:501–507.

Koushik, S. V., Chen, H., Thaler, C., Puhl, H. L., III, and Vogel, S. S. 2006. Cerulean, Venus, and VenusY67C FRET reference standards. *Biophysical Journal* 91:L99–L101.

Koushik, S. V., and Vogel, S. S. 2008. Energy migration alters the fluorescence lifetime of Cerulean: Implications for fluorescence lifetime imaging Forster resonance energy transfer measurements. *Journal of Biomedical Optics* 13:031204.

Lakowicz, J. R. 1999. *Principles of fluorescence spectroscopy.* New York: Kluwer Academic/Plenum Publishers.

Lakowicz, J. R., Gryczynski, I., Gryczynski, Z., Danielsen, E., and Wirth, M. J. 1992. Time-resolved fluorescence intensity and anisotropy decays of 2,5-diphenyloxzole by two-photon excitation and frequency-domain fluorometry. *Journal of Physical Chemistry* 96:3000–3006.

Lakowicz, J. R., Szmacinski, H., Nowaczyk, K., Berndt, K. W., and Johnson, M. 1992. Fluorescence lifetime imaging. *Analytical Biochemistry* 202:316–330.

Marushchak, D., and Johansson, L. B. 2005. On the quantitative treatment of donor–donor energy migration in regularly aggregated proteins. *Journal of Fluorescence* 15:797–803.

McClain, W. M. 1972. Polarization dependence of three-photon phenomena for randomly oriented molecules. *Journal of Chemical Physics* 57:2264–2272.

Nagai, T., Ibata, K., Park, E. S., Kubota, M., Mikoshiba, K., and Miyawaki, A. 2002. A variant of yellow fluorescent protein with fast and efficient maturation for cell-biological applications. *Nature Biotechnology* 20:87–90.

Periasamy, A., and Day, R. N., eds. 2005. *Molecular imaging: FRET microscopy and spectroscopy.* New York: Oxford University Press.

Perrin, F. 1926. Polarization de la lumiere de fluorescence: Vie moyenne des molecules dans l'etat excite. *Journal de Physique et le Radium V* 7:390–401.

Piston, D. W., and Rizzo, M. A. 2008. FRET by fluorescence polarization microscopy. *Methods in Cell Biology* 85:415–430.

Rao, M., and Mayor, S. 2005. Use of Forster's resonance energy transfer microscopy to study lipid rafts. *Biochimica et Biophysica Acta* 1746:221–233.

Rizzo, M. A., and Piston, D. W. 2005. High-contrast imaging of fluorescent protein FRET by fluorescence polarization microscopy. *Biophysical Journal* 88:L14–16.

Rizzo, M. A., Springer, G. H., Granada, B., and Piston, D. W. 2004. An improved cyan fluorescent protein variant useful for FRET. *Nature Biotechnology* 22:445–449.

Runnels, L. W., and Scarlata, S. F. 1995. Theory and application of fluorescence homotransfer to melittin oligomerization. *Biophysical Journal* 69:1569–1583.

Scott, T. W., Haber, K. S., and Albrecht, A. C. 1983. Two-photon photoselection in rigid solutions: A study of the B2u → A1g transition in benzene. *Journal of Chemical Physics* 78:150–157.

Sharma, P., Varma, R., Sarasij, R. C., Ira, Gousset, K., Krishnamoorthy, G., et al. 2004. Nanoscale organization of multiple GPI-anchored proteins in living cell membranes. *Cell* 116:577–589.

Soleillet, P. 1929. Sur les parametres caracterisant la polarisation partielle de la lumierer dans les phenomenes de fluorescence. *Annales de Physique (Paris)* 12:23–97.

Suhling, K., Siegel, J., Lanigan, P. M., Leveque-Fort, S., Webb, S. E., Phillips, D., et al. 2004. Time-resolved fluorescence anisotropy imaging applied to live cells. *Optics Letters* 29:584–586.

Tanaka, F., and Martaga, N. 1979. Theory of time-dependent photo-selection in interacting fixed systems. *Photochemistry and Photobiology* 29:1091–1097.

Tanaka, F., and Martaga, N. 1982. Dynamic depolarization of interacting fluorophores. Effect of internal rotation and energy transfer. *Biophysical Journal* 39:129–140.

Thaler, C., Koushik, S. V., Puhl, H. L., Blank, P. S., and Vogel, S. S. 2009. Structural rearrangement of CaMKIIa catalytic domains encodes activation. *Proceedings of the National Academy of Sciences USA* 10.1073/pnas.0901913106.

Valeur, B. 2002. *Molecular fluorescence.* Weinheim: Wiley–VCH.

van der Meer, B. W. 2002. Kappa-squared: From nuisance to new sense. *Journal of Biotechnology* 82:181–196.

Varma, R., and Mayor, S. 1998. GPI-anchored proteins are organized in submicron domains at the cell surface. *Nature* 394:798–801.

Vogel, S. S., Thaler, C., and Koushik, S. V. 2006. Fanciful FRET. *Sci STKE* 2006: re2.

Volkmer, A., Subramaniam, V., Birch, D. J., and Jovin, T. M. 2000. One- and two-photon excited fluorescence lifetimes and anisotropy decays of green fluorescent proteins. *Biophysical Journal* 78:1589–1598.

Wallrabe, H., and Periasamy, A. 2005. Imaging protein molecules using FRET and FLIM microscopy. *Current Opinion in Biotechnology* 16:19–27.

Wang, X. F., Periasamy, A., and Herman, B. 1992. Fluorescence lifetime imaging microscopy (FLIM): Instrumentation and applications. *Critical Reviews in Analytical Chemistry* 23:369–395.

Weber, G. 1952. Polarization of the fluorescence of macromolecules. I. Theory and experimental method. *Biochemical Journal* 51:145–155.

Yan, Y., and Marriott, G. 2003. Fluorescence resonance energy transfer imaging microscopy and fluorescence polarization imaging microscopy. *Methods in Enzymology* 360:561–580.

Yeow, E. K., and Clayton, A. H. 2007. Enumeration of oligomerization states of membrane proteins in living cells by homo-FRET spectroscopy and microscopy: Theory and application. *Biophysical Journal* 92:3098–3104.

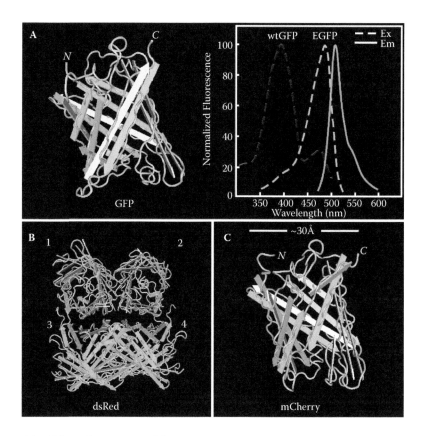

FIGURE 3.1　The β-barrel structure of the FPs.

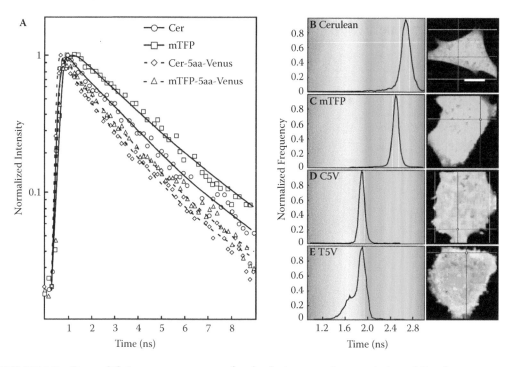

FIGURE 3.3　Donor lifetime measurements for the fusion proteins consisting of Cerulean or mTFP linked to Venus.

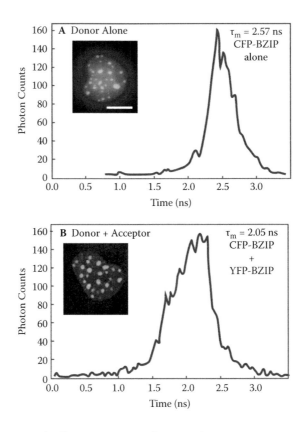

FIGURE 3.4   FRET-FLIM of cells co-expressing donor and acceptor.

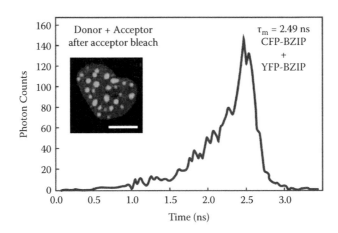

FIGURE 3.5   FRET-FLIM of cells co-expressing donor and acceptor after acceptor photobleaching.

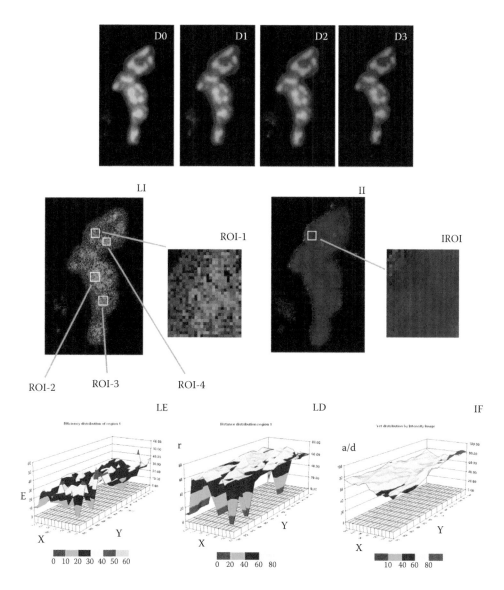

FIGURE 4.6   Demonstration of CFP-YFP-C/EBPα protein behavior in the foci of a living cell nucleus during the protein interaction process.

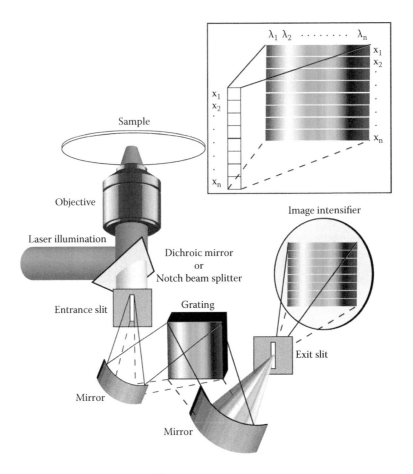

FIGURE 5.6 A schematic of the spectral FLIM setup.

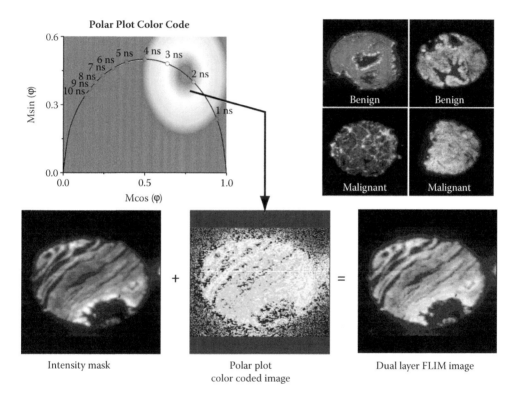

FIGURE 5.7 Dual layer FLIM images where an intensity image is used to mask the color coded lifetime image.

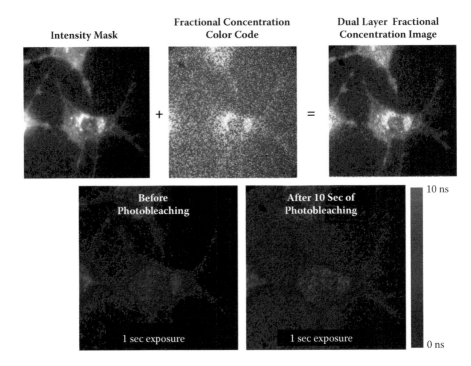

FIGURE 5.8 Top: Overlay of the intracellular locations and concentrations of PpIX (protoporphyrin IX) monomer and dimer. Bottom: Dual layer fluorescence lifetime images of PpIX in live 3T3 cells incubated with ALA.

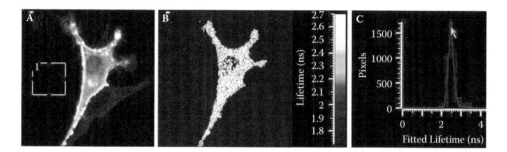

FIGURE 6.14 Intensity (A), lifetime (B) image and the lifetime histogram (C) of a GPI-GFP stained Her14 cell.

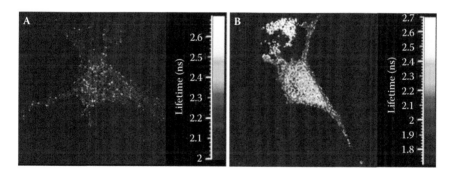

FIGURE 6.15 Lifetime image of Her14 cell stained with GPI-GFP and GM1-CTB-Alexa594 before (A) and after extraction (B) of cholesterol from the lipid rafts.

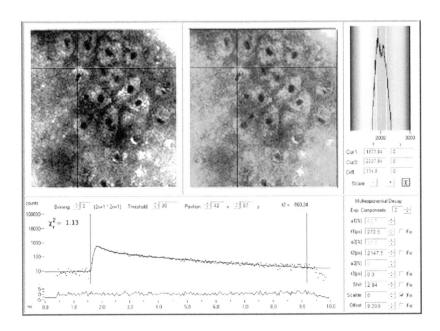

FIGURE 7.2   Fluorescence intensity image, color-coded FLIM image, and τ_2 distribution of skin autofluorescence.

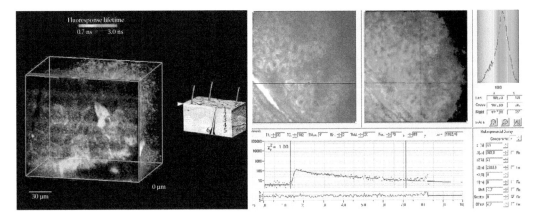

FIGURE 7.8   Typical autofluorescence decay curves and FLIM images of normal Caucasian human skin.

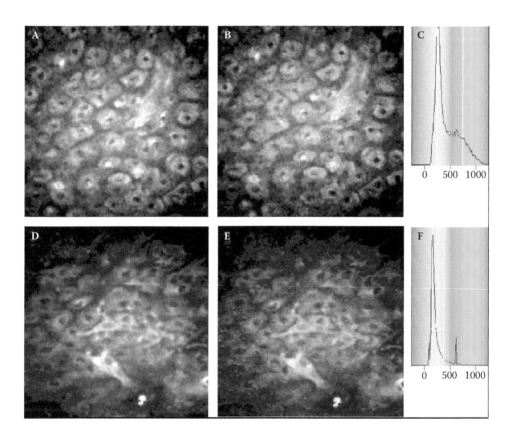

FIGURE 7.10   False-color-coded FLIM images of nevi and melanoma.

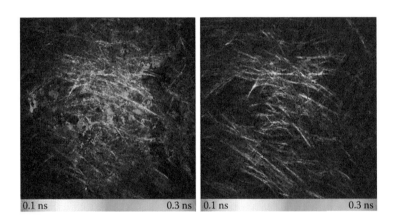

FIGURE 7.12   False-color-coded FLIM images of autofluorescence and SHG signals.

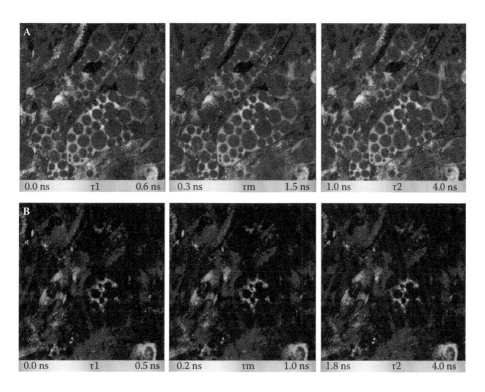

FIGURE 7.14 False-color-coded FLIM images of cells after adipogenic differentiation at 750nm (A) and 900 nm (B) excitation.

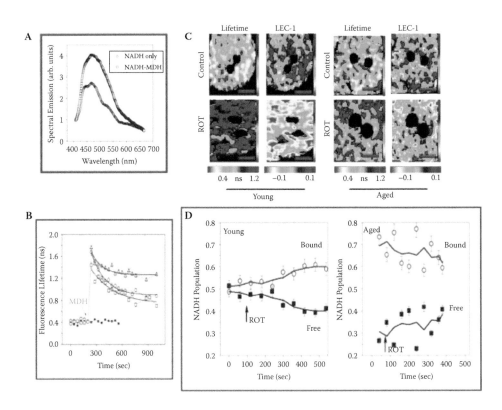

FIGURE 8.2   Imaging dynamic compartmentalization of free and bound fractions of NADH.

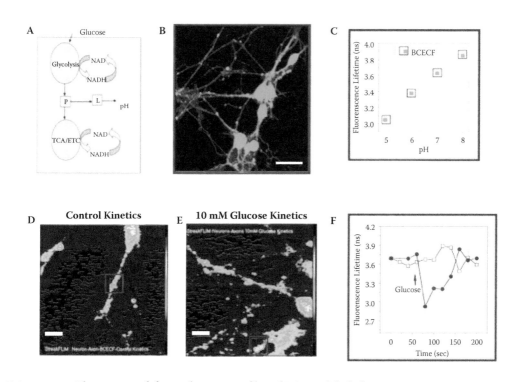

A Glucose

Glycolysis

NAD

NADH

P → L → pH

TCA/ETC

NAD

NADH

B

C BCECF

D Control Kinetics

E 10 mM Glucose Kinetics

F Glucose

FIGURE 8.4 Fluorescence lifetime kinetic profiles of BCECF labeled neurons.

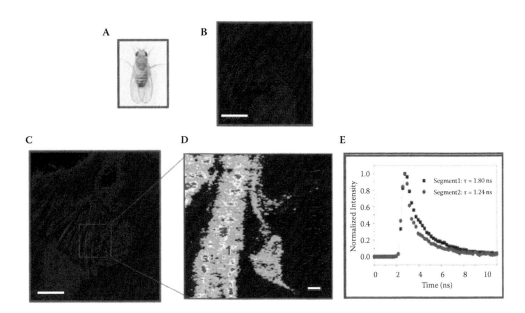

FIGURE 8.5 Fluorescence lifetime imaging for monitoring spatially-resolved, differential metabolic activity in live animals.

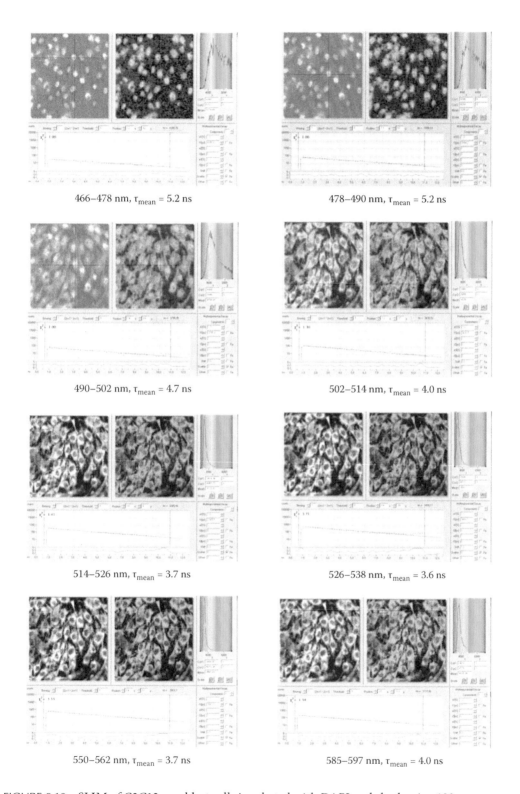

466–478 nm, τ_{mean} = 5.2 ns

478–490 nm, τ_{mean} = 5.2 ns

490–502 nm, τ_{mean} = 4.7 ns

502–514 nm, τ_{mean} = 4.0 ns

514–526 nm, τ_{mean} = 3.7 ns

526–538 nm, τ_{mean} = 3.6 ns

550–562 nm, τ_{mean} = 3.7 ns

585–597 nm, τ_{mean} = 4.0 ns

FIGURE 9.10   SLIM of C2C12 myoblast cells incubated with DAPI and rhodamine 123.

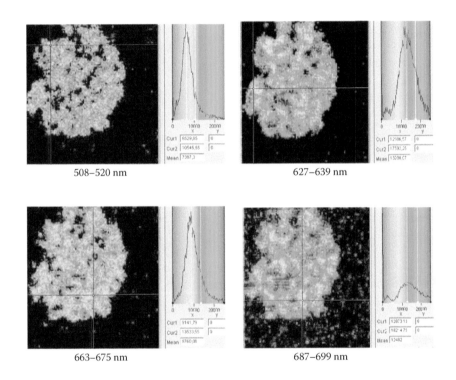

508–520 nm

627–639 nm

663–675 nm

687–699 nm

FIGURE 9.11   SLIM of HepG2 cells demonstrated within different spectral ranges.

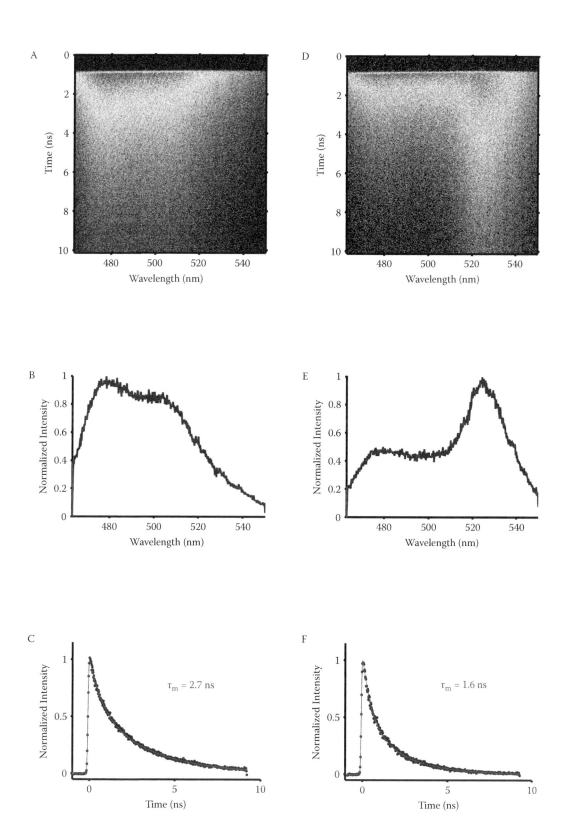

FIGURE 9.12 Analysis of HEK293 cells streak FLIM images.

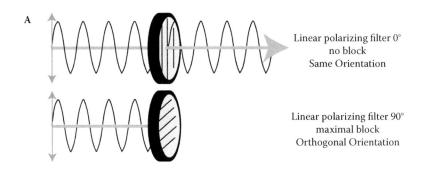

Linear polarizing filter 0°
no block
Same Orientation

Linear polarizing filter 90°
maximal block
Orthogonal Orientation

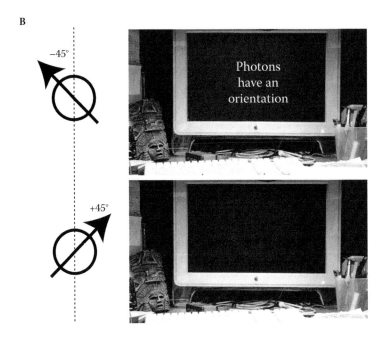

FIGURE 10.1   Detecting polarization.

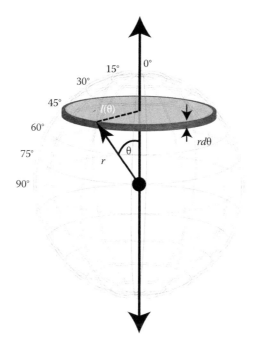

FIGURE 10.3   Random dipole orientation distributions.

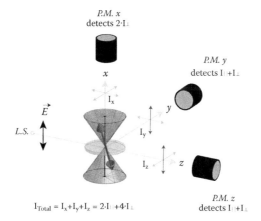

FIGURE 10.5   Detecting polarization on a microscope.

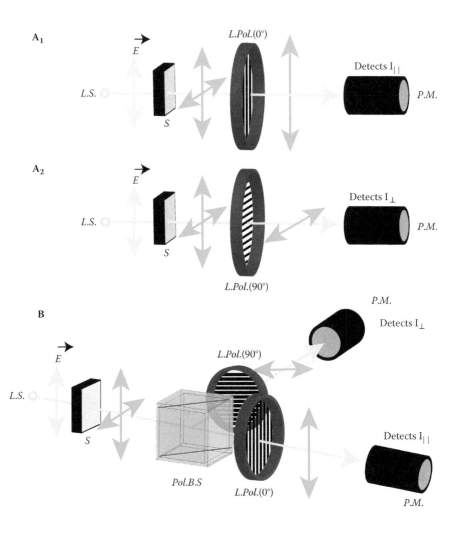

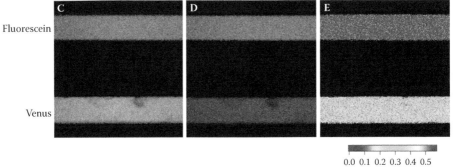

FIGURE 10.6 Separating $I_{\parallel}$ and $I_{\perp}$.

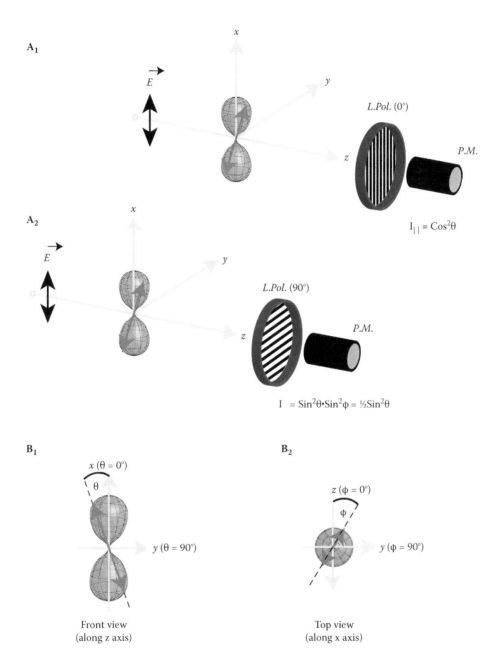

FIGURE 10.7   The probability of detection through a polarizing filter.

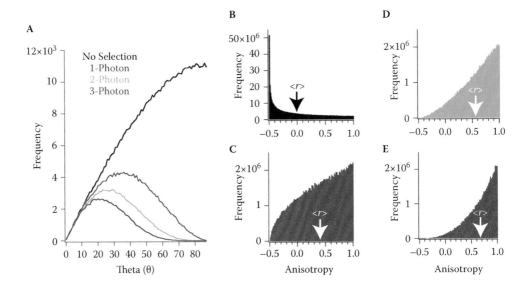

FIGURE 10.8   The fundamental anisotropy value of an isotropic population.

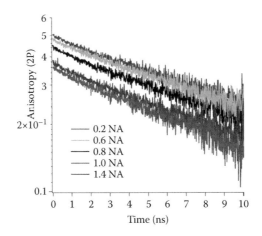

FIGURE 10.10   The influence of numerical aperture on anisotropy.

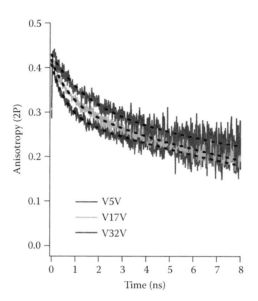

FIGURE 10.13   Depolarization due to FRET.

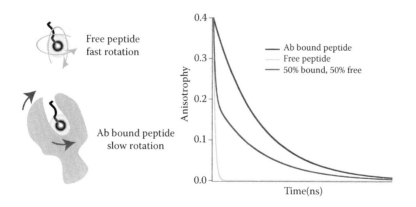

FIGURE 10.15   Using anisotropy to measure phosphorylation.

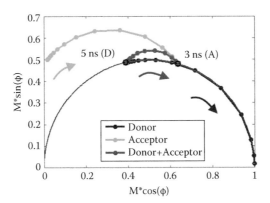

FIGURE 11.3 The polar plot of the fluorescence signals of the donor, the acceptor, and the donor and acceptor together.

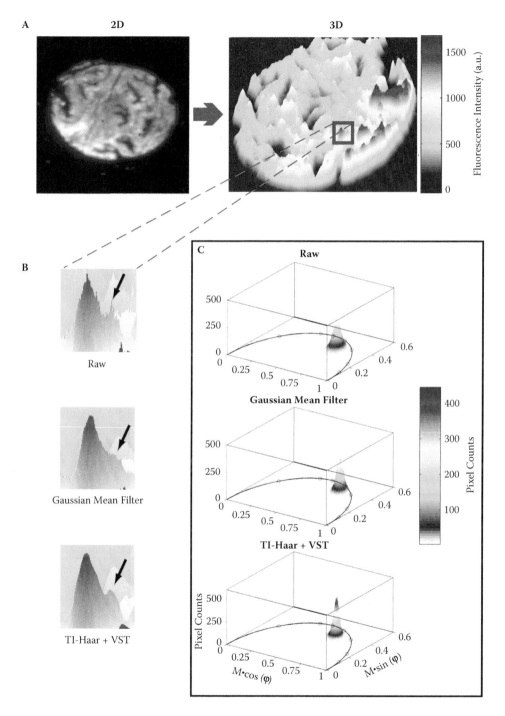

FIGURE 11.6 Demonstration of the signal-dependent noise removal image analysis for FLIM, using a fluorescence intensity image.

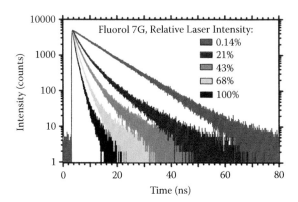

FIGURE 12.4 Intensity decays measured in TCSPC system for different excitation intensity.

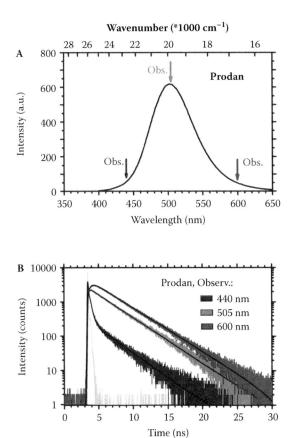

FIGURE 12.5 Intensity decays measured for different observation wavelengths.

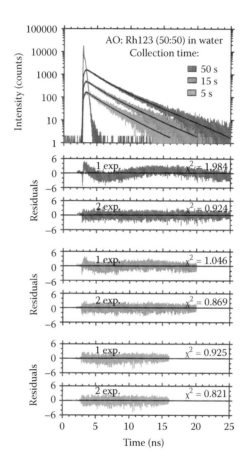

FIGURE 12.7   Intensity decays for mixture of AO and Rh123 and residual distribution for different data collection times.

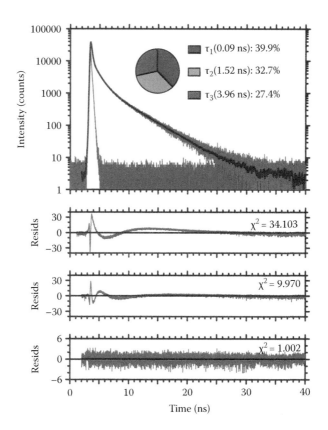

FIGURE 12.11 Mixture of three dyes (ErB, PxB, Rh123) and recovered parameters.

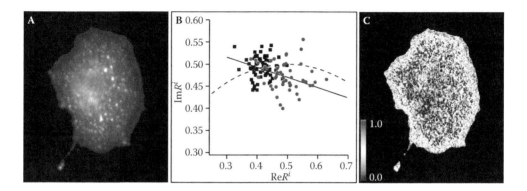

FIGURE 13.1   Global analysis of FRET data.

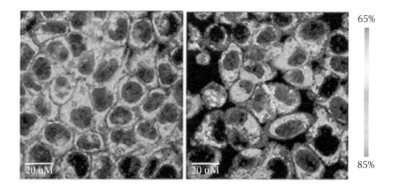

FIGURE 14.1   Autofluorescence lifetime images recorded from untreated (left) and synchronized with hydroxyurea HeLa cells (S-phase).

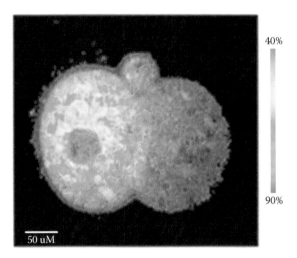

FIGURE 14.2   Lifetime distribution of metabolic activity in a mouse embryo at the stage of two blastomers in the NADH spectrum band.

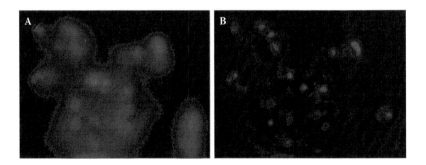

FIGURE 14.4   Time-gated image of HPV DNA to remove the autofluorescence signal from CaSki cells.

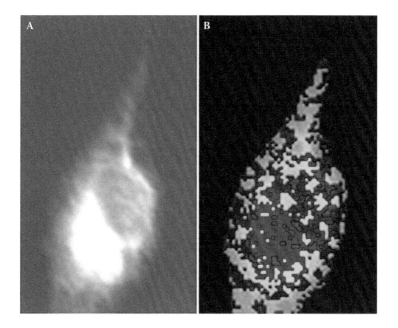

FIGURE 14.5   FLIM analysis of the presenilin 1(PS1) protein in Alzheimer's disease.

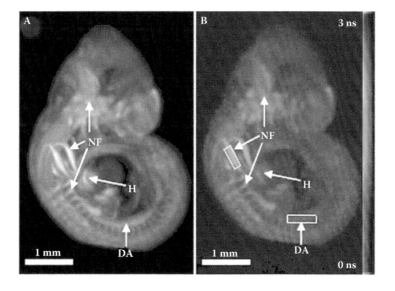

FIGURE 14.6   Intensity and lifetime image of a mouse embryo.

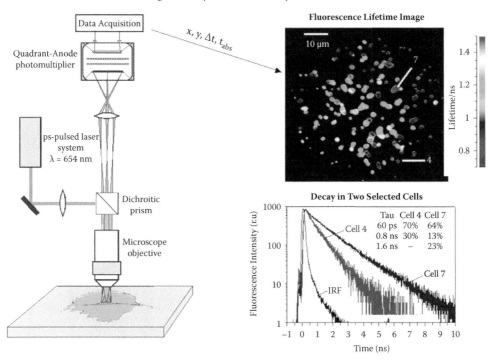

Widefield FLIM (single photon counting)
Living Cells of Cyanobacterium Acaryochloris Marina

Data Acquisition

$x, y, \Delta t, t_{abs}$

Quadrant-Anode
photomultiplier

ps-pulsed laser
system
$\lambda = 654$ nm

Dichroitic
prism

Microscope
objective

Fluorescence Lifetime Image

10 µm

7

4

Lifetime/ns

1.4

1.2

1

0.8

Decay in Two Selected Cells

Cell 4

Tau	Cell 4	Cell 7
60 ps	70%	64%
0.8 ns	30%	13%
1.6 ns	–	23%

Cell 7

IRF

Fluorescence Intensity (r.u)

1000

100

10

1

−1 0 1 2 3 4 5 6 7 8 9 10

Time (ns)

FIGURE 14.7 Schematic of the FLIM set up and the fluorescence dynamics in individual cells of
A. *marina*.

3

Data Analysis

General Concerns of FLIM Data Representation and Analysis

Frequency-Domain Model-Free Analysis

Yi-Chun Chen, Bryan Q. Spring, Chittanon Buranachai, Bianca Tong, George Malachowski, and Robert M. Clegg

11.1 INTRODUCTION

In the first part of this chapter we develop several mathematical expressions describing fluorescence lifetime measurements, and show the close correspondence between the time and frequency domains (Lakowicz 1999; Valeur 2002, 2005). We then discuss how the underlying physical parameters are deciphered from the measured FLIM experiments and present some techniques of image analysis and display that are peculiar to FLIM.

Calculating the time-dependent fluorescence $F(t)_{meas}$ for different modes of measurement, Box 11.1 describes the basic convolution integral that is our starting point for the description of all time-dependent fluorescence experiments. Equation 11.1 simply says that the fluorescence signal at time t depends on all excitation events previous to t (causality), and the fluorescence response is the linear sum (integral) of all the prior excitation events. Equation 11.2 states that if fluorophores are excited instantaneously at time t', then the fluorescence will decay exponentially as $(t - t')$ (Clegg and Schneider 1996; Clegg, Schneider, and Jovin 1996). The defining parameters in the fluorescence decay are the amplitudes, $F_{0,i}$ and the lifetimes τ_i (see Box 11.1). The amplitudes carry information concerning the number of fluorophores that have been excited, and each lifetime provides valuable physical information about all the participating pathways of de-excitation for that particular fluorescence species, as described in Chapter 1.

BOX 11.1

For the case where all fluorophores are initially excited by the excitation light, the basic relation of the time-dependent measured fluorescence $F(t)_{meas}$ to the form of the excitation event is through a finite convolution integral (Birks 1970; Clegg and Schneider 1996; Clegg et al. 1996; Dern and Walsh 1963; Schneider and Clegg 1997).

$$F\left(t\right)_{meas} = \int_0^t E\left(t'\right) F_\delta\left(t-t'\right) dt' \tag{11.1}$$

$E(t')$ is the time dependence of the excitation event; this is usually a repetitive waveform and can vary from a pure sinusoidal wave or a repetitive square wave to a repetitive series of pulses of short duration. $F_\delta(t - t')$ is the *fundamental fluorescence response* of the fluorescent sample to a single delta function excitation pulse, $E_\delta(t')$. t' is the time that $E_\delta(t')$ arrives to excite the sample, and t is the time, following t', that the measurement is made. By delta function pulse, we mean simply a light pulse that is very short compared to any rate of fluorescence decay. If every fluorescing species is excited directly by $E_\delta(t')$, then the decay of every separate component of fluorescence can be described by the *fundamental fluorescence response, expressed as an exponential decay.*

$$F_\delta\left(t-t'\right)_{meas} = F_\delta\left(t-t'\right) = F_0 \exp\left(-\left(t-t'\right)/\tau\right) \tag{11.2}$$

Many components may decay exponentially, either from the same chemical species in different environments or from different chemical species. In that case, *the fundamental fluorescence response is a sum of exponential decays,* each weighted with corresponding amplitudes and each with a different lifetime.

$$F_\delta\left(t-t'\right)_{meas} = \sum_i F_{\delta,i}\left(t-t'\right) = \sum_i F_{0,i} \exp\left(-\left(t-t'\right)/\tau_i\right) \tag{11.3}$$

The inverse lifetimes, $1/\tau_i$, are the sum of all the rate constants of de-excitation of all the pathways leading out of the excited state of species (see Chapter 1). For noninteracting fluorophores (see later discussion for cases of excited-state reactions, such as FRET), each decay component contributes individually to the signal. For simplicity in the following discussion, we consider only one fluorescence component, with the understanding that the measured signal is the sum of all components.

For those with an engineering background, the procedure of convolution in Equation 11.1 is simply an expression of a *causal recursive filter* in the terminology of electronic measurement or communications engineering. This means that the filter only acts on past occurrences that influence the signal. An *instrument response function* is the temporal response of the measuring instrumentation to a delta function pulse signal and is also expressed as such a convolution. The instrument response function is determined by the experimenter separately from the experiment. A lucid discussion of the basics of such filters can be found in a book by Hamming (1989). That is, the fluorescence response of Equation 11.1, $F(t)_{meas}$, is subsequently further convoluted with the instrument response function of the recording device.

Because the convolution of the fluorescence signal with the instrument response function is specific to every individual instrument, we will not consider it expressly in this chapter. But one should be aware that the fluorescence signal will be affected by the instrumentation (see Section 11.2). An analogous effect arises when observing the product of an excited-state reaction in Section 11.7.1.2, where the signal (the fluorescence response of the donor) is convoluted with the response function of the acceptor (which, in this case, is analogous to a measurement device; see Box 11.8). The mathematics is identical.

There are different methods for calculating the measured fluorescence response, $F(t)_{meas}$, described in Equation 11.1. The most straightforward way is to calculate the convolution integral directly, knowing the form of $E(t')$. This is a very general procedure and can be carried out for any form of excitation. To demonstrate different approaches found in the literature, we will derive analytical expressions for $F(t)_{meas}$ using the convolution integral directly, as well as several other methods. The excitation event is always a repetitive wave form, with a definite period, T; that is, the form of the excitation pulse is identical in every period of the excitation pulse train. Because the signal is acquired and averaged over many periods, we are not interested in transient terms that arise immediately after the excitation pulse train is started and decay to zero in the time range of the longest fluorescence decay. We are interested in the steady-state time-dependent terms, which are identical in every period. We can analyze $F(t)_{meas}$ by following the time progress directly or by expressing $F(t)_{meas}$ in terms of its corresponding frequency parameters.

11.2 TIME DOMAIN ASSUMING VERY SHORT EXCITATION PULSES

Because the *fundamental fluorescence response* $F_\delta(t - t')$ (Equation 11.3) is a sum of exponential relaxations in response to a delta function excitation pulse, $E_\delta(t')$, we first consider the case of a repetitive series of identical excitation pulses that are very short compared to any of the fluorescence lifetimes (we will call these *delta function pulses*, or δ-*pulses* (Bracewell 1978; Brigham 1974; Valeur 2002, 2005).) With the advent of lasers with femto- to picosecond pulse durations, this is now routinely possible, and it is a standard way of exciting a FLIM sample (see Chapters 6 and 7). When the excitation pulses are much shorter than any of the fluorescence lifetimes, the pulses can be considered to be δ-pulses and $F(t - t')_{meas} = F_\delta(t - t')$; in this case, there is no need for deconvolution of the excitation event. If we know $E(t')$ and it cannot be considered a δ-pulse, all that is necessary to simulate $F(t)_{meas}$ is to carry out the convolution integral in Equation 11.1.

However, as mentioned in the introduction, even if the excitation is a δ-pulse, the *instrument response* must be taken into account. The instrument response is convoluted into the fluorescence signal in a manner similar to that for the convolution integral of Equation 11.1. The true fundamental fluorescence response, Equations 11.2 and 11.3, is only recovered if the instrument response is also deconvoluted from the measured signal. The instrument response is measured by recording a δ-pulse emission signal, such as light scattering, Raman, or an exceptionally rapid fluorescence decay. The instrument response convolution will not be considered in this chapter, but one must be aware of it (see Chapters 6 and 7). The following discussion will assume a perfect data acquisition system.

BOX 11.2

The measured fluorescence decay in each period arising from a single fluorescent species excited by a repetitive train of very short excitation pulses (δ-pulses) can be calculated. The pulses are repeated with period T. The measured signal is represented by the fluorescence response within one time period after a steady-state response is reached.

$$F(t)_{meas} = \int_0^t \left[E_\delta(t')\right] F_\delta(t-t')dt' = \int_0^{NT+\Delta t}\left[F_0 \sum_{n=0}^{n=N}\delta(t'-nT)\right] F_\delta(t-t')dt'$$

$$= F_0 \sum_{n=0}^{n=N}\exp\left(-(NT+\Delta t - nT)/\tau\right) = F_0 \exp\left(-(NT+\Delta t)/\tau\right)\sum_{n=0}^{n=N}\exp\left(nT/\tau\right)$$

$$= F_0 e^{-(NT+\Delta t)/\tau}\frac{1-e^{(N+1)T/\tau}}{1-e^{T/\tau}} = F_0 e^{-\Delta t/\tau}\frac{\left(e^{-(NT)/\tau}-e^{T/\tau}\right)}{1-e^{T/\tau}}$$

$$\simeq F_0 e^{-\Delta t/\tau}\frac{-e^{T/\tau}}{1-e^{T/\tau}} = F_0 e^{-\Delta t/\tau}\left[\frac{1}{1-e^{-T/\tau}}\right]$$

(11.4)

F_0 is the number of photons emitted per unit time or, for single molecules, the probability of detecting a photon per unit time. It is proportional to the number of molecules excited by each excitation δ-pulse. The value of F_0 depends on the absorption constant of the fluorophore and the intensity of the pulse. The measurement is made for times between $NT < t < (N+1)T$ and Δt, the time following the beginning of the last pulse, $t = (NT + \Delta t)$. N is, of course, continually increasing. $F(t)_{meas} = [F_0/(1 - \exp(-T/\tau))]\exp(-\Delta t/\tau)$ is measured in each period and is measured only for times $\Delta t \leq T$. The full decay is not recorded if $T \lesssim 5\tau$. After the system reaches a steady-state response, the signals are averaged (or the photon-counting process is repeated many times) to give the final recorded signal.

The convolution integral for a single fluorescence exponential component excited by a series of δ-pulses, commonly known as a "Dirac comb" (Bracewell 1978; Brigham 1974) (assuming the transient terms have decayed to zero), is given in Box 11.2 (Figure 11.1a). Every component of the sum in Equation 11.3 can be analogously handled.

Thus, according to Box 11.2, for repeated δ-pulses, the amplitude of the *fitted* fluorescence response (that is, the pre-exponential amplitude) will be, according to Equation 11.4 (Elder, Schlachter, and Kaminski 2008),

$$\text{Pre-exponential amplitude} = F_0 \big/ \left(1-\exp\left(-T/\tau\right)\right) \tag{11.5}$$

The pre-exponential amplitude is larger than the amplitude due to an isolated single δ-pulse of magnitude F_0; that is, only when $T > 5\tau$ does the signal decay as $F_0\exp(-\Delta t/\tau)$; the latter is the response to a *single* short pulse. If $T < 5\tau$, there is some excited state left from the previous pulses, raising the pre-exponential amplitude of the decay function. At first this may seem paradoxical, but the shorter the period is, the greater is the buildup of the excited state. When $T < 5\tau$, the complete relaxation of the excited state has not taken place; therefore, the amplitude of the *observed* fraction of the fluorescence relaxation in every

A

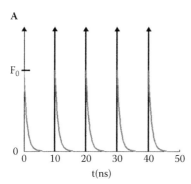

B

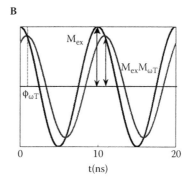

FIGURE 11.1 Excitations (black lines) at 100 MHz repetition rate and the fluorescence emissions (gray lines) of the fluorophores with 1 ns lifetime. (A) A Dirac comb excitation and the exponential time decays of the fluorescence signal. F_0 is the number of photons emitted per unit time. (B) A cosine wave excitation with modulation fraction M_{ex} and the resulting fluorescence emission, which has a modulation fraction and a phase delay.

period is smaller than the pre-exponential factor (see Equation 11.7). It should be said that T is rarely (if ever) less than 10 ns (this corresponds to a pulse frequency of 100 MHz).

As a check of the equations, we calculate the *observed steady-state (S.S.) time-independent DC signal,* which is simply the time average of steady-state signal over one period (the average of intensity × time) divided by the period, which is proportional to the DC intensity of fluorescence:

$$\text{S.S. DC fluorescence intensity} = \frac{F_0}{T}\int_0^T \exp(-\Delta t/\tau)\big/\big(1-\exp(-T/\tau)\big)d\Delta t$$

(11.6)

$$= -\frac{F_0\tau}{T}\Big[\exp(-\Delta t/\tau)\Big]_0^T\Big/\big(1-\exp(-T/\tau)\big)=\frac{F_0\tau}{T}$$

Thus, the time-independent intensity increases as T decreases. This is understandable because the intensity is the number of fluorescence photons per unit time and number of δ-pulses per unit time increase as T decreases. We have implicitly assumed that the ground state is not significantly depleted and therefore that every pulse excites the same number of fluorophores.

Also interesting is the *measured magnitude of the signal change in one period* at steady state (that is, from $\Delta t = 0$ to $\Delta t = T$, which is not the same as the pre-exponential decay). The amplitude of this signal change is

$$\text{recorded signal change over one period} = F_0\frac{\exp(-t/\tau)_{t=0}-\exp(-t/\tau)_{t=T}}{\big(1-\exp(-T/\tau)\big)}=F_0 \quad (11.7)$$

Odd as it may seem at first, this is the same as the fluorescence response to a single δ-pulse. However, this is not too surprising when we remember that the pre-exponential factor increases according to $1/(1 - \exp(-T/\tau))$ as T decreases.

If more than one fluorescent component is present with different lifetimes, then the fundamental fluorescence response is a sum of exponentials (see Equation 11.3), $F_\delta(t - t')_{meas}$ = $\sum_i F_{0,i} \exp(-(t - t')/\tau_i)$, where it is usually assumed that $t' = 0$, and every pre-exponential amplitude is related to the repetition period as Equation 11.5.

If the duration of the repetitive pulses is not short compared to the fluorescence relaxation times (that is, if they are not δ-pulses), then the shape of the pulses must be known, and one must deconvolute $E(t')$ from $F(t)_{meas}$ to obtain $F_\delta(t - t')_{meas} = \sum_i F_{0,i} \exp(-(t - t')/\tau_i)$ (see Equation 11.1) (Elder et al. 2008). The deconvoluted data are convenient for applying fitting algorithms in order to determine the species-specific pre-exponential factors, which are proportional to the radiative rate constants, the absorption coefficients, and the relative concentrations of the different components, and the lifetimes. Conversely, one can numerically convolute a known $E(t')$ with an assumed model of the fluorescence response $F_\delta(t)$ and fit the functional result of the convolution directly to the data. A very readable and critical account of numerous methods for analyzing exponential decays, with many references, is given by Istratova and Vyvenko (1999).

If the excitation pulse is not (infinitely, or at least significantly) short compared to the fluorescence relaxation times, the period of time that the excitation light is on must also be taken into account (see Marriott et al. 1991, which is an account of delayed luminescence imaging microscopy—DLIM, where a quantitative account of the buildup time of the luminescence is described; Birks 1970). This is not important when using 50–100 fs excitation pulses and recording pico- to nanosecond responses, which is now the common way of exciting the sample for measurements in the time domain.

11.3 FREQUENCY DOMAIN

Several procedures have been described to derive the equations corresponding to the measured fluorescence when the fluorescence response is expressed and analyzed as a signal in the frequency domain (Bailey and Rollefson 1953; Birks 1970; Birks and Dawson 1961; Birks and Little 1953; Clegg and Schneider; Clegg et al. 1996; Gratton, Jameson, and Hall 1984). Because frequency-domain methods are not familiar to many researchers, we present a few ways of deriving the frequency-domain fluorescence response. The basics of frequency-domain methods of data acquisition in FLIM are essentially identical to methods used in AM and FM radio communication, radar (Goldman 1948; Lawson and Uhlenbeck 1950), electric circuit analysis (Millman and Halkias 1972), dielectric dispersion (Bottcher and Bordewijk 1978; Cole and Cole 1941; Hill et al. 1969; Jonscher 1983; Raju 2003; von Hippel 1954), material relaxation processes (Mueller 1953), rheology (Kim, Lee, and Lee 2007), and acoustics and ultrasound (Feng and Isayev 2004). Similar methods were described very early by Rayleigh in his treatise *Theory of Sound* in 1894 (Rayleigh 1945).

11.3.1 Calculating $F(t)_{meas}$ Directly from the Convolution Integral

Just as in the time domain, we can directly carry out the convolution integral. We consider $E(t')$ to be a repetitive wave form that can theoretically have any shape. Because $E(t')$ is repetitive, it can be expanded as a Fourier series (see Box 11.3 and Figure 11.1b).

BOX 11.3: FOURIER SERIES EXPANSION OF AN ARBITRARY REPETITIVE EXCITATION PULSE

Because the excitation light is applied as a repetitive function in time, $E(t') = E(t' + T)$, we can expand it in a Fourier series (Bracewell 1978; Brigham 1974; Butz 2006; Byron and Fuller 1969; Davis 1993; Franklin 1949; Tolstov 1962):

$$E(t') = \sum_n \left[\frac{1}{T} \int_0^T E(t) e^{-2\pi j\left(\frac{n}{T}\right)t} dt \right] e^{j2\pi\left(\frac{n}{T}\right)t'} = \sum_n \left[\frac{1}{T} \int_0^T E(t) e^{-jn\omega_T t} dt \right] e^{jn\omega_T t'}$$

where

$$n\omega_T = 2\pi\left(\frac{n}{T}\right)$$

We have written this in the form of the complex exponential.

The real part of $E(t')$ is what corresponds to our measurement, which can be obtained using a Euler identity (Brown and Churchill 1996; Hamming 1989), $e^{j\varphi} = \cos\varphi + j\sin\varphi$. For a pure sinusoidal perturbation, the only terms of the Fourier series are the $n = 0,1$ terms. For $n = 0$ and $n = 1$, the frequency is the repetition frequency. $E(t')$ can be written as

$$E_{\omega_T}(t) = F_0 \left[1 + M_{ex} \cos\left(\frac{2\pi}{T}t\right) \right] = F_0 \left[1 + M_{ex} \cos(\omega_T t) \right]$$

F_0 is proportional to the time average number of fluorophore molecules that are excited at the instant of the excitation application—that is, at time t. Note that we have expressed $E_{\omega_T}(t)$ in terms of the excited fluorophore. This represents the instantaneous excitation of the fluorophore. F_0 is the time-independent steady-state (DC) level of fluorescence and M_{ex} is defined as the modulation fraction of the excitation light; $M_{ex} \leq 1$. Note in the preceding sum that there is always a zero frequency term, F_0, because there can be no negative intensity of the excitation light.

The frequency of each sinusoidal component of the Fourier series of $E(t')$ is an integer multiple of the frequency of the repetition period, $1/T$. All the Fourier components are sinusoidal functions of time with radial frequencies $n\omega_T = 2\pi(n/T)$, $n = 1, 2, 3....$ The coefficients of a Fourier series expansion of a particular excitation waveform furnish the relative amplitudes of the different frequency components of the excitation perturbation (Butz 2006; Dern and Walsh 1963; Tolstov 1962). Therefore, for demonstration purposes and without any loss of generality, we can derive the fluorescence response for just one particular frequency component of a sinusoidal $E(t')$. For simplicity, we choose the $n = 1$ cosine component with the basic repetition frequency $f = \omega_T/2\pi = 1/T$ and the $n = 0$ component (see Box 11.3). The result of convoluting this cosine wave with the fundamental fluorescence relaxation for a single fluorescent component is shown in Box 11.4.

The modulation fraction of the time-dependent term

$$M_{\omega_T} = \frac{1}{\sqrt{1 + (\omega_T \tau)^2}}$$

BOX 11.4

Here we calculate the convolution integral of a sinusoidal excitation wave form with the fundamental fluorescence response of a single fluorescence component. The fractional depth of excitation modulation is M_{ex} (see Box 11.3), so the excitation wave train, including the time-independent term, can be represented as

$$E_{\omega_T}(t) = F_0\left[1 + M_{ex}\cos\left(\frac{2\pi}{T}t\right)\right]$$

First, we calculate the convolution integral of the $\cos\left(\frac{2\pi}{T}t\right)$ part of $E_{\omega_T}(t)$,

$$F(t)_{meas} = \int_0^t M_{ex}F_0\cos\left(\frac{2\pi}{T}t'\right)F_\delta(t-t')dt' = \int_0^t M_{ex}F_0\cos(\omega_T t')\exp\left(-(t-t')/\tau\right)dt'$$

$$= \frac{1}{2}\int_0^t M_{ex}F_0\left(\exp(j\omega_T t') + \exp(-j\omega_T t')\right)\exp\left(-(t-t')/\tau\right)dt'$$

where we have used $\cos\varphi = (1/2)(e^{j\varphi} + e^{-j\varphi})$.
By carrying out the integrations, the following can be derived:

$$F(t)_{meas} = M_{ex}F_0\frac{\tau}{1+j\omega_T\tau}e^{j\omega_T t} = M_{ex}F_0\tau e^{j\omega_T t}\frac{1}{\sqrt{1+(\omega_T\tau)^2}}e^{-j\tan^{-1}\omega_T\tau} = M_{ex}F_0\tau e^{j\omega_T t}M_{\omega_T}e^{-j\phi_{\omega_T}}$$

We have defined the demodulation factor of fluorescence, M_{ω_T}, and the phase delay, ϕ_{ω_T} of the $n = 1$ (basic repetition frequency) component, $\omega_T \equiv 2\pi/T$. The real part of this expression is the final form of the time-dependent term:

$$\left[F(t)_{meas}\right]_{real} = M_{ex}F_0\tau M_\omega\cos(\omega_T t - \phi_{\omega_T}) \tag{11.8}$$

The time average of the $\cos(\omega_T t - \phi_{\omega_T})$ signal in Equation 11.8 is, of course, zero. The time-independent steady-state DC contribution is

$$\int_0^t F_0\exp\left(-(t-t')/\tau\right)dt' = \tau F_0 e^{-t/\tau}\left[e^{t'/\tau}\right]_0^t = \tau F_0 \tag{11.8'}$$

Therefore, the total signal, Equation 11.8 + Equation 11.8', is

$$\left[F(t)_{meas}\right]_{real} = F_0\tau\left(1 + M_{ex}M_\omega\cos(\omega_T t - \phi_{\omega_T})\right) \tag{11.8''}$$

and the phase delay $\phi_{\omega_T} = \tan^{-1}\omega_T\tau$ of the fluorescence signal are related to the experimentally measured parameters (for a single component, the τs are identical); see Chapter 12. Note that, in deriving Equation 11.8, we have assumed that the phase of the excitation sinusoid $\varphi_{ex} = 0$. If this is not the case, then the result should be written as

$$\left[F(t)_{meas}\right]_{real} = F_0\tau\left[1 + \frac{M_{ex}}{\sqrt{1+(\omega_T\tau)^2}}\cos\left(\omega_T t - \phi_{ex} - \phi_{\omega_T}\right)\right]$$

$$= F_0\tau\left[1 + M_{ex}M_{\omega_T}\cos\left(\omega_T t - \phi_{ex} - \phi_{\omega_T}\right)\right]$$

(11.9)

If there is only a single fluorescence component with a single lifetime, then the lifetimes (as in Box 11.4) determined from the measurement of either M_{ω_T} or ϕ_{ω_T} are the same. If more than one lifetime component contributes to the fluorescence signal, the two parameters show different values of τ (see Equation 11.3).

Also note that the amplitude of the frequency-domain expression has the factor $F_0\tau$ (Equations 11.8 and 11.9), whereas the time-domain amplitude (Equations 11.4 and 11.5) just has the term F_0. Also, the expression for the steady state fluorescence intensity (Equation 11.6) has the term $F_0\tau/T$. In both the frequency domain and the steady state fluorescence intensity, the signal is averaged over the time course of the repetition period, and this averaging introduces the factor τ (essentially from the integration). The time-domain measurement measures individual time points following the excitation pulse, and then each time point is averaged independently over all the pulses; therefore, the time-domain expression does not pick up the factor τ. For multiple lifetimes, this factor is important because it directly affects the contribution of a fluorescence component to the intensity.

Exciting the sample with a sinusoidal modulated light source is a common way to carry out frequency-domain FLIM. In the next section we demonstrate how the short-pulse time-domain signal is correlated to the corresponding frequency domain.

11.3.2 Calculating $F(t)_{meas}$ from the Finite Fourier Transform of the Repetitive δ-Pulse Result

We can also calculate the coefficients of the Fourier series expansion of $F(t)_{meas}$ given in Equation 11.4, which is the fluorescence time response to a repetitive series of δ-pulses see (Box 11.5). This will give us the frequency representation of the fluorescence response to a series of δ-pulses (Valeur 2005).

Notice that the time-independent term (Equation 11.10′) is the same as that calculated directly from the convolution with the repetitive δ-pulse (Equation 11.6). This is expected. On the other hand, the expression for an $n = 1$ frequency representation (finite Fourier transform) of the fluorescence response to repetitive δ-pulses of Box 11.3, which is given by Equation 11.10, differs by a factor of $1/T$ from Equation 11.8.

Equation 11.8 is calculated by convoluting the fundamental fluorescence response $F_\delta(t - t')$ with $\cos(2\pi/Tt')$. That is, in Box 11.4, throughout every period the excitation light varies continuously as a sinusoidal function; whereas in Box 11.5 we are dealing with a repetitive δ-pulse. (We have assumed for the comparison the same maximum F_0 value in both cases; realistically, this will not be the case.) Thus, the time-averaged time signal in the case of the repetitive δ-pulse is reduced by the factor of $1/T$ compared to the repetitive $\cos(\omega_T t)$ excitation wave. Of course, the coefficients of the $n = 1, 2, 3\ldots$ Fourier components

BOX 11.5

The nth Fourier coefficient (the *time-dependent term* with the repetition frequency, $\omega = 2\pi/T$, of the fluorescence response, $F(t)_{meas}$, to a repetitive series of δ-pulses, given in Box 11.2) is

$$\frac{1}{T}\int_0^T \frac{F_0}{\left(1-\exp\left(-T/\tau\right)\right)} e^{-\Delta t/\tau} e^{-2\pi jn\frac{1}{T}\Delta t}\,d\Delta t = \frac{1}{T}\frac{F_0}{\left(1-\exp\left(-T/\tau\right)\right)}\int_0^T e^{-\left(\frac{1}{\tau}+2\pi jn\frac{1}{T}\right)\Delta t}\,d\Delta t$$

$$=\frac{1}{T}\frac{-\tau F_0}{\left(1-\exp\left(-T/\tau\right)\right)}\frac{1}{1+2\pi jn\left[\frac{\tau}{T}\right]}\left[\exp\left(-\left(\frac{1}{\tau}+2\pi jn\frac{1}{T}\right)\Delta t\right)\right]_0^T$$

$$=\frac{1}{T}\frac{-\tau F_0}{\left(1-\exp\left(-T/\tau\right)\right)}\frac{1}{1+2\pi jn\left[\frac{\tau}{T}\right]}\left[\exp\left(-\left(\frac{1}{\tau}+2\pi jn\frac{1}{T}\right)T\right)-1\right] \qquad (11.10)$$

$$=\frac{1}{T}\frac{F_0\tau}{1+2\pi jn\left[\frac{\tau}{T}\right]}=\frac{1}{T}\frac{F_0\tau}{1+jn\omega_T\tau}=\frac{1}{T}\frac{F_0\tau\left(1-jn\omega_T\tau\right)}{1+\left(n\omega_T\tau\right)^2}$$

$$=\frac{1}{T}\frac{F_0\tau}{\sqrt{1+\left(n\omega_T\tau\right)^2}}\exp\left(-j\tan^{-1}n\omega_T\tau\right)=\frac{1}{T}F_0\tau M_{n\omega_T}\exp\left(-j\phi_{n\omega_T}\right)$$

Therefore, the time dependence of the $n=1$ term is $\frac{1}{T}F_0\tau M_{\omega_T}\exp\left(-j\left(\phi_{\omega_T}+\omega_T\Delta t\right)\right)$.

We need the real part of the last expression, $\left[1/T\right]F_0\tau M_{\omega_T}\cos\left(-\left(\omega_T\Delta t+\phi_{\omega_T}\right)\right)$.

As in Box 11.4, $M_{\omega_T}=\dfrac{1}{\sqrt{1+\left(\omega_T\tau\right)^2}}$, and $\phi_{\omega_T}=\tan^{-1}\omega_T\tau$.

The coefficient of the *time-independent* Fourier term is

$$\frac{1}{T}\int_0^T \frac{F_0}{\left(1-\exp\left(-T/\tau\right)\right)}e^{-\Delta t/\tau}\,d\Delta t=\frac{1}{T}\frac{-F_0\tau}{\left(1-\exp\left(-T/\tau\right)\right)}\left[e^{-\Delta t/\tau}\right]_0^T=\frac{F_0\tau}{T} \qquad (11.10')$$

(the measured amplitudes of these components) are also reduced by $1/T$, as we would expect. As T increases, many fewer photons are emitted per unit time for the case of repetitive δ-pulses.

The preceding comparison (see Boxes 11.3 and 11.5) emphasizes the equivalence of representing time-dependent fluorescence signals in terms of a temporal progress curve of a series of exponential time decays or in terms of the phase and modulation fractions of the corresponding Fourier frequency components. At first sight, there is no obvious continuous sine wave characteristic of a pulse train of δ-pulses. However, using Fourier techniques, we can clearly represent the fluorescence signal in terms of modulation fractions and phase delays of the different frequency components of the time-domain fluorescence response.

This is, of course, true for any repetitive excitation wave form because any repetitive wave form can be expanded in a Fourier series. We have deliberately chosen the extreme case of delta function excitation pulses for demonstration and comparison in order to emphasize this correspondence. For those acquainted with Fourier techniques, this is obvious; the frequency representation of the signal is just the Fourier transform of the time-domain signal or, equivalently, the exponential time representation of the signal is the finite Fourier transform of the frequency-domain signal. The usual frequency-domain representation assumes a sinusoidal excitation perturbation as in Box 11.4.

11.3.3 Calculating the frequency response from the Convolution Theorem of Fourier Transforms

The convolution theorem of Fourier transforms says that the Fourier transform of the convolution of two functions, $FT\{f_1 * f_2\}$, is the multiplication of the Fourier transforms of the two separate functions; that is, $FT\{f_1 * f_2\} = FT\{f_1\}FT\{f_2\}$. In the previous section, we calculated the $FT\{f_1 * f_2\}$ directly in Box 11.5 and showed that, except for a factor of $1/T$, the signal for $n = 1$ was the same as that calculated directly from the convolution of a sinusoidal

$$E_\omega\left(t'\right) = F_0\left[1 + M_{ex}\cos\left(\frac{2\pi}{T}t'\right)\right]$$

with $F_\delta(t - t')$ (Box 11.4).

The factor of $1/T$ in Equation 11.10′ is not in Equation 11.8′ simply because the sinusoidal perturbation $E_\omega(t')$ is continuous during the whole period, T, and therefore for such an excitation the signal strength does not decrease as the period length is increased. For completeness and to demonstrate the power and convenience of the convolution theorem for calculating the frequency-domain expression, we show in Box 11.6 that both ways of calculating the frequency representation—$FT\{f_1 * f_2\}$ or $FT\{f_1\}FT\{f_2\}$—result in the same expressions for the frequency domain as in Box 11.5, when $E_\omega(t')$ is a series of δ-pulses, as in Box 11.4.

Comparison of the results in Box 11.6 with Equations 11.10 and 11.10′ shows that the frequency-domain expressions for the infinite comb of δ-pulses using $FT\{f_1\}FT\{f_2\}$ (Box 11.6) is identical to that of $FT\{f_1 * f_2\}$ (Box 11.5). The convolution theorem is a very important relationship because it simplifies greatly the calculation of the frequency-domain expression representing $F(t)_{meas}$ for any form of $E_\omega(t')$. One can avoid long convolution calculations (Equation 11.1) by calculating the frequency representation of $F(t)_{meas}$ using the convolution theorem.

Equation 11.11″ shows that a train of very short pulses (δ-pulses) can also be used to obtain the higher harmonic frequencies; that is, one is not limited to only one frequency. All the higher frequencies are also available for analysis. With the right acquisition method and electronics, one can obtain a complete frequency dispersion (all frequencies, $2\pi k/T$, $k = 0, 1, 2, 3, 4,\ldots$) of the fluorescence response (Gratton et al. 1984, 2003; Lakowicz 1999; Valeur 2005).

BOX 11.6

The expression for the Fourier transform of the series of δ-pulses is

$$FT\left\{F_0\sum_{n=-\infty}^{n=\infty}\delta(t'-nT)\right\}=\frac{F_0}{T}\sum_{k=-\infty}^{k=\infty}\delta\left(\omega-\frac{2\pi k}{T}\right)$$

This equality can be shown as follows. $\sum_{n=-\infty}^{n=\infty}\delta(t'-nT)$ is expanded in a Fourier series:

$$\sum_{n=-\infty}^{n=\infty}\delta(t'-nT)=\frac{1}{T}\sum_{k=-\infty}^{k=\infty}\exp\left(\frac{j2\pi kt}{T}\right)$$

Therefore, the Fourier transform of the series of δ-pulses is

$$\int_{-\infty}^{+\infty}\frac{F_0}{T}\sum_{k=-\infty}^{k=\infty}\exp\left(\frac{j2\pi kt}{T}\right)\exp(-j\omega t)dt=\frac{F_0}{T}\sum_{k=-\infty}^{k=\infty}\delta\left(\omega-\frac{2\pi k}{T}\right) \qquad (11.11)$$

This final equality follows from the orthogonality of the sinusoidal functions.

The FT of the fundamental fluorescence response of a single fluorescence component excited with a series of δ-pulses is

$$FT\left\{e^{-t/\tau}\right\}_{\text{for } t>0}=\frac{\tau}{1+j\omega\tau} \qquad (11.11')$$

Multiplying the two Fourier transforms (Equations 11.11 and 11.11′) together results in

$$FT\{f_1\}FT\{f_2\}=\frac{1}{T}\sum_{k=-\infty}^{k=\infty}\frac{F_0\tau}{1+j\omega\tau}\delta\left(\omega-\frac{2\pi k}{T}\right)=\frac{1}{T}\sum_{k=-\infty}^{k=\infty}\frac{F_0\tau}{1+j\left(k\frac{2\pi}{T}\right)\tau} \qquad (11.11'')$$

The $k = 0$ term is $F_0\tau/T$ and the $k = 1$ term (for the fundamental repetition frequency $\omega = 2\pi/T$) is

$$\frac{1}{T}\frac{F_0\tau}{1+j\omega\tau}=\frac{1}{T}\frac{F_0\tau}{1+j\frac{2\pi}{T}\tau}$$

All terms of the sum in Equation 11.11″ with $k > 1$ represent the higher frequency harmonics.

11.4 ANALYSIS OF THE MEASURED DATA, $F(T)_{MEAS}$, AT EVERY PIXEL

When the signal in the time domain is analyzed, one often derives $F_\delta(t)_{meas}$ from the acquired data $F(t)_{meas}$ (Equation 11.3). This is carried out by deconvoluting $E(t')$ from $F(t)_{meas}$ to derive $F_\delta(t)_{meas}$. Then, a variety of regression algorithms are employed to fit $F_\delta(t)_{meas}$ to a series of decaying exponentials. The regression analysis can be carried out effectively the same way that analysis of single-channel (cuvette) fluorescence lifetime data has been handled for many years (Brochon et al. 1990; Cundall and Dale 1983; Grinvald and Steinberg 1974; Istratova and Vyvenko 1999). Many different numerical algorithms and stringent criteria for statistical reliability have been extensively worked out for single-channel lifetime data in order to extract accurate lifetimes and relative amplitudes. Strict limits can be set on the quality of the experimental data, such as the number of photons collected or the intensity of fluorescence. Essentially, the same statistical criteria for balancing signal-to-noise requirements and reliability apply to all the different modes of FLIM.

In the frequency domain, the signal does not have to be deconvoluted from the excitation wave form. As described earlier, the repetitive signal is parsed into separate frequency Fourier components, and each frequency is an integer multiple of the repetition frequency. Each frequency component is orthogonal to all others and can be analyzed separately. In general, the fitting procedures available for analyzing the sinusoidal components are very fast because noniterative digital Fourier techniques can be employed; however, other techniques are also used, especially if the time or phase acquisition points are not evenly spaced. This convenience arises because sines and cosines are orthogonal to each other, whereas exponential decays with real exponents are highly nonorthogonal (but see the section on Chebyshev polynomial fitting). Chapters 12 and 13 explain in detail some specific methods of fitting data; in this chapter, we mention some major points in order to compare the time and frequency domains.

11.5 REMARKS ABOUT SIGNAL-TO-NOISE CHARACTERISTICS OF TIME- AND FREQUENCY-DOMAIN SIGNALS: COMPARISON TO SINGLE-CHANNEL EXPERIMENTS

The general attributes and signal-to-noise requirements are identical for the time and frequency domains. For simplicity, we compare the signal-to-noise characteristics of single-channel and FLIM measurements with time-domain measurements. But, the analysis in this section applies to frequency domain as well as the time domain (we have shown that the two methods are essentially equivalent). For the calculation, the method of measurement is assumed to be photon counting.

Two prominent differences between FLIM and single-channel data (cuvette measurements), especially when acquiring data by photon counting, are the signal-to-noise level of the data and the number of separate time records recorded. We make a simple calculation to emphasize these differences and to stress the challenges involved for making accurate FLIM measurements (see also Duncan et al. 2004). FLIM records a time series for every pixel, and FLIM images have up to $N_{pixel} = 10^6$ pixels. N_{count} is defined as the number of photons at the maximum of the decay that must be recorded for every time series in order to achieve the desired signal-to-noise level. For high accuracy and discrimination, N_{count} may be required to be 10^5 in order to differentiate two lifetimes; this is normal for single-channel (cuvette) experiments (Brochon et al. 1990; Cundall and Dale 1983; Grinvald and Steinberg 1974; Istratova and Vyvenko 1999).

The number of time intervals recorded at each pixel can be 10^3; this differs appreciably depending on the instrumentation, but we assume this number—although for single-channel experiments, this number is often up to 10^4. In order to achieve a very high signal-to-noise level, one must average over longer times to realize these high N_{count} counts per pixel. If the accuracy requires $N_{count} = 10^5$, then we require the following separate light pulses per pixel (each count is a detected photon and requires a separate excitation pulse):

$$10^5 \int_0^{1000} \exp\left(-x/200\right)dx = -200 \cdot 10^5 \left[\exp\left(-x/200\right)\right]_0^{1000} = 2 \cdot 10^7$$

For photon counting, only one photon can be detected per pulse for each separate measurement channel. Assuming that a photon is detected for every excitation pulse (for single-channel experiments, one photon is observed per 100 pulses in order to avoid pulse overlap), for $N_{pixel} = 10^6$, this is a total of 2×10^{13} excitation pulses, minimum, if every pixel is measured separately. With modern hardware, the pulse rate can be approximately 10^8 per second. Thus, a total experiment would take 2×10^5 seconds, 3,333 minutes, or 55 hours. This assumes that a single photon is detected for every pulse; if the photon detection efficiency is 1% per pulse, then the time of acquisition would be 100 times longer.

For this calculation, we have chosen an extreme case with the stringent precision usually set for a single-channel lifetime measurement. We have assumed that we want this high signal-to-noise level at every pixel, in order to achieve the highest spatial resolution. Obviously, the procedure just outlined is not applicable for a reasonable FLIM measurement. But the point is that the time of acquisition for such high temporal and spatial accuracy, together with acquiring the data sequentially at every pixel, would be exceptionally long for a FLIM measurement.

Analyzing the data of a time-domain lifetime experiment is typically carried out by fitting the time progress of the relaxation with a discrete sum of exponential functions or a continuous distribution of exponentials. This type of numerical regression requires the proposition of a model; that is, one must decide whether the signal consists of discrete exponentials, or, if a distribution is assumed, the shape of the distribution must be chosen. The particular exponential model selected depends on the underlying molecular dynamics and interactions and on the experience of the investigator.

The parameters to be determined are typically the fluorescence lifetimes and the fractional amplitudes, but they can be directly the particular molecular photophysical model chosen by the researcher. The parameters are usually ascertained by iteratively minimizing the variance between the data and the fit. The available standard iterative numerical methods are quite accurate, but can be time consuming. Such detailed analyses are, by and large, the intent of single-channel cuvette-type experiments. However, for FLIM data with up to 10^6 pixels, the analysis time can become burdensome.

One of the hallmarks of fluorescence measurements is the sensitivity of the lifetime to the environment of the fluorophore. Unless the sample contains only one or very few fluorescing components and the environment is homogeneous (such as in a cuvette experiment), the interpretation of a fluorescence response as a series of a few individual lifetime components that can be determined with high accuracy is dubious at best. This is the case for most FLIM measurements made on biological cells or in tissue. Therefore, it is essential to remember the limitations for making statistically accurate interpretations of FLIM data in order not to overinterpret the data.

11.6 FLIM EXPERIMENTS: CHALLENGES, ADVANTAGES, AND SOLUTIONS

The discussion in the previous section presents clearly the challenges for implementing FLIM experiments. Great numbers of detected photons are required if highly accurate lifetime determinations capable of determining the number of lifetime components and discriminating closely lying lifetimes are desired. In addition, if the necessary spatial resolution is on the order of individual pixels, each pixel in the area of interest must have a sufficient number of counts. There is no essential obstacle against acquiring and analyzing every pixel of FLIM data in the same manner and to the same accuracy as single-channel cuvette experiments; however, the time required to do this is usually prohibitively long. If the highly accurate determinations of fluorescence lifetimes described in the previous section are necessary for answering specific, detailed questions of some molecular mechanism, probably FLIM is not the right choice unless the image data are also important.

On the other hand, usually the primary objective of a FLIM investigation is quite different from single-channel measurements. One is still interested in determining the lifetime-resolved information as accurately and robustly as possible. However, in FLIM, the structure of some object (e.g., in a cell) is investigated, and the morphology of the image is a major concern. Thus, the scientific questions asked are analogous to normal intensity fluorescence imaging; one is interested in correlating the spectroscopic information with different locations in the imaged object. By acquiring lifetime-resolved fluorescence in addition to the intensity, considerable quantitative information is available that aids in the identification and differentiation of fluorophores.

If the overall rate of fluorescence decay (or something equivalent—see, for instance, the polar plot discussion) can be determined accurately and reproducibly, changes in this *average rate* with time can be followed at different locations. Such high signal-to-noise ratios are often not necessary to determine spatial differences, and it may not even be of major interest to determine the actual lifetimes of the individual components. It may also be that the interesting lifetimes are far enough apart so that such high signal-to-noise ratios are not required to differentiate multiple lifetimes and determine their fractional amplitudes, even if the separate decays are not pure exponentials. Additional quantitative information is often available from FLIM (even with relatively few detected photons per pixel) that is not easily obtained from intensity measurements. For instance, this is the case for Förster resonance energy transfer (FRET).

If the lifetime of a donor fluorophore is known in the absence of an acceptor, it is easy to locate positions in the image with faster lifetime decay, indicating increased efficiency of energy transfer. With FLIM, because lifetimes are independent of the concentration, complicated control experiments and multiple wavelengths (which may be difficult to align in the image) are not required. FRET is probably the most common application of FLIM, and in many cases FLIM is the most reliable determination of the efficiency of FRET. FLIM is also an excellent way to discriminate objects in an image that have different lifetimes but similar emission wavelengths; that is, the image contrast is improved. A common application of FLIM is the elimination of background fluorescence, such as intrinsic fluorescence or unbound fluorophore where the bound and unbound fluorophores exhibit different lifetimes. Many variations have been developed for acquiring and analyzing FLIM data. Chapters 4, 7, 12, and 13 describe details of these methods.

11.7 HOW FLIM CIRCUMVENTS THE DATA DELUGE

Several obvious ways for FLIM measurements to avoid time-consuming data acquisition and data analysis procedures are to

- collect data at lower signal-to-noise levels (that is, average fewer excitation repetitions; Chapter 12);
- collect fewer time points per pixel (in either the time or frequency domain; Chapter 4);
- collect fewer pixels or, equivalently, average many pixels into superpixels (called "binning"; Chapters 4, 5, and 12);
- avoid time-consuming iterative fitting procedures (accomplished in the frequency domain automatically; this chapter);
- simplify the data form before fitting (such as taking the logarithm), which will avoid iterative regression (Chapter 4);
- use fitting procedures that are not iterative (such as the Chebyshev method; this chapter);
- analyze the data directly—not trying to determine the lifetimes per se, but rather using equivalent descriptions of the fluorescence response (such as the "polar plot" described in this chapter);
- vary parameters such as wavelengths of excitation or emission to gain additional information (Chapter 9); and
- assume that the fluorescence relaxation is identical throughout the image (or parts of the image) and fits the data globally (Chapter 13).

Indeed, FLIM enthusiasts have come up with many variations that provide insight into particular samples, as well as simplify and perfect the analysis. Of course, as discussed in the previous section, one usually assumes that the objective of a FLIM experiment is not to provide the possibility of the highly accurate and detailed photophysical information at every pixel, which is the purview and the raison d'être of single-channel fluorescence lifetime measurements. As a matter of fact, the information sought from FLIM experiments is usually very different.

Other chapters discuss the time-domain methods of acquisition and analysis in detail. The remainder of this chapter will concentrate on some recent methods of analysis not included in the other chapters:

- the introduction of image analysis methods into FLIM analysis (wavelets and denoising);

- model-free analysis of FLIM data (polar plots);

- analyzing the fluorescence response of fluorophores not excited solely directly by the excitation pulse (such as observing the resultant components of excited-state reactions, e.g., the acceptor fluorescence in FRET);

- the development of noniterative fitting of the fluorescence response (the use of Chebyshev and Laguerre polynomials; Malachowski, Clegg, and Redford 2007); and

- the combination of the model-free polar plots with fluorescence emission spectra.

11.7.1 Polar Plots of Frequency-Domain Data (Model-Free Analysis)

11.7.1.1 Polar Plot Description of Fluorescence Directly Excited by Light Pulses

The polar plot (Buranachai and Clegg 2007; Buranachai et al. 2008; Colyer, Lee, and Gratton 2008; Holub et al. 2007; Redford and Clegg 2005; Redford et al. 2005) is also referred to as a "phasor plot" (Digman et al. 2007), an "A/B plot" (Clayton, Hanley, and Verveer 2004; Hanley and Clayton 2005), or as a plot of the Fourier transform pairs (Berberan-Samtos 1991). It is a way to display the frequency characteristics (modulation fraction and phase) of signals that are described in the time domain by exponential decays; that is, $F_\delta(t - t')_{meas} = \Sigma_i F_{0,i} \exp(-(t - t')/\tau_i)$ (see Equations 11.2 and 11.3).

Polar plots and similar frequency-domain characterizations of data have been used extensively to analyze and display dielectric dispersion data (Cole and Cole 1941; Hill et al. 1969; Jonscher 1983; von Hippel 1954) and calculate the frequency characteristics of electronic circuits (Millman and Halkias 1972). They are applicable to all physical systems excited with repetitive perturbations. Remember that $F_{0,i}$ (Equation 11.3) represents the fluorescence amplitude at $t = 0$ of species i, following a short pulse. $F_{0,i}$ is directly proportional to the concentration of the ith fluorescent species, its radiative rate constant, and absorption coefficient. All exponential terms contribute linearly and separately to the fundamental fluorescence response, $F_\delta(t - t')_{meas}$. Each term contributes separately and additively to the measured repetitively modulated fluorescence, $F(t)_{meas}$ (see Box 11.4). In Box 11.7, we derive the expression for the measured normalized fluorescence signal from a solution with multiple fluorophores, all excited by the same excitation wave form.

Equation 11.15 (see Box 11.7) is the general normalized expression for multiple discreet fluorescence components excited directly by the repetitive excitation wave form. We have only considered the fundamental frequency component, which is the repetition frequency, $\omega_T = 2\pi/T$. Higher harmonics, in case the excitation is not a pure sinusoid, are just the higher frequency Fourier components. For each of the higher harmonics, the only change to Equation 11.15 is to replace $\omega_T = 2\pi/T$ by $n\omega_T = n2\pi/T$. We will consider only the fundamental frequency component, but for generality we use the general frequency designation, ω. Each of the terms in the summation of the last expression in Equation 11.15 represents the contribution to the normalized signal of a discreet fluorescent species with a particular lifetime, τ_i (normalized to $\Sigma_i F_{0,i} \tau_i$). It is clear from this equation that every fluorescence

BOX 11.7: NORMALIZING THE FREQUENCY-DOMAIN SIGNAL

We derived in Box 11.4, Equation 11.8, the expression for a sinusoidal excitation waveform with radial frequency ω, and a *single*-exponential relaxation component:

$$F(t)_{meas} = F_0\tau\left[1 + \frac{Mex}{1+j\omega\tau}e^{j\omega t}\right] = F_0\tau\left[1 + e^{j\omega t}\frac{Mex}{\sqrt{1+(\omega\tau)^2}}e^{-j\tan^{-1}\omega\tau}\right]$$

Remember that $j = \sqrt{-1}$. For a series of exponentials, the expression for the frequency (phase and modulation fraction) mode of expressing the data is therefore simply

$$F(t)_{meas} = \left[\sum_i F_{0,i}\tau_i + \sum_i \frac{MexF_{0,i}\tau_i}{1+j\omega\tau_i}e^{j\omega t}\right]$$

$$= \left[\sum_i F_{0,i}\tau_i + e^{j\omega t}\sum_i \frac{MexF_{0,i}\tau_i}{\sqrt{1+(\omega\tau_i)^2}}e^{-j\tan^{-1}\omega\tau_i}\right] \tag{11.12}$$

Expressing this in real terms (see Box 11.4),

$$\left[F(t)_{meas}\right]_{real} = \left[\sum_i F_{0,i}\tau_i + \sum_i F_{0,i}\tau_i\frac{Mex}{\sqrt{1+(\omega\tau_i)^2}}\cos(\omega t - \phi_{\omega,i})\right]$$

$$= \left[\sum_i F_{0,i}\tau_i + \sum_i F_{0,i}\tau_iMexM_{\omega,i}\cos(\omega t - \phi_{\omega,i})\right] \tag{11.13}$$

The first sum, $\sum_i F_{0,i}\tau_i$, is the time-independent (steady-state intensity) signal, $F_{meas,ss}$. The fractional contribution of species i to the measured steady-state intensity is

$$\left[\text{fractional steady-state intensity of species i}\right] \equiv \alpha_i = \frac{F_{0,i}\tau_i}{\sum_i F_{0,i}\tau_i} \tag{11.14}$$

where, of course, $\sum_i \alpha_i = 1$.

Thus, if we normalize the signal $F(t)_{meas}$ to the time independent steady-state intensity, the normalized signal in complex notation is

$$\frac{F(t)_{meas}}{F_{meas,ss}} = 1 + Mex\sum_i\frac{\alpha_i}{1+j\omega\tau_i}e^{j\omega t} = 1 + e^{j\omega t}Mex\sum_i\frac{\alpha_i}{\sqrt{1+(\omega\tau_i)^2}}e^{-j\tan^{-1}\omega\tau_i}$$

$$\tag{11.15}$$

$$= 1 + e^{j\omega t}Mex\sum_i\frac{\alpha_i}{\sqrt{1+(\omega\tau_i)^2}}\left[\cos(\phi_{i,\omega}) + j\sin(\phi_{i,\omega})\right]; \quad \phi_{i,\omega} = \tan^{-1}\omega\tau_i$$

lifetime component has the exact same frequency as the excitation modulation. The time dependent term (proportional to $e^{j\omega\tau}$) is divided by the modulation fraction of excitation, Mex. Then the modulation fraction and the phase of each fluorescence component are represented by the following (see Equation 11.15 and Box 11.4):

$$\frac{\alpha_i}{1+j\omega\tau_i} = \frac{\alpha_i}{\sqrt{1+\left(\omega\tau_i\right)^2}}\left[\cos\left(\phi_{i,\omega}\right)+j\sin\left(\phi_{i,\omega}\right)\right] \tag{11.16}$$

$\phi_{i,\omega} = \tan^{-1}\omega\tau_i$. $\dfrac{\alpha_i\cos\left(\phi_{i,\omega}\right)}{\sqrt{1+\left(\omega\tau_i\right)^2}}$ is the ith fluorescence component *in phase* with the sinusoi-

dal excitation, and $\dfrac{\alpha_i\sin\left(\phi_{i,\omega}\right)}{\sqrt{1+\left(\omega\tau_i\right)^2}}$ is the ith fluorescence component $\pi/2$ *out of phase* with the excitation.

11.7.1.1.1 Polar Plot of a Single Lifetime Component   First, we assume that only one fluorescence component is present; then, $\alpha_i = \alpha = 1$. This is the case of Equation 11.8. For a single component,

$$\frac{1}{1+j\omega\tau} = \frac{1}{\sqrt{1+\left(\omega\tau\right)^2}}\left[\cos\left(\phi_\omega\right)+j\sin\left(\phi_\omega\right)\right] = M_\omega\left[\cos\left(\phi_\omega\right)+j\sin\left(\phi_\omega\right)\right] \tag{11.17}$$

Equation 11.17 is just a complex number, $a + jb$, and complex numbers can be plotted on an Argand diagram, or polar axis (Brown and Churchill 1996). That is, the real part, $a = M_\omega\cos(\phi_\omega)$, is the value plotted on the x-axis, and the complex part, $b = M_\omega\sin(\phi_\omega)$, is the value plotted on the y-axis. It is well known and easy to show that, for all values of τ and ω, *every point* of Equation 11.17 lies on a universal semicircle (universal meaning for all τ and ω) with the center at (x = 0.5, y = 0) and radius 0.5. This plot is called the "polar plot" (Figure 11.2).

According to Chapter 5, the measured time varying fluorescence signal is always a sinusoidal signal with the same frequency as the excitation, and the modulation fraction $M_{meas,\omega}$ and the phase $\phi_{meas,\omega}$ of the measured normalized fluorescence relative to the excitation modulation depth (Mex) and phase (ϕex) can be represented in complex form as

$$F_{meas,\omega} = M_{meas,\omega}\left[\cos\left(\phi_{meas,\omega}\right)+j\sin\left(\phi_{meas,\omega}\right)\right] \tag{11.18}$$

Comparing Equations 11.17 and 11.18, we see that, for a single component, $M_{meas,\omega} = M_\omega$ and $\phi_{meas,\omega} = \phi_\omega$. Therefore, if we make a polar plot of the measured data, $x_{meas} = M_{meas,\omega}\cos(\phi_{meas,\omega})$ against $y_{meas} = M_{meas,\omega}\sin(\phi_{meas,\omega})$, the points must lie on the semicircle of the polar plot if only one fluorescence component is present. This is diagnostic for a signal from a single fluorescence lifetime component.

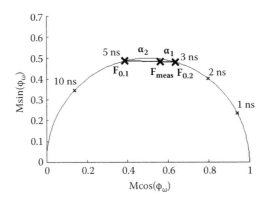

FIGURE 11.2 The polar plot of fluorescence lifetime components by a 40 MHz sinusoidal excitation. Single lifetime components lie on the semicircle; the shorter lifetimes are on the right side and the longer lifetimes on the left. The measured two lifetimes component F_{meas} lies inside the semicircle, and its fractional relative distances between the two corresponding fluorescence components $F_{0,1}$ and $F_{0,2}$ are α_2 and α_1, respectively.

The polar plot of frequency-domain data is made directly from the measured data, $M_{meas,\omega}$ and $\varphi_{meas,\omega}$. It is not necessary to fit the data to determine the lifetimes. The plot is visually diagnostic and model independent. If the frequency or the lifetime is varied, all the points of such an experiment will lie on the semicircle if only a single component is present. When the frequency is known, the lifetime can simply be read off the plot from the position on the semicircle. In this sense, the plot is said to be universal.

11.7.1.1.2 Polar Plot with Multiple Lifetime Components   If multiple components are present, then we use Equation 11.15, which we rewrite as

$$\frac{F(t)_{meas}}{F_{meas,ss}} = 1 + e^{j\omega t} \text{Mex} M_{meas,\omega} \left[\cos\left(\phi_{meas,\omega}\right) + j\sin\left(\phi_{meas\omega}\right) \right]$$

(11.19)

$$= 1 + e^{j\omega t} \text{Mex} \sum_i \frac{\alpha_i}{1 + j\omega\tau_i} = 1 + e^{j\omega t} \text{Mex} \sum_i \alpha_i M_{\omega,i} \left[\cos\left(\phi_{i,\omega}\right) + j\sin\left(\phi_{i,\omega}\right) \right]$$

where we remember that

$$\sum_i \alpha_i = 1, \ M_{\omega,i} = 1 \Big/ \sqrt{1 + \left(\omega\tau_i\right)^2}$$

and

$$\phi_{i,\omega} = \tan^{-1}\omega\tau_i$$

Just as in the case of only one lifetime component (Equation 11.17), we can think of each term of the summation in Equation 11.19 as being a vector (after dividing the $e^{j\omega\tau}$ term by Mex):

$$(x_i, y_i) = \alpha_i \left(M_{\omega,i} \cos(\phi_{i,\omega}), M_{\omega,i} \sin(\phi_{i,\omega}) \right)$$

Therefore,

$$(x_{meas}, y_{meas}) \equiv M_{meas,\omega} \left[\cos(\phi_{meas,\omega}) + j \sin(\phi_{meas\omega}) \right]$$

$$= \sum_i \alpha_i M_{\omega,i} \left[\cos(\phi_{i,\omega}) + j \sin(\phi_{i,\omega}) \right] \equiv \sum_i \alpha_i (x_i, y_i)$$

(11.20)

With more than one fluorescence lifetime component, the polar plot of (x_{meas}, y_{meas}) no longer lies on the semicircle. Each *separate* vector (x_i, y_i) would lie on the semicircle (without the fractional contribution factors, α_i). However, because $\alpha_i < 1$ and $\sum_i \alpha_i = 1$, it can easily be shown (Clayton et al. 2004; Jameson, Gratton, and Hall 1984; Redford and Clegg 2005) that the resultant vector, $(x_{meas}, y_{meas}) = \sum_i \alpha_i (x_i, y_i)$, lies inside the semicircle. This is a very useful diagnostic. If the (x_{meas}, y_{meas}) points lie inside the semicircle, there must be more than a single lifetime component. Also, if the (x_{meas}, y_{meas}) points lie outside the semicircle, either the model represented by Equation 11.3 (which says that the fluorescent components are independent and are all excited directly by the repetitive excitation pulse) is not valid (e.g., observing a fluorescent product of an excited-state reaction; see Section 11.7.1.2) or an artifact is in the measurement.

An interesting case arises for two-fluorescence components. Then, $(x_{meas}, y_{meas}) = \alpha_1(x_1, y_1) + \alpha_2(x_2, y_2)$. In this case, the points (x_{meas}, y_{meas}) for all possible values of the $\alpha_{1,2}$, $x_{1,2}$, and $y_{1,2}$ must lie on a straight line between the points (x_1, y_1) and (x_2, y_2), both of which lie on the semicircle (Clayton et al. 2004; Hanley and Clayton 2005; Redford and Clegg 2005). The fractional relative distance between (x_{meas}, y_{meas}) and (x_1, y_1) is α_2, and the fractional relative distance between (x_{meas}, y_{meas}) and (x_2, y_2) is α_1. The α_i, which can be read directly from the polar plot points, are the fractional intensities of the two corresponding fluorescence components (α_i was derived earlier, $\alpha_i = \tau_i F_{0,i} / \sum_i \tau_i F_{0,i}$). If we know the lifetimes of the two individual components, from the position of (x_{meas}, y_{meas}) on the straight line we can also calculate the species fractional population of each of the two fluorescence components (this latter ratio is $F_{0,i} / \sum_i F_{0,i}$).

This can all be derived from the measurement of one point on the polar plot without ever performing an iterative fit to relaxation data to determine the corresponding intensities. Also, if we know one lifetime, say τ_1 (and therefore (x_1, y_1)), and we determine from experiment (x_{meas}, y_{meas}), we can determine (x_2, y_2) by simply extending the straight line to the intersection with the semicircle, and this will give us τ_2. Of course, we also automatically have α_1 and α_2.

11.7.1.2 Polar Plot of Fluorescence from a Product Species of an Excited-State Reaction

The previous section was a general discussion of polar plots for the case where every contributing fluorescent component is excited directly by the excitation light. In this case, the convolution integral of Equation 11.1 predicts the time dependence of the fluorescence relaxation. However, if a fluorophore in an excited state (which is excited by the excitation pulse) reacts with another molecule in the ground state to produce an excited-state product, and if we observe the emission from the excited product of the excited-state reaction, then Equation 11.1 must be extended. This is exactly what takes place in FRET. The energy from the donor molecule, D, which is in the excited state, is transferred nonradiatively to a nearby acceptor, A, in the ground state. We will use this as an example of an excited-state reaction. A detailed description of FRET can be found in Chapter 2, and the general rate equations leading to the lifetime of fluorescence are discussed in Chapter 1.

If we observe the donor fluorescence in a FRET experiment, then the fluorescence time dependent relaxation is described by Equations 11.2 and 11.3; that is, all observed fluorescence components progress in time as decaying exponentials. The fluorescence signal expressed in terms of the frequency domain will behave as Equations 11.12–11.14. The lifetime of the donor is simply decreased by the additional FRET pathway of deactivation. However, the donor still decays exponentially.

On the other hand, the excitation of the acceptor in FRET takes place through the transfer of energy from the donor, rather than through direct excitation by the excitation light pulse. Therefore, if the acceptor fluorescence is observed, the time dependence of the acceptor fluorescence can no longer be expressed as decaying exponentials, following excitation through a light pulse. The fundamental fluorescence response of the acceptor (which is a simple exponential decay) is convoluted with the donor population decay through the energy transfer. The probability of transfer per unit time (the rate) is proportional to the population concentration of the excited donor, which itself is decaying exponentially in time.

Theoretical treatments for observing the time-dependent fluorescence signal of a fluorescent product species of an excited-state reaction have been known for a longer time (see the book by Birks, 1970, for a general treatment of the time- and frequency-domain experiments). Several accounts of expressions are available in the literature specifically for the time domain (Berberan-Santos and Martinho 1990; Birks 1968) and for the frequency domain (Lakowicz and Balter 1982a, 1982c). We show a short derivation that is suitable for our discussion of the polar plot (see also Forde and Hanley 2006).

The time dependence of the acceptor fluorescence (the excited-state reaction product) is derived in Box 11.8 for the usual case of sinusoidal excitation and a single component of donor and a single rate of energy transfer. These equations can easily be generalized. We calculate the double convolution for a sinusoidal modulation of the excitation light, Equation 11.21, because we want to discuss the behavior of the polar plot.

As expected, Equations 11.21–11.23 show that the measured fluorescence signal of the acceptor can be expressed as a modulated sinusoidal signal at the same frequency as the excitation light. The total modulation is $M_{ex}M_{DA}M_A$, and the total phase delay is $(\varphi_{DA} + \varphi_A)$ (Equation 11.23). It is informative to compare Equations 11.22 and 11.23 to Equation 11.9, assuming that $\varphi_{ex} = 0$ (that is, we set the phase of the excitation as phase zero). For convenience, we reproduce Equation 11.9 here with

$$\phi_{ex} = 0 : \left[F(t)_{meas} \right]_{real} = F_0 \tau \left[1 + M_{ex} M_{\omega_T} \cos\left(\omega_T t - \phi_{\omega_T} \right) \right]$$

M_{meas} and φ_{meas} are defined as the measured modulation and phase. We see that the effect of the successive nature of the excitation from the excited donor to the acceptor,

BOX 11.8

We express the excitation pulse train in complex notation as $E\left(t''\right) = 1 + M_{ex}e^{j\omega_T t''}$. M_{ex} is the fractional modulation of the excitation. The fundamental fluorescence decay of the donor (in the presence of the acceptor) is $F_\delta(t' - t'')_{DA} = F_{0,DA}\exp(-(t' - t'')/\tau_{DA})$. τ_{DA} is the fluorescence lifetime of the donor in the presence of the acceptor, which is shorter than the donor lifetime in the absence of the acceptor, τ_D. $t' - t''$ is the time between the excitation of the donor with the excitation wave at time t'', which would be observed at time t'. The fundamental fluorescence decay of the acceptor is $F_\delta(t - t')_A = F_{0,A}\exp(-(t - t')/\tau_A)$. τ_A is the fluorescence lifetime of the acceptor. $F_{0,A}$ is proportional to the extent of energy transfer. Therefore, we express the double convolution as

$$F_A(t) = \int_0^t \left[\int_0^{t'} E\left(t''\right) F_\delta\left(t' - t''\right)_{DA} dt'' \right] F_\delta\left(t - t'\right)_A dt' \tag{11.21}$$

$$= \int_0^t \left[\int_0^{t'} \left(1 + M_{ex}\exp\left(j\omega_T t''\right)\right) F_{0,DA}\exp\left(-\left(t' - t''\right)/\tau_{DA}\right) dt'' \right] F_{0,A}\exp\left(-\left(t - t'\right)/\tau_A\right) dt'$$

The integrations are straightforward, just as in Box 11.4. Also, just as in that box, we disregard the transient terms that decay in times comparable to the lifetimes of the D and A (remember that the data acquisition is a long-time average over many periods of repetition). The steady-state result is

$$F(t)_A = \tau_{DA}\tau_A F_{0,DA}F_{0,A}\left\{1 + M_{ex}\frac{\exp\left(j\omega_T t\right)}{\left(1 + j\omega_T \tau_{DA}\right)\left(1 + j\omega_T \tau_A\right)}\right\}$$

$$= \tau_{DA}\tau_A F_{0,DA}F_{0,A}$$

$$+ M_{ex}\frac{\tau_{DA}F_{0,DA}}{\left(1 + j\omega_T \tau_{DA}\right)}\frac{\tau_A F_{0,A}}{\left(1 + j\omega_T \tau_A\right)}\exp\left(-j\left(\tan^{-1}\left(\omega_T \tau_{DA}\right) + \tan^{-1}\left(\omega_T \tau_A\right)\right)\right)\exp\left(j\omega_T t\right)$$

$$= \tau_{DA}\tau_A F_{0,DA}F_{0,A} + M_{ex}\left[\tau_{DA}F_{0,DA}M_{DA}\right]\left[\tau_A F_{0,A}M_A\right]\exp\left(-j\left(\phi_{DA} + \phi_A\right)\right)\exp\left(j\omega_T t\right)$$

Changing from complex to real notation, we finally have

$$\boxed{F(t)_A = \tau_{DA}\tau_A F_{0,DA}F_{0,A} + M_{ex}M_{DA}M_A\left(\tau_{DA}F_{0,DA}\right)\left(\tau_A F_{0,A}\right)\cos\left(\omega_T t - \phi_{DA} - \phi_A\right)} \tag{11.22}$$

$$M_{DA} = \frac{1}{\sqrt{1 + \left(\omega_T \tau_{DA}\right)^2}}\ ;\ M_A = \frac{1}{\sqrt{1 + \left(\omega_T \tau_A\right)^2}}\ ;\ \phi_{DA} = \tan^{-1}\left(\omega_T \tau_{DA}\right);\ \phi_A = \tan^{-1}\left(\omega_T \tau_A\right).$$

The normalized fluorescence response (see Box 11.7) is Equation 11.21 divided by the average DC signal $\tau_{DA}\tau_A F_{0,DA}F_{0,A}$.

$$\boxed{F(t)_{A,norm} = 1 + M_{ex}M_{DA}M_A\cos\left(\omega_T t - \phi_{DA} - \phi_A\right)} \tag{11.23}$$

the modulation of the acceptor in Equations 11.22 and 11.23, is $M_{meas} = M_{ex}M_{DA}M_A$. If the donor fluorescence is observed alone, the total modulation fraction would be

$$M_{meas} = M_{ex}M_{\omega_T} \equiv M_{ex}M_{DA}$$

Also, the phase when observing the acceptor is $\varphi_{meas} = (\varphi_{DA} + \varphi_A)$, rather than just $\phi_{meas} = \phi_{\omega_T} \equiv \phi_{DA}$, which would be the phase when observing only the donor fluorescence. Also, instead of $F_0\tau \equiv F_{0,DA}\tau_{DA}$ for the donor DC fluorescence, we have $\tau_{DA}\tau_A F_{0,DA}F_{0,A}$ for the DC averaged acceptor fluorescence intensity. Remember that, for the polar plot, we normalize the fluorescence signal by dividing by the steady-state DC signal. Therefore, we divided Equation 11.22 by $\tau_{DA}\tau_A F_{0,DA}F_{0,A}$ to achieve the normalized fluorescence response in Equation 11.23.

When observing the acceptor (only), the points of the polar plot, $M_{meas}\cos(\varphi_{meas})$ versus $M_{meas}\sin(\varphi_{meas})$, will not lie inside the semicircle (see also Section 11.7.2.2). For successive excitation (e.g., FRET observing *only* the acceptor fluorescence), the plotted values of the polar plot will always lie *outside* the semicircle (see Figure 11.3). This is diagnostic of a successive excitation event, such as in an excited-state reaction, or FRET. When measuring FRET, the polar plot depends on whether we observe the donor (points inside the semicircle) or the acceptor (points outside the semicircle).

If we are at a wavelength to observe both the donor and acceptor fluorescence, the plot is more complex, although it is still diagnostic (see Section 11.7.2.2). If the observed fluorescence is from both the acceptor and the donor (it is very difficult to observe the acceptor fluorescence without a contribution of the donor), then the positions of the polar plot points depend strongly on the relative signal strengths. An example of this is shown in Figure 11.3. Thus, in general, we can make diagnostic conclusions *without trying to fit lifetimes;* we simply plot directly the observed parameters of the frequency domain experiment (that is, the modulation and the phase of the fluorescence signal).

11.7.2 Combining Spectra and Polar Plots

Different fluorescent molecules have distinctive emission spectra. Therefore, spectral information is useful in discriminating fluorescence species. All FLIM experiments select wavelengths of excitation and emission, and this is usually accomplished by the use of band-pass or cutoff filters. However, much more information is available by collecting data over a range of emission wavelengths. Previous work has shown the advantages of incorporating a spectrograph to disperse the emission wavelengths in time-resolved fluorescence measurements (Lakowicz and Balter 1982b; Lakowicz et al. 1984) and into FLIM instrumentation (Siegel et al. 2001; Becker, Bergmann, and Biskup 2007; Bednarkiewicz, Bergmann, and Biskup 2008; Chorvat and Chorvatova 2006; De Beule et al. 2007; Hanley, Arndt-Jovin, and Jovin 2002; Ramadass et al. 2007; Rueck et al. 2007). Here we demonstrate the advantages of combining the polar plot analysis with FLIM data where multiple emission wavelengths are acquired simultaneously at every pixel.

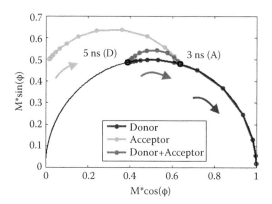

FIGURE 11.3 **(See color insert following page 288.)** The polar plot of the fluorescence signals of the donor alone (blue dots), the acceptor alone (green dots), and donor and acceptor together (red dots). The original lifetimes of the donor and the acceptor (which means the fluorophores are excited directly by light and there is no FRET occurring) are 5 and 3 ns, as indicated on the figure. When the donor-to-acceptor separation distance changes from twice the Förster distance to half the Förster distance, the acceptor-only fluorescence signal moves clockwise toward the original acceptor lifetime, and the donor-only signal moves along the semicircle to a smaller lifetime. Together, the donor and acceptor signal moves from the original lifetime of the donor to the original lifetime of the acceptor, as the FRET efficiency increases from ~0 to ~100%. The arrows show the direction of lifetime signal change as the FRET efficiency increases.

11.7.2.1 Two Different Noninteracting Fluorophores

We consider two different molecules with disparate spectra. The measured time-independent spectra can be represented as

$$F(\lambda)_{meas} = F_{0,\alpha}F_\alpha(\lambda) + F_{0,\beta}F_\beta(\lambda) = \alpha(\lambda)F(\lambda)_{meas} + \beta(\lambda)F(\lambda)_{meas} \qquad (11.24)$$

where $F_\alpha(\lambda)$ and $F_\beta(\lambda)$ are the normalized emission spectra of two different molecules, $\alpha(\lambda) + \beta(\lambda) = 1$. The spectra are normalized by setting the integration of the measured fluorescence spectra over wavelength equal to 1; that is,

$$\int_0^\infty F_\alpha(\lambda)d\lambda = 1 \text{ and } \int_0^\infty F_\beta(\lambda)d\lambda = 1$$

$\alpha(\lambda)$ and $\beta(\lambda)$ are defined as the fractional intensity of the fluorophores.

The measured lifetime data at every wavelength λ (and at every pixel) can be expressed in terms of the frequency-domain presentation as Equation 11.25, where we consider the practical case that the fluorescence lifetimes are independent of wavelength:

$$F(t)_{meas,\lambda} = F_{0,\alpha}F_\alpha(\lambda)F_\alpha(t) + F_{0,\beta}F_\beta(\lambda)F_\beta(t)$$

$$= F_{0,\alpha}F_\alpha(\lambda)[1 + M_\alpha\cos(\omega_T t - \phi_\alpha)] + F_{0,\beta}F_\beta(\lambda)[1 + M_\beta\cos(\omega_T t - \phi_\beta)] \qquad (11.25)$$

$$= F(\lambda)_{meas}\left\{[\alpha(\lambda)+\beta(\lambda)] + [\alpha(\lambda)M_\alpha\cos(\omega_T t - \phi_\alpha) + \beta(\lambda)M_\beta\cos(\omega_T t - \phi_\beta)]\right\}$$

The preceding equations show that if we have prior knowledge of the emission spectrum of both components, the fluorescence percentage composition $\alpha(\lambda)$ and $\beta(\lambda)$ at every wavelength can be identified through spectral linear unmixing (the average time-independent fluorescence values recorded at each pixel). The resulting values of $\alpha(\lambda)$ and $\beta(\lambda)$ are then used globally over all wavelengths in the lifetime calculation to provide statistically more reliable results. Conversely, $\alpha(\lambda)$ and $\beta(\lambda)$ can be determined by first fitting the lifetime data at every wavelength of every pixel, keeping the lifetimes constant for all wavelengths, globally at every pixel. Then Equation 11.24 can be used to construct the emission spectra of both molecules.

The individual spectrum is determined from the density of points on the polar plot straight line; from the position of the points on the straight line, we know the fractional contribution from each component (and we know the wavelength at which each point was measured). Spectral FLIM is especially helpful to distinguish molecules having almost overlapping spectra, in which case it is difficult to separate the fluorescent signals. The additional spectral information increases the reliability of the lifetime determinations.

In Figure 11.4, numerically simulated spectral FLIM data are analyzed and presented on the polar plot. The fluorescence signal of two noninteracting molecules with different spectra and lifetimes (5 and 3 ns) is simulated (Figure 11.4a, b). White noise has been added to the signal. In Figure 11.4(a), $\alpha(\lambda)$ and $\beta(\lambda)$ are acquired from prior knowledge of the two spectra, and then the resulting values of $\alpha(\lambda)$ and $\beta(\lambda)$ are used in a least-squares calculation to determine lifetime values, which are 5 and 3 ns (also indicated on Figure 11.4b). As has been shown in Section 11.7.1.1.2, all the polar plot points from a sample with just two emitting species with two separate lifetimes must lie on the straight line between the lifetime positions of the two molecules individually. Therefore, all the data with two components taken at different wavelengths of emission must lie on this straight line.

The only thing that changes is the fractional contribution of the two lifetime components at the different emission wavelengths. Therefore, the position on the straight line is different for every wavelength because this fraction depends on the emission wavelength. We can also use the polar plot data in another way: If we know the lifetimes of the two separate components alone, we can get $\alpha(\lambda)$ and $\beta(\lambda)$ and then rebuild the two spectra. The indicated calculations can be shown to produce quite accurate estimates of the lifetimes and/or the spectra.

We emphasize that the most insightful aspect about projecting the complete spectral FLIM data from two independent fluorophores on a polar plot is that the data points of the polar plot projection of the FLIM data at all the different wavelengths lie on the line connecting the two lifetimes (as should be if there are only two contributing fluorescence

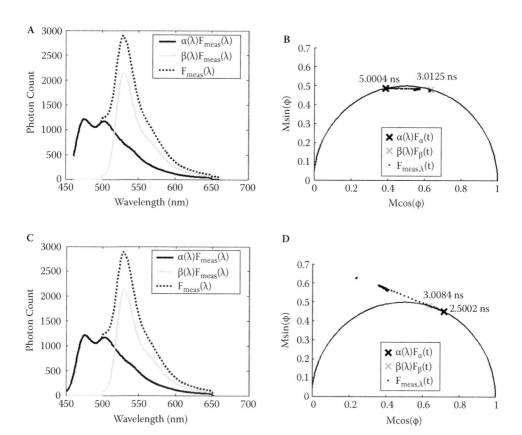

FIGURE 11.4   Spectral-FLIM analysis for two simulated data. (A) The simulated spectra $F(\lambda)_{meas}$ (the black dotted line) of two noninteracting fluorophores. The prior knowledge of the emission spectra of both components (the black and the gray lines) is applied to calculate the fluorescence percentage composition $\alpha(\lambda)$ and $\beta(\lambda)$ at every wavelength. (B) The values of $\alpha(\lambda)$ and $\beta(\lambda)$ acquired from (A) are then used globally over all wavelengths in the lifetime calculation. The black dots are the simulated lifetime data over the wavelengths, and white noise has been added to the signal. (C) The simulated spectra $F(\lambda)_{meas}$ (the black dotted line) of both donor and acceptor observed simultaneously. Similarly, the prior knowledge of the emission spectra is used to calculate the values of $\alpha(\lambda)$ and $\beta(\lambda)$. (D) The lifetime data of the donor and acceptor when they are separated by their corresponding Förster distances. The straight line connecting the data points over the wavelengths intersects the semicircle at two points: the acceptor lifetime in the absence of FRET and the donor lifetime in the presence of FRET. Again, these two values can be calculated from $\alpha(\lambda)$ and $\beta(\lambda)$ estimated from (C).

components, each with a wavelength independent lifetime). This provides an immediate visualization and determination of the two lifetime values without any need to fit the lifetimes. The separate contributing lifetimes just correspond to the two intersections of this line with the semicircle (see Section 11.7.1.1.2).

The points at all the different wavelengths lie on the straight line because the values of $F_{\alpha}(\lambda)$ and $F_{\beta}(\lambda)$ vary at different wavelengths; however, the lifetimes of the two species do not change. Therefore, the position of the polar plot data moves along the line depending on the relative fractional values of $F_\alpha(\lambda)$ and $F_\beta(\lambda)$. The position on the straight line is dependent on the wavelength of the emission spectrum at which each individual FLIM determination is made, and this depends on the underlying spectra of the two independent species. A straight line with the properties just given is also diagnostic for two lifetime components.

11.7.2.2 FRET: Observing Donor and Acceptor Fluorescence Simultaneously

Spectral FLIM is also very insightful for visualizing directly the lifetimes of a FRET pair from the polar plot—that is, without fitting the data. As shown in Section 11.7.1.2, in the presence of FRET, when the *acceptor fluorescence only* is measured, all points on the polar plot will lie outside the semicircle; the positions of these points, as shown in Figure 11.3, will lie on a particular curve. However, when measuring the phase and modulation of fluorescence of a FRET signal at wavelengths where *both the acceptor and the donor contribute to the measured fluorescence,* the points can lie either inside or outside the semicircle, depending on the fractional intensities and the efficiency of fluorescence.

For a particular value of FRET efficiency, the data points taken at different emission wavelengths when observing the emission of both the acceptor and the donor will lie on a straight line; the locations on this straight line depend on the wavelength of emission. That is, the straight line is determined by measuring the phase and modulation of the FRET sample at many different emission wavelengths (with differing overlapping fractional contributions from the donor and acceptor).

An extension of this straight line intersects the universal semicircle at two locations (see Figure 11.4d): (1) the location of the polar plot point corresponding to the acceptor lifetime in the absence of FRET, τ_A; and (2) the location on the polar plot semicircle for the donor lifetime in the presence of the acceptor, τ_{DA} (this is the lifetime of the donor corresponding to the particular value of the FRET efficiency). As we know from Section 11.7.1.1.1, when only the donor fluorescence is measured (which is the common way of measuring FRET with FLIM), all points on the polar plot must lie on the semicircle for all FRET efficiencies (see Figure 11.3). As the FRET increases, the value of τ_{DA} decreases and the point on the semicircle of the polar plot moves clockwise. Also, τ_{DA} for a particular FRET efficiency is the same for all donor emission wavelengths.

Extending the straight line in the other direction intersects the point of the polar plot where one would be observing *only the acceptor* in the presence of FRET for the particular FRET efficiency. This point lies outside the semicircle because when *only* the acceptor fluorescence is observed, as we discussed earlier, it must lie outside the semicircle. This latter point is difficult to determine experimentally (without fitting the data to a model involving lifetimes to separate τ_A and τ_{DA}) because the donor emission spectrum almost always overlaps the acceptor fluorescence. However, we do not need to determine this point in order to determine the FRET efficiency.

The interpretation of the intersecting point on the semicircle for τ_A is easy to understand. In the limit of 100% transfer, the acceptor is excited by extremely rapid transfer from the donor, as though the acceptor is excited by the excitation light directly. Therefore, the phase and modulation of this latter intersection with the semicircle is a limit point for the acceptor fluorescence for 100% transfer, which is identical to that if the acceptor were excited by the light directly. The FLIM polar plot data approach this point as the FRET efficiency increases (the curve following increasing FRET efficiency is not linear; see Figure 11.3).

Equation 11.26 demonstrates that the three points just discussed corresponding to a *particular* FRET efficiency and observing both the acceptor and donor fluorescence lie on a common straight line in the polar plot. The slope *dy/dx* of the line connecting these three data points is exactly the same as each of the two straight lines:

1. the straight line connecting polar plot points corresponding to the acceptor lifetime in the presence of FRET ($M_{DA}M_A\cos(\varphi_{DA} + \varphi_A)$, $M_{DA}M_A\sin(\varphi_{DA} + \varphi_A)$) and the donor lifetime in the presence of FRET ($M_{DA}\cos(\varphi_{DA})$, $M_{DA}\sin(\varphi_{DA})$); and

2. the straight line connecting polar plot points corresponding to the acceptor lifetime in the absence of FRET ($M_A\cos(\varphi_A)$, $M_A\sin(\varphi_A)$) and the donor lifetime in the presence of FRET ($M_{DA}\cos(\varphi_{DA})$, $M_{DA}\sin(\varphi_{DA})$).

As shown in the polar plot section, $M_{\omega,i}\cos(\varphi_{i,\omega})$ and $M_{\omega,i}\sin(\varphi_{i,\omega})$ can be expressed respectively as

$$\left(1\Big/\sqrt{1+\left(\omega\tau_i\right)^2}\right)\left(1\Big/\sqrt{1+\left(\omega\tau_i\right)^2}\right)=1\Big/\left(1+\left(\omega\tau_i\right)^2\right)$$

and

$$\left(1\Big/\sqrt{1+\left(\omega\tau_i\right)^2}\right)\left(\omega\tau_i\Big/\sqrt{1+\left(\omega\tau_i\right)^2}\right)=\omega\tau_i\Big/\left(1+\left(\omega\tau_i\right)^2\right)$$

We therefore obtain the following for the value of the slope $(\omega^2\tau_{DA}\tau_A - 1)/(\omega(\tau_{DA} + \tau_A))$:

$$\frac{dy}{dx}=\frac{M_{DA}M_A\sin\left(\phi_{DA}+\phi_A\right)-M_{DA}\sin\left(\phi_{DA}\right)}{M_{DA}M_A\cos\left(\phi_{DA}+\phi_A\right)-M_{DA}\cos\left(\phi_{DA}\right)}$$

$$=\frac{M_A\sin\left(\phi_A\right)-M_{DA}\sin\left(\phi_{DA}\right)}{M_A\cos\left(\phi_A\right)-M_{DA}\cos\left(\phi_{DA}\right)} \tag{11.26}$$

$$=\frac{\omega^2\tau_{DA}\tau_A-1}{\omega\left(\tau_{DA}+\tau_A\right)}$$

For each different efficiency of FRET, a corresponding straight line will connect the described three points, and every straight line will intersect the semicircle at the point on the polar plot corresponding to τ_A (of course, τ_A does not depend on FRET). For different FRET efficiencies, the point on the semicircle for τ_{DA} will change; consequently, the straight line will change its slope according to the FRET efficiency. The straight line just tips like a see-saw with the fulcrum the point on the polar plot corresponding to τ_A.

This "straight line" property is useful. For instance, one can measure the phase and modulation at multiple wavelengths throughout the emission spectrum, which contains contributions from both the acceptor and donor fluorescence, and the resulting straight line will intercept the semicircle at the point corresponding to τ_{DA}. Therefore, τ_{DA} can be determined *without* fitting any lifetimes, without worrying about selecting only the donor fluorescence, and without specifying a model. Measuring the phase and modulation of the total fluorescence contributed by both the donor and acceptor and constructing the polar plot are especially useful for high efficiencies of FRET, where the donor fluorescence is weak. From the τ_{DA} intercept, one can easily determine the efficiency of energy transfer simply by using the relation $E = 1 - (\tau_{DA}/\tau_D)$ (τ_D is the lifetime of the donor in the absence of FRET and is assumed known). Then, Equations 11.22 and 11.25 can be applied to distinguish donor and acceptor and determine the fractional contributions from which the FRET efficiency can also be derived.

Even if a fraction of the acceptor is excited directly (i.e., not through FRET), the total fluorescence signal of the FLIM experiment does not change the slope or the location of this straight line, but rather the position of the points on the straight line for the points acquired when detecting the acceptor and donor fluorescence at different wavelengths. When the fluorescence of a FRET sample is measured, the preceding method is still valid. Varying the relative fraction of the acceptor that is excited directly (by exciting at different wavelengths) will also produce points on this straight line from which τ_{DA}, and therefore the efficiency of transfer, can be determined.

In Figure 11.4(C, D), the simulated spectral FLIM data include the donor and acceptor signals. The two molecules in the simulation are separated by their corresponding Förster distance; therefore, the lifetime of the donor is quenched from 5 to 2.5 ns. The data points on the polar plot when measuring only the acceptor fluorescence, which is excited through FRET, fall outside the semicircle, as discussed previously (the lifetime of acceptor excited directly is 3 ns and is shown on the plot). Just as in the analysis of spectral FLIM polar plot data for two noninteracting fluorophores, the FRET pair spectrum is used to resolve the two lifetimes on the semicircle. It can also be shown that the spectra can be calculated from the lifetimes determined from the polar plot (not shown in figure).

11.8 WAVELETS AND DENOISING

11.8.1 Why Use This Image Analysis?

The raw data of frequency-domain homodyne FLIM is a set of full-field fluorescence intensity images (typically three to eight images) collected for various phase shifts between the detector gain and the excitation light wave forms (see Chapter 5). Thus, image analysis techniques may be readily implemented to the individual fluorescence phase-selected images. For instance, image analysis can be used to filter out noise and background fluorescence from the raw intensity data at each phase before the fluorescence lifetimes are calculated; therefore, both image quality and fluorescence lifetime accuracy are improved, provided that this filtering does not change the signal.

We have found two image analysis techniques, in particular, to be very valuable for enhancing the analysis of our video-rate FLIM images:

- Wavelet transforms are useful for selecting objects according to their spatial morphology and, likewise, for separating fluorescence lifetime components that originate from different spatial morphologies. Both the fluorescence intensity and fluorescence lifetimes of unwanted background components can be removed from the data using wavelet-based filtering of spatial morphologies.

- An advanced image analysis technique, originally designed exclusively for removing photon noise from intensity images, has been adapted to noise removal for homodyne FLIM.

It is generally useful to apply these compatible techniques in parallel. For the purpose of providing a general introduction, we avoid mathematical and technical details of the image analysis; rather, we describe fundamental principles of the techniques and demonstrate how they are used for FLIM analysis. Comprehensive publications that provide mathematical proofs, algorithm pseudocode, and much more detail are referenced throughout the following discussion.

11.8.2 Wavelet Transforms for Discriminating Fluorescence Lifetimes Based on Spatial Morphology

11.8.2.1 What Is a Wavelet Transform?

Wavelets have only recently been applied in fluorescence imaging (see Bernas et al. 2006 and references therein). There are many different wavelet methods and transforming functions (Starck, Murtagh, and Bijaoui 1998). Wavelets can be used for one-, two-, and three-dimensional data. In imaging, wavelets are most often used for denoising (Nowak and Baraniuk 1999; Starck and Bijaoui 1994), denoising before later image processing (Boutet de Monvel, Le Calvez, and Ulfendahl 2001), image compression (Bernas et al. 2006; Grgic, Grgic, and Zovko-Cihlar 2001; Shapiro 1991; Shapiro, Center, and Princeton 1993), location of dominant effects in an image (Olivo-Marin 2002), edge detection (Willett and Nowak 2003), and for object detection before particle tracking (Genovesio et al. 2006). Wavelets have not been applied to FLIM measurements until recently. In this section, we provide a general discussion of the wavelet transform and discuss how wavelets are specifically advantageous for FLIM.

The wavelet transform is a multiscale analysis scheme. Wavelet transforms extract features within an image by selective filtering of localized space/scale characteristics of images. It is a way to detect spatial frequency characteristics of image data at a local scale. It is instructive to contrast wavelets with the more familiar Fourier analysis. Two-dimensional Fourier image analysis reports on spatial frequencies in an image on a global scale; the transform is a function of frequency only. That is, the Fourier transform of an image produces a power spectrum of the frequency components on a global scale, but does not keep a record of the spatial locations of the frequency components.

Windowed Fourier transformation of images tries to analyze localized regions of signals or images, but suffers from drastic edge effects, so the transform is still only a function of the spatial frequency (Walker 1997). An important distinction is that the wavelet transform intrinsically has a continuous "window" size for resolution of the full spatial frequency spectrum; this is a major advantage over the fixed window size of Fourier analysis. However, in practice, digital wavelet transforms with an exponential scale resolution are sufficient for many applications—as opposed to more cumbersome large "continuous" scale resolution.

The wavelet transform achieves this flexibility because wavelets are localized functions. In contrast to the infinite-range sinusoidal functions comprising Fourier analysis, wavelets are generalized local basis functions that can be stretched (dilated) and translated with a flexible resolution in both space and frequency. The fundamental distinction is that wavelet transformations have units of both space and frequency (scale). Stretched wavelet basis functions correlate with spatial resolution of low-frequency scale components—coarse features; contracted wavelet basis functions correlate with high spatial frequencies and fine structure.

The dilation/contraction property of the wavelet sets the spatial extent, and the translation property locates the position of the wavelet application within the image. Thus, the wavelet transform produces a matrix of coefficients; *each* pixel of the image is assigned coefficients for *each* space/scale frequency. The wavelet coefficient matrix is obtained by translating the wavelet across the image in the horizontal, vertical, and diagonal directions for various degrees of wavelet dilation/stretching. Thus, the spatial orientation of objects can also be discerned with wavelets. This application of wavelets is known as wavelet multiresolution decomposition, which more generally refers to decomposing a signal into its spatial or temporal frequency components.

Formally, wavelet analysis involves two sets of localized functions: a scaling function and an associated wavelet function. As described by S. G. Mallat (1989), the scaling function can be seen as a low-pass filter, whereas the wavelet is similar to a band-pass filter. In Figure 11.5, the example image is passed through a series of dilated wavelets together with the associated scaling functions; this analysis acts similarly to a "filter bank" (Hong

FIGURE 11.5 An example of wavelet multiresolution decomposition applied to a fluorescence intensity image of a nuclear spread under oil. Chromatin (extended strings and loops) and nuclear bodies (circular spots of intensity) are visible. The raw image (upper left corner) is decomposed into approximations (A1–A6) and details (D1–D6). At each level of the decomposition, the approximation is split into an approximation and detail image. The detail images combine the horizontal, vertical, and diagonal details together. To recompose the entire image, the details are all added to the final approximation (i.e., total image = D1 + D2 + D3 + D4 + D5 + D6 + A6). The chromatin contains higher frequency components that show up in the details (D2–D4), while the nuclear bodies largely do not appear in the detail images until levels five and six. Further decomposition (not shown), eventually leads to a flat approximation image that is the dark current of the CCD. The wavelet analysis is performed by a set of functions in MATLAB's® Wavelet Toolbox 4 (The Mathworks, Inc., Massachusetts) based on a dyadic scale using biorthogonal wavelets (bior3.7). (Image courtesy of Dr. Snehal B. Patel and Prof. Michel Bellini, Department of Cell and Developmental Biology, University of Illinois at Urbana-Champaign.) The grayscale bar indicates the fluorescence intensity in arbitrary unit (a.u.).

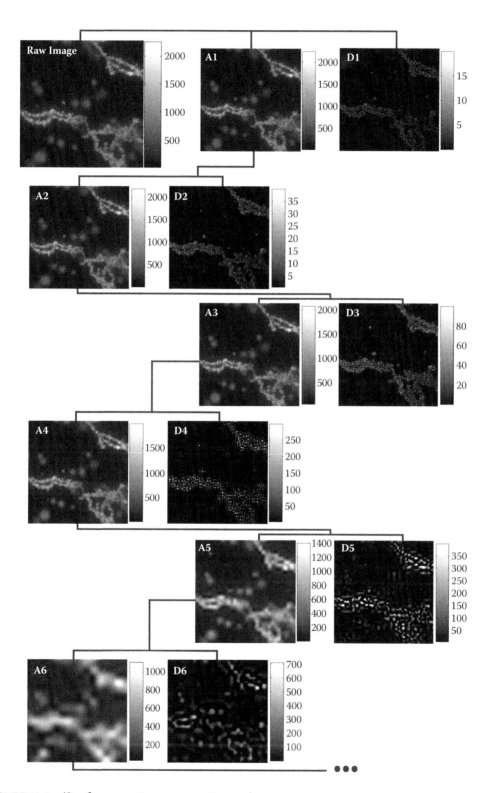

FIGURE 11.5 (See figure caption on opposite page).

1993). As the image data are iterated through the image filter bank, the wavelet algorithm selects morphologies in the image with scales of successively narrower spatial frequency band pass and lower spatial resolution (less detail), decomposing the original image into multiresolution "approximation" and "detail" coefficient matrices.

For a particular scale, the approximation is the low-frequency filtered image and the detail is the high-frequency components. For the subsequent iteration, the approximation is partitioned into approximation and detail images. Thus, the low-pass approximation image at some iteration of the wavelet procedure will be an image that has been filtered of high frequencies.

Using all the coefficients derived through this process, the original image can be recreated exactly by applying the inverse wavelet transform. Figure 11.5 illustrates this process for a biologically relevant fluorescence image. The differences between two low-pass scaled images select the morphologies in the image with spatial frequencies in the differential band-pass region. Using the coefficients in this spatial frequency band pass, the inverse transform creates an image with the selected morphologies (corresponding to the selected localized spatial frequencies). The band-passed images are then used for the FLIM analysis. For example, in Figure 11.5, the high-frequency elements of chromatin structure could be automatically selected and separated from the lower frequency nuclear bodies.

11.8.2.2 Applications of Wavelets to Homodyne FLIM

The efficacy of the wavelet transforms for analyzing FLIM images has been reported (Buranachai et al. 2008). The wavelet multiresolution process can automatically select localized morphologies in images based on their spatial frequencies. Regions with different spatial frequency characteristics can be removed, even when they overlap spatially with the regions of interest. This is especially powerful for FRET imaging when broadly dispersed, diffuse, low spatial frequency background fluorescence is detrimental.

Even though FLIM is able to distinguish locations based on the fluorescence lifetime, better results are often obtained if background can be subtracted before the phase and modulation analysis is carried out—even with confocal images. We emphasize the fact that not only is the background intensity removed, but also the contribution of the background lifetime component to the lifetime signal is removed (Buranachai et al. 2008). The wavelet transform is applied to each of the phase selected fluorescence intensity images before the FLIM data are analyzed to extract the phase and modulation values. Spatial morphologic decomposition results in robust reduction or often complete elimination of background fluorescence lifetime contributions and other unwanted fluorescence, thus leaving only the image locations of interest, without the interfering lifetime components.

The contribution of background can be especially detrimental for estimating accurate lifetimes. Even with confocal measurements, background light can arise from reflections and scattering of fluorescence from the surrounding material; this scattering is low, but because the FLIM signal emanating from a sample is independent of the intensity, this can lead to confusing artifacts. In many cases, wavelets eliminate this background contribution, and the result is a better spatial definition of the components with specific lifetimes.

11.8.3 Denoising Homodyne FLIM Data

11.8.3.1 Sources of Noise for Homodyne FLIM

An important factor to improve fluorescence lifetime determinations is the reduction of random photon and instrument noise; these are often a major contribution to the fluorescence signal. Homodyne FLIM systems include an image intensifier for modulating the emitted fluorescence at every pixel of the image simultaneously. That is, the photon gain of the intensifier is modulated at radio-frequencies. The output of the intensifier is focused onto a charge-coupled device (CCD) camera. The CCD is operated in charge accumulation mode for homodyne FLIM, as opposed to photon-counting mode. In charge accumulation mode, the detector substantially degrades the photon statistics of the fluorescence signal.

Much of the noise is due to the variability in each of the photoelectron multiplication steps within the microchannel plates (MCPs) of the image intensifier. The fluorescence light incident on the intensifier cathode is converted into photoelectrons, which are injected into the MCP. These electrons collide with the MCP channel walls and multiple electrons are ejected in each collision. The signal-dependent instrument noise is known as the "exponential pulse height distribution," an exponential probability function (Sandel and Broadfoot 1986). The pulse height distribution (also used to describe photomultiplier tube noise in analog mode) refers to the fact that the number of electrons ejected per collision inside the MCP is variable and a multitude of such events result in the exponential probability function. Thus, the major sources of noise are the unavoidable Poisson noise of the initial photo detection (always present when detecting photons) and the detector noise discussed before.

Following the photon gain through the electron multiplication process, the photoelectrons are converted back to photons by impinging onto a phosphor screen. The image of the phosphor screen is collected by a CCD camera and digitized. The noise in these final steps can be shown to be relatively small, and it is not included in the theoretical consideration of the noise. Rather, the focus is on the photon gain steps described previously, which amplify the noise of prior steps. The mathematical foundation for interdependent noise propagation for cascading systems has been worked out and has been discussed for intensified CCD (ICCD) detectors (Frenkel, Sartor, and Wlodawski 1997).

11.8.3.2 Removal of Signal-Dependent Noise: TI-Haar Denoising

Signal-dependent noise is not a trivial matter. Its removal requires partitioning the image to take into account local intensities. A significant challenge is to accomplish this and preserve the image detail—that is, without blurring (oversmoothing) the data. Thus, signal-dependent noise (e.g., photon noise) requires alternative techniques to take into account the rise in noise with signal level. Nowak and Willet have addressed this issue with a technique termed "fast Haar translational invariant denoising (TI-Haar)" that is designed specifically for photon-limited medical images and removal of Poisson noise (Willett and Nowak 2003, 2004).

This technique is based on Haar wavelets—square-shaped wavelets—to partition the image and fit the individual pieces of the image locally. In analogy to wavelets, TI-Haar is a multiresolution analysis of the noise, using constant-square or gradient-square (called platelets; Willett and Nowak 2003) basis functions. This procedure is not a normal filter of high-frequency noise, but rather is specifically aimed at the characteristics of Poisson noise (of photon detection). The great advantage of this multiresolution approach is that it does not round the edges of sharp spatial transitions and does not blur the image in regions of sufficient signal-to-noise ratio to warrant preservation of features.

The core of the TI-Haar algorithm is a penalized maximum likelihood estimator to guide aggregate decisions of the square partitions. For each iteration of the algorithm, the likelihood that the measured data of a local region of the image can be attributed to a constant square (representing the estimate of the true mean signal) is computed and the complexity of the particular partition (the penalty) is also computed. In the next iteration, in a coarse-to-fine procession, that region of the image is further portioned and the penalized likelihood is compared to the previous step. If the resulting partition does not increase the penalized maximum likelihood criteria, then the partition is thrown out. Thus, a complex partition of a local region of the image (meaning fine-scale squares to fit local variation in intensity) requires sufficient likelihood that the mean intensity truly varies at that location. It follows that low signal-to-noise regions are fit to larger squares because the intensity profile in these regions is due to noise alone.

Finally, once the partition has been optimized, the terminal nodes of the partition tree are fused to construct the estimate of the true intensities of the image. This type of analysis proves powerful for removing noise without loss of spatial resolution (Figure 11.6).

11.8.3.3 TI-Haar Denoising Improves Homodyne FLIM Accuracy

As described before, the video-rate FLIM system requires a different statistical basis to account for the instrument noise in addition to the removal of the Poisson distributed photon noise. Theoretical considerations demonstrate that the final noise distribution for the homodyne FLIM system is a signal-dependent Gaussian distribution. Fortunately, Nowak and Willet developed a "Gaussian mode" noise intended for removing noise following the normal distribution; however, this mode is intended specifically for signal-independent noise with a variance of unity. Hence, in order to apply TI-Haar to homodyne FLIM data, it is necessary to carry out a variance stabilizing transform (VST) (Prucnal and Saleh 1981), which transforms the image into signal-independent noise with a variance of unity. Following denoising, the inverse VST can be applied to the data to return the image to its familiar signal level.

FIGURE 11.6 **(See color insert following page 288.)** Demonstration of the signal dependent noise removal image analysis for FLIM, using a fluorescence intensity image. (A) A gray-scale fluorescence intensity image of a prostate tissue microarray core from a FLIM data set. Also shown is a three-dimensional fluorescence intensity image of the same core, using a color scale to represent the intensity amplitudes. (B) A zoom-in view of a region of the thee-dimensional intensity image shows noise in the data. The same zoom view is shown after application of a Gaussian mean filter, which calculates a weighted average (by averaging each pixel's intensity with its neighbors) and for the TI-Haar + VST custom routine we developed for our FLIM setup. The black arrows indicate an intensity peak that is "averaged out" by the Gaussian mean filter, while it remains intact in the TI-Haar + VST image. (C) Polar plot histograms of the entire fluorescence intensity image are shown; all pixels are included in the analysis. The TI-Haar + VST routine substantially tightens the distribution on the polar plot without blurring the image. The Gaussian mean filter also tightens the distribution to some extent, as expected, but it ruins the structure of the image.

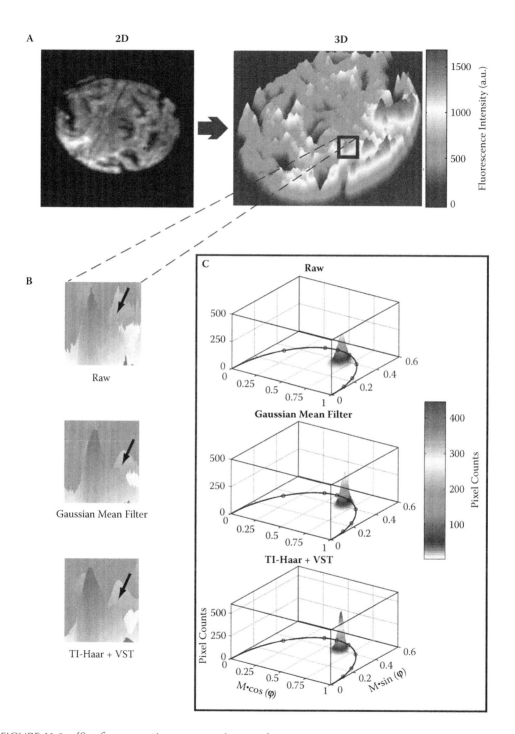

FIGURE 11.6   (See figure caption on opposite page.)

In brief, we carry out the denoising routine using the TI-Haar in Gaussian mode, rather than Poisson mode. In Figure 11.6, we refer to this final noise removal process as "TI-Haar + VST." The VST is carried out based on empirical data collected with our FLIM instrument—a signal–noise curve carried out for the exact same instrument settings as the data to be denoised. A comprehensive discussion of the empirical determination of the signal–noise curve and VST calculation for our FLIM setup is provided in a recent publication (Spring and Clegg, forthcoming).

11.8.4 The Future of Wavelet and Denoising Image Analysis for Homodyne FLIM

Wavelet transforms provide improved recognition and spatial clarity of structures to improve the reliability of the structural identification, which is very useful for coupling the FLIM analysis to the morphology of the region of interest. The wavelet analysis plays a distinctive and unique role in FLIM; not only are the local spatial frequencies selected, as in a normal intensity image, but the lifetime components are also selected in a way unique to the wavelet analysis in conjunction with FLIM (see Figure 11.7). This has major consequences and is a new major advance in FLIM technology.

The TI-Haar + VST image analysis also has many exciting implications for FLIM. Removal of noise from the intensity images means that the FLIM data processing will result in more accurate calculations of the fluorescence lifetimes. This noise removal is accomplished by adapting TI-Haar to the noise probability distribution of the FLIM setup. This method of noise reduction enhances the sensitivity of homodyne, video-rate FLIM and broadens its applicability. This should be particularly helpful when rapid FLIM imaging of multiple fluorophores or fluorescent species, with subtle differences in fluorescence lifetimes, is needed. Moreover, longer integration times may be prohibitive for some dynamic FLIM experiments, so denoising could play a critical role for achieving higher imaging rates while preserving the accuracy of the fluorescence lifetime data.

Neither of the image analysis methods is prohibitively time costly. The wavelet analysis is very fast (less than a minute for a typical FLIM data set) the noise-removal analysis is more significantly time expensive (a few minutes for a typical FLIM data set) using a 2.80 GHz Intel CPU with 704 MB of RAM. Figure 11.7 and its legend provide a simple example showing the simultaneous application of wavelets in conjunction with the denoising algorithm.

FIGURE 11.7 Wavelet background subtraction and TI-Haar denoising applied in combination to simulated FLIM data. Poisson noise has been added to the image. (A) A fluorescence intensity image from a FLIM data set includes a circle (a high-frequency structure). A fluorescent species with a 10 ns fluorescence lifetime resides within the circle. A 1 ns background fluorescence lifetime is everywhere (i.e., the circle is floating within the background such that the pixels inside the circle include both the 10 ns and 1 ns fluorescence lifetime components). (B) A polar plot histogram of the raw image data. Broad distributions of pixels fall at the 1 ns mark on the semicircle (background pixels) and on the line between the 10 ns and 1 ns mark (pixels within the circle). (C) A single, tight peak at 10 ns is found following wavelet background subtraction (to remove the 1 ns background) and Poisson denoising (TI-Haar in Poisson mode). Thus, the background fluorescence intensity and lifetime and the Poisson noise are all completely removed.

A   Inside the circle: 10 ns fluorescence lifetime
and background fluorescence with 1 ns lifetime

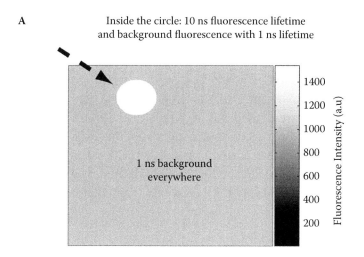

B   Polar plot histogram:
Raw data

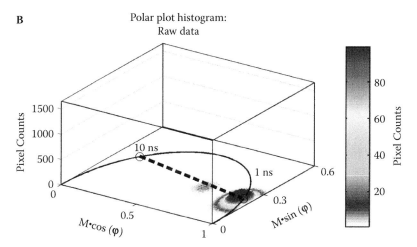

C   Polar plot histogram:
TI-Haar and wavelets combined

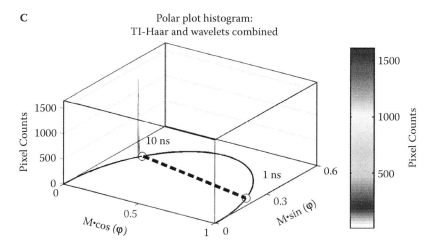

FIGURE 11.7   (See caption on opposite page).

11.9 NONITERATIVE DATA REGRESSION (CHEBYSHEV AND LAGUERRE POLYNOMIALS)

11.9.1 Noniterative Data Regression

FLIM images consist of hundreds of thousands pixels, and if the time of analysis is at a premium, analytical techniques are required that can be carried out rapidly and accurately. The traditional ways of analyzing data, such as exponential decays, involve iterative numerical techniques; if they are carried out on so many pixels, these techniques take an inordinate amount of time. This section introduces a noniterative transform method that is rapid and accurate and avoids false minima. The method has been presented in detail in previous publications by Malachowski (2005; Malachowski et al. 2007); here we summarize the background and indicate the advantages. Examples can be found in the Malachowski references.

11.9.2 Convexity in Modeling and Multiple Solutions

In this section, the concepts of numerical fitting and modeling and the notion of convexity are introduced in the mathematical process of fitting functions. The transformation to Hilbert spaces (Box 11.10) demonstrates that such a transformation eliminates the problems of numerical fitting, which require starting conditions, iteration, and sifting through multiple local solutions.

Figure 11.8(a) shows an attempt to fit a decaying harmonic. The hills and valleys are numerous in the χ^2 function (see Box 11.9 and Figure 11.8(B). The problem of using

---

BOX 11.9: CONVENTIONAL FITTING NUMERICAL TECHNIQUES

The Levenberg–Marquardt algorithm is a numerical minimization of a χ^2 metric of the form

$$\chi^2 = \sum_{j=1}^{N} (y_j - f(t_j, v))^2$$

where v is a vector of parameters defining a general modeling function $f(t_j, v)$, and y_j and t_j are pairs of data values and sample points, respectively. The purpose is to find the best fit of the function $f(t_j, v)$ to the data set of points y_j.

The algorithm establishes the steepest descent down the χ^2 function from a set of starting values and the fastest rate of convergence by employing the Levenberg–Marquardt equation for each step in the convergence (Levenberg 1944; Marquardt 1963). The steepest descent algorithm delivers a vector of values used to alter the current parameter estimations. The equation explicitly is

$$(J^T J + \lambda diag(J^T J))\delta v = J^T (y - f(v))$$

where J is the Jacobian matrix (Kaplan 1984) of the function $f(t_j, v)$ and δv is a vector to increment or decrement the vector of parameter values v such that a lower value of χ^2 is obtained. This process iterates until the lowest χ^2 is obtained and the parameters of the best fit to the data by the model function are determined.

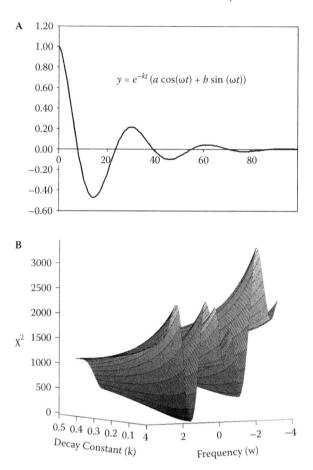

FIGURE 11.8 (A) Decaying harmonic; (B) χ^2 fit with exponentials.

numerical methods to find the global minimum is clearly demonstrated by the proliferation of local minima. The number of these minima can be large and numerical algorithms must hunt for the smallest χ^2.

The gradient-descent technique is based on the observation that if a real-valued function F (the χ^2 function) is defined and differentiable in a neighborhood of a point p, then the function decreases *fastest* in the direction of the negative gradient of F at that point. The assumption is that for a small enough computational step size γ along the gradient, $\gamma > 0$, the smallest χ^2 can be found. With this observation in mind, one starts with a guess for a local minimum of F and considers the sequence such that the process terminates in a reasonable number of iterations. The sequence should converge to the desired local minimum. The value of the computational step size γ is allowed to change at every iteration.

Gradient descent has problems with pathological functions such as the Rosenbrock function (a nonconvex function used as a test problem for optimization algorithms). The Rosenbrock function has a narrow curved valley that contains the minimum; the bottom of the valley is very flat. The curved, flat valley implies a slowly varying Jacobian variety, indicating that the optimization is zigzagging slowly with small step sizes toward the minimal solution.

11.9.2.1 Formulation of Modeling as a Dynamic System

BOX 11.10: DISSERTATION ON DYNAMIC SYSTEMS

Fitting data to single or multiple exponentials is equivalent to the regression of data against a general linear autonomic (self-defining) equation. As an example, the fitting of two exponentials with rate constants k_1, k_2 is equivalent to performing a regression of a second-order differential equation (this equation has as solution exponentials, which depend on the constraints and bounds of the problem):

$$\frac{d^2y}{dt^2} + x_1 \frac{dy}{dt} + x_2 y = 0$$

where $x_1 = k_1 + k_2$, $x_2 = k_1 k_2$

This problem has been approached previously (Prony 1795; Simon 1987), but t is compromised because the formulation is subject to noise resulting from the derivative terms, which inherently increase the noise component of the original data. This can be avoided by integrating the equation to second order:

$$\int \int \left(\frac{dy}{dt^2}^2 + x_1 \frac{dy}{dt} + x_2 y \right) d^2t = y + x_1 \int y dt + x_2 \int \int y d^2t + a_0 + a_t t = 0$$

The integral terms in y can be determined by a "Hilbert transform" to an orthogonal space of polynomials. The right-hand side of the equation vanishes for these types of functions and the integral form becomes a simpler algebraic equation:

$$d_j + x_1 d_{1j} + x_2 d_{2j} = 0$$

where $d_j = H_j(y)$, $d_{1j} = H_j(\int y dt)$, and $d_{2j} = H_j(\int \int y d^2 t)$. H_j is the transform into a Hilbert space of orthogonal functions (they behave similarly to orthogonal x,y,z vectors). The ideal functions for this transform are the discrete Chebyshev polynomials, though other forms such as Laguerre or Laplace polynomials can be used. The transform H_j in the case of the Chebyshev polynomials is the integral

$$H_j = \frac{1}{\pi} \int_{-1}^{1} \frac{y(t)}{\sqrt{(1-t^2)}} dt$$

The transform of the integrals of the input function $y(t)$ may be obtained from the set of d_j by linear and finite combinations of the d_j (Malachowski and Ashcroft 1986).

11.9.2.2 Solution to Convexity in a Hilbert Space

A Hilbert space is a multidimensioned space of functions that have a property similar to the orthogonal axes of a Euclidian space, where a point can be defined by a composition of coordinates along each axis. A Hilbert space is the composition of an arbitrary function as a set of orthogonal functions. Infinite sets of functions can describe a Hilbert space. The classic example is the Fourier transform, where a function can be described as a composition of sine and cosine terms that are the multidimensional components of the Hilbert space.

The idea here is to take a discrete function and characterize it by a series of discrete orthogonal Chebyshev polynomials. The Chebyshev polynomial transform produces coefficients of the continuous Chebyshev polynomials of the first kind, which are specific for the particular function being fitted (similar to the coefficients of a Fourier series). These polynomials have particular advantages for our case and have been described by Abramovitz and Stegun (1972) regarding orthogonal polynomials in general. The problem is then reduced to solving an algebraic equation, instead of fitting with an iterative method.

BOX 11.11: THE SOLUTION TO NONCONVEXITY

Convexity is an important issue in fitting methodologies. A convex function has only one minimum, and others may have many. As mentioned previously, the problem with numerical techniques to determine the best fit of a set of data to a defined function is that many solutions are possible. These iterative methods are described in the previous sections of this chapter (see Box 11.9), as well as in other chapters. Fitting exponentials—or a combination of exponentials and harmonics—is a very difficult numerical problem when there are multiple functions (exponentials) to be fit to the data. However, the transformation of the problem via a Hilbert transform removes the multiple solutions characterized by hills and valleys in the χ^2 space. The time domain fitting regression of a double exponential defined by the differential equation in Box 11.10 is then converted to

$$d_j + x_1 d_{1j} + x_2 d_{2j} = 0$$

As indicated in Box 11.10, derivatives are removed and no starting conditions need be applied; there is only one solution and it is directly accessible analytically (Figure 11.9 shows the χ^2 fit with Hilbert space-transformed decaying harmonic of Figure 11.8(A).

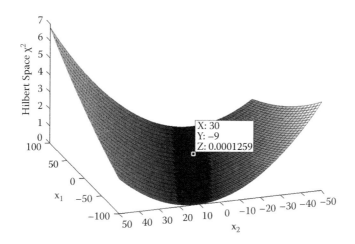

FIGURE 11.9   Chi-square fit with Hilbert space-transformed exponentials of Figure 11.8(A).

11.9.2.3 Error Evaluation

BOX 11.12: THE MONTE CARLO ESTIMATE OF ERROR SENSITIVITY

The Monte Carlo method is used to simulate the fitting of a time series of data points with ideal representations, where the data are contaminated with noise. The purpose is to determine the extent of error in the returned parameters by determining the spread of the parameters for a particular fitting scenario. The fitting of complex functions makes the determination of error difficult to estimate and the Monte Carlo method provides a very effective method to estimate errors. The method is as follows:

Create an ideal data set y_j and add random noise so that $y_j \rightarrow y_j +$ noise. Then calculate the χ^2 for each ideal data set plus noise:

$$\chi^2 = \sum_{j=1}^{N} (y_j - f(t_j, v))^2$$

Calculate the fit by minimizing χ^2 with the methods described before and create histograms of the parameter results. Iterations can be made such that the distributions of fitted values are smooth and accurate estimates of the error in those parameters can be made.

With increasing numbers of functions, such as exponentials being fitted and increasing noise, the ability of any algorithm to recover the rate constants and amplitude becomes more difficult because the fitting parameters dissolve into each other rendering their value to be meaningless (see Figure 11.10). The meaning of fitting parameters vanishes with increasing noise. Numerical techniques become intractable and the analytic techniques described above demonstrate the methods that can be used to determine the useful parameters that are delivered for a particular fitting. To see an application of the above discussion, see a recent paper (Malachowski et al. 2007) where this technique is used to take into account exponentially decaying signal due to photolysis of homodyne frequency domain FLIM data. The improvement in the speed of fitting when compared to an interative method to account for the exponential decay of the FLIM signal is dramatic.

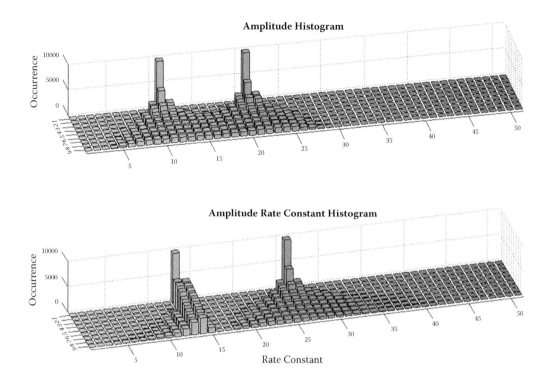

FIGURE 11.10   Top: Amplitude histograms showing the spread of content as noise increases. Bottom: Rate constant histograms showing the spread of the parameter estimation as noise increases.

REFERENCES

Abramovitz, M., and Stegun, I. A. 1972. *Handbook of mathematical functions*. New York: Dover Publications.

Bailey, E. A., and Rollefson, G. K. 1953. The determination of the fluorescence lifetimes of dissolved substances by a phase shift method. *Journal of Chemical Physics* 21:1315–1322.

Becker, W., Bergmann, A., and Biskup, C. 2007. Multispectral fluorescence lifetime imaging by TCSPC. *Microscopy Research and Technique* 70:403–409.

Bednarkiewicz, A., Bouhifd, M., and Whelan, M. P. 2008. Digital micromirror device as a spatial illuminator for fluorescence lifetime and hyperspectral imaging. *Applied Optics* 47:1193–1199.

Berberan-Samtos, M. N. 1991. The time dependence of rate coefficients and fluorescence anisotropy for non-delta production. *Journal of Luminescence* 50:83–87.

Berberan-Santos, M. N., and Martinho, J. M. G. 1990. Kinetics of sequential energy-transfer processes. *Journal of Physical Chemistry* 94:5847–5849.

Bernas, T., Asem, E., Roinson, J., and Rajwa, B. 2006. Compression of fluorescence microscopy images based on the signal-to-noise estimation. *Microscopy Research and Technique* 69:1.

Birks, J. B. 1968. Energy transfer in organic systems VI. Fluorescence response functions and scintillation pulse shapes. *Journal of Physics B (Proceedings of the Physical Society)* 1:946–957.

Birks, J. B. 1970. *Photophysics of aromatic molecules*. London: Wiley.

Birks, J. B., and Dawson, D. J. 1961. Phase and modulation fluorometer. *Journal of Scientific Instruments* 38:282–295.

Birks, J. B., and Little, W. A. 1953. Photo-fluorescence decay times of organic phosphors. *Proceedings of the Physical Society Section A* 66:921–928.

Bottcher, C. J. F., and Bordewijk, P. 1978. *Theory of electric polarization.* New York: Elsevier.

Boutet de Monvel, J., Le Calvez, S., and Ulfendahl, M. 2001. Image restoration for confocal microscopy: Improving the limits of deconvolution, with application to the visualization of the mammalian hearing organ. *Biophysical Journal* 80:2455–2470.

Bracewell, R. N. 1978. *The Fourier transform and its applications.* Tokyo: McGraw–Hill Kogakusha, LTD.

Brigham, E. O. 1974. *The fast Fourier transform.* Englewood Cliffs, NJ: Prentice Hall.

Brochon, J.-C., Livesey, A. K., Pouget, J., and Valeur, B. 1990. Data analysis in frequency-domain fluorometry by the maximum entropy method—Recovery of fluorescence lifetime distributions. *Chemistry and Physics Letters* 174:517–522.

Brown, J. W., and Churchill, R. V. 1996. *Complex variables and applications.* New York: McGraw–Hill, Inc.

Buranachai, C., and Clegg, R. M. 2007. Fluorescence lifetime imaging in living cells. In *Fluorescent proteins: Methods and applications.*, ed. J. Rothnagel. Towata, NJ: Humana Press.

Buranachai, C., Kamiyama, D., Chiba, A., Williams, B. D., and Clegg, R. M. 2008. Rapid frequency-domain FLIM spinning disk confocal microscope: Lifetime resolution, image improvement and wavelet analysis. *Journal of Fluorescence.* Epub ahead of print: http://www.ncbi.nlm.nih.gov/pubmed/18324453

Butz, T. 2006. *Fourier transformation for pedestrians.* New York: Springer.

Byron, F. W., and Fuller, R. W. 1969. *Mathematics of classical and quantum physics.* Reading, MA: Addison–Wesley Publishing Company.

Chorvat, D., Jr., and Chorvatova, A. 2006. Spectrally resolved time-correlated single photon counting: A novel approach for characterization of endogenous fluorescence in isolated cardiac myocytes. *European Biophysics Journal* 36:73–83.

Clayton, A. H. A., Hanley, Q. S., and Verveer, P. J. 2004. Graphical representation and multicomponent analysis of single-frequency fluorescence lifetime imaging microscopy data. *Journal of Microscopy* 213:1–5.

Clegg, R. M., and Schneider, P. C. 1996. Fluorescence lifetime-resolved imaging microscopy: A general description of the lifetime-resolved imaging measurements. In *Fluorescence microscopy and fluorescent probes,* ed. J. Slavik, 15–33. New York: Plenum Press.

Clegg, R. M., Schneider, P. C., and Jovin, T. M. 1996. Fluorescence lifetime-resolved imaging microscopy. In *Biomedical optical instrumentation and laser-assisted biotechnology,* ed. A. M. Verga Scheggi, S. Martellucci, A. N. Chester, and R. Pratesi, 325, 143–156. Dordrecht, the Netherlands: Kluwer Academic Publishers.

Cole, K. S., and Cole, R. H. 1941. Dispersion and absorption in dielectrics. *Journal of Chemical Physics* 9:341.

Colyer, R. A., Lee, C., and Gratton, E. 2008. A novel fluorescence lifetime imaging system that optimizes photon efficiency. *Microscopy Research and Technique* 71:201–213.

Cundall, R. B., and Dale, R. E., eds. 1983. *Time-resolved fluorescence spectroscopy in biochemistry and biology.* NATO ASI series. Series A, life sciences. New York: Plenum.

Davis, H. F. 1993. *Fourier series and orthogonal functions.* New York: Dover.

De Beule, P., Owen, D. M., Manning, H. B., Talbot, C. B., Requejo-Isidro, R., Dunsby, C., et al. 2007. Rapid hyperspectral fluorescence lifetime imaging. *Microscopy Research and Technique* 70:481–484.

Dern, H., and Walsh, J. B. 1963. Analysis of complex waveforms. In *Physical techniques in biological research: Electrophysiological methods, part B,* ed. W. L. Nastuk, VI, 99–218. New York: Academic Press.

Digman, M. A., Caiolfa, V. R., Zamai, M., and Gratton, E. 2007. The phasor approach to fluorescence lifetime imaging analysis. *Biophysical Journal: Biophysical Letters* 107:L14–L16.

Duncan, R. R., Bergmann, A., Cousin, M. A., Apps, D. K., and Shipston, M. J. 2004. Multidimensional time-correlated single photon counting (TCSPC) fluorescence lifetime imaging microscopy (FLIM) to detect FRET in cells. *Journal of Microscopy* 215:1–12.

Elder, A., Schlachter, S., and Kaminski, C. F. 2008. Theoretical investigation of the photon efficiency in frequency-domain fluorescence lifetime imaging microscopy. *Journal of the Optical Society of America A* 25:452–462

Feng, W., and Isayev, A. I. 2004. Continuous ultrasonic devulcanization of unfilled butyl rubber. *Journal of Applied Polymer Science* 94:1316–1325.

Forde, T. S., and Hanley, Q. S. 2006. Spectrally resolved frequency domain analysis of multifluorophore systems undergoing energy transfer. *Applied Spectroscopy* 60:1442–1552.

Franklin, P. 1949. *An introduction to Fourier methods and the Laplace transformation.* New York: Dover.

Frenkel, A., Sartor, M., and Wlodawski, M. 1997. Photon-noise-limited operation of intensified CCD cameras. *Applied Optics* 36:5288–5297.

Genovesio, A., Liedl, T., Emiliani, V., Parak, W., Coppey-Moisan, M., and Olivo-Marin, J. 2006. Multiple particle tracking in 3-D+ t-microscopy: Method and application to the tracking of endocytosed quantum dots. *IEEE Transactions on Image Processing* 15:1062–1070.

Goldman, S. J. 1948. *Frequency analysis, modulation and noise.* New York: Dover.

Gratton, E., Breusegem, S., Sutin, J., Ruin, Q., and Barry, N. P. 2003. Fluorescence lifetime imaging for the two-photon microscope: Time-domain and frequency-domain methods. *Journal of Biomedical Optics* 8:381–390.

Gratton, E., Jameson, D. M., and Hall, R. 1984. Multifrequency phase and modulation fluorometry. *Annual Review of Biophysics and Bioengineering* 13:105–124.

Grgic, S., Grgic, M., and Zovko-Cihlar, B. 2001. Performance analysis of image compression using wavelets. *IEEE Transactions on Industrial Electronics* 48:682–695.

Grinvald, A., and Steinberg, I. Z. 1974. On the analysis of fluorescence decay kinetics by the method of least-squares. *Analytical Biochemistry* 59:583–593.

Hamming, R. W. 1989. *Digital filters.* New York: Dover.

Hanley, Q. S., Arndt-Jovin, D. J., and Jovin, T. M. 2002. Spectrally resolved fluorescence lifetime imaging microscopy. *Applied Spectroscopy* 56:155–166.

Hanley, Q. S., and Clayton, A. H. A. 2005. AB-plot assisted determination of fluorophore mixtures in a fluorescence lifetime microscope using spectra or quenchers. *Journal of Microscopy* 218:62–67.

Hill, N. E., Vaughan, W. E., Price, A. H., and Davies, M. 1969. *Dielectric properties and molecular behavior.* New York: van Nostrand Reinhold Company.

Holub, O., Seufferheld, M. J., Gohlke, C., Govindjee, Heiss, G. J., and Clegg, R. M. 2007. Fluorescence lifetime imaging microscopy of *Chlamydomonas reinhardtii:* Nonphotochemical quenching mutants and the effect of photosynthetic inhibitors on the slow chlorophyll fluorescence transients. *Journal of Microscopy* 226:90–120.

Hong, L. 1993. Multiresolutional filtering using wavelet transform. *IEEE Transactions on Aerospace and Electronic Systems* 29:1244–1251.

Istratova, A. A., and Vyvenko, O. F. 1999. Exponential analysis in physical phenomena. *Review of Scientific Instruments* 70:1233–1257.

Jameson, D. M., Gratton, E., and Hall, R. D. 1984. The measurement and analysis of heterogeneous emissions by multifrequency phase and modulation fluorometry. *Applied Spectroscopy Reviews* 20:55–106.

Jonscher, A. K. 1983. *Dielectric relaxation in solids.* London: Chelsea Dielectrics Press.

Kaplan, W. 1984. *Advanced calculus,* 98–99, 123, 238–245. Reading, MA: Addison–Wesley.

Kim, H., Lee, H., and Lee, J. W. 2007. Rheological properties of branched polycarbonate prepared by an ultrasound-assisted intensive mixer. *Korea–Australia Rheology Journal* 19:1–5.

Lakowicz, J. R. 1999. *Principles of fluorescence spectroscopy.* New York: Kluwer Academic/ Plenum Publishers.

Lakowicz, J. R., and Balter, A. 1982a. Analysis of excited-state processes by phase-modulation fluorescence spectroscopy. *Biophysical Chemistry* 16:117–132.

Lakowicz, J. R., and Balter, A. 1982b. Differential-wavelength deconvolution of time-resolved fluorescence intensities. A new method for the analysis of excited-state processes. *Biophysical Chemistry* 16:223–240.

Lakowicz, J. R., and Balter, A. 1982c. Theory of phase-modulation fluorescence spectroscopy for excited-state processes. *Biophysical Chemistry* 16:99–115.

Lakowicz, J. R., Gratton, E., Cherek, H., Maliwal, B. P., and Laczko, G. 1984. Determination of time-resolved fluorescence emission spectra and anisotropies of a fluorophore–protein complex using frequency-domain phase-modulation fluorometry. *Journal of Biological Chemistry* 259:10967–10972.

Lawson, J. L., and Uhlenbeck, G. E., eds. 1950. *Threshold signals*. New York: Dover.

Levenberg, K. 1944. A method for the solution of certain nonlinear problems in least squares. *Quarterly of Applied Mathematics* 2:164–168.

Malachowski, G., and Ashcroft, R. 1986. The Chebyshev transformation linearizes the fitting of exponential, Gaussian and binding function data. *Proceedings of the Australian Society of Biophysics*, 10th annual meeting, Canberra, Australia.

Malachowski, G. C. 2005. Analytical solutions to the modeling of linear dynamic systems, accompanying paper. *Journal of Microscopy*.

Malachowski, G. C., Clegg, R. M., and Redford, G. I. 2007. Analytic solutions to modeling exponential and harmonic functions using Chebyshev polynomials: Fitting frequency-domain lifetime images with photobleaching. *Journal of Microscopy* 228:282–295.

Mallat, S. 1989. A theory for multiresolution signal decomposition: The wavelet representation. *IEEE Transactions on Pattern Analysis and Machine Intelligence* 11:674–693.

Marquardt, D. 1963. An algorithm for least-squares estimation of nonlinear parameters. *SIAM Journal on Applied Mathematics* 11:431–441.

Marriott, G., Clegg, R. M., Arndt-Jovin, D. J., and Jovin, T. M. 1991. Time-resolved imaging microscopy. Phosphorescence and delayed fluorescence imaging. *Biophysical Journal* 60:1374–1387.

Millman, J., and Halkias, C. C. 1972. *Integrated electronics: Analog and digital circuits and systems*. New York: McGraw–Hill Book Company.

Mueller, F. H., ed. 1953. *Das Relaxationsverhalten der Materie 2. Marburger Discussionstagung. Sonderausgabe der Kolloid-Zeitschrift*. Marburg: Dietrich Steinkopff.

Nowak, R., and Baraniuk, R. 1999. Wavelet-domain filtering for photon imaging systems. *IEEE Transactions on Image Processing* 8:666–678.

Olivo-Marin, J.-C. 2002. Extraction of spots in biological images using multiscale products. *Pattern Recognition* 35:1989–1996.

Prony, B. G. R. 1795. Essai éxperimental et analytique: Sur les lois de la dilatabilité de fluides élastique et sur celles de la force expansive de la vapeur de l'alkool, à différentes températures. *Journal de l'École Polytechnique* 1:24–76.

Prucnal, P., and Saleh, B. 1981. Transformation of image-signal-dependent noise into image-signal-independent noise. *Optics Letters* 6:316–318.

Raju, G. G. 2003. *Dielectrics in electric fields*. New York: Marcel Dekker.

Ramadass, R., Becker, D., Jendrach, M., and Bereiter-Hahn, J. 2007. Spectrally and spatially resolved fluorescence lifetime imaging in living cells: TRPV4–microfilament interactions. *Archives of Biochemistry and Biophysics* 463:27–36.

Rayleigh, J. W. S. 1945. *The theory of sound (1894)*. New York: Dover.

Redford, G. I., and Clegg, R. M. 2005. Polar plot representation for frequency-domain analysis of fluorescence lifetimes. *Journal of Fluorescence* 15:805–815.

Redford, G. I., Majumdar, Z. K., Sutin, J. D. B., and Clegg, R. M. 2005. Properties of microfluidic turbulent mixing revealed by fluorescence lifetime imaging. *Journal of Chemical Physics*. 123:224504–224510.

Rueck, A., Huelshoff, C., Kinzler, I., Becker, W., and Steiner, R. 2007. SLIM: A new method for molecular imaging. *Microscopy Research and Techniques* 70:485–492.

Sandel, B., and Broadfoot, A. 1986. Statistical performance of the intensified charged-coupled device. *Applied Optics* 25:4135–4140.

Schneider, P. C., and Clegg, R. M. 1997. Rapid acquisition, analysis, and display of fluorescence lifetime-resolved images for real-time applications. *Review of Scientific Instruments* 68:4107–4119.

Shapiro, J., Center, D., and Princeton, N. 1993. Embedded image coding using zerotrees of wavelet coefficients. *IEEE Transactions on Signal Processing* (see also *IEEE Transactions on Acoustics, Speech, and Signal Processing*) 41:3445–3462.

Shapiro, J. M. 1991. Embedded image coding using zerotrees of wavelet coefficients. *IEEE Transactions on Signal Processing* 41:3445–3462.

Siegel, J., Elson, D. S., Webb, S. E. D., Parsons-Karavassilis, D., Lvque-Fort, S., Cole, M. J., Lever, M. J., French, P. M. W., Neil, M. A. A., Jukaitis, R., Sucharov, L. O., and Wilson, T. 2001. Whole-field five-dimensional fluorescence microscopy combining lifetime and spectral resolution with optical sectioning. *Optics Letters* 26:1338–1340.

Simon, W. 1987. *Mathematical techniques for biology and medicine.* New York: Dover Publications.

Spring, B. Q. and Clegg, R. M. (to appear). Image analysis of denoising full-field frequency domain fluoresence lifetime images. *Journal of Microscopy.*

Starck, J., Murtagh, F., and Bijaoui, A. 1998. *Image processing and data analysis: The multiscale approach.* Cambridge, England: Cambridge University Press.

Starck, J.-L., and Bijaoui, A. 1994. Filtering and deconvolution by the wavelet transform. *Signal Processing* 35:195–211.

Tolstov, G. P. 1962. *Fourier series.* New York: Dover.

Valeur, B. 2002. *Molecular fluorescence: Principles and applications.* Weinheim: Wiley-VCH.

Valeur, B. 2005. Pulse and phase fluorometries: An objective comparison. In *Fluorescence spectroscopy in biology,* ed. O. S. Wolfbeis, M. Hof, R. Hutterer, and V. Fidler, 3, 30–48. Berlin: Springer.

von Hippel, A. R. 1954. *Dielectrics and waves.* New York: Wiley.

Walker, J. 1997. Fourier analysis and wavelet analysis. *Notices of the AMS* 44:658–670.

Willett, R., and Nowak, R. 2003. Platelets: A multiscale approach for recovering edges and surfaces in photon-limited medical imaging. *IEEE Transactions on Medical Imaging* 22:332–350.

Willett, R., and Nowak, R. 2004. Fast multiresolution photon-limited image reconstruction. *IEEE International Symposium on Biomedical Imaging: Macro to Nano, 2004* 1192–1195.

CHAPTER 12

Nonlinear Curve-Fitting Methods for Time-Resolved Data Analysis

Ignacy Gryczynski, Rafal Luchowski, Shashank Bharill, Julian Borejdo, and Zygmunt Gryczynski*

12.1 INTRODUCTION

Time-resolved fluorescence spectroscopy and recently time-resolved fluorescence micros-copy have proven to be powerful technologies for studying macromolecular interactions and dynamics in biological systems on the subpicosecond to millisecond timescale. In complex biological systems including membranes, nuclei, or even entire cells, the fluo-rescence decay kinetics can reveal detailed information about intrinsic relaxation mecha-nisms, macromolecular interactions, conformational changes, and dynamics of complex molecular processes. Depending on the excitation type, the instrumental methods for mea-suring fluorescence intensity decays (fluorescence lifetimes) traditionally are divided into two dominant methods: time domain (O'Connor and Philips 1984; Birch and Imhof 1991; Demas 1983; Chapter 6) and frequency domain (Gratton and Limkeman 1983; Gratton et al. 1984; Lakowicz 1999; Chapter 5).

In time-domain technology, the sample is excited with a pulse of light. The width of the pulse is made as short as possible and is preferably much shorter than the fluores-cence lifetime of the sample. The time-dependent intensity decay is measured following the excitation pulse. In the frequency-domain or phase-modulation method, the sample is excited with intensity modulated light. Typically, a sine wave modulation of light is used for excitation. The intensity modulated excitation forces the fluorescence inten-sity to respond at the same modulation frequency. Because of the finite fluorescence

---

* We dedicate this chapter to Professor Enrico Bucci for his contribution to the field of data analysis in hemoprotein systems.

lifetime, the fluorescence intensity is delayed in time relative to the excitation. In effect, the delayed response results in a phase shift and demodulation of the signal compared to the excitation.

Technological improvements, such as picosecond solid-state laser diodes and new fast detectors (microchannel plate photomultipliers and avalanche photodiodes), together with advancements in electronics, enabled many new practical applications of time- and frequency-domain measurements. Speed of data acquisition in time- and frequency-domain methods has been greatly increased, allowing in certain cases, almost real-time time-resolved microscopy imaging. In effect, time-resolved measurements of fluorescence can now be applied to the study of a variety of chemical and biochemical processes in cells and tissue.

As technology progresses, the precision and quality of time-resolved data collected in spectroscopic or imaging experiments quickly increase and routine time-resolved measurements now reveal much fundamental information about biological systems. Obtained data are relatively complex and the interpretation of fluorescence decays is very demanding. In spite of tremendous technological progress, multiple problems arise from the fact that the instrument response time is in many cases comparable to the fluorescence lifetime, masking the true fluorescence response of the system. To obtain a pure fluorescence response of the system (decays free from instrumental distortions), one must apply analytical procedures allowing deconvolution of experimental data from the instrument perturbation; this can be a complex and lengthy process.

The purpose of this chapter is to familiarize the reader with the basics of data analysis, show how time-resolved data can be analyzed and interpreted, and present and discuss the limits of fitting procedures. Because typical time-resolved experiments are very advanced, we also want to focus attention on possible errors that may lead to misinterpretation of experimental data.

12.2 BACKGROUND

The goal of numerical analysis of fluorescence decay kinetics is to determine the analytic expression and the corresponding numerical values of the physical parameters that properly describe the intensity decay process. The concept of the most common method of least-squares analysis (Grinvald and Steinberg 1974; Good et al. 1984; Johnson and Fresier 1985; Johnson 1994) is to assume a plausible functional form for the decay data and to adjust the values of the parameters until the statistically best fit is obtained between the experimental data and the calculated decay function. The goodness of the fit is usually judged by the sum of the weighted squares of the residuals, which is supposed to be a minimum.

Generally, time-resolved data are relatively complex and the interpretation of fluorescence decays often requires knowledge of the analytical expression for the decay function. A fundamental problem for time-domain methods arises from the fact that the fluorescence lifetimes are frequently comparable to the excitation pulse width and instrument response function, and this greatly complicates data analysis. The typical instrument response function (IRF) usually cannot be neglected, and to obtain decays free from instrumental distortion, one must use analytical methods to separate the pure undistorted fluorescence intensity decay that is convolved with the excitation pulse (Ware et al. 1973; Grinvald and

Steinberg 1974; O'Connor, Ware, and Andre 1979; Straume, Frasier-Cadoret, and Johnson 1991).

Such deconvolution procedures of experimental data are complex and are frequently lengthy processes. The purpose of deconvoluting fluorescence decays is to free them from sources of distortion, such as the finite rise time, the width, and the decay of the excitation source, as well as to remove distortions introduced by the detector and timing apparatus. Typical deconvolution methods are divided into two groups: those requiring an assumption of the functional form of the undistorted system response function (assumption of the decay law) and those that directly give the undistorted system decay law without any assumption about the decay law.

Over the years many numerical methods have been proposed for the analysis of time-resolved fluorescence decays. Some of them, like the Fourier (Munro and Ramsay 1968) and Laplace (Gafni, Modlin, and Brand 1975; Ameloot 1992) transform methods do not require prior knowledge of the undistorted system response function. However, they frequently suffer from instabilities due to the random fluctuations inevitably present in real data. Today, a much more widely used method of deconvolution by convolution (Helman 1971; Ware, Doemeny, and Nemzek 1973) involves the assumption of the exact functional form of the undistorted fluorescent system response. The principles of this method have been discussed by various authors (Knight and Seliger 1971; Ware et al. 1973; Demas and Crosby 1970; Grinvald and Steinberg 1974; Isenberg et al. 1973; Small 1992; Johnson 1994). This deconvolution method usually assumes a simple exponential form for the undistorted system response function.

12.3 METHODS

The obvious reason for carrying out the experiment and data analysis is to recover the actual undistorted system response function and subject it to examination by various physical models. All linear and nonlinear least-squares data analyses are based on a series of assumptions that must be strictly obeyed in order to obtain valid results. One has to remember that the fundamental goal of the parameter estimation technique (fit) is to obtain an estimate of parameter values that have the highest probability of being correct. Before describing the basics of the mathematics involved in parameter estimation, we need to introduce some commonly used terminology.

12.3.1 Basic Terminology and Assumptions

Today, most time-resolved fluorescence measurements are made using time-domain or frequency-domain technology. For a time-domain experiment, the data set consists of a series of observations of the fluorescence response and the time course of the lamp intensity (for both, the data are typically acquired by counting photons) collected as a function of time (in time-separated channels). For the frequency domain, the measured data set consists of a series of observations of amplitude modulation (demodulation) and the phase shift of emitted light as a function of modulation frequency of the excitation light. For a mathematical description, we refer to a data set as a series of observations, X_i and Y_i, where the subscript refers to each individual data point within the data set.

We call the quantities whose values we can control by appropriate adjustments of the instrumentation *independent variables*. The independent variable for a time-domain experiment is time (t) and for a frequency-domain experiment it is usually the frequency (ω) of excitation light modulation. We define the independent variable, X_i.

Those quantities whose values are actually measured by the experimental protocol we call the *dependent variables*. For the time-domain experiment, the temporal response of the fluorescence intensity (number of collected photons) is the dependent variable. For the frequency-domain experiment, the amplitude of modulation and the phase shift are the dependent variables.

For clarity, we present the independent and dependent variables in two dimensions as a series of scalar quantities (X_i and Y_i). However, the mathematical procedures apply equally when X_i and Y_i are vectors in multidimensional space. Some excellent reviews can be found in Straume et al. (1991) and Johnson and Faunt (1992). A good example for the vector representation is a frequency-domain measurement where two dependent variables (amplitude of the modulation and the phase shift) can appropriately be handled as a two-dimensional vector.

The *fitting function* is the mathematical relationship between the independent and dependent variables:

$$Y_i = G(\alpha, X_i) + \rho_i \qquad (12.1)$$

where α is a vector of parameters of the fitting function, $G(\alpha, X_i)$, and ρ_i is the experimental uncertainty vector that describes the difference between an individual data point, Y_i, and the fitting function obtained for the optimal values of α as expressed by Equation 12.1.

The overall goal of data analysis of the given set of data, X_i and Y_i, is to evaluate a set of parameters, α, of the fitting function, $G(\alpha, X_i)$, that have the maximum likelihood (highest probability) of being correct. Furthermore, the obtained function for these optimal parameters must be a reasonable description of the actual experimental data in the absence of experimental uncertainties. In the preceding decades, several regression algorithms have been developed to extract the best satisfying parameter values numerically. It can be proven mathematically that a typical least-squares analysis correctly estimates the parameter values if the data (both the function $G(\alpha, X_i)$ and the distribution of experimental uncertainties–errors) satisfy a series of assumptions:

- All of the uncertainties of the experimental data are only attributed to the dependent variables, Y_i.

- The experimental uncertainties of the dependent variable Y_i obey a Gaussian distribution centered at the correct value of the dependent variable.

- There are no unknown/uncontrollable systematic uncertainties in either the independent, X_i, or dependent, Y_i, variables.

- The assumed fitting function, $G(\alpha, X_i)$, is the correct mathematical description of the experimental data.

- The number of data points is sufficient to yield a good random sampling of the random experimental uncertainty.

- Each data point must be an independent observation.

In practice, the first assumption means that the values of the independent variables are known with much greater precision than the values of the dependent variables and the mathematical fitting may ignore experimental error in the independent variable. Because of this assumption, it is convenient to plot the independent variables on the X-axis (abscissa). The second assumption requires that the distribution of experimental uncertainties must follow a Gaussian distribution. The third assumption requires that the experimental procedure have no systematic errors. An example of systematic uncertainty could be a constant time shift between the lamp intensity profile and the observed fluorescence intensity induced by color effect in a photomultiplier. We will later discuss in detail the effect of systematic errors.

The fourth assumption states that the fitting function, $G(\alpha, X_i)$, correctly describes the phenomena being investigated. Good examples can be the decay curve of a system undergoing normal radiative and nonradiative decay along with a time-dependent spectral shift due to a solvent-solute relaxation process (later we will present the example of Prodan, showing the drastic solvent effect) or fluorescence decay in the presence of diffusion and Forster resonance energy transfer (FRET). In both cases, a typical exponential decay law may not be obeyed.

The fifth assumption requires a sufficient number of data points. The exact number of sufficient data points to ensure a good statistical sampling of the random experimental uncertainties is not clearly defined and usually differs for each set of experimental conditions. In general, a time-domain experiment in a cuvette collects a number of data points sufficient for adequate data analysis (typically, over 100,000 counts per decay). However, this is usually time consuming and an optical microscope cannot always be realized. Many modern frequency-domain systems collect data at multiple frequencies and generate an adequate number of data points for a proper least-squares analysis. The final assumption requires that each experimental point be independent and not influence any other experimental point.

If the experimental set of data obeys all six assumptions, it can be demonstrated that the method of least-squares analysis will yield parameters values, α, with the highest probability of being correct. Overall, both time and frequency domains satisfy the assumptions of least-squares analysis.

12.3.2 Least-Squares Analysis

In a typical least-squares analysis, one starts by assuming that a selected model properly describes the experimental data. The most common representation of the fluorescence intensity decay, $I(\alpha, t)$, as a function of time, t, is given by the following exponential function (Lakowicz 1999; Gryczynski 2005):

$$I(\alpha, t) = I_0 \, e^{-\frac{t}{\tau}} \tag{12.2}$$

where I_0 is the (initial) fluorescence intensity at time zero and τ is the fluorescence lifetime.

12.3.2.1 Time Domain

For time-domain (photon-counting) types of measurement such as time-correlated sin-gle-photon counting (TCSPC) (O'Connor and Philips 1984; Birch and Imhof 1991), the fluorescence intensity is directly related to the number of molecules in the excited state. In photon counting, the time-dependent number of counts per unit time, $N(\alpha, t)$, is

$$N(\alpha,t) = N_0\, e^{-\frac{t}{\tau}} \tag{12.3}$$

In many cases, the fluorescence intensity decays are longer than the typical duration of the excitation pulse and the time response of the instrument (detector/electronics). However, for time-dependent analysis one should remember that the observed time-dependent signal is a convolution of undistorted fluorescence intensity decay with the excitation profile and the detection response (IRF). The corresponding fitting function, $G(\alpha, X_i)$, is then a convolution integral of the true fundamental fluorescence response, $N(\alpha, t)$, and the instrument response function (IRF), also called lamp function, $L(t)$:

$$G(\alpha, t) = \int_0^{X_i} L(t)\, N(\alpha, X_i - t)\, dt \tag{12.4}$$

where the integration over the time, t, is from time 0 to X_i, corresponding to the time value of the particular data point. The vector of fitted parameters, α, consists of two values: the initial (time zero) number of counts, N_0, and fluorescence lifetime, τ. The IRF as a function of time must be experimentally determined for each measurement.

Consequently, one must solve the integral Equation 12.4 that involves the undistorted luminescence system response function (undistorted intensity decay or fundamental fluorescence response) and the independently measured time-dependent IRF. The analysis of fluorescence intensity decays has to consider the finite duration of the IRF, which contains pulse duration and the response time of the detection system.

Importantly, the IRF must have identical profiles in both measurements, a requirement that can be difficult to achieve experimentally with high accuracy. We can mathematically (numerically) correct the different number of counts (intensity) of the fluorescence and IRF signals, but a change of shape is usually very difficult to correct. For example, the observation (emission) wavelength for the fluorescence measurement and the IRF measurement are typically different, and the color effect in the detector may not produce only a temporal shift of the detected pulse; more importantly, the effective shapes of IRF for longer and shorter wavelengths may differ (Bebelaar 1986). In the experimental section, we present examples of how to assess the pulse shape change due to wavelength differences.

We assume that the decay law in Equations 12.2 and 12.3 is the fitting function $G(\alpha, X_i)$ (Equation12.1). A least-squares analysis involves differentiation only with respect to the components of the vector α. After differentiation, the integrals in Equation 12.4 will not involve an unknown fitting parameter and can thus be evaluated. The calculated time-dependent intensity decay, $I_F^{cal}(t_i)$ (in our case, $N_F^{cal}(t_i)$), can be calculated from the convolution integral (Equation 12.4):

$$N_F^{cal}(t_i) = \int_0^{t_i} L(t_i - t) N_0 e^{-t/\tau} \, dt \tag{12.5}$$

The calculated fluorescence intensity decay depends on two fitted parameters: N_0 and τ. If $N_F^{ob}(t_i)$ is the measured number of counts (observed intensities), the residual ρ_i for the ith data point is given by

$$\rho_i = N_F^{ob}(t_i) - N_F^{cal}(t_i) \tag{12.6}$$

The deconvolution problem reduces to obtaining the coefficients (fitted parameters) that minimize residuals ρ_i for all of the data points. The quality of the fit can then be judged by the magnitude of the sum of the weighted squares of the residuals as defined by

$$\Phi = \sum_{i=1}^n w_i (N_F^{ob}(t_i) - N_F^{cal}(t_i))^2 \tag{12.7}$$

where t_i is the time at the ith time interval for which a measurement or calculation is made (for photon counting, it typically corresponds to the channel number), and w_i is the weighting factor given to the square deviation between $N_F^{cal}(t_i)$ and $N_F^{ob}(t_i)$ at the ith time increment. The weighting factor, w_i, generally depends upon the experimental setup and is related to the corresponding data point variance, s_i^2:

$$s_i^2 = \left[\frac{N_F^{ob}(t_i) - N_F^{cal}(t_i)}{\sigma_i} \right]^2 \tag{12.8}$$

by the following relation:

$$w_i = \frac{1/s_i^2}{(1/n) \sum_{i=1}^n (1/s_i^2)} \tag{12.9}$$

where σ_i is the standard deviation of the random experimental uncertainty for a particular data point.

For photon-counting techniques, the statistical noise typically obeys Poisson statistics, approaching a Gaussian distribution at large numbers of counts. The standard deviation, σ_i, is proportional to the square root of the number of photons. The overall goodness of fit for all data points is the sum of the corresponding variances, usually called chi square (χ^2):

$$\chi^2 = \sum_{i=1}^{n} s_i^2 = \sum_{i=1}^{n} \left[\frac{N_F^{ob}(t_i) - N_F^{cal}(t_i)}{\sigma_i} \right]^2 = \sum_{i=1}^{n} \frac{(N_F^{ob}(t_i) - N_F^{cal}(t_i))^2}{N_F^{ob}} \qquad (12.10)$$

Chi square depends on the number of data points and is not a convenient parameter to judge the goodness of fit. A much more convenient way is to use the value of reduced chi square, χ_R^2, which represents the standard weighted least squares for the set of experimental data points and is given by

$$\chi_R^2 = \frac{1}{NDF} \sum_{i=1}^{n} \chi_i^2 = \frac{1}{NDF} \sum_{i=1}^{n} \left[\frac{N_F^{ob}(t_i) - N_F^{cal}(t_i)}{\sigma_i} \right]^2 \qquad (12.11)$$

where NDF is the number of degrees of freedom defined as the number of independent experimental observations (for the time domain, the number of data points), n, minus the number of fitting parameters (number of floating parameters), p; $NDF = n - p$.

12.3.2.2 Frequency Domain

For frequency-domain measurements, there are two dependent variables (amplitude modulation and phase shift) and two fitting functions, $G(\alpha, X_i)_{am}$ and $G(\alpha, X_i)_{ph}$ and two fitting functions should be considered (Gratton et al. 1984):

$$G(\alpha, X_i)_{am} = [D(\alpha, \omega)^2 + N(\alpha, \omega)^2]^{1/2} \quad and \quad G(\alpha, X_i)_{ph} = \frac{1}{\tan[N(\alpha, \omega) / D(\alpha, \omega)]} \qquad (12.12)$$

where X_i is the excitation light modulation frequency, ω, and $N(\alpha, \omega)$ and $D(\alpha, \omega)$ are sine and cosine intensity transforms:

$$N(\alpha, \omega) = \frac{\int_0^{\infty} I(\alpha, t) \sin \omega t \, dt}{\int_0^{\infty} I(\alpha, t) \, dt} \quad and \quad D(\alpha, \omega) = \frac{\int_0^{\infty} I(\alpha, t) \cos \omega t \, dt}{\int_0^{\infty} I(\alpha, t) \, dt} \qquad (12.13)$$

For frequency-domain experiments, the vector of fitted parameters, α, has only two lifetime elements: one for amplitude modulation, called τ_m, and one for phase shift, called τ_p. The initial intensity amplitude, I_0, has been canceled by the integral normalization. The calculated frequency-dependent values of the phase angle ($\phi_{cal\omega}$) and the demodulation ($m_{cal\omega}$) are given by

$$\tan(\varphi_{cal\omega}) = \frac{N_\omega}{D_\omega} \quad and \quad m_{cal\omega} = \sqrt{(N_\omega^2 + D_\omega^2)} \tag{12.14}$$

By analogy to time-domain representation, one may define the reduced chi square, χ_R^2, that defines the goodness of fit:

$$\chi_R^2 = \frac{1}{NDF} \sum_\omega \left[\frac{\varphi_\omega - \varphi_{cal\omega}}{\delta\varphi} \right] + \frac{1}{NDF} \sum \left[\frac{m_\omega - m_{cal\omega}}{\delta\omega} \right] \tag{12.15}$$

where *NDF* is the number of degrees of freedom that, in the frequency domain, is twice the number of frequencies minus the number of variable parameters. Because the number of variable frequencies is usually below 30, the values of uncertainties in phase, $\delta\phi$, and modulation, δm, cannot be estimated directly from the Poisson statistic. These values are defined by the instrumentation prior to the experiment or measured during the experiment from multiple determinations for each point.

The quality of the fit is judged based on the values of χ_R^2; for an appropriate model and for a random noise distribution, its value is supposed to be close to 1.

12.3.3 Least-Squares Parameter Estimation

The nonlinear least-squares minimization is the preferred method of data analysis in time-resolved fluorescence measurements. Typically, the fitting process can be performed by a number of numerical procedures—for example, Gauss–Newton (Stroume et al. 1991) or the Nelder–Mead 1965 simplex procedure. The goal of these numerical procedures is to find a set of values for the fitting parameters, α, that best describe the experimental data set. It is not our intent to discuss numerical methods in this chapter; numerous detailed descriptions are available (Bevington 1969; Ware et al. 1973; Grinvald and Steinberg 1974; O'Connor et al. 1979; Livesey and Brochon 1987; Straume et al. 1991; Johnson 1994). Instead of detailed discussion of numerical procedures, we will focus on how to interpret the results of data analysis and how to judge the results of the analysis.

In the analysis of time-resolved data, one should consider different fits to the experimental data. Each fit is judged by the χ_R^2 value. It is intuitive that the value of χ_R^2 decreases for a model with more adjustable parameters. It is important to consider how large a change in the χ_R^2 value is adequate to justify the acceptance of a different model or of increasing the number of parameters. Typically, it is not simple to answer this question based only on mathematics. In many cases, one has to depend on experience and on knowledge of the sample. Because a random error distribution is assumed, one can statistically predict the probability of obtaining a given value of χ_R^2. For interested readers, we recommend more detailed descriptions where many mathematical considerations have been discussed

(Grinvald and Steinberg 1974; Straume et al. 1991; Johnson 1994). Many of the considerations have been adequately described and typical probability values have been tabularized (Bevington 1969; Bard 1974; Montgomery and Peck 1982; Lakowicz 2005).

In many cases, reliance on only statistical considerations can be misleading and may frequently result in data overinterpretation. One has to remember that χ_R^2 is just a relative number and its absolute value can be strongly affected by instrumental (systematic) errors. Usually, decay analysis is performed first with the simplest model (for example, a one-exponent model) and then with more complex models. Which model should be accepted? Only large change in the value of χ_R^2—twofold or larger—can be easily accepted. Smaller changes in the χ_R^2 value should be interpreted with great caution, and decisions should be based on prior understanding of the system under investigation.

12.3.4 Diagnostics for Quality of Curve-Fitting Results

It is important to know how one can interpret the quality of curve fitting (least-squares analysis) in order to judge the correctness of the assumed model. Various models (mathematical expressions) may yield reasonable fits to the set of experimental data points (reasonable value of χ_R^2), so the sum of the squares of the residual in Equation 12.6, Φ, is at minimum with the magnitude determined by the intrinsic noise of the experiment (Grinvald and Steinberg 1974).

However, not all models truly reflect physical processes. Ideally, the experimentally observed points, $N^{ob}(t_i)$, should be randomly distributed around the model function that describes the fluorescence intensity decay (which is proportional to the number of molecules in excited state, $N^{cal}(t_i)$). On many occasions, the wrong models will betray themselves by nonrandom scattering of experimental points around the time course of the mathematical function. In this case, the randomness of the experimental distribution can be judged from the autocorrelation function, $C(t_j)$, of the weighted residuals, which can be defined as (Grinvald and Steinberg 1974)

$$C(t_j)=\frac{\dfrac{1}{m}\displaystyle\sum_{i=1}^{m}\sqrt{w_i}\,\Delta_i\sqrt{w_{i+j}}\,\Delta_{i+j}}{\dfrac{1}{n}\displaystyle\sum_{i=1}^{n}w_i\,\Delta_i^2} \qquad (12.16)$$

where $\Delta_i = N^{cal}(t_i) - F^{ob}(t_i)$ represents the deviation in the ith data point, n is the number of experimental points (number of channels in TCSPC), and m is the number of terms in the numerator. The index j can assume the values 1, 2,..., $(n - m)$ and in general practice the value for m should be greater than $n/2$ because smaller values for m may introduce improper averaging for $C(t_j)$.

12.3.5 Uncertainty of Curve-Fitting Procedures

The analysis of experimental data by a least-squares procedure will not only evaluate the parameters, α, that have the maximum likelihood of being correct, but also should evaluate a reasonable range of error for the estimated parameters. This will provide a measure of the overall precision of the estimated parameters. Unfortunately, there are no general methods for estimating the reliability of the extracted values by nonlinear least-squares procedures. Even if the uncertainties are reported by most fitting programs, they are not always adequately considering parameter correlation and they are smaller than the actual uncertainties. We would like to warn experimentalists not to trust completely to the degree of precision reported by data analysis software.

The most common way to determine the range of estimated parameters is to examine the changes in value of χ_R^2 when one parameter is varied around its originally estimated value. This procedure is typically called support plane analysis. The idea is to fix one parameter to a value different from its best estimate and than rerun the least-squares analysis allowing the other parameters to adjust to a new minimum that yields a new value of χ_R^2. We call this new value for the fixed parameter value $\chi_R^2(fp)$ and the minimum value of χ_R^2 for all parameter floating (which is the smallest value) we call $\chi_R^2(min)$. The routine is repeated for larger and larger changes in the value of the fixed parameter, each time allowing all other parameters to adjust to a new minimum.

The procedure is repeated until the $\chi_R^2(fp)$ value for a given value of the parameter exceeds acceptable value. Typically, the range of parameter values is expanded until the new value $\chi_R^2(fp)$ exceeds the value predicted by F-statistics for a given number of parameters (p), the number of degrees of freedom (NDF), and the chosen probability level, P. P describes the probability that the value of the F-statistic (ratio of chi squares for two fits) is due to random error in the data and can be related to the standard deviation of the experiment. For a one standard-deviation confidence interval, the value of P is approximately 67% and for a two standard-deviation confidence interval the value of P increases to about 95%. We express the following (Box 1960; Bevington 1969; Straume et al. 1991):

$$F = \frac{\chi_R^2(fp)}{\chi_R^2(min)} = 1 + \frac{p}{NFD} F(p, NFD, P) \tag{12.17}$$

where $F(p, NDF, P)$ is the F-statistic value with p parameters and NDF with a probability of P. The typical values for the F-statistic can be found in Montgomery and Peck (1982) and Lakowicz (2005).

The supporting plane analysis is usually a convenient way to represent uncertainty in the values of determined parameters graphically.

12.4 EXAMPLES

12.4.1 How to Analyze Experimental Data

Before proceeding with data analysis, one should critically review experimental conditions. In standard experiments, two sources of errors are typically found. One is statistical (random) distribution of experimental errors related to the precision of instrumentation, and the second is the systematic errors generated by experimentalists, instruments, or unknown or unexpected physical or chemical processes occurring in the system. As we discussed earlier, for statistical error, the data analysis and data diagnostic procedures should yield a high probability that the obtained parameters are proper and adequately reflect the physical system. The accuracy of the obtained parameters strictly depends on the number of collected data points and the number of parameters one wants to determine. Later, we will present an example of how the number of collected data points affects the accuracy of obtained parameters.

12.4.2 Systematic Errors

First, we discuss various types of systematic errors. Such errors are much more difficult to detect and are the frequent sources of data misinterpretation. They may dramatically affect data analysis and on many occasions lead to the completely wrong solution. Unfortunately, numerical analysis does not always indicate that a systematic error is present in the measurement and the error should be eliminated before acquiring experimental data.

In typical time-resolved experiments, multiple potential errors are frequently overlooked by the experimentalist. As the time resolution of the instrumentation increases (ability to measure faster times) and one attempts to detect very fast decays, one should consider even the effects generated by the various "passive" optical elements. At the same time, instruments are more advanced and include more active components, and cumulative effects cannot be ignored. Even the simplest system typically uses optical filters to separate excitation from emission. One has to remember that absorption and emission spectra are typically very well separated and the wavelengths for emission and for excitation can sometimes be 100 nm shifted.

In this section, we will present a few examples of possible experimental errors and briefly discuss how these errors influence results. Our goal is to alert the reader to the most common systematic experimental errors, which affect time-resolved fluorescence measurements and mislead a numerical analysis.

12.4.2.1 Light Delay

The wavelength difference between the excitation and emission spectra (Stoke's shift) is considered a very convenient property of fluorescence that allows easy separation of the fluorescence signal from the excitation wavelength simply by use of optical filters or monochromators. Typical optical filters are 2–4 mm thick and it is common to use a combination of several filters. Also, to balance the signal intensity for scattering and emission, it is convenient to use neutral density filters for both fluorescence and scattering measurements. In this case, one should realize that a difference of a few millimeters in filter thickness may result in a significant pulse shift. For example, 1 mm of glass, which has a typical refractive index of 1.5, will delay light about 1.3 ps and a difference of 4 mm in overall thickness (average single-glass filter) results in a delay of over 5 ps. This seems relatively small, but it can have a significant effect on the measurement.

To demonstrate this effect, we used the FT 200 system from Picoquant GmbH (http://www.picoquant.com/) equipped with a multichannel plate photomultiplier (MCP PMT) detector and a 470 nm pulsed laser diode, also from Picoquant. The typical response time

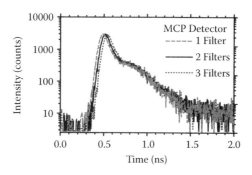

FIGURE 12.1   Instrument response function (IRF) measured through one, two, and three glass filters.

of MCP PMT is in the order of 30 ps (Gryczynski and Bucci 1993) and the pulse width of the laser diode is about 90 ps. Figure 12.1 shows the measured scattering response (IRF) observed with one , two, and three glass filters, respectively (3, 6, and 9 mm total thickness, respectively) in the emission path. A significant pulse shift is clearly visible.

This small shift will not affect long (nanosecond) fluorescence lifetimes, but when short decay components, as in FRET experiments, are measured, it may produce some problems. Such a pulse shift can easily be accounted for in the time-domain data analysis by using the pulse shift function (if the problem is identified), but it could be a significant problem for frequency-domain experiments. The temporal pulse shift affects measured phase, but does not change modulation. For a discussion of the possible effects of pulse delays and straight scattering, see Gryczynski and Bucci (1993).

12.4.2.2   Color Effect in the Detector

In a typical time-resolved experiment, the sample is excited with one wavelength and the fluorescence is observed with a longer wavelength. Under many circumstances, the wavelength difference can be very significant, over 50 nm. Frequently, when long pass filters are used, the wavelength difference could be over 100 nm. As mentioned earlier, the different response of typical detectors to different optical wavelengths may result in very difficult problems in the interpretation of time-resolved data. To demonstrate how significant this effect can be, we present a simple experimental approach for assessing the difference in temporal response of various detectors due to the wavelengths' differences (Carraway et al. 1985).

For this we use the Raman scattering of water that is easy to detect in any laboratory. Raman scattering of water is shifted about 3.3 kK (Keiser = 1 cm^{-1}) compared to the frequency of the excitation beam. For the excitation wavelength of 470 nm, the Raman scattering is detected at 556 nm. This extent of wavelength shift is also typical for many fluorophores. Raman scattering is known to be an instantaneous process (zero time delay) and preserves the temporal shape of the original pulse (the signal is linearly proportional to the intensity). We report results for this experiment using the three most common detectors: MCP PMT and regular PMT in TCSPC (FT 200 and FT 100 Picoquant GmbH, respectively) systems and a PerkinElmer detector in our MT 200 microscope system (Picoquant GmbH).

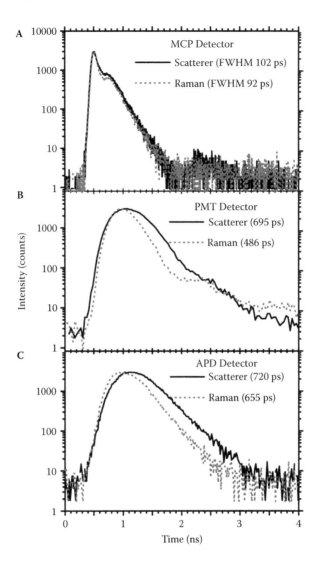

FIGURE 12.2   Different IRF responses measured at 470 and 556 nm (Raman) for MCP PMT, PMT, and APD.

Figure 12.2 shows the scattering and Raman scattering responses measured with MCP PMT (A), PMT (B), and PerkinElmer (C). The first thing that can be noticed is the significantly different IRF for the last two detectors. This difference is generated by the different response time of the detector and electronic system. As expected, the color effect for MCP PMT is minimal, practically undetectable. This effect is quite significant for a PerkinElmer detector and a regular PMT.

It is interesting to note that, in this wavelength range, the response function is typically narrower for longer wavelengths (Raman scattering). The change in the bandwidth of the IRF depends on the wavelength range and may have a significant effect on the results of data fitting, especially when short lifetime components are considered. One can see that in both frequency-domain and time-domain lifetime measurements, using scattering as a reference, a significant problem arises for both PerkinElmer and PMT detectors. Unfortunately, correcting for this effect is very difficult. The only solution is to use a fluorescence standard with a similar spectral profile as the reference. Using a reference fluorophore as a standard is easy to realize and commonly used in the frequency domain, but most time-domain systems are not ready for this.

12.4.2.3 Polarization Effect

Most fluorescence measurements are now made with polarized excitation. Typical laser excitation is intrinsically polarized. Even if unpolarized light is used for excitation, the light is isotropically polarized around the direction of propagation, and molecules are not excited if their transition dipoles lie in the same direction as the excitation beam direction. This induces only limited photoselection and the expected anisotropy for the nonpolarized excitation beam is 0.2—half of the 0.4 for perfectly polarized excitation (Gryczynski, Gryczynski, and Lakowicz 1999; Lakowicz 2005).

The depolarization processes may significantly influence the collected intensity decays, so polarizer orientation on the emission should be very carefully controlled. As an example, we consider acridine orange in propylene glycol (PG). The correlation time for this molecule in PG is comparable to its fluorescence lifetime. Figure 12.3 presents results of fluorescence lifetimes measured with vertically polarized excitation and different orientations of polarizer on observation (emission). Figure 12.3 shows the measured intensity decays for magic angle (A), VV (B), and VH (C) observations. The fit in Figure 12.3(A) for proper magic angle condition (where all rotational motions are canceled; Gryczynski et al. 2005) yields satisfactory results ($\chi_R^2 = 1.002$) already for a single lifetime component of 3.56 ns.

Figure 12.3(B) shows the intensity decay as seen through the vertically oriented polarizer. In this case, the single-exponential fit gives poor results, with clear evidence of systematic error seen in the residuals' distribution (middle panel). Figure 12.3(B) shows a two-exponential fit that gives satisfactory value of χ_R^2. The detailed results of the fit are presented in Figure 12.3(B). However, the second short component (1.54 ns) is clearly an artifact.

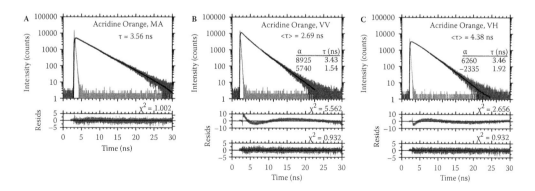

FIGURE 12.3 Fluorescence intensity decay of Acridine Orange in propylene glycol measured at (A) magic angle, (B) vertical, and (C) horizontal polarizer orientation.

Figure 12.3(C) presents the same experiment, but now performed with horizontal orientation of the observation polarizer. The single-exponential fit (Figure 12.3B) again gives an unexpected value of χ_R^2 and residuals' distribution. A careful reader will notice that the residual distributions for the vertical and horizontal observations are opposites showing oscillations that are highly correlated. The two-exponential fit (Figure 12.3C) produces an excellent fit to the data, but now one component has a negative pre-exponential factor. This is a typical result for an orthogonal polarization of observation; however, in this case we know that this is not a physically acceptable result.

Looking at the results in Figure 12.3 makes it possible to realize how easy it is to make such a mistake for an unknown sample and overinterpret the experimental results. Unfortunately, in all three cases (delay of the pulse, color effect, and polarization), there is no simple way to judge (diagnose) if such a mistake has been made. Usually this requires a highly experienced investigator and the use of advanced approaches like global analysis of multiple experiments (as discussed in the following chapter).

In this section, we have discussed these effects in the time domain, which is very intuitive. Nevertheless, the same errors apply to frequency-domain measurements, although the observed effects are not as intuitive to present and interpret.

12.4.2.4 Pileup Effect

In the final example of instrument-associated errors, we want to focus attention on a problem frequently encountered in TCSPC technology. In an effort to speed up the detection process, sometimes we can overload the data collection. One has to remember that the working principle of TCSPC technology is detecting only one photon in the measurement cycle (the first detected photon stops/starts the clock).

Because we want to collect photons through the entire decay without any bias, the requirement is that the probability for detecting a fluorescence photon in any cycle is low (1% or less). In practice, this means that we detect a photon only once in 100 or more cycles. This ensures that the probability of detecting the photon at a time following the excitation pulse is the same as that for immediately after the excitation and depends only on the number of available photons. If this assumption is broken, one preferentially detects only the early photons and has practically no chance to detect the late photons.

This effect has been discussed many times in detail (Birch and Imhof 1994; Lakowicz 1999) and we only want to demonstrate how this affects collected decay curves. For this demonstration, we selected Fluorol 7G in ethanol. Figure 12.4 shows the measured intensity decays for different signal strengths (different excitation intensities). The collection time was adjusted to maintain the number of counts detected at the same level, and the number of detected counts is sufficient to obtain very reliable fitting results. For the lowest intensity, as expected, the analysis clearly reveals single-exponential decay. As we increase the intensity of the excitation light, abnormalities in the decay are observed. As seen in Figure 12.4, for excitation intensities that are too high, the fluorescence decay exhibits dramatically different results, which obviously lead to completely different fluorescence lifetimes.

This is a very important effect that puts significant constraints on TCSPC in FLIM applications. We typically should limit the repetition rate, usually to a maximum of 40 MHz. A

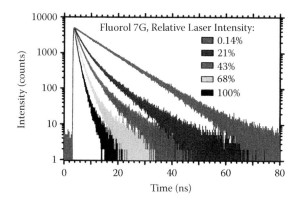

FIGURE 12.4 **(See color insert following page 288.)** Intensity decays measured in TCSPC system for different excitation intensities. Data collection time was adjusted to keep constant number of counts in the peak.

higher repetition rate results in insufficient time separation between the pulses for proper measurements of the fluorescence lifetime. Proper measurement should have essentially zero probability to detect photons emitted from a molecule that was excited by the previous pulse. In practice, we generally recommend that a pulse-to-pulse separation should be at least more than four to five times the longest intensity decay (longest lifetime), which for the fluorescence lifetime of 4 ns gives 20 ns (the corresponding laser repetition rate should be less than 50 MHz).

To achieve a reasonable resolution, 10^5 photons or more should usually be collected throughout the entire decay (Birch and Imhof 1991). A simple calculation shows that in this perfect condition, one will need ~250 ms per pixel. For a typical image of 10^6 pixels, the scanning time becomes ~4,000 min, which is completely unrealistic. In practice, we usually sacrifice the accuracy (number of counts) within the single pixel and use multiple pixels to calculate the intensity decays. This allows dropping the measurement time to a single minute per image. It is important to note that the frequency domain does not suffer from the pileup problem (as long as we do not saturate the system) and allows faster FLIM measurement.

12.4.2.5 Solvent Effect

The previous three examples referred to errors introduced by the instrumentation or improper use of optical elements. In this section, we want to focus attention on possible errors arising from the sample itself. For this experiment, we selected the popular membrane probe 6-propionyl-2-dimethylaminonaphthalene (Prodan) (Lakowicz 1999). Prodan is characterized by significant dipole moment changes between ground and excited states and is consequently very sensitive to the surroundings and an excellent probe of the local environment. But it is important to remember that this change is associated with very strong spectral relaxation (due to the physical relaxation of the solvent or matrix surrounding the fluorophore, which arises from a change in the molecular dipole when passing from the ground to the excited state)—a process that frequently occurs in a timescale comparable to the fluorescence lifetime.

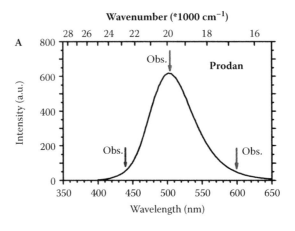

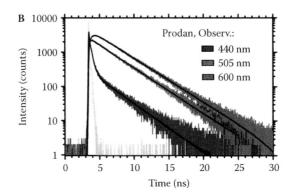

FIGURE 12.5 **(See color insert following page 288.)** Top: emission spectrum of Prodan. Bottom: intensity decays of Prodan measured for different observation wavelengths.

Figure 12.5 (top) shows a typical emission spectrum of Prodan in propylene glycol. The spectrum has an almost perfect bell shape and looks like it represents a single species. Figure 12.5 (bottom) shows the fluorescence intensity decays for three different observation wavelengths: short (440 nm), intermediate (505 nm), and long (600 nm). Even the least experienced investigator will realize that these decays are dramatically different. The short wavelength has a high contribution of the short lifetime component due to the quick relaxation process. On the other hand, the long wavelength has the large contribution of the "pumping" component typical for an excited-state reaction and similar to that observed for horizontal polarization in the previous example. The intermediate part of the spectrum is a pure intensity decay characterized by single-exponential decay, as expected from Prodan.

12.4.3 Analysis of Multiexponential Decays

In this section, we present some examples of how to approach data analysis in order to resolve fluorescence lifetime components properly. We think that the best way to understand analysis of time-resolved data is to consider some representative results. For this we will present some original examples of simple model systems representing how precise one may expect the results from data analysis to be. For more examples, see Lakowicz 1999.

It is intuitive that the quality of the fit will depend on the number of data points. In the time domain, it will depend on the number of channels and the total number of collected photons; in frequency domain, it will depend on the number of frequencies and time averaging for each frequency. Usually, when the error is purely statistical, a good fit can clearly reveal a single-exponential decay even for a limited signal level (limiting number of counts). However, this becomes much more problematic when one tries to resolve two or more lifetime components. This problem becomes very significant when the fluorescence lifetimes are not very well separated (closely spaced) and the number of collected photons is limited.

12.4.3.1 Effect of the Signal Level

For the first example, we selected a disodium fluorescein (uranin) in buffer solution (pH ~ 10) well characterized by a single-exponential decay of about 3.95 ns. We wanted to test how experimental error depends on the fluorescence signal and how this may alter the obtained results. Figure 12.6 shows the measured intensity decays and measured corresponding IRF using the Picoquant system FT-200 and a solid-state laser diode excitation (470 nm). To change the fluorescence signal, we used simple dilution of the sample. For the highest concentration (0.5 μM) in Figure 12.6 (left), we obtained a superb decay that was fitted with one exponential with excellent χ_R^2. A 100-fold dilution lowers the signal and the data collection time has to be increased to maintain the same total number of counts. The data fit in Figure 12.6(B) resulted in a very good fit and yielded a fluorescence lifetime of 3.94 ns, which is in excellent agreement with the expected lifetime of 3.95 ns.

Next, the 100-fold dilution further lowered the signal and we again had to increase collection time by 10-fold. This is now clearly visible from the level of dark counts (counts contributed by the background) in Figure 12.6 (right). Again, the fit to the experimental data is quite good and the result is very satisfactory (3.92 ns). We want to stress that, for the experiment performed in a cuvette, we can significantly increase data collection time without risking damaging the sample. However, this procedure can become useless for imaging because increasing the collection time to minutes or even seconds per pixel extends the total time needed for collecting one image to unrealistic periods and could also cause photobleaching of the fluorophores.

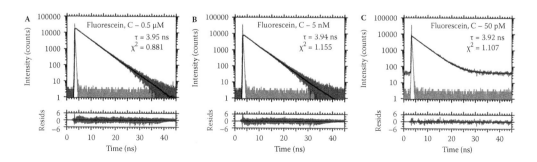

FIGURE 12.6   Fluorescence intensity decays of fluorescein at different concentrations.

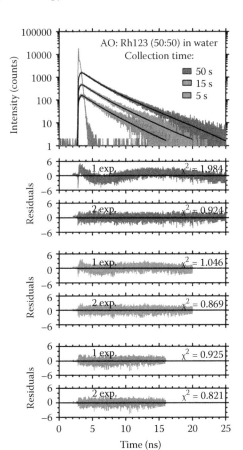

FIGURE 12.7 **(See color insert following page 288.)** Intensity decays for mixture of AO and Rh123 and residual distribution for different data collection times.

It is clearly evident that, even for very weak signals (we want to stress that difference in signal level between the first and last sample is 10,000-fold), the random error obviously increases, but this does not significantly affect the obtained results when a single lifetime component is measured. The situation becomes much more complicated when one wants to resolve two or more closely spaced fluorescence lifetime components.

To demonstrate this, we selected two dyes, rhodamine (Rh123) and rhodamine acridine orange (AO) in water, for which fluorescence lifetimes are 3.98 and 1.81, respectively (the individual fits are presented later). Figure 12.7 shows the measurement of the intensity decay for a 50:50 mixture of AO and Rh123. The intensity was adjusted to collect a sufficient number of counts in maximum over 50 s. Remember that one should be careful with the signal level in time-domain TCSPC experiments because a signal that is too high could lead to the so-called pileup effect discussed earlier.

Then we remeasured the intensity decays for shorter collection times. Figure 12.7 shows the results of the individual fits for a declining number of collected data points. Note that during the experiment we collected data up to 40 ns. However, the data analysis considers shorter times (25, 20, and 16 ns). For data fittings, it is important to consider experimental

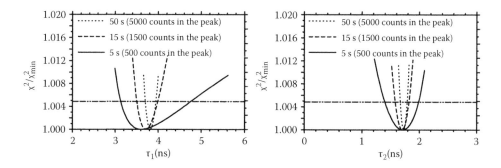

FIGURE 12.8   Lifetime χ_R^2 surfaces for the mixture AO and Rh123 for individual analysis of intensity decays for different collection times (different number of counts).

points significantly above the background level. If, for example, for a shortest acquisition time (5 s) we extend the range for data fitting to 25 ns, this will artificially lower the χ^2 value. This is because a big part of the decay is within the background level (instrument noise) and this will contribute to lowering the overall χ_R^2 value.

Figure 12.8 presents the results of supporting plane analysis for different acquisition times (different number of collected data points). This method is a very convenient way to represent the results of the fit graphically (Straume et al. 1991). As seen in the figure, for a high number of counts, the resolution for two components is very good. However, when the number of collecting photons is lowered, the confidence intervals quickly increase and resolving the components becomes increasingly problematic. The horizontal line on the graph represents the 97% confidence level.

12.4.3.2 Two and Three Components of Intensity Decays

To test the possibility of resolving multiple lifetime components, we selected three dyes (erythrosine B [ErB], phloxin B [PxB], and rhodamine 123 [Rh123]), which individually display single-exponential decay in water solution. Figure 12.9 shows the intensity decays and fitting results for the ErB with a fluorescence lifetime of about 89 ps, practically comparable to the response function of the instrument. The PxB has a lifetime of 1.072 ns and the Rh123 a lifetime of 3.981 ns. All three dyes can be excited with a 470 nm laser diode. Figure 12.9 (top) shows the emission spectra of each dye excited by a 470 nm wavelength. We selected the observation (emission) wavelength at 556 nm because this is where emission spectra have an isosbestic point. Fluorescence intensity decays of individual dyes are shown in the lower panel of Figure 12.9.

Next, we prepared various mixtures of two individual dyes. First, we selected two dyes, ErB and PxB, which have about 10-fold difference in fluorescence lifetimes. Figure 12.10A (left) shows the results of fluorescence lifetime measurements for the 50:50 mixture. As expected, the single component fits the data poorly with a large error (data not shown). Addition of the second component in the fit dramatically improves the quality of the fit (Figure 12.10A, left), and now we can clearly see two components, at 0.096 ns and 1.085 ns, that correspond very well to the expected values. Addition of the third component does not produce significant improvement in the χ_R^2 or error distribution. Figure 12.10(B, C; left)

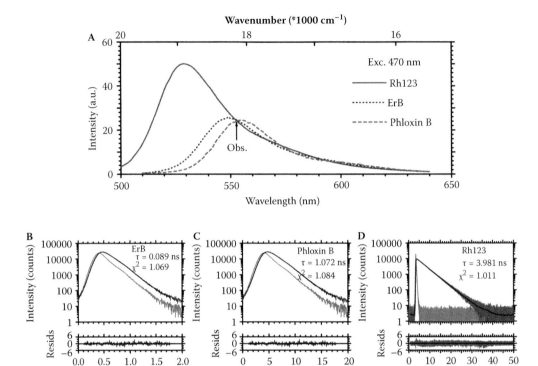

FIGURE 12.9 Emission spectra (top) and fluorescence intensity decays (bottom) measured for individual dyes ErB, PxB, and Rh123.

shows the fits obtained for 20:80 and 80:20 mixtures. Still, two lifetime components can be clearly resolved if a sufficient number of counts is collected. This is a very encouraging result because both lifetimes are relatively short.

Figure 12.10 (right) shows the results obtained for mixtures of ErB and Rh123 in three proportions (50:50, 10:90, and 90:10). The individual lifetimes of 0.089 ns and 3.981 are now very different and even the small 10% contribution is very well detected by data analysis within 15% error. The two component fits yield very good χ_R^2 values and addition of a third component does not produce noticeable improvement in χ_R^2. The measured values of fluorescence lifetimes are very close to expected values and the corresponding amplitudes are within 20% of expected values. This is proving that, for a larger difference in fluorescence lifetimes, the resolution quickly gets better.

Finally, we prepared a 1:1:1 mixture of all three dyes. The results are shown in Figure 12.11. Fits with one and two components are unacceptable (see the residual and the chi square value in the figure). The addition of the third component significantly improves the fit. We know that the system has three lifetime components and the results are in very good agreement with our expectations. Addition of a fourth component in the fit does not produce a significant improvement in the χ_R^2 value. The recovered fluorescence lifetimes are very close to the expected values and recovered amplitudes are within 20% error. This result shows that if an experiment is performed carefully and sufficient numbers of

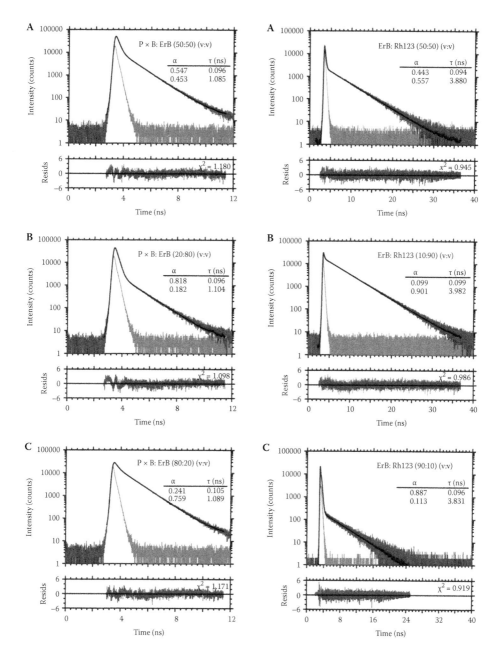

FIGURE 12.10 Intensity decays and recovered parameters for different mixtures of PxB and ErB (left) and ErB and Rh123 (right).

counts have been collected, it is not difficult to detect three lifetime components with high precision.

We want to stress that when we attempted to perform experiments with various mixtures, we did not expect such good results. Because the length of the fluorescence lifetime and its amplitude are strongly correlated parameters, these results exceeded our expectations. Typically, limiting the number of collected points results in a declining resolution and

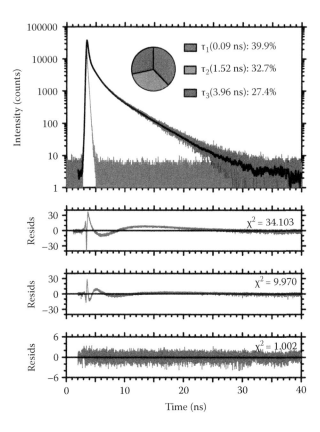

FIGURE 12.11   **(See color insert following page 288.)** Mixture of three dyes (ErB, PxB, and Rh123) and recovered parameters.

inability to recover parameters properly. But even with sufficient numbers of collected data points, resolving two not-well-separated components from a single measurement could be problematic.

In this case, prior information about the relations between the fractions or fluorescence lifetimes can be used in data analysis. Knowing the process that regulates changes in fractions (like temperature that will thermodynamically shift the equilibrium) and applying the global analysis approach (Beechem et al. 1991) may significantly improve results. See the studies of pH dependence of time-resolved fluorescence of myoglobin and hemoglobin in Gryczynski, Lubkowski, and Bucci (1995, 1997).

12.4.3.3 Fluorescence Lifetime Distribution: Biological Examples

Most biological systems are very complex and single or multiple exponents may not truly reflect the real situation. For example, a macromolecule (protein, DNA, RNA) may exist in multiple states (conformations), and the equilibrium between the states depends on many physical parameters like temperature or pH. Also, a typical biological system is a very dynamic system that is constantly changing between multiple conformations.

Such conformational alternations are usually very slow compared to the fluorescence lifetime. Thus, it is not a surprise that in biological systems one may expect many slightly differing lifetime components representing different conformational states. In such situations, we could expect a distribution of possible lifetime components around some mean value. Depending on the specimen, we can have a Lorentzian or Gaussian distribution for amplitudes of various lifetime components (Acala et al. 1987).

In the case of Lorentzian types of distribution, the total fluorescence intensity decays are represented by the sum of individual decays weighted by the amplitudes with distribution:

$$\alpha(t) = \frac{A_i}{\pi} \frac{\Gamma_i/2}{(\tau - \bar{\tau}_i)^2 + (\Gamma_i/2)^2} \tag{12.18}$$

and the experimental data can be fitted to the functional form of intensity decay in the following form:

$$I(t) = \int_0^\infty \alpha(t) e^{-t/\tau} d\tau \tag{12.19}$$

Here, A_i is the amplitude of the ith component, $\bar{\tau}_i$ is the central value of the ith distribution, and Γ_i is the full width at the half maximum (FWHM). Typically, the use of continuous distribution $\alpha(\tau)$ minimizes the number of floating parameters in the fitting algorithms.

We want to stress that, based on a single experiment, it is very difficult to judge if the results represent a lifetime distribution or just two or three lifetime components. Typically, one would use the global analysis approach that is strongly supported by the physical model or some other experimental data to accept the fluorescence lifetime distribution fit. There are numerous examples for lifetime distributions (Gryczynski et al. 1997; Lakowicz 1999; Bharill et al. 2008).

In this section, we present a simple example of N-acetyl-l-tryptophanamide (NATA). This is a very well known tryptophan system that may adopt some limited conformational states. Figure 12.12 shows the intensity decay measured for NATA with a 295 nm LED excitation and 365 nm observation (emission). The figure shows the single-exponent fit, which is already good. Addition of the second component slightly improves the fit. Comparing the differences in Figure 12.12(A, B) the evidence for accepting two components' fit is rather weak. Figure 12.12(C) shows the same intensity decay, but now fitted using the single-exponential Lorentzian distribution model. The fit is further improved and we propose to accept the distribution model. However, we want to stress that, mathematically, the difference in the chi square values is too small to accept the distribution model unequivocally.

We have already shown the solvent effect and how it influences the intensity decays. Also, solvent polarity may significantly influence the fluorescence spectra and intensity decays of polar molecules. A good example is the 2-aminopurin (2AP) that we recently studied (Ballin et al. 2008; Bharill et al. 2008). The excitation forces a charge separation in 2AP that causes an increase of overall dipole moment of the molecule. Accordingly, a dipole–dipole interaction between 2AP and polar solvent molecules will lower the excitation energy of 2AP, resulting in a red-shifted emission.

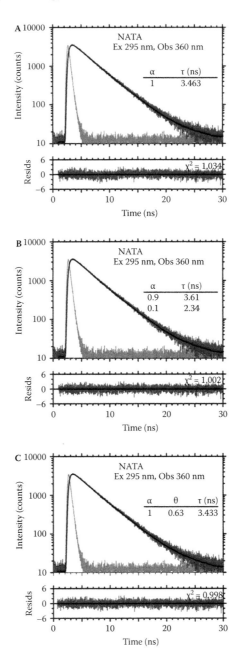

FIGURE 12.12   Fluorescence intensity decay of NATA. Analysis with one exponential (A), two exponentials (B), and Lorentzian distribution (C).

Mixtures of polar and nonpolar solvents are a good system to study the effect of solvent polarity on the spectral shift. Generally, the steady-state spectral shift is small and has limited sensitivity to be used for the detailed interpretation of the effect. Alternatively, the intensity decays using a Lorentzian lifetime distribution model may reflect the average number of molecules emitting in different microenvironments and the range of mobility and polarity in each environment.

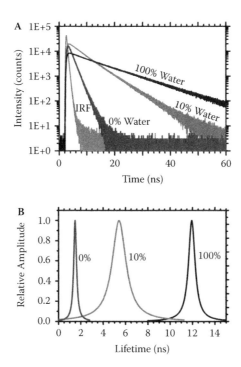

FIGURE 12.13 (A) Fluorescence intensity decays of 2AP in dioxane, dioxane + 10% water, and water. (B) Recovered lifetime distribution.

We studied the 2AP fluorescence intensity decays in dioxane/water mixtures. After excitation, the fluorophore remains unrelaxed in the Franc–Condon state (F) or it loses its energy to the environment and relaxes to a lower energy state (R). The two states have their own characteristic lifetime decays. Typically, the unrelaxed (F) state exhibits a shorter lifetime than one in the R state because F decays through both emission and relaxation. Studies of 2AP fluorescence revealed that it undergoes solvent relaxation and two states emit approximately at 350 nm (F) and 410 nm (R).

Figure 12.13 shows the fluorescence intensity decays and fits for different dioxane/water mixtures. The emission from the (R) state (410 nm), as expected, was well resolved by a single exponential. The intensity decay observed for the (F) state (350 nm) was rather poorly approximated by the single exponential. All intermediate states were not suitable for fitting with one or two single-exponential decays. In contrast, 2AP intensity decay from all mixtures can be very well fitted with unimodal Lorentzian distribution. However, the distributions are dramatically different in the sense of FWHM. Figure 12.13 shows the recovered distributions for different mixtures. See Bharill et al. (2008) for detailed studies of observed effects.

12.5 SUMMARY

Time-resolved fluorescence measurements are moderately complex and frequently require advanced methods for data analysis. The results of proper time-resolved measurements may supply extensive basic information about the system being investigated. A good example is FRET, which can reveal detailed information about different macromolecular interactions, arrangements, and dynamics of conformational changes.

In this chapter, we presented basics of the least-squares data analysis method in the time-domain and frequency-domain techniques. We briefly discussed possible problems that may emerge during the time-resolved data collection. Using an example of a time-domain approach in the TCSPC method, we presented how systematic experimental errors may impact obtained data sets. Also, we presented the effect of solvent relaxation that may exhibit different decay kinetics for different observation wavelengths. Finally, we discussed some examples of multiexponential decays and the limits of resolution as a function of collected data points and number of detected photons. Generally, for a sufficient number of counts, the resolution of well spaced two and three lifetime components can be successful. However, in the microscope imaging setup, these methods may suffer due to an insufficient number of counts in each pixel.

REFERENCES

Alcala, J. R., Gratton, E., and Prendergast, F. G. 1987. Resolvability of fluorescence lifetime distributions using phase fluorometry. *Biophysical Journal* 51 (4): 587–596.

Ameloot, M., 1992. Laplace deconvolution of fluorescence decay surfaces. *Methods in Enzymology* 210:279–304.

Ballin, J. D., Prevas, J. P., Bharill, S., Gryczynski, I., Gryczynski, Z., Wilson, G. M. 2008. Local RNA conformational dynamics revealed by 2-aminopurine solvent accessibility. *Biochemistry* 47:7043–7052.

Bard, Y. 1974. *Nonlinear parameter estimation.* New York: Academic Press.

Bebelaar, D. 1986. Time response of various types of photomultipliers and its wavelength dependence in time-correlated single photon counting with an ultimate resolution of 47 ps FWHM. *Review of Scientific Instruments* 57:1116–1125.

Beechem, J. M., Gratton, E., Ameloot, A., Knutson, J. R., and Brand, L. 1991. The global analysis of fluorescence intensity and anisotropy decay data: Second-generation theory and program. In *Topics in fluorescence spectroscopy, vol. 2: Principles,* ed. J. R. Lakowicz, 241–305. New York: Plenum Press.

Bevington, P. R. 1969. *Data reduction and error analysis for the physical science.* New York: McGraw–Hill.

Bharill, S., Sarkar, P., Ballin, J. D., Gryczynski, I., Wilson, G. M., and Gryczynski, Z. 2008. Fluorescence intensity decays of 2-aminopurine solutions: Lifetime distribution approach. *Annals of Biochemistry* 377:141–149.

Birch, D. J. S., and Imhof, R. E. 1991. Time-domain fluorescence spectroscopy using time-correlated single photon counting. In *Topics in fluorescence spectroscopy, vol. I, Techniques,* ed. J. R. Lakowicz, 1–95. New York: Plenum Press.

Box, G. E. P. 1960. Fitting empirical data. *Annals of the New York Academy of Sciences* 86:792–816.

Carraway, E. R., Hauenstein, B. L., Demas, J. N., and DeGraff, B. A. 1985. Luminescence lifetime measurements. Elimination of phototube time shifts with the phase plane method. *Annals of Chemistry* 57:2304–2308.

Demas, J. N., and Crosby, G. A. 1970. Photoluminescence decay curves: An analysis of the effects of flash duration and linear instrumental distortions. *Analytical Chemistry* 42:1010–1017.

Gafni, A., Modlin, R. L., and Brand, L. 1975. Analysis of fluorescence decay curves by means of the Laplace transformation. *Biophysical Journal* 15:263–280.

Good, H. P., Kallir, A. J., and Wild, U. P. 1984. Comparison of fluorescence lifetime fitting techniques. *Journal of Physical Chemistry* 88:5435–5441.

Gratton, E., Jameson, D. M., and Hall, R. D. 1984. Multifrequency phase and modulation fluorometry. *Annual Review of Biophysics & Bioengineering* 13:105–124.

Gratton, E., and Limkeman, M. 1983. A continuously variable frequency cross-correlation phase fluorometer with picosecond resolution. *Biophysical Journal* 44:315–324.

Grinvald, A., and Steinberg, I. Z. 1974. On the analysis of fluorescence decay kinetics by the method of least squares. *Analytical Biochemistry* 59:583–598.

Gryczynski, Z., and Bucci, E. 1993. Design and application of a new optical cell for measuring weak fluorescent emission with time resolution in the picosecond time scale. *Biophysical Chemistry* 48:31–38.

Gryczynski, Z., Gryczynski, I., and Lakowicz, J.R. 2005. Basics of fluorescence and FRET. In *Molecular imaging. FRET microscopy and spectroscopy,* ed. A. Periasamy and R. N. Day, 21–56. New York: Oxford University Press.

Gryczynski, Z., Lubkowski, J., and Bucci, E. 1995. Heme-protein interactions in horse heart myoglobin at neutral pH and exposed to acid investigated by time-resolved fluorescence in the pico- to nanosecond time range. *Journal of Biological Chemistry* 270:19232–19237.

Gryczynski, Z., Lubkowski, J., and Bucci, E. 1997. Fluorescence of myoglobin and hemoglobin. *Methods of Enzymology* 78:538–569.

Helman, W. P. 1971. Analysis of very fast transient luminescence behavior. *International Journal for Radiation Physics and Chemistry* 3:283–294.

Isenberg, I., Dyson, R. D., and Hanson, R. 1973. Studies on the analysis of fluorescence decay data by the method of moments. *Biophysical Journal* 13:1090–1115.

Johnson, M. L. 1994. Use of least-squares techniques in biochemistry. *Methods in Enzymology* 240:1–22.

Johnson, M. L., and Faunt, L. M. 1992. Parameter estimation by least squares method. *Methods in Enzymology* 210:1–37.

Johnson, M. L., and Fresier, S. G. 1985. Nonlinear least-squares analysis. *Methods in Enzymology* 117:301–342.

Knight, A. E. W., and Selinger, B. K. 1971. The deconvolution of fluorescence decay curves. A non-method for real data. *Spectrochimica Acta Part A* 27:1223–1234.

Lakowicz, J. R. 1999. *Principles of fluorescence spectroscopy,* 2nd ed. New York: Plenum Press.

Livesey, A. K., and Brochon, J. C. 1987. Analysis of distribution of decay constants in puls-fluorimetry using a maximum entropy method. *Biophysical Journal* 52:693–706.

Montgomery, D. C., and Peck, E. A. 1982. *Introduction to linear regression analysis.* New York: John Wiley & Sons.

Munro, I. H., and Ramsay, I. A. 1968. Instrumental response time corrections in fluorescence decay measurements. *Journal of Physics E* 1:147–148

Nedler, J. A., and Mead, R. 1965. A simplex method for function minimization. *Computer Journal* 7:308–313.

O'Connor, D. V., and Philips, D. 1984. *Time-correlated single photon counting.* New York: Academic Press.

O'Connor, D. V. O., Ware, W. R., and Andre, J. C. 1979. Deconvolution of fluorescence decay curves. A critical comparison of techniques. *Journal of Physical Chemistry* 83:1333–1343.

Small, E. W. 1992. Method of moments and treatment nonrandom error. *Methods of Enzymology* 210:237–279.

Straume, M., Frasier-Cadoret, S. G., and Johnson, M. L. 1991. Least-square analysis of fluorescence data. In *Topics in fluorescence spectroscopy, vol. 2: Principles,* ed. J. R. Lakowicz, 177–240. New York: Plenum Press.

Ware, W. R., Doemeny, L. J., and Nemzek, T. L. 1973. Deconvolution of fluorescence and phosphorescence decay curves. A least-square method. *Journal of Physical Chemistry* 77:2038–2048.

CHAPTER 13

Global Analysis of Frequency Domain FLIM Data

Hernan E. Grecco and Peter J. Verveer

13.1 INTRODUCTION

Fluorescence lifetime imaging microscopy (FLIM) has become a common tool to image the fluorescence lifetimes of fluorophores in cells, providing information about their state and immediate molecular environment. The fluorescence lifetime is sensitive to environmental conditions such as pH (Carlsson et al. 2000) and excited-state reactions such as Förster resonance energy transfer (FRET; Clegg 1996; Chapters 1 and 9)—properties that can be exploited to resolve physiological parameters in the cell. As a result, FLIM has become a valuable tool to investigate cellular processes in intact, living cells (Bastiaens and Squire 1999; Bastiaens and Pepperkok 2000; Bastiaens et al. 2001). The best example of this is the use of FLIM to detect quantitatively protein–protein interactions via FRET between fluorescence proteins such as the green fluorescence protein (GFP) and its many spectral variants.

The fluorescence lifetime is defined as the average time the fluorophore spends in the excited state after absorption of a photon. Fluorescence lifetimes can be measured with various approaches that can be broadly divided in time-domain and frequency-domain methods (Bastiaens and Squire 1999). These methods and standard approaches to analyze data acquired with these instruments are described in the various chapters of this book. In this chapter, we describe the global analysis methods developed for frequency-domain data (Clayton, Hanley, and Verveer 2003; Verveer, Squire, et al. 2000; Verveer et al. 2001a, 2001b; Verveer and Bastiaens 2003).

Global analysis is an approach to increase the accuracy of a data fit by analyzing data sets from multiple experiments simultaneously (Beechem 1992). By exploiting the prior knowledge that some of the experimental parameters are always the same, such a simultaneous fit can be much improved compared to individual fits of the data sets. This idea can be applied to FLIM data because in many cases the intrinsic fluorescence lifetimes of the fluorophores are fixed quantities (Beechem et al. 1983; Knutson, Beechem, and Brand 1983; Verveer,

Squire, et al. 2000). We will explain how this approach is implemented in the case of single-frequency FLIM data and how it is used to analyze FRET-FLIM data quantitatively.

13.2 FOURIER DESCRIPTION OF FLIM DATA

Frequency domain data are acquired as a function of a sinusoidal excitation signal, and they can therefore be treated with Fourier transform techniques for the quantitative analysis of FRET-FLIM data. Here we will discuss the *homodyne* FLIM method, which is currently most widely used in wide-field frequency domain setups. In the homodyne detection scheme, the detector is modulated at a frequency that is equal to the frequency of modulation of the excitation signal (see Chapters 5 and 11).

The decay kinetics of a fluorescent sample can generally be described by the response to a delta function as a sum of multiple exponentials with fluorescence lifetimes equal to τ_q:

$$D(t) = \sum_{q=1}^{Q} \frac{\alpha_q}{\tau_q} \exp(-t/\tau_q) \tag{13.1}$$

where α_q is the fraction of fluorescence emitted by the qth fluorescent species. Therefore, the sum of the fractional fluorescence α_q is normalized to one: $\sum_{q=1}^{Q} \alpha_q = 1$. Box 13.1 explains how homodyne frequency domain FLIM methods can be analyzed with Fourier methods. If the fundamental frequency is given by $\omega = 2\pi/T$, we can write the normalized fluorescence response as follows:

$$F_N(\phi) = 1 + \sum_{\substack{n=-\infty \\ n\neq 0}}^{\infty} \sum_{q=1}^{Q} \frac{M_n \alpha_q \exp\left(j\left(\arctan(n\omega\tau_q) - n\phi\right)\right)}{\sqrt{1 + (n\omega\tau_q)^2}} \tag{13.2}$$

where the complex number M_n depends on the Fourier coefficients of the excitation and the detection modulation signals. $F_N(\phi)$ is a function of a phase shift ϕ, which is used to sample $F_N(\phi)$ systematically. In order to have a measurable change of $F_N(\phi)$ as a function of ϕ, $n\omega\tau$ must be in the order of unity; that is, for lifetimes in the order of nanoseconds, modulation frequencies in the order of tens of megahertz must be chosen.

Equation 13.2 provides an expression for the normalized fluorescence in response to an arbitrary periodic excitation. The values of τ_q and α_q can be recovered from the Fourier coefficients of Equation 13.2, given calibration values of M_n. In classical frequency-domain systems, the modulation of both excitation and detection are chosen to be sinusoidal, and all harmonics $n > 1$ are equal to zero.

However, the modulation signals may be chosen to be nonsinusoidal, in which case significant higher harmonics will be present that, in principle, can also be utilized (Squire, Verveer, and Bastiaens 2000). Because most devices (modulators and detectors) are unable to reach high modulation on a sinusoidal-only waveform, many frequency FLIM systems are designed such that higher harmonics present are present. However, these are usually

BOX 13.1: FOURIER ANALYSIS OF FLIM DATA

We can write the excitation signal as a Fourier series:

$$E(t) = E_0 \left(1 + \sum_{\substack{n=-\infty \\ n\neq 0}}^{\infty} E_n \exp(-jn\omega t) \right)$$

where E_0 is real and E_n is complex for $n > 0$. The fluorescence response to the excitation is given by convolving $E(t)$ with the fluorescence decay kinetics given by Equation 13.1:

$$F(t) = F_T \int_{-\infty}^{t} E(u) D(t-u) du$$

where F_T is the total fluorescence intensity. The result can be written as a Fourier series:

$$F(t) = F_T E_0 \left(1 + \sum_{\substack{n=-\infty \\ n\neq 0}}^{\infty} \sum_{q=1}^{Q} \frac{E_n \alpha_q \exp(-jn\omega t)}{1 - jn\omega\tau_q} \right)$$

In *homodyne* FLIM approaches, a modulated detector is employed with a modulation signal $G(t)$, which is also periodic with a fundamental frequency ω. The detected signal is then given by a multiplication of the modulation signal $G(t)$ with the fluorescence signal $F(t)$. The result has many high-frequency cross-terms involving multiples of ω. If the detector has a slow response that integrates out the high frequencies, only a constant signal remains:

$$F(t) = F_T E_0 G_0 \left(1 + \sum_{\substack{n=-\infty \\ n\neq 0}}^{\infty} \sum_{q=1}^{Q} \frac{E_n G_n \alpha_q}{1 - jn\omega\tau_q} \right)$$

where G_n are the Fourier coefficients of $G(t)$. By systematically varying the phase of the modulation signal $G(t)$, we obtain the fluorescence signal as a function of the systemic delay in $G(t)$ that is also in the form of a Fourier expansion:

$$F(\phi) = F_T M_0 \left(1 + \sum_{\substack{n=-\infty \\ n\neq 0}}^{\infty} \sum_{q=1}^{Q} \frac{M_n \alpha_q \exp(-jn\phi)}{1 - jn\omega\tau_q} \right)$$

$$= F_T M_0 \left(1 + \sum_{\substack{n=-\infty \\ n\neq 0}}^{\infty} \sum_{q=1}^{Q} \frac{M_n \alpha_q \exp\left(j\left(\arctan(n\omega\tau_q) - n\phi\right)\right)}{\sqrt{1 + (n\omega\tau_q)^2}} \right)$$

where $M_0 = E_0 G_0$ and $M_n = E_n G_n \exp(jn\varphi)$. The phase φ is usually selected by controlling a delay in the modulation $G(t)$. This means that, generally, G can be written as $G = |G| \exp(jn(\psi - \varphi))$, where ψ is a constant and φ is determined by the controllable delay. Thus, we can write $M = E|G| \exp(jn\psi)$, which is a constant that can be calibrated.

weak compared to the first harmonic and not used for data analysis. As long as the signal is sampled accurately (i.e., if the number of points sampled is more than twice the number of significant harmonics), these higher harmonics do not pose any problems.

13.3 GLOBAL ANALYSIS OF FLIM DATA

Global analysis is an approach to improve data analysis by simultaneously fitting multiple data sets that are acquired under varying experimental conditions to a common model (Beechem 1992). In a conventional fitting approach, each data set is fit independently and, for all parameters of the model, a value is estimated for each data set. In a global analysis approach, the experimental conditions are chosen such that some of the model parameters are the same in all experiments. These parameters are still unknown, but at least it is known that their values must be the same in all data sets. This prior knowledge can be exploited in a global fit, which analyzes all data sets simultaneously while constraining the subset of invariant parameters to single values for all data sets. It was shown that the global analysis approach leads to significant improvements in the data fit and it has since become a standard approach for fitting biochemical and biophysical data (Beechem 1992).

This approach was applied first to lifetime data acquired with fluorometer setups that collect data in a single measurement point, usually in solutions (Beechem et al. 1983; Knutson et al. 1983). By collecting fluorescence lifetime data under different conditions (i.e., by changing such parameters as temperature or solvent conditions), multiple data sets were acquired. Depending on the experimental conditions, some of the parameters, such as the fluorescence lifetimes, could be assumed to be invariant, and a global fit of all the data led to an improved estimation of the parameters.

The idea of global analysis can be applied to FLIM data to resolve multiple molecular species with different fluorescence lifetimes with increased accuracy and precision in each pixel of the image. In this case, it is not necessary to change experimental parameters such as temperature. Instead, it is assumed that the sample is composed of two or more species with given fluorescence lifetimes that do not vary over the sample. Each pixel in the FLIM image can be treated as a separate data set, where the relative concentrations of the molecular species will be different in each pixel, but their fluorescence lifetimes will be invariant over the image or even over multiple FLIM images. Thus, global analysis methods can be applied using the knowledge that the fluorescence lifetimes are the same in all pixels of all data sets.

This approach was applied to single-frequency FLIM data. It was shown that this provides sufficient information to resolve a biexponential decay model and, moreover, that the accuracy and precision of the estimated parameters could be increased significantly compared to a nonglobal analysis where the each pixel is assumed to have independent lifetime values (Verveer, Squire, et al. 2000). It was subsequently shown by Clayton et al. (2003) and later confirmed by others (Esposito, Gerritsen, and Wouters 2005; Redford and Clegg 2005) that this is a linear problem, making the calculation simple and efficient. The mathematical background of this approach is described in detail in Box 13.2 and a general protocol given in Box 13.3. Although the restriction to two lifetimes may seem limiting, it covers a number of important biological applications, such as FRET-FLIM. However, for global analysis to be applicable to such a FRET-FLIM assay, a number of assumptions must hold; these are described in more detail in Box 13.4.

13.4 APPLICATION TO FRET-FLIM DATA

Global analysis of FLIM data has been used successfully in biological applications to determine the fraction of interacting molecules with FRET assays quantitatively (Ng et al. 2001; Reynolds et al. 2003; Rocks et al. 2005; Verveer, Wouters, et al. 2000; Xouri et al. 2007); we provide an example of this in Figure 13.1. In a FRET-FLIM application, the lifetime of one fluorophore—the *donor*—is measured. If the donor-tagged molecule interacts with another fluorophore-tagged molecule—the *acceptor*—the excitation energy of the donor can be transferred to the acceptor fluorophore, provided the conditions for FRET are met (Clegg 1996). As a result of FRET, the fluorescence lifetime of the donor is decreased to a lower value. If the binding geometry of the complex is relatively constant, the fluorescence lifetime of the complex will also be constant in the sample.

Thus, in this case we can describe the system as a mixture of two molecular species: free donor and donor/acceptor complex; each has unique fluorescence decay characteristics. If the individual decays are monoexponential, then the fluorescence kinetics of the sample can be described by a biexponential model. The fluorescence lifetimes are the same in each pixel, and only the relative fraction of each species varies; therefore, global analysis can be applied.

Provided the FLIM system is properly calibrated for the instrument response, no additional measurements with reference samples are needed in this case. This is an advantage over intensity-based FRET methods that require separate calibration samples (Gordon et al. 1995). Moreover, unlike intensity-based FRET methods, the recovered lifetimes yield an estimate of the quantum yield of free donor and of donor/acceptor complexes, making it possible to renormalize the estimated fractional fluorescence to true relative concentrations (Verveer, Squire, et al. 2000; Verveer et al. 2001a).

Figure 13.1 shows the application of global analysis to FRET data obtained from samples from a well-established biological FRET-FLIM assay (Verveer, Wouters, et al. 2000; Wouters and Bastiaens 1999). MCF7 cells expressing the epidermal growth factor receptor (EGFR) tagged with yellow fluorescent protein (YPF) were fixed and incubated with Cy3.5-labeled PY72, a generic antibody against phosphotyrosine. Panel A shows the YFP fluorescence of one cell from a sample that was fixed 5 min after stimulation with soluble EGF. Panel B shows a phasor plot of a subset of the data, and the result of the fit to the accepted data points is shown by the dotted line. Panel C shows the corresponding relative fractions of phosphorylated receptor, demonstrating that most of the receptor is active near the periphery of the cell, as expected at such short stimulation times and in agreement with previous work (Verveer, Wouters, et al. 2000; Wouters and Bastiaens 1999).

13.5 DISCUSSION AND OUTLOOK

Global analysis of fluorescence lifetime imaging microscopy data is a method to fit biexponential models to single-frequency FLIM data. Unless further prior knowledge is available, this is normally not possible with single-frequency FLIM data because only two measurements (phase and modulation) are available in each pixel; this is not sufficient to estimate

BOX 13.2: GLOBAL ANALYSIS OF BIEXPONENTIAL FLIM DATA

We focus on the application of global analysis to fitting a biexponential model to a set of FLIM data. We use a superscript i to indicate different FLIM data sets (i.e., different pixels in one or more FLIM images). We assume that the two fluorescence lifetimes τ_1 and τ_2 are the same in each data set, but that the relative fraction α^i may be different in each data set i. We rewrite Equation 13.2 as

$$F_N^i(x) = 1 + \sum_{\substack{n=-\infty \\ n \neq 0}}^{\infty} M_n R_n^i \exp(-jn\phi)$$

where

$$R_n^i = \sum_{q=1}^{Q} \alpha_q^i R_{n,q}$$

and

$$R_{n,q} = (1 - jn\omega\tau_q)^{-1}$$

In general, only the first harmonic is used for the global analysis, and therefore we drop the subscript n. For the special case of a biexponential decay we can then write explicitly

$$R^i = (1 - \alpha^i)R_1 + \alpha^i R_2$$

where we substituted $\alpha_2^i = \alpha^i$ and $\alpha_1^i = 1 - \alpha^i$ (because $\alpha_1^i + \alpha_2^i = 1$), which means that α^i corresponds to the fractional fluorescence emitted by the species with a lifetime equal to τ_2. If M_1 is known, the R^i follow from the first harmonic Fourier coefficients of the data. The global analysis then proceeds by first estimating the quantities R_1 and R_2 and thereby the fluorescence lifetimes τ_1 and τ_2. The fractions α_i in each pixel can then be estimated afterward using the previously recovered lifetimes. The procedure to estimate τ_1, τ_2, and α_i is most easily understood graphically using a phasor plot (Clayton et al. 2003; Jameson, Gratton, and Hall 1984; see Chapter 11 for a general discussion of the phasor plot). In the literature, this plot has also been called an AB-plot (Clayton et al. 2003) or a polar plot (Redford and Clegg 2005).

The graph shows a phasor plot of the imaginary part of R^i against its real part (see Figure 13.2). It can be seen that all points, in the absence of noise, fall on the straight line that passes through R_1 and R_2, a consequence of the fact that R^i is a linear combination of R_1 and R_2. All points derived from monoexponential curves will fall on a half-circle (Clayton et al. 2003). Because R_1 and R_2 represent monoexponential curves, they must individually fall on this half-circle and can be found by finding the intersections of the straight line with the half-circle. Based on this, it can be derived that the offset u and the slope v of the line $\mathrm{Im}R^i = u + v\mathrm{Re}R^i$ can be used to calculate the two fluorescence lifetimes (Clayton et al. 2003):

$$\tau_{1,2} = \frac{1 \pm \sqrt{1 - 4u(u+v)}}{2\omega u}$$

To find the two fluorescence lifetimes in the presence of noise, a straight line is fit to a plot of $\mathrm{Im}R^i$ against $\mathrm{Re}R^i$ (for instance, using a least-squares fit), and the lifetimes are estimated from the estimated offset and slope. Given the fluorescence lifetimes, the fluorescence fraction α^i can then be found by projecting R^i on the fitted line and calculating the projected distance to R_1, which gives (Verveer and Bastiaens 2003):

$$\alpha^i = \frac{\omega\left(\tau_1 + \tau_2\right)\mathrm{Re}\,R^i + \left(\omega^2\tau_1\tau_2 - 1\right)\mathrm{Im}\,R^i - \omega\tau_2}{\omega\left(\tau_1 - \tau_2\right)}$$

BOX 13.3: GLOBAL ANALYSIS PROTOCOL

The steps for global analysis are as follows:

1. Acquire multiple standard frequency domain FLIM data sets, each consisting of a series of images taken at different phase settings (see Chapter 5).
2. Calculate an average fluorescence image by averaging all phase images in the FLIM series.
3. Find the features of interest in the fluorescence image by standard segmentation techniques, such as intensity thresholding.
4. Find a small region representative for the background, and correct each image in the FLIM series by subtracting the average value within that region.
5. Collect the FLIM sequences corresponding to the pixels of interest selected in step 3 and calculate the first Fourier coefficients by a standard Fourier transform method.
6. Correct the Fourier coefficients obtained in step 5 using calibration values obtained from an independent standard (see Chapter 5).
7. Calculate a linear fit of the imaginary parts as a function of the real parts of the calibrated Fourier coefficients obtained in step 6, using a standard linear fitting approach such as linear least-squares.
8. Estimate the fluorescence lifetimes τ_1 and τ_2 from the offset u and slope v, fitted in step 7, according to Box 13.2.
9. For each pixel in the FLIM data set, calculate the relative fraction of fluorescence α^i according to Box 13.2, using the fluorescence lifetimes τ_1 and τ_2 estimated in step 8. This yields the fluorescence fraction corresponding to the species with lifetime τ_2. In the case of a FRET assay, this should be the shortest lifetime estimated; swap the estimated lifetime values if necessary.
10. Renormalize the relative fraction of fluorescence to obtain the relative concentration, using the estimated lifetime values of step 8 by

$$c^i = \frac{\alpha^i \tau_1}{\tau_2 + \left(\tau_1 - \tau_2\right)\alpha^i}$$

Display c^i after appropriate contrast stretching and application of a suitable color table.

Some of these steps, such as the calculation and calibration of the Fourier coefficients, are part of the normal analysis routine of frequency-domain FLIM data. The global analysis procedure can therefore be easily incorporated into the normal analysis flow for FLIM data.

BOX 13.4: PITFALLS IN GLOBAL ANALYSIS

For global analysis to be successful, the usual caveats for FLIM, such as the need to avoid or correct photobleaching artifacts, apply; however, in addition, a number of conditions need to be fulfilled. First of all, the main assumption of global analysis is that some of the parameters that need to be fit to the data are invariant. In the case of global analysis of frequency-domain FLIM data, it is assumed that the fluorescence lifetimes of all species are the same in each pixel of the data, and that only the relative amount of each species varies.

This assumption is equivalent to the assumption that the fluorescence lifetime of each species is not influenced by environmental conditions. In the case of FRET data, it means that the only process that should change the lifetime of the donor should be the transfer of excited-state energy to the acceptor and, moreover, that the FRET process should be the same everywhere. This is a condition that can be fulfilled in good approximation for the case of two partners that bind tightly in a fixed geometry because FRET efficiency depends on distance and orientation.

The second assumption that must be fulfilled in current single-frequency FLIM global analysis methods is that each species should have monoexponential fluorescence decay kinetics. This condition is not always fulfilled; for instance, the commonly used donor CFP has a distinct biexponential decay and should therefore not be used in a global analysis assay if the assumption is made that CFP is monoexponential. (This problem is not unique to global analysis. Generally, any two-component fit of FLIM data with CFP as a donor will not yield quantitative results, unless its multiexponentiality is taken into account explicitly.) The assumption of monoexponentiality is not a fundamental limit of the global analysis method, but rather is due to the use of single-frequency FLIM. GFP has been used in global analysis FRET-FLIM assays even though its decay kinetics shows a slight biexponential behavior. In practical biological applications, this was shown to have only a minor effect, effectively introducing a small bias in the estimation of the relative concentrations of complexes.

As a rule of thumb, a fluorophore can be considered monoexponential for this type of analysis if the lifetimes derived independently from phase and modulation of the FLIM data match within the experimental uncertainty. The best fluorescent protein so far in this respect has been YFP, which has decay kinetics that for all practical purposes are monoexponential. Also, the second partner in a two-component system must be monoexponential; in the case of a FRET system, this means that the donor should still have monoexponential kinetics if bound to an acceptor. This condition can be fulfilled if a single acceptor binds in a tight fixed geometry, or it can be approximately fulfilled if multiple acceptors bind in a configuration such that the resulting distribution of FRET efficiencies and thereby of the lifetimes is very narrow.

The conditions of invariance and monoexponentiality can be checked to some extent using samples with only one of the species (i.e., donor-only samples in the case of a FRET system). With frequency-domain systems, standard FLIM analysis can be used to assess the variations of the lifetimes over the image, and the extent to which the fluorophore is monoexponential can be evaluated by comparing the lifetimes derived from phase and modulation.

The last requirement for successful global analysis is less strict, but it affects the quality of the result. If more data sets are available for the simultaneous fit, with a broad variety in relative concentrations of the different species, the result will be more precise and accurate. Global analysis methods will have difficulties if there is little variation in the sample. For instance, if there is no FRET at all, global analysis is meaningless because no second lifetime is present. The variation in the set of samples in a FRET assay can, however, often be extended experimentally—for instance, by using different stimuli to induce more or less binding, by adding measurements of a sample without acceptor, or by photobleaching the acceptor.

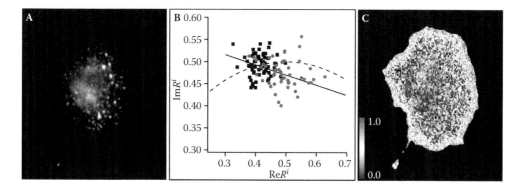

FIGURE 13.1 **(See color insert after page 288.)** Global analysis of FRET data. (A) EGFR-YFP fluorescence. (B) Phasor plot of a subset of the data; black squares: donor-only samples; red circles: donor/acceptor samples. The straight line represents a linear fit through the data. Estimated lifetimes were 2.4 and 1.0 ns. (C) Estimated relative concentrations of phosphorylated EGFR-YPF.

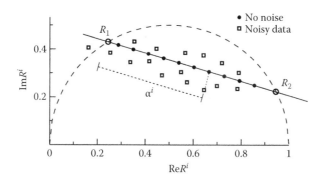

FIGURE 13.2 Illustration of the phasor plot. See the text for details.

three unknowns (two lifetimes and a relative fluorescence fraction). With global analysis, the number of measurements is twice the number of pixels (N) exceeding the number of unknowns ($N + 2$). In addition, global analysis methods allow the simultaneous analysis of multiple FLIM data sets, thus increasing the precision and accuracy of the results by eliminating degeneracy on the parameter landscape.

An important advantage for the analysis of FRET data, which does not exist for intensity-based FRET methods, is that the relative concentrations of interacting species can be determined quantitatively. Ultimately, these are the quantities of interest for researchers in biology who are not interested in physical parameters such as fluorescence lifetimes, but rather wish to answer questions such as, "How much of my protein is interacting with this other protein?"

Global analysis should always be applied with care. It is essential that the data be understood well with the help of standard analysis methods before application of global analysis. However, as soon as it is clear that the investigated system fulfills the requirements for global analysis, data can be analyzed routinely to acquire quantitative information that is not available otherwise, such as the relative fraction of binding molecules.

The methods described in this chapter are designed for frequency-domain FLIM systems. However, there is no reason that the principle of global analysis cannot be applied to time-domain FLIM methods. Indeed, such a method has been reported for confocal photon-counting FLIM systems (Pelet et al. 2004), although to our knowledge this has not been applied yet in a biological study.

Current global analysis methods are limited to two-component systems and restricted to monoexponential chromophores. These restrictions have not inhibited the successful use of global analysis in relevant biological questions (for example, see Ng et al. 2001; Reynolds et al. 2003; Rocks et al. 2005; Verveer, Wouters, et al. 2000; Xouri et al. 2007). Future developments in FLIM instrumentation and in data analysis methods are expected to widen further the range of biological systems where this approach can be applied successfully.

13.6 SUMMARY

In this chapter, we explained the use of global analysis to analyze frequency-domain fluorescence lifetime imaging microscopy (FLIM) data. We described the basis of the method and its application to fluorescence resonance energy transfer (FRET) assays, which we illustrated with an example of an established FRET assay. We discussed the conditions that must be met by the FLIM data to be amendable for analysis with global analysis. If this is the case, global analysis can be applied to determine the relative concentrations of interacting molecules quantitatively, which is ultimately the parameter of interest in many biological applications.

APPENDIX 13.1: METHODS

Cell Preparation

MCF7 breast tumor cells (American Type Culture Collection, ATCC, Manassas, Virginia) were grown in DMEM (Sigma-Aldrich Biochemie GmbH, Hamburg, Germany) supplemented with 10% FCS and 2 mM glutamine (Sigma-Aldrich Biochemie GmbH, Hamburg, Germany). Cells were transiently transfected with an EGFR-YFP plasmid using Lipofectamine 2000 (Invitrogen, Carlsbad, California). Cells were allowed to express for 24 h, after which 1 mg/mL of the antibiotic G418 (Serva Electrophoresis Gmbh, Heidelberg, Germany) was added in order to select stable transfectants. Before sample preparation, transfected cells were seeded on glass-bottom Lab-Tek chambers (Nalge Nunc International, Rochester, New York), grown overnight, and then starved for 5 h in DMEM supplemented with 0.5% FCS. The cells were stimulated with 100 ng/mL of EGF (Cell Signaling Technology, Inc. Danvers, Massachusetts) for 5 min, fixed with 4% paraformaldehide in PBS, washed three times with TBS for 5 min, permeabilized with 0.1% TX-100 for 5 min, and washed twice in PBS. The samples were then incubated with 15 μg/mL of PY72, a generic antibody against phosphotyrosine (In vivo Biotech Service, Hennigsdorf, Germany) labeled with Cy3.5. Donor-only samples were prepared by skipping this last step.

Fluorescence Lifetime Imaging Microscopy

The FLIM system is based on a fully motorized fluorescence microscope (IX-81, Olympus, Germany) with a computer-controlled stage (Corvus-2, Märzhäuser, Germany). A multiline Argon Laser (Innova I305C, Coherent, Germany) modulated at 80 MHz using an acousto-optic modulator (SWM-804AE1-1, Intraaction, Bellwood, Illinois), is coupled into the microscope through the upper backport. The wavelength is selected using an acousto-optic tuneable filter (AOTF.nC-VIS, AA Optoelectronic, Orsay, France), which is also used as a shutter. The fluorescence is collected with a 100X 1.4 NA oil immersion objective through a 530LP dichroich together with a 538/25 emission filter and detected by an intensified camera CCD (PicoStar HR 12, LaVision, Germany), using 2×2 hardware binning. The gain of the camera is modulated at the same frequency as the excitation. The signals used to drive the AOM and the intensifier are generated using two synchronized function generators (2023A, Aeroflex, Plainview, New York). To record a FLIM sequence, the phase of the detector modulation is systematically shifted, and an image is taken at each phase setting.

REFERENCES

Bastiaens, P. I., and Pepperkok, R. 2000. Observing proteins in their natural habitat: The living cell. *Trends in Biochemical Science* 25:631–637.

Bastiaens, P. I., and Squire, A. 1999. Fluorescence lifetime imaging microscopy: spatial resolution of biochemical processes in the cell. *Trends in Cell Biology* 9:48–52.

Bastiaens, P., Verveer, P. J., Squire, A., and Wouters, F. 2001. Fluorescence lifetime imaging microscopy of signal transduction protein reactions in living cells. In *New trends in fluorescence spectroscopy. Applications to chemical and life sciences,* ed. B. Valeur, 297–301. New York: Springer.

Beechem, J. M. 1992. Global analysis of biochemical and biophysical data. *Methods in Enzymology* 210:37–54.

Beechem, J. M., Knutson, J. R., Ross, J. B. A., Turner, B. W., and Brand, L. 1983. Global resolution of heterogeneous decay by phase/modulation fluorometry: Mixtures and proteins. *Biochemistry* 22:6054–6058.

Carlsson, K., Liljeborg, A., Andersson, R. M., and Brismar, H. 2000. Confocal pH imaging of microscopic specimens using fluorescence lifetimes and phase fluorometry: Influence of parameter choice on system performance. *Journal of Microscopy* 199:106–114.

Clayton, A. H. A., Hanley, Q. S., and Verveer, P. J. 2003. Graphical representation and multicomponent analysis of single-frequency fluorescence lifetime imaging microscopy data. *Journal of Microscopy* 213:1–5.

Clegg, R. M. 1996. Fluorescence resonance energy tranfer. *Fluorescence Imaging Spectroscopy and Microscopy* 137:179–251.

Esposito, A., Gerritsen, H. C., and Wouters, F. S. 2005. Fluorescence lifetime heterogeneity resolution in the frequency domain by lifetime moments analysis. *Biophysical Journal* 89:4286–4299.

Gordon, G. W., Chazotte, B., Wang, X. F., and Herman, B. 1995. Analysis of simulated and experimental fluorescence recovery after photobleaching. Data for two diffusing components. *Biophysical Journal* 68:766–778.

Jameson, D. M., Gratton, E., and Hall, R. 1984. The measurement and analysis of heterogeneous emissions by multifrequency phase and modulation fluorometry. *Applied Spectroscopy Review* 20:55–106.

Knutson, J. R., Beechem, J. M., and Brand, L. 1983. Simultaneous analysis of multiple fluorescence decay curves: A global approach. *Chemistry and Physics Letters* 102:501–507.

Ng, T., Parsons, M., Hughes, W. E., Monypenny, J., Zicha, D., Gautreau, A., et al. 2001. Ezrin is a downstream effector of trafficking PKC-integrin complexes involved in the control of cell motility. *EMBO Journal* 20:2723–2741.

Pelet, S., Previte, M. J. R., Laiho, L. H., and So, P. T. C. 2004. A fast global fitting algorithm for fluorescence lifetime imaging microscopy based on image segmentation. *Biophysical Journal* 87:2807–2817.

Redford, G. I., and Clegg, R. M. 2005. Polar plot representation for frequency-domain analysis of fluorescence lifetimes. *Journal of Fluorescence* 15:805–815.

Reynolds, A. R., Tischer, C., Verveer, P. J., Rocks, O., and Bastiaens, P. I. H. 2003. EGFR activation coupled to inhibition of tyrosine phosphatases causes lateral signal propagation. *Nature Cell Biology* 5:447–453.

Rocks, O., Peyker, A., Kahms, M., Verveer, P. J., Koerner, C., Lumbierres, M., et al. 2005. An acylation cycle regulates localization and activity of palmitoylated Ras isoforms. *Science* 307:1746–1752.

Squire, A., Verveer, P. J., and Bastiaens, P. I. 2000. Multiple frequency fluorescence lifetime imaging microscopy. *Journal of Microscopy* 197:136–149.

Verveer, P. J., and Bastiaens, P. I. H. 2003. Evaluation of global analysis algorithms for single frequency fluorescence lifetime imaging microscopy data. *Journal of Microscopy* 209:1–7.

Verveer, P. J., Squire, A., and Bastiaens, P. I. 2000. Global analysis of fluorescence lifetime imaging microscopy data. *Biophysical Journal* 78:2127–2137.

Verveer, P. J., Squire, A., and Bastiaens, P. I. 2001a. Improved spatial discrimination of protein reaction states in cells by global analysis and deconvolution of fluorescence lifetime imaging microscopy data. *Journal of Microscopy* 202:451–456.

Verveer, P. J., Squire, A., and Bastiaens, P. I. 2001b. Frequency-domain fluorescence lifetime imaging microscopy: A window on the biochemical landscape of the cell. In *Methods in cellular imaging*, ed. A. Periasamy, 273–292. Oxford: Oxford University Press.

Verveer, P. J., Wouters, F. S., Reynolds, A. R., and Bastiaens, P. I. H. 2000. Quantitative imaging of lateral ErbB1 receptor signal propagation in the plasma membrane. *Science* 290:1567–1570.

Wouters, F. S., and Bastiaens, P. I. 1999. Fluorescence lifetime imaging of receptor tyrosine kinase activity in cells. *Current Biology* 9:1127–1130.

Xouri, G., Squire, A., Dimaki, M., Geverts, B., Verveer, P. J., Taraviras, S., et al. 2007. Cdt1 associates dynamically with chromatin throughout G1 and recruits Geminin onto chromatin. *EMBO Journal* 26:1303–1314.

4

Applications

CHAPTER 14

FLIM Applications in the Biomedical Sciences

Ammasi Periasamy and Robert M. Clegg

14.1 INTRODUCTION

Fluorescence lifetime measurements acquired with a spectrofluorometer (nonimaging) or light microscope (imaging) are vital tools in biological and clinical sciences because of the valuable information contained in fluorescence lifetime. Fluorescence is inherently a dynamic phenomenon; the steady-state intensity averages the fluorescence signal over the decisive time when the fluorescence is dynamically interacting and communicating on a molecular scale with its immediate environment (usually less than 10 ns). All physical and chemical events that transpire within the lifetime of the excited state of the fluorophore, such as solvent relaxation, rotational freedom of the probe, Förster resonance energy transfer, other excited-state reactions, and dynamic quenching, affect the dynamic response of an electronically excited fluorophore.

Usually these events can be observed directly by following the fluorescence emission on that time scale. However, steady-state fluorescence measurements integrate the fluorescence signals over this critical timescale, so extra assumptions are required for their interpretation that are not required for the interpretation of lifetime-resolved measurements. Thus, lifetime resolution provides extra parameters that are often critical for correctly interpreting fluorescence measurements, especially in the environment of complex biological systems. Of course, FLIM provides spatial resolution in addition to this temporal information and, just as in single-channel measurements, the lifetime resolution provided by FLIM greatly extends the capabilities of existing fluorescence imaging techniques. Different lifetime contributions can be distinguished, increasing contrast of the image and providing better separation of multiple components and many times better elimination of autofluorescence or other unwanted fluorescent background.

As explained in various chapters in this book, each molecule has its distinct lifetime value. Because the lifetime is sensitive to the environment of the molecule and most FLIM experiments are carried out on ensembles of molecules, the measured lifetimes at any location usually reflect a distribution of lifetimes that depends on the heterogeneity of the molecular environment within the smallest observed volume. The majority of lifetime distributions in biological systems is around 1–10 ns; however, as mentioned earlier, in some applications, longer lifetime fluorophores prevail in the range of microseconds to milliseconds. These long-lifetime probes are used for various applications to allow elimination of autofluorescence as described in this chapter. On the other hand, autofluorescence can also be a useful tool to study some biological processes.

As covered in the various chapters, a number of methodologies or devices are available to measure lifetimes in an image. Often, the data acquisition appears simple, provided one understands the fundamentals of the measurement. Also, nowadays many different instruments are commercially available, so assembling the instrumentation and carrying out the primary data acquisition often is not a problem. In the 1990s, no commercial instruments existed and, consequently, lifetime usage was minimal. Now, various commercial units make it relatively simple to collect lifetime data. However, the difficult part of lifetime measurements is in the analysis and interpretation of the data. It can consequently be a challenge for biologists to take full advantage of FLIM data acquisition and processing.

In this chapter, we provide historic background about FLIM and selected applications not listed in the other chapters.

14.2 A BRIEF HISTORICAL JOURNEY THROUGH THE DEVELOPMENT OF LIFETIME-RESOLVED IMAGING

Luminescence lifetime-resolved imaging in a microscope began quite some time ago; the first examples were the measurement of phosphorescence. Because the emission lifetime of phosphorescence (or delayed fluorescence) (Phipson 1870; Pringsheim 1949) is orders of magnitude longer than prompt fluorescence (tens of microseconds up to seconds for phosphorescence, compared to nanoseconds or subnanoseconds for fluorescence), phosphorescence was the first time-resolved luminescence to be observed, as well as the first instance of lifetime-resolved imaging.

All the early phosphorescence microscopes were essentially constructed similarly to the Becquerel phosphoroscope (Becquerel 1868, 1871): a mechanical rotating disc (or two synchronous discs) or a rotating cam with slits adjusted so that the emitted luminescence was observed repetitively only after the excitation light illuminated the sample. The phosphorescence can be observed easily, even in the presence of scattering and prompt fluorescence, because it is delayed by a time that is many times the lifetime of the prompt fluorescence. Because of the time delay required to separate prompt fluorescence and scattering from the delayed luminescence signal, this is lifetime-resolved imaging. With such relatively simple mechanical instrumentation, very weak phosphorescence from an image can be observed, even in the presence of high intensities of prompt fluorescence, and many ingenious experiments have been carried out for over 200 years.

It is difficult to say when the first time-resolved phosphorescence images (as opposed simply to observing the phosphorescence decay) were observed without the attempt to localize the source of emission within a spatially extended sample (even though one can look directly at an image with a simple Becquerel phosphoroscope). However, recognition as the inventor of the first imaging phosphorescence microscope is usually given to Newton Harvey (Harvey and Chase 1942), followed by later studies with microscopes capable of making lifetime-resolved phosphorescence images from biological material (Polyakov, Rosanov, and Brumberg 1966; Vialli 1964; Zotikov 1982; Zotikov and Polyakov 1975, 1977). All these microscopes were basically designed according to the Becquerel phosphoroscope.

More recent phosphorescence microscope instruments, based on CCD (charge coupled device) image data acquisition, appeared in the late 1980s and early 1990s; these instruments were used to determine the phosphorescence decay time (Beverloo et al. 1990; Jovin et al. 1989, 1990; Marriott et al. 1991). Similar instruments have since been constructed and applied to measure delayed luminescence (that is, with lifetimes longer than microseconds) in biological samples. This type of lifetime-resolved microscopy is also very useful when using lanthanides, which have lifetimes of hundreds of microseconds to milliseconds (Vereb et al. 1998). Delayed luminescence imaging is usually not considered under the label of FLIM, but it is still an active field of research.

Fluorescence lifetime measurements in the nanosecond time range in optical microscopes were first carried out by Venetta in the frequency domain as early as 1959 (Chapters 5 and 11). In Venetta's instrument, the spatial location within an image that one wanted to study was selected by focusing an intermediate image in the microscope onto an opaque surface (a mirror) with a small circular hole, which selected the part of the image impinging onto the detector. The detected fluorescence was thereby limited to a small selected area of the original image. The light was modulated at about 5–10 MHz via a standing ultrasonic grating, and the modulated fluorescence signal was phase detected using a homodyne technique (see Chapter 5).

The excitation light was detected with one photomultiplier (PM) and fluorescence with a second photomultiplier. The amplification levels of the two photomultipliers were modulated at the same frequency; the phase of the modulation on the reference PM was adjusted until the two signals were 180° out of phase. The nondelayed timing of the fluorescence channel was calibrated by detecting the excitation light with both PM channels. By comparing the phase of the fluorescence signal relative to a scattering signal, the phase—and therefore the lifetime—of the emitted fluorescence could be estimated (see Chapter 5). The instrument was capable of dissecting the image into areas of interest and can therefore be classified as an imaging fluorescence lifetime instrument.

After proving the precision of the associated modulation electronics and the accuracy of the data acquisition, lifetime measurements were carried out on fluorochromes bound to the nuclei of tumor cells, as well as autofluorescence of biological tissue samples. Interestingly, this development was apparently far ahead of the general interest of the scientific community for measuring fluorescence lifetimes in microscopic samples and was even developed before most of the phosphorescence microscopes were constructed. Venetta's instrument was developed at the same time as the frequency domain instrumentation of nonimaging fluorescence lifetime measurements (Birks and Little 1953; Birks and Dawson 1961).

Almost 20 years elapsed before other fluorescence lifetime measurements in microscopes, with imaging capabilities, were attempted by comparing decay curves from an analog simulator with the fluorescence decay on an oscilloscope (Loeser and Clarck 1972; Loeser et al. 1972), using Ortec photon counting instrumentation (Arndt-Jovin et al. 1979), and with a homemade digital averager (Andreoni et al. 1975, 1976; Bottiroli 1979). As one would expect, the measurement time was very long in order to acquire satisfactory signal-to-noise ratio for multiexponential analysis for photon-counting methods where this was possible. The delay in the development of FLIM was probably partially due to the lack of commercial electronics. The data acquisition electronics was either homemade or not meant for weak signals from a microscope compared to cuvette-based nonimaging fluorescence lifetime measurements. Light sources were also not convenient lasers, but rather a pulsed nitrogen lamp (Arndt-Jovin et al. 1979) or a nitrogen-pumped tunable dye laser (Bottiroli 1979).

In addition, the possibility of quantifying optical measurements was only beginning to interest most researchers dealing with biological problems. By the late 1970s, the field of nanosecond fluorescence lifetime measurements (nonimaging) was well underway, and commercial instrumentation (in both the time and frequency domains) was available; also, highly sophisticated data analysis was being seriously developed and was available. Researchers in general (that is, not only the aficionados) became aware of the advantages and importance of incorporating lifetime-resolution in their fluorescence measurements.

In spite of these technical developments and increased interest in acquiring fluorescence lifetimes, as well as the exciting biological applications of time-resolved nonimaging fluorescence, other than the few pioneering attempts mentioned previously, the field of FLIM did not start in earnest until the late 1980s. This timing coincided with the ready availability of advanced CCD cameras, image intensifiers, and scanning confocal microscopes. As is often the case, it is not easy to distinguish the desire of researchers in making certain experiments and the ability to do so that is often limited by readily available instrumentation. Therefore, it is not surprising that the "dawn of FLIM" took place in a few research laboratories with ready expertise in time-resolved fluorescence, fast electronics, and usually a strong interest in biological problems; a review with references to many of the subsequent publications from many research groups is available (van Munster and Gadella 2005). However, immediately following the initial publications and announcements at international meetings, FLIM attracted the attention of many laboratories; eventually, industry became convinced of the utility of and the market for FLIM.

14.3. AUTOFLUORESCENCE LIFETIME IMAGING OF CELLS

A fluorescence lifetime measurement from endogenously fluorescent elements of cells and tissues has established a new fast emerging field of applications with the potential of practical use in medicine. The majority of information thus derived originates from reduced nicotinamide adenine dinucleotide (NADH) and its phosphate forms (NAD(P)H) along with flavins and flavoproteins. These fluorescing molecules are participants of cellular metabolism at which the energy-rich molecules are oxidized in cytoplasm (glycolysis) and mitochondria (oxidative phosphorylation) to CO_2 and H_2O with the release of the energy, which is reused by cells mainly for synthesis and motion. As the principal electron donor in glycolytic and oxidative energy metabolism, NAD(P)H is thus a convenient noninvasive fluorescent probe of the metabolic state.

Traditionally, spectrofluorimetric studies characterized the metabolic dynamics of the total NAD(P)H concentration. However, it has been pointed out that the reaction velocity of a given intracellular NADH-linked dehydrogenase depends on the concentration of locally available NADH—that is, the local concentration of free NADH (Williamson, Lund, and Krebs 1967). Additionally, given the mechanism of the oxidative phosphorylation, the ratio of the free and bound forms of NADH may serve as a good indicator of cellular metabolic activity and hypoxic states. Aside from the biochemical analysis, which requires extraction of pyridine nucleotides, fluorescence lifetime is the only technique currently available that provides a noninvasive way to assess the free/bound ratio of NADH. The short lifetime τ_1 at the range of ~300–500 ps is usually attributed to the free form of NADH and the long lifetime τ_2 (~2,500–3,000 ps) to the bound form (Vishwasrao et al. 2005). Many enzymes bind to NADH in the metabolic pathway, and as favored metabolic pathways shift with cancer progression, the distribution of NADH binding sites changes. The fluorescence lifetime of protein-bound NADH changes depending on the enzyme to which it is bound. This suggests that changes in metabolism with cancer development can be probed by the lifetime of protein-bound NADH.

The relation of NADH free and bound fractions' ratio to metabolic state makes FLIM a prospective technique for the diagnosis of a range of diseases that affect production and utilization of energy in cells—cancer, Alzheimer's disease, Parkinson's disease, hypoxemia, ischemia, etc. Thus, one of the hallmarks of carcinogenesis is a shift from cellular oxidative phosphorylation to cellular glycolysis for ATP production (the Warburg effect) (Gulledge and Dewhirst 1996; Warburg 1930). This phenomenon in its turn leads to the increase of the free/bound ratio, which has been utilized to diagnose precancerous changes in epithelia by means of FLIM (Skala 2007). In addition to the study of cancers, the approach has been shown as informative for studies of ischemia. FLIM of NADH also has a potential in treatment optimization: Recent studies have demonstrated that the coenzyme lifetimes change crucially at the apoptosis as opposed to the deleterious for the cell environment necrosis (Wang 2008). This contrast in NADH lifetime dynamics can be applied for the apoptosis-favored dosimetry in cancer treatment—in particular, in photodynamic therapy.

With the increasing role of the autofluorescence lifetime and future prospects of its application in medical diagnostics, a detailed characterization of NADH fluorescence lifetime dynamics in normal states is required for the proper assessment of pathologies. A series of such studies of NADH fluorescence lifetime perturbations under normal conditions has been conducted by the researchers from the Institute of Biophotonics, National Yang-Ming University, Taiwan. Among these is the characterization of the autofluorescence lifetime dynamics at different cell densities in a culture. Ghukasyan et al. have conducted measurements in NAD(P)H bands from cervical cancer cells (HeLa line) under physiological conditions (37°C, 5% CO_2) maintained by a microscope cage incubator.

The results appeared to be well correlated with the dynamics of cell culture growth. Little to no change in lifetime parameters was observed during the first 2 days after plating—the period of recovery after stress and restoration of normal metabolic levels. As the cells proceed to the exponential growth phase—the ratio, a_1/a_2, of the short and long lifetimes—correspondingly (free/bound ratio) exhibit a decrease. These values slightly grow and then enter a plateau; the cell culture growth slows down and enters the confluency state. The changes have also been found to be location specific: The cells in the middle of the colonies exhibit a broad variation of the pre-exponential factors' ratio and lifetime values, whereas at the edge there is a significant trend of increased a_1 values, reflecting a larger portion of free NAD(P)H molecules at this location.

Earlier studies have reported the dependence of the metabolism on cell density. Oxygen consumption, net lactate production, ATP-content, NAD-content, and NAD-redox potential have been shown to decrease with the increasing cell density; the phenomenon has been attributed mostly to cell–cell contacts rather than to proliferation (Bereiter-Hahn, Munnich, and Woiteneck 1998). The results obtained by the NYMU group demonstrate that the lifetime dynamics in this cell culture cannot be attributed solely to the levels of oxygen available, as was demonstrated earlier for MCF-10A (Bird et al. 2005). The real cause for such dynamics is a subject of further studies.

The energy demands of cells change as they go through the cell cycle stages. The cell cycle is generally considered to consist of four basic phases: the gap 1 phase (G1-phase), the DNA synthesis phase (S-phase), the gap 2 phase (G2-phase), and the mitotic phase (M-phase). Upon duplication of the DNA content of the nucleus, cells enter the G2-phase, followed by mitosis. As cells go through the cycle, the biosynthetic activity levels change and thus the demand for energy. Ghukasyan et al. tested whether these changes in activity are correlated with the NADH fluorescence lifetime via the synchronization of cell cultures. Synchronized with hydroxyurea, a well-known agent to arrest cell cycles at the S-phase, HeLa cells exhibited well distinguishable changes in NADH fluorescence lifetime as compared to the control cells. The most prominent shift was observed in changes of the pre-exponential factors. Despite the fact that the majority of the signal from the images still exhibited 80% of a_1 (similar to control cells), another peak at 75% appeared, which is well illustrated by the color coding in Figure 14.1. Change in metabolic activity of dividing cells has also been demonstrated on mouse embryos. Fluorescence lifetime measurements of NAD(P)H taken from the embryo at the stage of two blastomers exhibit different levels of metabolic activity, as shown in Figure 14.2.

14.4 PAP SMEAR DETECTION USING TIME-GATED LIFETIME IMAGING MICROSCOPY

This section describes how to remove the autofluorescence in cellular imaging by using the time-gated lifetime imaging method. To achieve an autofluorescence reduction, as shown in Figure 14.3, a mixture of two fluorescent compounds of varying lifetimes is excited with a short pulse of light. The excited molecules will emit fluorescence levels with a time dependence related to the length of the excited-state lifetime. For example, although both types of fluorescence decay follow an exponential curve, short-lived fluorescence dissipates to zero in nanoseconds, whereas long-lived fluorescence demonstrates lifetimes consisting of microseconds to 10 ms. The lifetime distribution due to autofluorescence from cells or tissue media as well as the microscope optical elements is on the order of <1 μs.

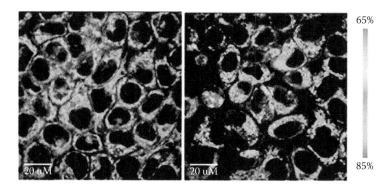

FIGURE 14.1   **(See color insert following page 288.)** Autofluorescence lifetime images recorded from untreated (left) and synchronized with hydroxyurea HeLa cells (S-phase). The color coding represents the contribution in percentage of the shorter lifetime, which for the given wavelength (450 nm) is usually attributed to the free NADH. Synchronization at the S-phase of the cell cycle with hydroxyurea results in the majority of the bound form, indicating increased level of metabolism resulting from the high demand in ATP. (Figure courtesy of Drs. Vladimir Ghukassyan and Fu-Jen Kao, National Yang Ming University, Taipei, Taiwan.)

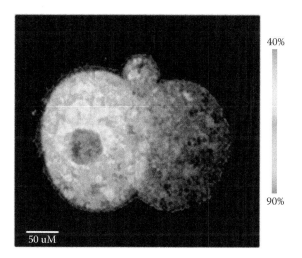

FIGURE 14.2   **(See color insert following page 288.)** Lifetime distribution of metabolic activity in mouse embryo; the mouse embryo is at the stage of two blastomers in the NADH spectrum band (456 ± 60). The color coding by the pre-exponential component a_1 (free NADH) shows that only one blastomer exhibits high metabolic activity with high concentration of NADH and the majority of it bound. (Figure courtesy of Drs. Vladimir Ghukassyan and Fu-Jen Kao, National Yang Ming University, Taipei, Taiwan.)

Thus, if no measurements are taken during the first 1–100 μs after laser pulse excitation, all short-lived fluorescence will decay to zero and long-lived fluorescence signals can be measured with very high sensitivity. As an example, the time-gated lifetime imaging microscopy technique enables the separation of true fluorescence emitted from HPV

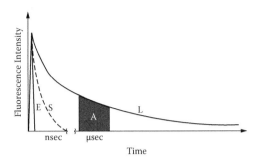

FIGURE 14.3 Illustration of autofluorescence (nanoseconds) and long lifetime (microseconds) fluorophore decays. (Periasamy, A. et al. 1995. *Microscopy and Analysis,* March 19–21.)

cDNA probes labeled with long-lifetime probes (microseconds to milliseconds) from autofluorescence (1–100 ns) of PAP absorption stains (Periasamy et al. 1995).

The availability of bioreagents such as monoclonal antibodies and nucleic acid probes has opened up new possibilities for the localization and analysis of proteins and nucleic acid sequences in cells, tissues, and chromosomes. Unfortunately, due to autofluorescence from solvents, solutes, cells tissues, fixed tissues, or clinical samples, autofluorescence effectively decreases the signal-to-noise ratio of detection. For example, the main assay of detection of cervical dysplasia and cervical cancer is the PAP smear. Although it is performed in large numbers worldwide, the PAP smear unfortunately detects only between 50 and 80% of abnormalities subsequently found by histological examination of biopsy specimens. In addition, screening of PAP smears is associated with a false negative rate of 5–10%, depending on how and where the screening is undertaken (Moscicki et al. 1991).

Substantial evidence has been presented that strongly associates human papillomavirus (HPV) infection with the development of cervical cancer. HPVs have been divided into high- and low-risk groups based on their association with benign or malignant lesions; the majority of malignant lesions are infected with the high-risk genotypes HPV-16 or HPV-18. Thus, the ability to detect high-risk HPV, if present, in cells of a PAP smear might lead to better discrimination of women at high risk for cervical disease. The most sensitive and reliable assays for detecting HPV are based on nucleic acid hybridization; however, neither radioactive nor nonradioactive enzymatic in situ hybridization allows quantification of HPV copy number at the single-cell level, and they suffer from inherent lack of linearity.

Fluorescence in situ hybridization (FISH), on the other hand, allows quantification of HPV copy level in single cells with very high sensitivity. The FISH assay for HPV detection that can detect as little as one copy of HPV/cell has been listed in the literature (Siadat-Pajouh, Ayscue, et al. 1994; Siadat-Pajouh, Periasamy, et al. 1994). Unfortunately, the assay requires the use of fresh cervicovaginal cell preparations and cannot be used on standard PAP smears because the absorption stains used by cytopathologists are intensely autofluorescent and interfere with the FISH signal. This autofluorescence has a defined lifetime.

As illustrated in Figure 14.4 (a simulated image—the original image can be seen in the literature; Periasamy et al. 1995), the removal of the autofluorescence signal from the HPV DNA signal in CaSki cervical cells was demonstrated by using FISH in combination with a

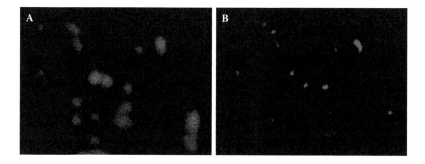

FIGURE 14.4 **(See color insert following page 288.)** Simulation of time-gated image to remove the autofluorescence signal. (A) HPV DNA signal of CaSki cells of normal fluorescence image (no gating). (B) Time-gated image collected after laser pulse excitation as illustrated in Figure 14.1. The autofluorescence was not recorded during microsecond time imaging and the punctuated DNA signal was very clear. (Periasamy, A. et al. 1995. *Microscopy and Analysis,* March 19–21.)

long-lifetime probe (Europium conjugated avidin) and time-gated images. The long-lifetime probe (lifetime is 700 µs) was excited by 355 nm and the emission was collected at 610 nm using a 40x oil 1.3 Nikon objective lens. The camera used for this data collection was a gated image intensifier CCD camera (Hamamatus Photonics, New York) and the gating of the camera was synchronized with the excitation laser pulse, as described in the literature (Periasamy et al. 1995).

Figure 14.4(A) shows the digitized prompt fluorescence image of the CaSki cells after hybridization with specific HPV probes. Punctate HPV DNA fluorescence signals were observed in the nuclei of these cell lines, which were surrounded by strong autofluorescence signals from the fixative, cytoplasm of the cells, and other microscope optical elements. Figure 14.4(B) illustrates the image obtained when the gated image intensifier is changed to the gating mode. The camera was turned on for a brief period of time (250 µs) 30 µs after laser pulse excitation. This time-resolved approach results in the disappearance of the autofluorescence and a many-fold increase in the sensitivity of detection compared to that when autofluorescence is present.

This kind of study demonstrates that a long-lifetime probe such as europium chelate can be used with FISH to provide a suitable marker for clinical imaging to detect HPV DNA signal in HPV infected cells. In addition to this application, previous studies have shown that europium fluorescence can be distinguished from that of an excess amount of a prompt decaying fluorescence substance (e.g., fluorescein, rhodamine 123, which may have decay times of <3–5 ns). Thus, multiparameter detection of cellular constituents based upon the characteristic lifetime of the probes is possible. This is particularly advantageous because many molecules exhibit similar spectra but drastically distinct lifetimes. In addition, a number of natural peptides are also fluorescent when excited with ultraviolet wavelength of light, resulting in autofluorescence. This inherent autofluorescence can be suppressed by using a phosphorescence probe and the time-gated microscopy technique. This technique has another advantage in that it allows the user to excite the sample with a brief pulse of excitation intensity for a fixed duration of time, instead of s continuous illumination that could potentially limit photodamage of the sample under study.

14.5 FLIM IN ALZHEIMER'S DISEASE

Alzheimer's disease (AD) is a progressive and fatal brain disease; about five million Americans are living with AD. This disease was first described by German physician Alois Alzheimer in 1906 (http://www.alz.org/alzheimers_disease_what_is_alzheimers.asp#history). Alzheimer's destroys brain cells, causing problems with memory, thinking, and behavior. The disease gets worse over time and it is fatal. Two abnormal structures, called plaques and tangles, are prime suspects for damaging and killing nerve cells in the brain. Plaques build up between nerve cells and contain deposits of a protein fragment called beta-amyloid. Tangles are based on another protein, called tau. Though most people develop some plaques and tangles as they age, those with AD tend to develop far more. Scientists believe that communication among nerve cells is somehow blocked and activities that cells need to survive disrupted.

Deposition of amyloid b (AB)-containing plaques in the brain is one of the major neuropathological hallmarks of AD. The final enzymatic step in generating AB via intramembranous cleavage of amyloid precursor protein (APP) is performed by the presenilin 1 (PS1)-dependent r-secretase complex (Fraser et al. 2000; Selkoe 2001). More than 100 mutations spread throughout the PS1 molecule are linked to autosomal dominant familial AD. Bacskai's group studied the different conformations of PS1 protein using a FRET-FLIM technique (Berezovska et al. 2005; Lleó et al. 2004). As shown in Figure 14.5, the conformation of PS1 distribution is clearly shown in the plasma membrane.

14.6 OPTICAL PROJECTION OF FLIM IMAGES OF MOUSE EMBRYO

Three-dimensional optical projections of biological images are an interesting tool in biological investigations (Dodt et al. 2007; Huisken et al. 2004; Sharpe et al. 2002). Recently, McGinty et al. (2008) described a quantitative fluorescence projection tomography that measures the fluorescence lifetime distribution in three dimensions. As described in their paper, this can be achieved by acquiring a series of wide-field, time-gated lifetime images at different relative time delays, with respect to a train of excitation pulses, at a number of projection angles.

A mouse embryo was fixed in formaldehyde and the neurofilament was labeled with an Alexa-488 conjugated antibody. The intensity image and the FLIM image were collected using a 485 nm excitation wavelength and the emission was collected at 515/25 nm. The three-dimensional time-integrated fluorescence intensity and FLIM image reconstruction of the mouse embryo are shown in Figure 14.6. The single exponential yielded an average fluorescence lifetime of the neurofilament at about 1,360 ps. The autofluorescence lifetime of the heart and dorsal aorta was about 1,030 ps. This demonstration provided a clear example for lifetime optical projection tomography to distinguish between extrinsic and intrinsic fluorescence signals without prior anatomical knowledge of the labeling sites. This kind of information cannot be obtained with steady-state fluorescence intensity imaging.

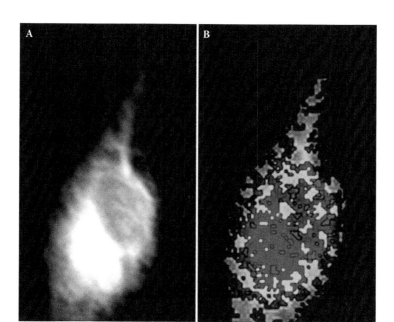

FIGURE 14.5 **(See color insert following page 288.)** FLIM analysis of the subcellular distribution of the presenilin 1(PS1) protein in different conformations. Chinese hamster ovary cells expressing human PS1 were immunostained with fluorescently labeled antibodies against PS1 N-terminus (NT, Alexa-488 donor fluorophore) and C-terminus (CT, Cy3 acceptor fluorophore). The proximity between the PS1 NT and PS1 CT was monitored with FRET using FLIM. The intensity image (A) shows PS1 expression (donor alone) throughout the cell. The pseudocolored FLIM image (B) displays different lifetimes of the donor within the cell. The conformation of PS1 changes as the protein matures and is trafficked to the plasma membrane. FLIM allows determination of the conformation of the peptide within cellular compartments. The pseudocolor is based on discrete color ranges for the lifetimes as follows: red: 100–1,000 ps; green: 1,001–1,800 ps; blue: 1,801–2,500 ps. A shorter lifetime represents closer proximity between the labeled PS1 termini. The lifetime of the Alexa-488 donor in the absence of Cy3 acceptor is ~2,400 ps. (Figure courtesy of Brian J. Bacskai, Massachusetts General Hospital, Boston.)

14.7 FULL-FIELD FLIM WITH QUADRANT DETECTOR

Eckert, Petrasek, and Kemnitz (2006) have applied a novel low-intensity, nonscanning fluorescence lifetime imaging microscopy technique to the investigation of fluorescence dynamics in living cells of the chlorophyll *d*-containing cyanobacteria *Acaryochloris marina* and in individual chloroplasts of moss leaves. The imaging technique is a nonscanning wide-field FLIM with time- and space-correlated single-photon counting (TSCSPC). The detector is a novel microchannel plate photomultiplier with quadrant anode (QA-MCP) (Lampton and Malina 1976). The time-resolved images are acquired under low-excitation conditions with parallel data acquisition. A schematic of the instrument is shown in Figure 14.7 (left).

Eckert et al. have demonstrated the capability of their wide-field, nonscanning TSCSPC fluorescence lifetime microscope to investigate the fluorescence dynamics of individual

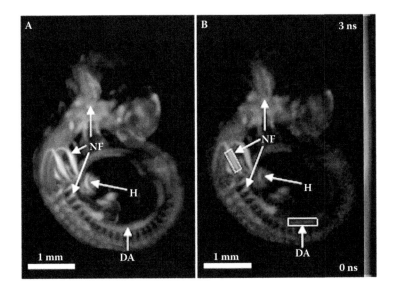

FIGURE 14.6 **(See color insert following page 288.)** Intensity and lifetime image of a mouse embryo. (A) Three-dimensional time-integrated fluorescence intensity and (B) three-dimensional fluorescence lifetime rendering of a mouse embryo. The Alexa-488 antibody labeled neurofilament (NF) was excited at 485 nm and obtained a consistent fluorescence lifetime of 1360 ± 180 ps. The shorter lifetime (1030 ± 135 ps) component corresponds to autofluorescence from H-heart and DA-dorsal aorta. (Permission from *Journal of Biophotonics* 1:390–394, 2008.)

living cells at average light intensity in a sample of less than 10 mW/cm^2 in order to avoid noteworthy detrimental photodynamic reactions. The shortest time for data acquisition is 2 min, and some of the data is taken over the duration of an hour. However, the method is capable of measuring pico- to nanoseconds with high photon-counting accuracy. This is not a scanning technique, but rather a full-field method that nevertheless has the statistical accuracy of normal photon-counting methodologies and allows accurate determinations of relatively close fluorescence lifetimes. An example is given in Figure 14.7, where the two lifetime decays from the two cells selected from a full field of many cells are shown. Eckert and colleagues stress that their technique yields high-quality data without significant photodamage due to the low level of excitation and method of acquisition. The authors were able to show that the fraction of "closed" PS II reaction centers increases with the measurement time due to the illumination of the cells by the excitation laser beam.

14.8 CONCLUSION

There is no question that the application of FLIM microscopy in the biomedical sciences will continue to increase in all directions, driven by development of advanced technologies. A growing number of researchers will be able to use this approach routinely. FLIM provides many advantages and its use is likely to expand; however, not every laboratory will have a FLIM facility or the required expertise or experience in rigorous FLIM data analysis. Further advances will arise from the convergence of the basic FLIM measurement with other technologies, such as Förster resonance energy transfer (FRET), dual-

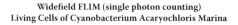

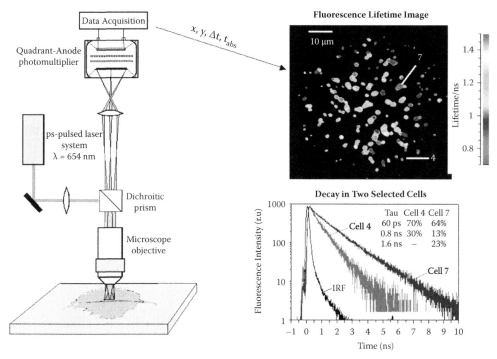

FIGURE 14.7 **(See color insert following page 288.)** Schematic of the FLIM setup and the fluorescence dynamics in individual cells of *A. marina*. See text for details. (Figure courtesy of H.-J. Eckert, Max Volmer Laboratory for Biophysical Chemistry, Technical University Berlin, Germany.)

color fluorescence correlation spectroscopy, image cross-correlation spectroscopy (ICCS), and others in the area of various biological and clinical applications.

New fluorophores—in particular, quantum dots—will expand the usefulness of FLIM qualitatively and quantitatively. The opportunities of additional mathematical modeling will lead to detailed insights into cellular dynamics. For example, in FLIM-FRET, using the efficiency of energy transfer and distance information, donor-to-acceptor ratios, knowledge of protein structure, etc.—plus data gleaned from other experimental methods—will produce interesting three-dimensional or four-dimensional information of the biological process in living cells and tissue.

At the time of this writing, many hundreds of publications employ FLIM (and this number is increasing rapidly), and over ten industrial firms offer some form of FLIM. The FLIM field is mature and the prospective user is now confronted with many different possibilities (and sometimes difficult choices), if he or she wishes to include FLIM in the laboratory repertoire. The reason for this book is to present the fundamentals of FLIM methodology, instrumentation, and data analysis to assist the prospective user of FLIM, as well as those interested in judging the many recent publications, in understanding many of the aspects of FLIM that are not emphasized in normal publications.

REFERENCES

Andreoni, A., Sacchi, C. A., Cova, S., Bottiroli, G., and Prenna, G. 1975. A study of the fluorescence pattern of chromosomes. In *Lasers in physical chemistry and biophysics,* ed. J. Joussot-Dubien, 413–424. Amsterdam: Elsevier Scientific.

Andreoni, A., Sacchi, C. A., Svelto, O., Longoni, A., Bottiroli, G., and Prenna, G. 1976. In *Proceedings of the Third European Electro-Optics Conference,* ed. H. A. Elion, 99, 258–270. Washington, D.C.: SPIE.

Arndt-Jovin, D. J., Latt, S. A., Striker, G., and Jovin, T. M. 1979. Fluorescence decay analysis in solution and in a microscope of DNA and chromosomes stained with quinicrine. *Journal of Histochemistry and Cytochemistry* 27:87–95.

Becquerel, E. 1868. *La lumiere. Ses causes et ses effets.* Paris: Firmin Didot.

Becquerel, E. 1871. Memoire sur l'analyse de la lumière émise par les composés d'uranium phosphorescents. *Annales de Chimie et Physique* 27:539.

Bereiter-Hahn, J., Munnich, A., and Woiteneck, P. 1998. Dependence of energy metabolism on the density of cells in culture. *Cell Structure and Function* 23:85–93.

Berezovska, Lleo, A., Herl, L. D., Frosch, M. P., Stern, E. A., Bacskai, B. J., et al. 2005. Familial Alzheimer's disease presenilin 1 mutations cause alterations in the conformation of presenilin and interactions with amyloid precursor protein. *Journal of Neuroscience* 25:3009–3017.

Beverloo, H. B., van Schadewijk, A., van Gelderen-Bode, S., and Tanke, H. J. 1990. Inorganic phosphors as new luminescent labels for immunocytochemistry and time-resolved microscopy. *Cytometry* 1:784–792.

Bird, D., Yan, L., Vrotsos, K. M., Eliceiri, K. W., Vaughan, E.M., Keely, P. J., White, J. G., and Ramanujam, N. 2005. Metabolic mapping of MCF10A human breast cells via multiphoton fluorescence lifetime imaging of the coenzyme NADH. *Cancer Research* 65 (19): 8766–8773.

Birks, J. B., and Dawson, D. J. 1961. Phase and modulation fluorometer. *Journal of Scientific Instruments* 38:282–295.

Birks, J. B., and Little, W. A. 1953. Photofluorescence decay times of organic phosphors. *Proceedings of the Physical Society, Section A* 66:921–928.

Bottiroli, G. P., Andreoni, A., Sacchi, C. A., and Svelto, O. 1979. Fluorescence of complexes of quinacrine mustard with DNA: Influence of the DNA base composition on the decay time in bacteria. *Photochemistry and Photobiology* 29:23–28.

Dodt, H. U., Leischner, U., Schierloh, A., et al. 2007. Ultramicroscopy: Three-dimensional visualization of neuronal networks in the whole mouse brain. *Nature Methods* 4:331–336.

Eckert, H. J., Petrasek, Z., and Kemnitz, K. 2006. Application of novel low-intensity nonscanning fluorescence lifetime imaging microscopy for monitoring excited-state dynamics in individual chloroplasts and living cells of photosynthetic organisms. *Proceedings of SPIE, Advanced Photon Counting Techniques* 6372:637207

Fraser, P. E., Yang, D., Yu, G., Levesque, L., et al. 2000. Presenilin structure, function and role in Alzheimer's disease. *Biochemica et Biophysica Acta* 1502:1–15.

Ghukasyan, V., and Kao, F. J. 2009. Monitoring cellular metabolism with fluorescence lifetime of reduced nicotinamide adenine dinucleotide. *Journal of Physical Chemistry C.* In Press.

Gulledge, C. J., and Dewhirst, M. W. 1996. Tumor oxygenation: A matter of supply and demand. *Anticancer Research* 162:741–749.

Harvey, E. N., and Chase, A. M. 1942. The phosphorescence microscope. *Review of Scientific Instruments* 13:365–368.

Huisken, J., Swoger, J., Del bene, F., Wittbrodt, J., and Stelzer, E. H. K. 2004. Optical sectioning deep inside live embryos by selective plane illumination microscopy. *Science* 305:1007–1009.

Jovin, T. M., Arndt-Jovin, D. J., Marriott, G., Clegg, R. M., Robert-Nicoud, M., and Schormann, T. 1990. Distance, wavelength and time: The versatile third dimensions in light emission microscopy. In *Optical microscopy for biology,* ed. B. Herman and K. Jacobson, 575–602. New York: Wiley–Liss.

Jovin, T. M., Marriott, G., Clegg, R. M., and Arndt-Jovin, D. J. 1989. Photophysical processes exploited in digital imaging microscopy: Fluorescence resonance energy transfer and delayed luminescence. *Berlin Bunsenges Physik Chemie* 93:387–391.

Lampton, M., and Malina, R. F. 1976. Quadrant anode image sensor. *Review of Scientific Instruments* 47:1360–1362.

Lleó, A., Berezovska, O., Herl, L., Raju, S., Deng, A., Bacskai, B. J., et al. 2004. Nonsteroidal anti-inflammatory drugs lower Abeta42 and change presenilin 1 conformation. *Nature Medicine* 10:1065–1066.

Loeser, C. N., and Clarck, E. 1972. Intercellular fluorescence decay time of aminonaphthalene sulfonate. *Experimental Cell Research* 72:485–488.

Loeser, C. N., Clarck, E., Maher, M., and Tarkmeel, H. 1972. Measurement of fluorescence decay time in living cells. *Experimental Cell Research* 72:480–484.

Marriott, G., Clegg, R. M., Arndt-Jovin, D. J. A., and Jovin, T. M. 1991. Time-resolved imaging microscopy: Phosphorescence and delayed fluorescence imaging. *Biophysical Journal* 60:1374–1387.

McGinty, J., Tahir, K. B., Laine, R., et al. 2008. Fluorescence lifetime optical projection tomography. *Journal of Biophotons* 1:390–394.

Moscicki, A. B., Palefsky, J. M., Gonzales, J., and Schoolnik, G. K. 1991. The association between human papillomavirus deoxyribonucleic acid status and the results of cytologic rescreening tests in young, sexually active women. *American Journal of Obstetrics and Gynecology* 165:67–71.

Periasamy, A., Siadat-Pajouh, M., Wodnicki, P., Wang, X. F., and Herman, B. 1995. Time-gated fluorescence microscopy for clinical imaging. *Microscopy and Analysis* March: 19–21.

Phipson, T. L. 1870. *Phosphorescence or the emision of light by minerals, plants and animals,* 210. London: L. Reeve & Co.

Polyakov, Y. S., Rosanov, Y. M., and Brumberg, E. M. 1966. The setting installation for studying the phosphorescence of micro-objects. *Cytologie SSSR* 8:677–681.

Pringsheim, P. 1949. *Fluorescence and phosphorescence.* New York: Interscience Publishers, Inc.

Selkoe, D. J. 2001. Presenilin, notch, and the genesis and treatment of Alzheimer's disease. *Proceedings of the National Academy of Sciences USA* 98:11039–11041.

Sharpe, J., Ahlgren, U., Perry, P., et al. 2002. Optical projection tomography as a tool for 3D microscopy and gene expression studies. *Science* 296:541–545.

Siadat-Pajouh, M., Ayscue, A. H., et al. 1994. Introduction of a fast and sensitive fluorescent in situ hybridization method for single copy detection of human papillomavirus (HPV) genome. *Journal of Histochemistry and Cytochemistry* 42:1503–1512.

Siadat-Pajouh, M., Periasamy, A., Ayscue, A. H., et al. 1994. Detection of human papillomavirus type 16/18 DNA in cervicovaginal cells by fluorescence-based in situ hybridization and automated image cytometry. *Cytometry* 15:245–257.

Skala, M. C., Riching, K. M., Bird, D. K., Gendron-Fitzpatrick, A., Eickhoff, J., Eliceiri, K. W. et al. 2007. *In vivo* multiphoton fluorescence lifetime imaging of protein-bound and free nicotinamide adenine dinucleotide in normal and precancerous epithelia. *Journal of Biomedical Optics* 12:024014.

van Munster, E. B., and Gadella, T. W. J. 2005. Fluorescence lifetime imaging microscopy (FLIM). In *Microscopy techniques (advances biochemical engineering/biotechnology),* ed. J. Rietdorf, 95, 143–175. Berlin: Springer–Verlag.

Venetta, B. D. 1959. Microscope phase fluorometer for determining the fluorescence lifetimes of fluorochromes. *Review of Scientific Instruments* 6:450–457.

Vereb, G., Jares-Erijman, E. A., Selvin, P. R., and Jovin, T. M. 1998. Temporally and spectrally resolved imaging microscopy of lanthanide chelates. *Biophysical Journal* 74:2210–2222.

Vialli, M. 1964. Indirizzi di ricerca nel campo della istoluminescenza. *Zeitschrift für wissenschaftliche Mikroskopie* 66:164–170.

Vishwasrao, H. D., Heikal, A. A., Kasischke, K. A., and Webb, W. W. 2005. Conformational dependence of intracellular NADH on metabolic state revealed by associated fluorescence anisotropy. *Journal of Biological Chemistry* 280:25119–25126.

Wang, H. W., Gukassyan, V., Chen, C. T., Wei, Y. H., Guo, H. W., Yu, J. S., and Kao, F. J. 2008. Differentiation of apoptosis from necrosis by dynamic changes of reduced nicotinamide adenine dinucleotide fluorescence lifetime in live cells. *Journal of Biomedical Optics* 13:054011.

Warburg, O. 1930. *The metabolism of tumors*. London: Constable.

Williamson, D. H., Lund, P. and Krebs, H. A. 1967. The redox state of free nicotinamide-adenine dinucleotide in the cytoplasm and mitochondria of rat liver. *Biochemical Journal* 103 (2): 514–527.

Zotikov, A. A. 1982. Some possibilities of the phosphorescence microscopy for the study of cells and chromatin. *Acta Histochemica* (suppl) 26:215–217.

Zotikov, A. A., and Polyakov, Y. S. 1975. Investigation of the phosphorescence of cells of various types with the phosphorescence microscope. *Izvestiya Akademii Nauk SSSR Seriya Bio. Eskaya* 2:277–280.

Zotikov, A. A., and Polyakov, Y. S. 1977. The use of the phosphorescence microscope for the study of the phosphorescence of various cells. *Microscopica Acta* 79:415–418.

Index

Printed and bound by CPI Group (UK) Ltd, Croydon, CR0 4YY

24/10/2024

01778288-0013